# Babylon to Voyager and Beyond

## A History of Planetary Astronomy

*Babylon to Voyager and Beyond* covers planetary research from the times of the Babylonians and Ancient Greeks through those of Kepler, Galileo and Newton to the modern era of space exploration. It outlines the key observational discoveries and theoretical developments in their historical context, covering not only the numerous successes but also the main failures.

Planetary astronomy has come a long way since the Babylonians used their extensive numerical skills to predict the positions of the moon and planets. That progress is the story of this book, ending, as it does, with the considerable discoveries of the space age, and the detection of planets around other stars. This unique account will appeal to both amateur and professional astronomers, as well as those with an interest in the history of science.

DAVID LEVERINGTON received his degree in Physics from Oxford University in 1963. Since then he has held a variey of senior positions in the space industry, including that of Technical Director at British Aerospace, Bristol, in the 1980s. During that time he was responsible, amongst other things, for the Giotto spacecraft that intercepted Halley's comet, and the Photon Detector Assembly and solar arrays for the Hubble Space Telecope. David Leverington is the author of many scientific papers on space science and astronomy, and has written two previous books: *A History of Astronomy from 1890 to the Present* (Springer, 1996) and *New Cosmic Horizons: Space Astronomy from the V2 to the Hubble Space Telescope* (Cambridge University Press, 2000). He obtained a PhD in 1997, and is included in *Who's Who in the World, 2000 edition* (Marquis, USA).

T0180483

# Babylon to Voyager and Beyond

## A History of Planetary Astronomy

DAVID LEVERINGTON

CAMBRIDGE
UNIVERSITY PRESS

CAMBRIDGE UNIVERSITY PRESS
Cambridge, New York, Melbourne, Madrid, Cape Town, Singapore, São Paulo

Cambridge University Press
The Edinburgh Building, Cambridge CB2 8RU, UK

Published in the United States of America by Cambridge University Press, New York

www.cambridge.org
Information on this title: www.cambridge.org/9780521808408

First published 2003
This digitally printed version 2007

*A catalogue record for this publication is available from the British Library*

*Library of Congress Cataloguing in Publication data*

Leverington, David, 1941–
  Babylon to Voyager and beyond: a history of planetary astronomy / David Leverington.
    p.  cm.
  Includes bibliographical references and index.
  ISBN 0 521 80840 5 (hardback) – ISBN 0 512 00461 6 (paperback)
  1. Planetology.  I. Title.
QB601.L59 2003
523.4–dc21    2002031557

ISBN 978-0-521-80840-8 hardback
ISBN 978-0-521-00461-9 paperback

# Contents

Contents

# Preface

Writing a history of astronomy, covering the whole period from the earliest written records to the present day, is a humbling experience. This is because one is reviewing the dedicated work of so many individuals who often operated in the most difficult circumstances, and because of the shear wealth of knowledge that has been accumulated. Key people undoubtedly stand out, like Aristotle, Copernicus, Galileo, Newton, Herschel, and so on, but none of these worked in a vacuum. Instead they built their ideas either consciously or subconsciously on those of their predecessors and contemporaries. Or as Newton put it,[1] 'If I have seen further it is by standing on the shoulders of giants'. Nevertheless, some philosophers, scientists and astronomers are undoubtedly key to our history, although who the key figures are today, in our era of multiauthored papers, is often hard to determine.

In any scientific enterprise it is difficult, if not impossible, to predict in advance which theories and ideas are likely to be the most productive, and which are plainly wrong. As a result, progress is usually achieved by taking six steps forward and three steps back, often finding oneself in a cul-de-sac in the process. Even some of the most distinguished astronomers have had ideas and beliefs which seem bizarre to us today. For example, Kepler was interested in astrology, Newton in alchemy, and Herschel believed that the sun was inhabited! But it is inevitable that great thinkers are likely to have their fair share of wrong ideas, if they are to push forward the frontiers of knowledge. Or as Edward Phelps said in rephrasing the old adage,[2] 'The man who makes no mistakes does not usually make anything'.

It is not possible nor sensible in a book of this size to try to discuss all the mistaken ideas and suspicious observations with which astronomy is littered. But neither is it correct to present the history of astronomy as one triumph after another, with no false steps in between. So I have tried to travel a middle way, outlining the key theories and ideas, as seen at the time they were made, even if they turned out to be wrong. Whilst it is easy to explain what was wrong with the earlier of these ideas, we naturally do not know which of our current ideas may be wrong, although some undoubtedly will

---

[1] Letter to Robert Hooke, 5th February 1676.
[2] Speech at the Mansion House, London, 24th January 1889.

be so. In fact, one or two of the current concepts described in the latter parts of this book may well need modifying, or even rejecting by the time that the book is published and read. But this history will not be the poorer for that. After all, it is in the nature of astronomical research that new things are found every day, which is what makes the subject so fascinating.

This book is written for the reader with a basic understanding of astronomy, whether as an amateur or professional. I have tried, therefore, to choose a set of units which would be acceptable to all these readers, some of whom may prefer imperial or cgs units, whereas others may prefer SI units where they exist.

Last, but not least, I would like to express my special thanks and appreciation to Richard Baum and Allan Chapman, who had the kindness and patience to read key parts of the text and suggest modifications to make the book more accurate and readable. To both of them I am most grateful, but if there are any errors of fact, misinterpretation or misunderstanding remaining, they are entirely mine.

# Introduction

Astronomy is like no other science, being partially associated, as it was in ancient times, with religion and superstition. It was studied by priests, philosophers, soothsayers and mystics, as well as the forerunners of today's men and women of science. In those early days, astronomy was seen as the most fundamental and important of the sciences, as it dealt with the motion of the sun, moon and planets across the sky, and their alignments, eclipses, and so on, to which the ancients attached great significance.

Today all that has changed in our cynical, materialistic world, where religion is struggling to survive, and where science is judged by most people on whether it is useful or not. On that basis, computer technology and genetic engineering are considered more important than trying to understand how the universe started, or whether there is life on other planets of the solar system or elsewhere in the universe. And yet, we are now beginning to understand how fragile and delicate the balance of nature is on earth, and to realise that a study of the other planets of our solar system could help us in this. Hopefully, such studies will help us to decide how to correct our harmful developments on earth before it is too late, and life, as we know it, becomes seriously compromised.

In 1600 Giordano Bruno suggested that there was a limitless number of planets around other stars supporting intelligent life. This was a challenge to the religious doctrine of the period, but, even today, the discovery of such intelligent life would have profound effects in some religious circles. Although it may be unlikely that we will discover such evidence in our lifetimes, we may find elementary lifeforms of some sort on Mars, Titan or Europa in the next few years or decades. Such a discovery would be yet one more twist to the tale featured in this book, which explores the vast increase in our knowledge of the solar system over recorded time, and which ends with the discovery of other planetary systems around other stars. In fact, by the time that you read this book, life may well have been discovered elsewhere in the universe. Who knows? That is the thrill of astronomy.

# Chapter 1 | THE ANCIENTS

## 1.1 Early astronomy

It is easy to ignore the skies these days when many of us live in light-polluted cities, rushing around without time to think, but things were very different for our distant ancestors. In those days, before the rise of the great civilisations of Babylon and Greece, for example, primitive peoples needed to know when to plant and harvest their crops, when to move their animals to winter pasture, and so on. They regulated their working and sleeping patterns according to the times of the rising and setting of the sun, and understood the regular cycle of the moon, but working out the seasons was more difficult. In the absence of any other information, they used their own experience, but this could sometimes prove unreliable, so a more structured guide to the seasons would be useful.

It was easy to measure a day or a lunar month, its so-called synodic period from one new moon to another, but how could a year and its seasons be determined? Some civilisations used midsummer's day as a baseline, when they knew that the noon-day sun was at its maximum altitude,[1] and worked out the seasons from that. Others used standing stones or stone circles, like Stonehenge, to measure the position of the rising or setting sun against the horizon on midsummer and midwinter's day. Yet others determined the seasons from the rising or setting of certain stars. The Egyptians, for example, knew when they were developing their calendar in about 3000 BC that the annual Nile floods occurred when Sirius, the brightest star in the sky,

---

[1] This is not so in equatorial regions, of course, where the seasons alternate between dry and rainy periods, rather than between the winter, spring, summer or autumn of higher latitudes.

became visible for the first time in the morning twilight just before dawn. On the other hand the Australian aborigines linked the start of spring to the visibility of the Pleiades star cluster in the evening sky. So the ancient civilisations became familiar with the movements of the sun, moon and stars in the heavens.

As time passed, more advanced civilisations started to use the sun and stars for navigation, which required a knowledge of their position in the sky with some accuracy. Other people began to ask not just where the sun, moon and stars were at a particular time of day or night, but what caused their movements through the heavens. The planets that were known to the ancients, that is Mercury, Venus, Mars, Jupiter and Saturn, were seen to move in the sky in a peculiar way which seemed to have a meaning. Total solar eclipses and comets were also observed from time to time, spreading alarm and despondency through many populations. So early astronomy became inextricably linked with religion and superstition.

Religion placed a different requirement on astronomy compared with agriculture, however. The start of the various seasons did not need to be known exactly for agricultural purposes, but religious observances demanded a precise calendar so that their rituals could be planned in advance. The problem is that there is not an integral number of days in a lunar month, nor is there an integral number of lunar months in a year, which complicated matters greatly.

One lunar (synodic) month, from new moon to new moon, is now known to be, on average, 29.53059 days, and one solar or tropical year,[2] as it is known, is 365.24220 days. So, one solar year is 12.3683 lunar months, or about 11 days longer than 12 lunar months. In fact, over 19 years there are almost exactly 235 lunar months, which is 7 months more than if there had been just 12 months in each of the 19 years. So, provided 7 extra, so-called 'intercalary' months are added over 19 years, the lunar phase on the first day of the year is almost exactly the same as it was 19 years previously. Such a discovery of this so-called 'Metonic cycle'[3] was made only gradually, by the empirical approach of adding an extra month, as required, to a year of 12 lunar months to correct for the observed agricultural seasons. Eventually the

---

[2] A year based on the seasons.

[3] This cycle was named after Meton, a Greek astronomer, who lived about 460–400 BC, and who was originally thought to have discovered it. The discovery was made more gradually, however, as mentioned above, but Meton was the first person to propose that it be adopted as the basis for a calendar.

Egyptian,[4] Babylonian and Jewish calendars were structured to have 12 years of 12 months and 7 years of 13 months in each 19 year period.

## 1.2 The Babylonians

The Babylonians of about 2000 BC, who lived in southern Mesopotamia,[5] had had to observe the western horizon to detect the new moon, and hence to officially start a new month. In the process they not only deduced its 29.5 day cycle, but also clearly observed Venus (called Nindaranna), which is often very bright in the west after sunset and in the east before sunrise. Venus' periods of visibility near the western and eastern horizons, and its periods when it could not be seen (because it is too close to the sun), have been clearly recorded in a text from Ashurbanipal's library of the seventh century BC, which was based on a text of about 1600 BC.[6]

The Assyrians, who came from northern Mesopotamia, became the most powerful force in Mesopotamia in about 800 BC, conquering Babylon and adopting a large part of their culture.[7] They studied the sky as a source of omens, however, rather than to improve their calendar, linking the planets to their gods like talismans. Mercury was linked to Nabu, Venus to Ishtar, Mars to Nergal, Jupiter to Marduk and Saturn to Ninib. Nabu was the god of wisdom, Ishtar the goddess of love, Nergal the god of war and the under-world, Marduk the god of Babylon and the chief god, and Ninib a solar deity responsible for the seasons.

Now that the positions of the planets in the sky were linked to omens, their movements were more closely studied by the Assyrians, as was that of the moon, in order to predict lunar eclipses which were considered to be of special significance. A record of previous omens, used by the priests to interpret new celestial alignments and positions, was contained in a collection of seventy tablets called the 'Enuma Anu Enlil', which dated back to the pre-conquest era of about 1400 BC.

---

[4] The Egyptians ran two calendars in parallel. The one described above was the astronomical calendar, but there was also an administrative calendar with a year of exactly 365 days, which was $1/4$ day shorter than in the astronomical calendar!

[5] Mesopotamia was the region of the Middle East which was based on the Tigris and Euphrates rivers. It is approximately equivalent to modern Iraq.

[6] These observations relate to the reign of King Ammisaduqa of Babylon who is variously thought to have reigned about 1702–1682 BC, 1646–1626 BC, or 1582–1562 BC, depending on whose chronology is used.

[7] The Assyrians first captured Babylon in 1234 BC but the occupation did not last long.

In 612 BC the Assyrian empire finally collapsed, and Babylon became the centre of a new empire under Nebuchadnezzar, but in 539 BC it became part of the Persian empire under Cyrus the Great. As time progressed the Babylonians became less and less interested in interpreting messages from the gods, as seen by planetary alignments and lunar eclipses, and became more interested in trying to see patterns in planetary and lunar movements to enable astronomical predictions to be made. Water clocks were used to measure time, and in a text of 523 BC the relative timings of sunrise and sunset, and moonrise and moonset, are recorded to an accuracy of about a minute. The Babylonians measured the positions of the planets in the sky relative to the stars, and deduced their synodic periods. So in the case of Jupiter, for example, they observed that the turning point in its movement amongst the stars was reached every 1.09 years, that is its synodic period, resulting in there being almost exactly 65 of Jupiter's synodic periods in 71 years. The Babylonians also recorded both partial and total lunar eclipses, and observed that the cycle of eclipses repeated itself almost exactly every 223 synodic months, a period now called a 'saros'. We now know that the reason for this is that eclipses can only take place when the sun–earth–moon line lies approximately along the line of nodes, which is where the inclined orbit of the moon intercepts the orbit of the earth around the sun. This line of nodes is not fixed in space, however, as it regresses (i.e. precesses backwards), allowing an alignment of the sun–earth–moon line every 223 synodic months.

Babylon was part of the Persian empire for about two hundred years, but in 331 BC it became part of the empire of Alexander the Great. Alexander's conquest resulted in the arrival of the Greek influence in the so-called Seleucid period which lasted until 247 BC. The Parthians then took control of the area, and successfully withstood invasion attempts by Rome, in particular, over the next three centuries. Fortunately, during all these changes the priests continued with their astronomical observations and analysis.

## The late Babylonian period

Our knowledge of astronomy in the late Babylonian period, covering the last three centuries BC, is contained in about 300 cuneiform tablets excavated from Babylon at the end of the nineteenth century, and from Uruk in southern Mesopotamia in the early twentieth century.

The Babylonian astronomers produced numerous extensive tables containing intricate calculations of the movements of the sun, moon and planets through the heavens. By now their positions were measured in ecliptical longitude and latitude coordinates.[8] The longitudes were recorded in one of twelve zodiacal signs, each of which covered 30° longitude, with their subdivisions being recorded in the sexagesimal notation based on units of sixty. So 28° 10′ 39$\frac{2}{3}''$ or 28°.177685 in our decimal system, based on powers of ten, becomes 28°,10,39,40 in the Babylonian system, based on powers of sixty.

Two systems were used in these cuneiform tablets to predict the movements of the sun and moon. So the first visibility of the crescent moon could be predicted every month, thus enabling the new month to be officially started.

The first analysis system was given in the earliest cuneiform fragments, dated about 170 BC. In these it was assumed that the sun moved along the ecliptic at 30° per synodic month for 194° of the ecliptic, and at $\frac{15}{16} \times 30°$ per month for the remaining 166°. This implied that there are 12.36889 or 12$\frac{83}{225}$ synodic months in a sidereal year (of 360°), or that there are 2783 synodic months in 225 sidereal years. Assuming a mean synodic month deduced by the Babylonians of 29.530594 days (29d 12h 44m 3$\frac{1}{3}$s, see later), gave a sidereal year of 365.260637 days or 365d 6h 15m 19s, which is only just over six minutes too long compared with the value known today.

The second analysis system was found in cuneiform tablets dating from about 130 BC. It assumed that the sun's velocity per synodic month varied linearly, in a zig-zag fashion, from 30° 1′ 59″ to 28° 10′ 39$\frac{2}{3}''$ and back again, changing at the rate of 0° 18′ per synodic month. This gave an average velocity of 29° 6′ 19$\frac{1}{3}''$ per synodic month, compared with the currently accepted value of 29° 6′ 20$\frac{1}{5}''$, which is remarkably close. Given the average solar velocity of 29° 6′ 19$\frac{1}{3}''$ per synodic month, it would take 12.368851 synodic months to complete a full 360° sidereal year.

At the rate of 0° 18′ per synodic month, however, it would take 12.369136 synodic months for the sun's velocity to go from maximum to minimum and back again, which is higher than the number of synodic months needed to traverse the 360° of a sidereal year. The time for the sun to go from maximum velocity to maximum velocity is now called the anomalistic year. The above estimate is 12;22,08,53,20 synodic months in Babylonian notation,[9] or 12$\frac{299}{810}$ synodic months using fractions, implying that there are 10,019 synodic months in 810 anomalistic years. At an average velocity of 29° 6′ 19$\frac{1}{3}''$ per synodic month, the sun would move 360° 0′ 29″.8 along the ecliptic in

---

[8] The ecliptic is the plane of the earth's orbit around the sun.

[9] The semicolon separates the full number from numbers less than unity.

an anomalistic year. So the Babylonians appear to have realised that the sun moves slightly more than 360° in going from maximum velocity to maximum velocity, the excess being 29".8, compared with the presently known value of 11".6.

The Babylonians, who were fascinated by numbers, saw the heavens in two dimensions, rather than the three dimensions envisaged by the Greeks. They were content to analyse the movements in the sky by arithmetic means, whereas the Greeks tried to explain the movements by recourse to geometrical figures. Although the Babylonians observed that the sun's motion along the ecliptic was not uniform, they did not seem to ask themselves why this was. We now know that this is because the earth's orbit around the sun is elliptical, with the earth moving faster when it is nearer to the sun at perihelion, and slower when it is furthest away at aphelion. As a result, the apparent solar velocity varies sinusoidally between these two extremes, rather than in the zig-zag fashion assumed by the Babylonians. We also now know that the orbit of the earth is not fixed in space, but it is moving such that the perihelion is precessing, thus explaining why the anomalistic year is not the same as the sidereal year.

In the second century BC, the Babylonians concluded that the average synodic month was 29d 12h 44m $3\frac{1}{3}$s or 29.530594 days long. It was also found that it could be as long as 29d 17h 57m $48\frac{1}{3}$s or as short as 29d 7h 30m $18\frac{1}{3}$s. The Babylonians analysed the moon's motion using a zig-zag function, with successive synodic periods, on the linear parts, increasing or decreasing by 1h 30m. This meant that it took 13.9444, $13\frac{17}{18}$ or $\frac{251}{18}$ synodic months for the moon to go from the longest synodic month to the shortest and back again. In this time the moon has orbited the earth $\frac{251}{18} + 1$ times, so $251 + 18 = 269$ anomalistic months is equivalent to 251 synodic months. So the average anomalistic month (from maximum to maximum orbital speed of the moon) was found to be 27.554569 days or 27d 13h 18m $34\frac{3}{4}$s, just 2.7 seconds from the value known today. Incidentally as $\frac{269}{251} \approx \frac{239}{223}$, a saros has almost exactly 223 synodic months or 239 anomalistic months.

As mentioned above, the main objective of these Babylonian analyses, which were mainly decoded in the early twentieth century by the Jesuit priest Jos. Schaumberger, was to predict when the crescent moon would first be visible each month. To do this, astronomers used any known lunar eclipse to provide the initial conditions. The angular separation of the sun and moon could then be calculated, knowing their various orbital velocities as determined above. The day of the year gave the inclination of the ecliptic to the horizon, and a knowledge of the moon's latitude

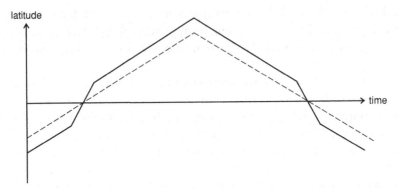

**Figure 1.1** Plot of the moon's latitude relative to the ecliptic, versus time, as used by the Babylonians to predict eclipses. The solid line shows the modified zig-zag function used.

relative to the ecliptic then enabled the visibility of the crescent moon to be determined.

The above Babylonian calculations were shown on extensive lunar tables, the most complete of which ran to eighteen columns of numbers, giving values of the various quantities used. These tables were produced not only to predict the visibility of the crescent moon, but also that of lunar eclipses. To do this, astronomers realised that it was vital to get a good estimate of the moon's latitude above or below the ecliptic, which is now known to be due to the moon's orbital plane being inclined at about 5° to the ecliptic. It was recognised by the Babylonians that a straightforward zig-zag function of latitude versus time would not be sufficient, so they used a modified form as shown in Figure 1.1. This enabled them to calculate an index which predicted not only when there would be a full or partial lunar eclipse, but also approximately how long it would last.

A slight diversion is now called for, outlining the movement of the planets as we now know them, before going on to describe the Babylonian observations.

An inferior planet, i.e. Mercury or Venus, that orbits the sun inside the orbit of the earth, has the apparent orbit around the sun as shown schematically in Figure 1.2 relative to a stationary earth. At point A the planet is first seen as a morning 'star', at what is called its heliacal rising. On subsequent days, it appears higher and higher in the sky before dawn until it reaches its greatest western elongation at point B, so-called because it is west of the sun in the sky (although it is seen in the eastern sky). On the following days the planet is seen to be lower and lower in the pre-dawn sky, until it is lost in the bright dawn sky after point C. Some time later at point D the planet

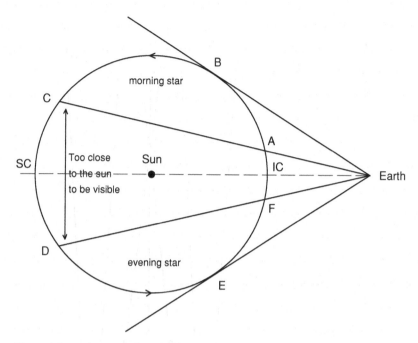

**Figure 1.2** A schematic showing the apparent orbit of an inferior planet relative to a stationary earth. (The real figure is not a circle, but the diagram illustrates the principle under discussion in the text.)

first starts to be seen low down in the evening sky just after sunset. Point E is its greatest eastern elongation, and point F is its last visible appearance in the evening sky at the planet's so-called heliacal setting. IC is called the planet's inferior conjunction and SC its superior conjunction. Because the orbits of Mercury, Venus and the earth are elliptical, the greatest elongations from the sun vary from about 18° to 28° for Mercury and from 45° to 47° for Venus.

We turn now to the superior planets that were known to the ancients, that is Mars, Jupiter and Saturn. Unlike the inferior planets, the superior planets can be seen at any time of night, depending on their orbital positions. When they reach their highest point in the sky, crossing the meridian at local midnight, they are 180° away from the sun, and are said to be at opposition. Whereas when they are directly in line with the earth and sun, but behind the sun, they are at conjunction.

The orbits of the superior planets, as seen from the earth, are complex, as their orbits are ellipses inclined to the ecliptic. They orbit the sun in the same direction as the earth with lower angular and linear velocities.

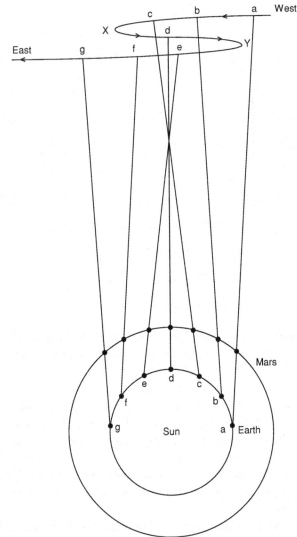

**Figure 1.3** Plot to show the apparent movement of Mars against the stars, as seen from a moving Earth. The first and second stationary points are labelled 'X' and 'Y'. Mars is at opposition at point 'd'.

As a result, if the positions of the superior planets are plotted against the stars night after night, they show a general movement to the east (see Figure 1.3).[10] Around opposition, however, they trace out an S-shaped curve, temporarily turning back on themselves at their first stationary

---

[10] Although Figure 1.3 is for Mars, the general configuration shown is the same for all the superior planets.

point. At the second stationary point the planets resume their easterly motion.

The Babylonians' planetary tables gave data on the heliacal rising and setting of the planets, and on the first and second stations (stationary points) and oppositions of the superior planets. The data on Jupiter are particularly extensive.

As in the case of the sun and moon above, Babylonian astronomers wished to predict the position of Jupiter in the sky based on their records of its position going back over many years. In their simplest arithmetical 'model' they assumed that Jupiter moved at 30° per synodic period when it is situated between 85° and 240° ecliptic longitude, and at 36° per synodic period over the remainder of the ecliptic. This gave the average movement in a synodic period (called a synodic arc) of 33° 8′ 45″, with 10.8611 synodic periods in a sidereal period, giving a sidereal period of 11.8611 years and a synodic period of 1.0921 years. These values of Jupiter's sidereal and synodic periods are within 0.01% of those used today.

The Babylonians also used a second, slightly more sophisticated model, assuming that Jupiter moved at 30° per synodic period for 120° of the ecliptic, 36° for 135°, and 33° 45′ for the transition regions of 53° and 52° between these two.

Finally, in a third alternative model they assumed that the movement of Jupiter varied linearly from 28° 15′ 30″ to 38° 2′ per synodic period and back again, in a zig-zag function over one sidereal period, changing at the rate of 1° 48′ per synodic period. The average movement was still 33° 8′ 45″, with $10.8611 = 10\frac{31}{36}$ steps or synodic periods per sidereal period, giving a sidereal period of $10\frac{31}{36} + 1 = \frac{427}{36} = 11.8611$ years.

Although these three models produced the same synodic and sidereal periods for Jupiter, they produced positions for the first and second stations of the planet that differed from the observed positions by up to almost 2° over the course of one sidereal period.

The Babylonians also produced similar data for Mercury, giving a synodic period of $\frac{480}{1513} = 0.31725$ years, Mars with a synodic period of $\frac{284}{133} = 2.1353$ years, and Saturn with a synodic period of $\frac{265}{256} = 1.0352$ years. Although these values compare very well with the values known today,[11] the predicted positions of the planets at their stations or heliacal risings or settings were, like those predicted for Jupiter, not all that accurate.

---

[11] The errors are all less than 0.01%.

## 1.3 **The Greeks**

As mentioned previously, the Babylonian astronomical observers were generally priests trying to predict the movement of heavenly objects using arithmetical techniques, whereas the Greeks were philosophers trying to understand the cosmos. There were other differences also. The Babylonian records do not give us the names of their observers, whereas we know the names of the Greek philosophers. Unfortunately, however, the Greeks used less durable materials for recording their work than the clay tablets of the Babylonians, and so most of the early Greek written material is lost, and we have to rely on second or third hand accounts, with the inevitable contradictions and confusion that that creates.

### Thales

Aristotle (384–322 BC) considered Thales of Miletus[12] (c. 625–547 BC) to be the founder of Greek natural philosophy. He was both a geometrician and philosopher, but he probably undertook no serious astronomical observations of his own, being more interested in theoretical ideas. Thales seems to have been the original absent-minded professor interested in 'higher things', if Plato (c. 427–347 BC) is to be believed. He tells the story that Thales was so engrossed in looking at the stars when he was out walking one night that he fell down a well. Or, as Plato explains it in his book *Thaetetus*, because Thales wanted to 'know what went on in the heaven, [he] did not notice what was in front of him, nay, at his very feet'.[13]

Thales hypothesised that the sun and stars were made of water, and that the earth was a flat disc that floated on water which evaporated to produce the air. There are also disputed Greek sources that say that Thales at some stage recognised that the earth was spherical, and understood the true cause of lunar and solar eclipses. Herodotus (c. 484–425 BC) claims that Thales predicted the solar eclipse that took place during a battle between the Lydians and the Medes, which is now thought to have taken place on 28th May 585 BC. This story must be treated with caution, however, as Herodotus says that Thales predicted the year of the eclipse, but not its day. This seems

---

[12] Miletus was a Greek colony on the west coast of Asia Minor (modern Turkey). It is just south of Ephesus.

[13] Translation by T. L. Heath in his *Greek Astronomy*, Dent, London, 1932, republished by Dover 1991 (afterwards referred to as Heath, *Greek Astronomy*), p. 1.

strange, as if Thales was able to predict solar eclipses, which at that time the Babylonians were not, he must have realised on approximately what days of the year they were possible.

## Anaximander

Anaximander (*c.* 610–545 BC), also of Miletus and one of Thales' pupils, thought that the earth was a cylindrical column surrounded by air. Its height was one-third of its breadth, and it floated in space at the centre of an infinite universe. Unlike Thales, however, he thought that, because the earth was at the centre of the universe, it was equidistant from all points on the celestial circumference, and so was naturally at rest, and did not need to be supported by water or anything else.

In Anaximander's scheme, a wheel or ring, with a diameter of 27 or 28 times the earth's diameter, surrounded the earth. This wheel had a hollow rim or tyre filled with fire, in which there was a hole the size of the earth. This hole is what we see as the sun.[14] From time to time this hole is closed, and then we see a total solar eclipse. Likewise, inside this wheel was another wheel, with a diameter 19 times that of the earth, again with a flame-filled tyre. A hole in this was the moon. Inside this was a sphere with a fire-filled surface, which had holes in it to produce points of light that were seen as the stars.

Anaximander proposed that the solar and lunar wheels and the stellar sphere all rotated around the earth at different rates. However, he did not, as far as we know, try to explain how we can see the sun and moon behind the stellar sphere, or why the sun was not eclipsed by the lunar wheel every time the sun crossed behind it. Although his model of the universe looks strange to us,[15] it was important as being the first one that we know that saw the universe as a 'mechanical' system.

We now know that the earth's equatorial plane is inclined at about $23\frac{1}{2}°$ to the plane of the earth's orbit around the sun (the ecliptic). But the projection of the earth's equatorial plane into space is seen as the celestial equator, so the ecliptic is inclined at about $23\frac{1}{2}°$ to the celestial equator; this angle

---

[14] There is some confusion about the size of the solar wheel, as Hippolytus says that Anaximander thought that it had a diameter of 27 times that of the moon's wheel, rather than 27 or 28 times the earth's diameter.

[15] Anaximander was closer to the mark in proposing that life started in the oceans, with the first animals being like sea urchins. They eventually left the water, adapting their structure to their new land-based environment.

being called the obliquity of the ecliptic. It can be measured in a number of ways, the simplest of which is by measuring the angle that the sun makes to the horizon at local noon on midsummer and midwinter's day, and dividing the difference by two. Such observations were probably first made by the Babylonians, Chinese or Egyptians, but the first Greek known to have made such an observation was either Anaximander or Pythagoras (c. 580–500 BC). Anaximander is also thought to have introduced the gnomon (a sundial with a vertical needle) into Greece from Babylon, which not only allowed him to measure the altitude of the sun, but also to find the dates of the equinoxes and solstices and hence deduce the length of the tropical year.

## Pythagoras

The name of Pythagoras is well known, so it is surprising that we have none of his original writings. Because of this, we are not sure what to attribute to him and what to attribute to his followers, but undoubtedly the basic philosophy espoused by them was due to him.

Pythagoras was born on Samos, an island off the coast of Asia Minor,[16] not far from Miletus, where Thales, Anaximander and a good many other Greek philosophers lived. A pupil of Anaximander and the mystic Pherekydes, Pythagoras travelled widely and probably visited Egypt and Babylon, like many other Greek scholars of his period. Then in about 530 BC, when he was about fifty, he left Samos for good and settled in Croton,[17] which was a Greek colony in southern Italy. There he set up a religious order and school of philosophers, sometimes known as the Pythagorean Brotherhood, that soon ruled the town and the surrounding area. The Brotherhood led a communal existence, sharing property and spending time in contemplation and reflection. Because of its political power, however, adjacent Greek states felt threatened. As a result, intervention by a Greek called Cylon caused Pythagoras to be banished towards the end of his life to Metapontum, another Greek colony in southern Italy. About sixty years later his school of philosophy was forcibly dispersed, and some of its members killed.

So Pythagoras was not just a mathematician and astronomer, but also a politician and religious leader. He even worked miracles, practised medicine,

---

[16] Asia Minor is now basically modern Turkey.

[17] The ancient Greek letter K or κ (kappa) is generally replaced by 'c' in proper names and 'k' in other words in English, although some writers replace it by 'k' in all words. So Croton is sometimes written Kroton, and the Ecphantus is sometimes written Ekphantus.

and went around the countryside giving religious sermons. He must have been quite a man.

The Pythagoreans believed in the harmony of nature and the purity of numbers, discovering, amongst other things, the pattern of standing waves on the plucked string of a musical instrument and the note that it produced. They were probably the first to appreciate that the earth is spherical,[18] and that the planets each move in separate orbits, all inclined to the celestial equator. In addition, the Pythagoreans were almost certainly the first Greeks to have realised that Phosphorus and Hesperus, the morning and evening stars, were one and the same astronomical body.

The early Pythagoreans introduced the doctrine of the 'Harmony of the Spheres' in which a non-spinning, spherical earth was surrounded by a series of concentric, crystalline spheres. The moon, sun, individual planets, and stars were each supported by a sphere, which revolved around the earth at different speeds. These spheres were thought to produce a musical sound, known as the 'Music of the Spheres', as they went past each other.

According to Pliny (23–79 AD), the musical notes were determined by the separations of the spheres as follows:

Earth to Moon, a tone
Moon to Mercury, a semitone
Mercury to Venus, a semitone
Venus to Sun, a minor third
Sun to Mars, a tone
Mars to Jupiter, a semitone
Jupiter to Saturn, a semitone
Saturn to the fixed stars, a minor third

This gives the scale, starting with C, of C, D, E flat, E, G, A, B flat, B, and D. A beautiful notion linking, as it does, the beauty of the sky with the magic of numbers and musical harmony.

## Philolaus

In the following century the Pythagorean Philolaus (c. 450–400 BC) devised a new cosmic model in which the earth was also moving, rather than being

---

[18] Very little is one hundred per cent certain about this period of Greek astronomy. Although the Pythagoreans are generally thought to have been the first to have proposed that the earth is spherical, some people attribute this to Parmenides of Elea (c. 504–450 BC).

static at the centre of the universe. He appears to have been influenced in his world picture by the fact that the moon always turns its same face to the earth. This led him to suggest that the earth orbits a central fire, called Hestia, once per day, in the earth's equatorial plane. He also suggested that the earth kept its inhabited part permanently facing away from the fire. This meant that the earth would have to rotate on its axis once per day, as it orbited Hestia, which was a revolutionary idea at the time. His model was even more revolutionary, as he proposed that there was also a counter-earth, or antichthon, on a line between the earth and the fire, also orbiting the central fire once per day, to shield the earth from its heat. This counter-earth was proposed, according to Aristotle, as it meant that there would be ten bodies in the universe, namely the earth, moon, sun, five planets, the sphere of fixed stars and the counter-earth; ten being a perfect number to the Pythagoreans.[19]

Outside the orbit of the earth, according to Philolaus, was the moon that was thought to orbit Hestia once every $29^1/_2$ days. Then came the sun, which took one year to complete an orbit, the planets, a sphere carrying the fixed stars, and finally a wall of fiery ether surrounding the universe. The latter, along with the central fire, provided the universe with light and heat. All the orbits, including that of the earth and counter-earth, were circular. The sun was not self-luminous, but was like a crystal ball, reflecting and scattering the light from the central fire and fiery ether, with an orbit inclined to that of the earth, thus explaining the seasons.

There is some confusion between the writers of antiquity when they come to describe the relative positions of the planets and the sun in Philolaus' universe. Some, including Plutarch (46–120 AD), state that Philolaus placed the orbits of Mercury and Venus between those of the moon and sun, whereas others say that he placed all five planets outside of the sun's orbit. He may well have done both, of course, at various times. Whatever is the case, the main point is that the Pythagoreans proposed schemes, including both an earth-centred and a non earth-centred universe, in which all the heavenly bodies were in orbit. This is very different from the relatively crude ideas of Thales, Anaximander and their ilk.

It had been known for a long time that there are more lunar eclipses than solar eclipses seen from a given point on the earth's surface. We now know that this is because the earth's shadow on the moon is larger than the moon's shadow on the earth, because of the relative sizes of these two bodies. This was not understood in fifth century BC Greece, however. Instead

---

[19] As ten is the sum of the first four whole numbers.

Philolaus explained the effect by supposing that lunar eclipses are caused by either the earth or counter-earth intercepting the sun's light en route to the moon, thus increasing the number of lunar eclipses, compared with solar eclipses.

Although Philolaus had required the earth to keep the same face to the central fire, as it orbited it once per day, he did not specifically mention that the earth spun on its axis. In fact, Hicetas of Syracuse seems to have been the first person to have done this. Unfortunately, little is known about Hicetas, except that he was a Pythagorean who lived at about the same time as Philolaus, and proposed that the earth spun on its axis at the centre of a geocentric (i.e. earth-centred) universe. It is not clear whether this idea superseded that of Philolaus amongst the Pythagoreans, or whether both existed in parallel. What is clear, however, is that Hicetas' geocentric universe, with a spinning earth, was adopted later in the fourth century BC by Ecphantus the Pythagorean and Heracleides of Pontus.

Heracleides[20] (c. 388–315 BC) was born at Heraclea in Pontus, on the Black Sea coast of Asia Minor, but emigrated to Athens where he came under the influence of Plato, Aristotle and the Pythagoreans. Whether he knew Plato and Aristotle is unclear, but he himself was an original thinker, having little time for academic tradition. His greatest contribution to the evolving world picture was in connection with Mercury and Venus, which are always very close to the sun in the sky. This led Heracleides to conclude that these two planets, unlike Mars, Jupiter and Saturn, actually orbited the sun as the sun orbited a spinning earth, all these orbits being circular. Given the misnomer 'The Egyptian System', this structure of the universe became accepted by a large number of people towards the end of the fourth century BC.

## Aristarchus

So far in the Pythagorean models of the universe, we have had those centred on a non-spinning or spinning earth and those centred on the central fire, Hestia, whilst Heracleides had proposed that Mercury and Venus orbit the sun as it goes round a spinning earth in a geocentric universe. Aristarchus (c. 310–230 BC) was to go one step further, and be the first to propose a sun-centred or heliocentric universe, in which the planets orbit the sun in the (correct) order of Mercury, Venus, Earth, Mars, Jupiter, and Saturn, with the moon orbiting the spinning earth.

---

[20] Sometimes latinised as Heraclitus.

Aristarchus, who was one of the last of the Pythagorean school, was born on Samos about 270 years after Pythagoras, and died in Alexandria about eighty years later. He and Archimedes knew each other, and it is from Archimedes that we have a clear statement in his book *The Sand Reckoner* (216 BC) of Aristarchus' cosmology. Archimedes says,[21] 'For he [Aristarchus] supposes that the fixed stars and the sun are immovable, but that the earth is carried round the sun in a circle which is in the middle of the course [i.e. the sun is at the centre of the circle[22]]; but the sphere of the fixed stars, lying with the sun round the same centre, is such a size that the circle, in which he supposes the earth to move, has the same ratio to the distance of the fixed stars as the centre of the sphere has to its surface'. The last half of the sentence is clearly nonsense, as Archimedes observed, as the centre of a sphere has no size. It is clear what Aristarchus meant, however, namely that the sphere of the fixed stars, centred on the sun, is at a very great distance from the sun and earth, as otherwise the effect of parallax would be seen as the earth orbited the sun.

Aristarchus' idea of an orbiting earth is also mentioned in Plutarch's book *On the Face in the Moon*, where he says that Aristarchus[23] 'supposed that the heavens stand still and the earth moves in an oblique circle at the same time as it turns round its axis'. Here Plutarch clearly shows that Aristarchus also believed that the earth spun on its axis.

We have some first-hand knowledge of Aristarchus' work, as we still have a copy of his book *On the Dimensions and Distances of the Sun and Moon*. In this he started with a number of premises, the most important of which is that the moon shines by reflected sunlight. He then used various ingenious measurements which enabled him to deduce the relative sizes and distances of the sun and moon from the earth. For example, he observed the moon at quadrature, when it is exactly half illuminated, and deduced that the sun–earth–moon angle was 87° at that time. This led him to conclude that the sun–earth distance was about 19 times (sec 87°) the moon–earth distance. Although this is over 20 times too small,[24] as the true angle is

---

[21] Translation by J. L. E. Dreyer in his *History of the Planetary Systems from Thales to Kepler*, Cambridge University Press, 1906, republished as *A History of Astronomy from Thales to Kepler*, Dover, 1953 (afterwards referred to as Dreyer, *A History of Astronomy from Thales to Kepler*), p. 137.

[22] This interpretation is given in Heath, *Greek Astronomy*, p. 106.

[23] Translation from Dreyer in his *A History of Astronomy from Thales to Kepler*, p. 138.

[24] Because of the difficulty in deciding when the moon is exactly half illuminated.

about 89° 50′, he had proved that the sun is much further away from the earth than is the moon. Since the angular sizes of the sun and moon, as seen from earth, are about the same, this meant that the sun was about 19 times the diameter of the moon, according to his measurements.

Observing the size of the earth's shadow on the moon during a lunar eclipse to be about double the diameter of the moon enabled Aristarchus to estimate that the earth is about $\frac{57}{20}$ times the diameter of the moon, implying that the sun's diameter is about $19 \times \frac{20}{57} = 6\frac{2}{3}$ times larger than that of the earth. Measuring the moon's angular diameter, and knowing that its linear diameter is $\frac{20}{57}$ or 0.35 times[25] that of the earth, enabled Aristarchus to deduce that the moon was about 25 earth diameters away. This is very close to the true value of about 30 earth diameters. So Aristarchus had estimated the size of the earth–moon system reasonably accurately, but his sun was about 20 times too close to the earth and so was about 20 times too small.

Strangely, as far as we know, Seleucus of Seleucia, in Mesopotamia, who lived in the second century BC, was the only person to adopt Aristarchus' ideas of a heliocentric universe, until Copernicus resurrected it about 1,700 years later. To understand why, we have to retrace our steps to Plato in the fourth century BC, to uncover the alternative cosmologies then under consideration.

## Plato

Plato (c. 427–347 BC) would have had little place in Gradgrind's[26] history of astronomy, were he to have written one, as Plato's contribution to the factual basis of astronomy is very small. However, he was to have a profound effect on the development of ideas regarding the structure of the universe for almost two thousand years. In about 387 BC he founded an Academy in Athens to encourage the systematic pursuit of philosophical and scientific ideas.

A great deal of Plato's work survives consisting of philosophical dialogues in which an idealised Socrates appears. Like Socrates, who was his tutor, Plato had little interest in the reality of nature preferring, instead, to examine the purity of ideas. Nevertheless his books covered a wide range of subjects from physics and mathematics to philosophy and politics, but his most interesting books, from an astronomical viewpoint, are *Timaeus* and the *Republic*.

---

[25] The true value is now known to be 0.27 times that of the earth.

[26] Gradgrind was the school teacher in Dickens' *Hard Times* who was only interested in facts, facts and more facts, rather than 'fanciful ideas'.

It is difficult to know how interpret the ideas expounded in Plato's books, as the descriptions given are allegorical and ambiguous. Sometimes he even seems to be deliberately vague, leaving scholars to continue to argue, even to this day, as to what he really meant and thought. That having been said, some of his ideas are clear. For example, in *Timaeus* he says,[27] 'And he gave the universe the figure which is proper and natural... Wherefore he turned it, as in a lathe, round and spherical, with its extremities equidistant in all directions from the centre, the figure of all figures most perfect and most like to itself... He allotted to it the motion which was proper to its bodily form... Wherefore, turning it round in one and the same place upon itself, he made it move with circular rotation...' He then goes on in the *Republic* to describe how the celestial bodies, which are fixed to one of eight concentric wheels, revolve around an axis passing through the earth at their centre. The order, from the earth, he took to be the moon, sun, Venus, Mercury, Mars, Jupiter, Saturn and the stars.[28] The outermost (stellar) circle moved rapidly in one direction, whilst the other celestial bodies moved slowly, relative to the stellar circle, in the opposite direction.

Plato recognised that Venus and Mercury behaved differently to the other planets, and he ascribed the same angular orbital rotation rates to them as to the sun. The moon was given the fastest rotation rate *relative to the stars*, followed in velocity by the sun, Venus and Mercury as a group, with the rates of Mars, Jupiter and Saturn being progressively less. He described the colour and intensity of all these celestial bodies, and stated that the moon shines by reflecting light from the sun. All the planets, plus the moon and sun, are said to be spherical, as that is the perfect shape. Finally, he is thought[29] to have defined the distances from earth of the moon, sun, Venus, Mercury, Mars, etc. as 1, 2, 3, 4, 8, 9 and 27, apparently by using two interlocking geometrical progressions namely, 1, 2, 4, 8 and 1, 3, 9, 27.

Plato's scheme was very crude as it did not explain the oscillations, as seen from the earth, of the inferior planets around the sun, nor the temporary reversal of direction of the superior planets at the stationary points. His lasting legacy was, however, his belief that the detailed movements of the

---

[27] Translation by Heath in his *Greek Astronomy*, pp. 49–50.

[28] It is not clear why Plato put Venus, rather than Mercury, after the sun. This could possibly have been because Venus does not move as far away from the sun in the sky as Mercury, or because it is brighter.

[29] This is the generally accepted interpretation of Plato's text, although there are other interpretations.

sun, moon and planets could be described using uniform circular motions, even if he was unable to find a suitable system himself. This idea was to dominate astronomical thinking for almost two thousand years.

## Eudoxus

One of Plato's pupils, Eudoxus (c. 408–355 BC) made a very bold attempt to solve the problem of defining a model of the universe based on the principle of uniform circular motion. Born in Cnidus in Asia Minor, Eudoxus attended Plato's Academy in Athens for two or three months at the age of twenty-three, and later went to Egypt for sixteen months. He then returned to Asia Minor, setting up his own Academy at Cyzicus. According to Seneca, Eudoxus studied planetary movements in Egypt, although there is no evidence that the Egyptians were experts in this field at that time, but he was probably drawn to observe the planets by their excellent climate. He also became versed in the intricacies of the solar and lunar cycles and their impacts on the calendar, as these had been studied by the Egyptians. Some time later Eudoxus appears to have been the first person to suggest that three 365 day years should be followed by a 366 day year, to keep the calendar in synchronisation with the year's natural length.[30] This was three hundred years before Julius Caesar implemented such a calendar.

According to Diogenes Laertius (fl. third century AD), Eudoxus was an expert geometer, astronomer, physician and legislator. He was reputed to have written parts of books V, VI and XII of Euclid's *Elements of Geometry*, but his main claim to fame in astronomy was his cosmological structure based on concentric spheres. This structure was originally explained in Eudoxus' book *On Velocities*, which is now lost. Aristotle (384–322 BC) mentioned the scheme briefly in his book *Metaphysics*, and his pupil Eudemus described it in more detail in the second book of his *History of Astronomy*, but that is now also lost. Sosigenes explained the theory, using Eudemus' book as the source, in yet another lost book, and finally in the sixth century AD Simplicius quoted extensively from Sosigenes in his book[31] that is now used as the main source of the theory.

---

[30] In 238 BC King Ptolemy III of Egypt was the first ruler to order that such a system be adopted, following advice from Aristarchus, but resistance to the proposal in Egypt was so great that it was not implemented.

[31] This was Simplicius' commentary on Book II of Aristotle's *De Caelo*.

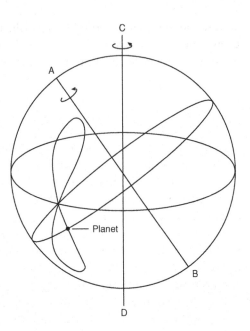

**Figure 1.4** The planet in this drawing, based on Eudoxus' system, is on the equator of an inclined sphere, whose poles A, B are embedded on the surface of a second sphere with a vertical axis CD. The two spheres rotate with equal and opposite speeds, as a result of which the planet moves in the figure-of-eight track shown. (From *Cambridge Illustrated History of Astronomy*, Michael Hoskin (ed.), 1997, p. 34, modified.)

Eudoxus found that he was able to qualitatively explain the oscillatory motion of a planet, assuming that the planet was on the equator of a sphere, which has its poles fixed on the surface of a second sphere. The two spheres were concentric, with the earth at their common centre, and they rotated uniformly with equal and opposite angular velocities about different poles. The composite motion of the two spheres produced a figure-of-eight planetary track called a hippopede (see Figure 1.4).

In order to explain the complete movement of the planets, Eudoxus used four spheres per planet, with the earth at their common centre. The first, outermost sphere rotated about the celestial poles in a day. The second, whose poles are the poles of the ecliptic, had those poles fastened on the surface of the first sphere, and rotated in the sidereal period for the superior planets or in one year for the inferior planets. Then there were two hippopede-producing spheres, fastened to the second sphere, to provide the backwards and forwards planetary motion.

We do not have all the numerical data that Eudoxus used to construct his spheres for each planet, but in 1875 Schiaparelli published a detailed reconstruction of the theory based on the planetary orbital parameters known to the Greeks of the time. This showed that, although the scheme worked qualitatively to produce the observed basic planetary motions, it was quantitatively deficient, particularly for Mars and Venus. In addition it could not explain the observed intensity variations of the planets, particularly of Mars, Venus and Mercury, as their distance from the earth remained constant in Eudoxus' theory.

Eudoxus' scheme for the moon consisted of three spheres. The outermost, like those for the planets, rotated once per day, the second rotated once every 18.6 years, which is the regression period for the moon's line of nodes, and the third rotated once every month of 27.2 days. Simplicius, for some reason, transposed the second and third spheres in his book, probably because they were also transposed in Aristotle's book *Metaphysics*, although it is strange that this error had not been picked up by Eudemus or Sosigenes. Eudoxus' correct theory produced a very satisfactory representation of the moon's motion, but the similar three sphere system for the sun was less successful. In all, Eudoxus' theory required twenty-seven spheres, consisting of four for each of the five planets, three for each of the moon and sun, and one for the stars. Although eight of these spheres served the same purpose of providing an identical daily movement, they were still treated as eight separate spheres.

It is not clear whether Eudoxus really thought that the sun, moon and planets were actually attached to transparent spheres, but it is thought unlikely. The scheme was probably intended as a mathematical concept, not as a model intended to mirror reality.

We are not certain what deficiencies were evident in Eudoxus' theory at the time, as the Greeks of his era were not great astronomical observers, nor how concerned he or his contemporaries were with these. However, Schiaparelli showed in the nineteenth century that Mars, Venus and the Sun had the largest discrepancies between their observed positions and those predicted by Eudoxus' theory.

## Callippus

About thirty years after Eudoxus published his theory, Callippus (c. 370–300 BC) produced a modified version by adding two spheres for each of the sun and moon, and one each for Mars, Venus and Mercury. No extra spheres were added for Jupiter and Saturn because, presumably, Callippus, like Schiaparelli over two thousand years later, found the agreement between theory and observation acceptable.

Callippus of Cyzicus was a pupil of Polemarchus who, in turn, had been a pupil of Eudoxus. He followed Polemarchus to Athens, where he stayed with Aristotle, and, with Aristotle's help, corrected and improved

Eudoxus' theory. Aristotle in his *Metaphysics* mentions the extra spheres added by Callippus, but does not say why they were added, nor what improvements were achieved. In turn, Simplicius says,[32] 'According to Eudemus, Callippus asserted that, assuming the periods between the solstices and equinoxes to differ to the extent that Euctemon and Meton held that they did, the three spheres in each case (i.e. for the sun and moon) are not sufficient to save [i.e. explain] the phenomena, in view of the irregularity which is observed in their motions. But the reason why he added the one sphere which he added in the case of each of the three planets Ares [Mars], Aphrodite [Venus], and Hermes [Mercury] was shortly and clearly stated by Eudemus.' Unfortunately we do not have a copy of Eudemus' book, so we are not aware of the latter reasons. We can only surmise that Callippus, and possible Eudoxus, had found the deficiencies too large for these three planets, as later demonstrated by Schiaparelli's reconstructions. Again Schiaparelli modified Eudoxus' system, by adding one sphere for each of these three planets in imitation of Callippus, and showed what improvements could be achieved. But how close his reconstruction was to Callippus' original in unknown.

Simplicius mentions that the periods between the solstices and equinoxes were found to vary by Euctemon and Meton. In about 430 BC these two astronomers did, in fact, find that the seasons, as measured between the solstices and equinoxes, varied from 93 days for spring, to 90 days for summer, 90 for autumn and 92 for winter. Eudoxus seems to have ignored this variation when he developed his theory about sixty years later, as he assumed that the annual motion of the sun is perfectly uniform. In about 330 BC Callippus measured the lengths of the seasons as 94, 92, 89 and 90 Days,[33] and it is known from modern calculations that the addition of two more spheres for the sun, as used by him, could explain this anisotropy. How accurately Callippus managed to correlate his theory with observation is unknown, however.

Simplicius' comment about the moon, in the above quotation, does not make sense, as the moon has nothing to do with the length of the seasons. Presumably he meant that the velocity of the moon in the sky is, like the sun, not uniform. Again the addition of two spheres can improve the correlation between theory and observation.

---

[32] Translation by T. L. Heath in his *Aristarchus of Samos; The Ancient Copernicus*, Clarendon Press, Oxford, 1913, republished by Dover, 1981, p. 213.

[33] These figures are, within rounding, the same as the modern values of 94.1, 92.2, 88.6 and 90.4 days.

## Aristotle

We turn now to one of the greatest of Greek philosophers, Aristotle (384–322 BC), who was born in the small Greek colony of Stagira (or Stagirus) in Macedonia. His father was court physician to Amyntas, king of Macedonia. When he was eighteen, Aristotle was sent to Plato's Academy in Athens, where he stayed until Plato's death some twenty years later. Unfortunately, there was a great deal of anti-Macedonian feeling at the time, as Amyntas' son, Philip, had just sacked the Greek city state of Olynthus. Also, because Aristotle had been born abroad, he could not own property in Greece, so he was ineligible to run the Academy. As a result, he left Athens when Plato's nephew Speusippus was made head of the Academy, and moved to Assus in Asia Minor, later moving to the island of Lesbos. A few years later, in about 342 BC, Aristotle was appointed by King Philip to tutor his son, Alexander, soon to become king (and later to be known as Alexander the Great). Then in 335 BC, at the age of forty-nine, Aristotle returned to Athens and set up the Lyceum institution as a rival seat of learning to Plato's Academy.

Aristotle's areas of interest were vast, covering logic, ethics, politics, rhetoric, psychology and biology, as well as astronomy. His approach to biology was practical, possibly influenced by his father's medical background, whereas in astronomy his approach was theoretical and strictly logical. Aristotle thought that he could sort out the structure of the universe purely by applying his logical skills, which were considerable, but this proved to be beyond him, like it had for all his predecessors. Knowing what we know now about the complexity of the universe, even just that part visible to the naked eye, this is hardly surprising.

Plato, like Philolaus and Empedokles (c. 494–434 BC) before him, thought that the universe consisted of four elements, namely earth, water, air and fire. Although all these elements were distributed throughout the universe, there were different proportions in some parts of the universe than others. So, for example, Plato saw the stars as being principally composed of fire. They moved in the ether, which was a specially pure form of air. Aristotle, on the other hand, assumed that the four basic elements occupied discrete regions in the so-called sub-lunary region below the moon's orbit. The element earth, for example, left to its own devices fell to the surface of the earth, water floated on the earth's surface, which was surrounded by air, and further out by fire. So Aristotle deduced that the order of elements from the centre of the earth was earth, water, air and fire. He also added a fifth element, called the 'quinta essentia' or 'ether', which was the most perfect of all the elements, and where natural motion was circular. The celestial bodies were, naturally, made of this the most perfect element.

Because all heavy things on earth try to get to the earth's centre, Aristotle reasoned that the exterior surface of the earth must be spherical, which is confirmed by the shape of the earth's shadow seen on the moon during a lunar eclipse. The moon is also spherical, as can be seen by the varying shape of the terminator (the line separating the illuminated from the non-illuminated part of the moon) as the month progresses. The earth cannot be very large because some of the stars seen in Egypt cannot be seen from Greece, and vice-versa. He also deduced that, as natural (i.e. non-forced) motion on the earth is up or down, the earth cannot spin around its axis, as that is contrary to the 'natural motion of its constituent parts'.

Aristotle saw motion as either circular around the central earth, or in a straight line towards or away from the earth, or a mixture of these. When a solid body is dropped on earth, it falls in a straight line to its home, the ground, where it will eventually become stationary, whereas flames rise upwards to their natural place above the earth. Change in the universe only occurs in the sub-lunar region of earth, water, air and fire, where we find comets, shooting stars, rainbows, thunderstorms and the like. The motion of the celestial bodies, however, is circular, invariable and eternal. Aristotle reasoned that the universe must be spherical, as that is the perfect shape. It must also be finite in size as it has a centre, namely the earth, and an infinite universe would have no centre.

So Aristotle's universe, which was perfectly spherically symmetric, was finite and spherical in shape, around a non-spinning earth at its centre. The celestial bodies were made of the most perfect element, ether, and they moved in circular, unchanging orbits. The earth, moon and all the celestial bodies were spherical in shape.

In detail, the structure of Aristotle's universe, which was based on that of Callippus, consisted of a series of concentric, transparent spheres, moving around the earth. As mentioned above, Eudoxus and Callippus probably saw their spheres as being mathematical devices, rather than real structures. Aristotle, on the other hand, saw his transparent spheres as real objects which were in contact with each other. The moving force for his system was transmitted from the outermost, celestial sphere, which was under Divine control, through the various planetary and solar spheres, to those of the moon. Aristotle was concerned, however, that if something were not done to alleviate matters, the complex motion of Saturn would be transmitted to Jupiter, as the spheres were in contact. Similarly, the complex Saturn and Jupiter motions would be transmitted to Mars, and so on. So he decoupled the motions of the various planets, the sun and the moon by introducing a number of additional 'unrolling' spheres, as he called them, between the spheres of Callippus.

In brief, Aristotle's system worked as follows.

Suppose that Callippus' four spheres for Saturn are labelled A, B, C, and D, with the outermost sphere A providing Saturn's basic daily motion, and with the innermost sphere, sphere D, supporting the planet. Inside these four spheres Aristotle added the 'unrolling' or 'reacting' spheres, which we will label D', C', and B', working inwards. D' rotates around the poles of D, with equal and opposite velocity, so D' cancels out the motion of D, as far as the next planet Jupiter is concerned (working inwards). As a result any point on D' will now move as if it is attached to C. Similarly, C' neutralises C by rotating about its same poles with equal and opposite velocity, and B' neutralises B in the same way. So any point on B' will move as if it is attached to A, and the first (outer) sphere of Jupiter will act as if all the spheres providing Saturn's unique motion did not exist.

Clearly there needed to be one less of these reacting spheres than the number of original or 'deferent' spheres, as Schiaparelli called them, for each of the planets and the sun. Aristotle did not add any reacting spheres for the moon, as there was nothing orbiting inside of it to be interfered with. As a result, in his universe Aristotle had one sphere for the stars, seven spheres (four deferent and three reactors) for each of Saturn and Jupiter, nine spheres (i.e. five plus four) for each of Mars, Mercury, Venus and the sun, and just the original five spheres for the moon, making a total of fifty-six spheres. Somewhat surprisingly, in view of the large number of spheres, Aristotle did not now eliminate the outer spheres for Jupiter, Mars, and so on, as their daily motions were now provided by Saturn's sphere A, to which their motion was locked.

Aristotle's theory was taught for a number of years after his death. But not only was it very complicated and inelegant, it still possessed the problems of Callippus' theory, probably the most important of which was that it could not explain the large brightness variations observed for Mars, Venus and Mercury over the course of one orbit. Nor could it explain the variations in the apparent size of the moon, as seen from earth, particularly from one annular solar eclipse to another. Autolycus, who lived shortly after Aristotle, tried to improve the model, but eventually it fell into disuse.

## 1.4 The Greeks in Egypt

Sometime around 3000 BC the Egyptians realised that the heliacal rising of Sirius, which was then on 22nd June, heralded the arrival of the Nile floods at their capital Memphis. This led them to devise a calendar linked to this heliacal rising, as the flooding of the Nile was the most important event in their agricultural year. Unfortunately, by about 2500 BC the predictive quality of Sirius had disappeared, as the date of its heliacal rising had moved so it was then after the start of the flood at Memphis owing, as we now know, to the precession of the equinoxes. As a result it became very important to the Egyptians to devise a reliable calendar so that the date of the flooding of the Nile could be anticipated.

The obvious astronomical objects to use in devising such a calendar were the sun and moon, so much of the early astronomical work in Egypt was devoted to trying to link the solar and lunar cycles in a reliable way. Consequently, by about the fourth century BC the Egyptians discovered a linkage, when they found that there were 309 synodic months in 25 of their 365 day years, which gave a synodic month of 29.53074 days. This compares with that deduced from the Metonic cycle, that was known in Greece and Babylon at the time, of 29.53192 days, or with the true value of 29.53059 days that the Babylonians and Greeks were not to deduce for a further two centuries.

We have already come across Alexander the Great as a pupil of Aristotle, and as the conqueror of the Persian empire in 331 BC. Although his and his successors' occupation of Babylon lasted over eighty years, it had only a limited effect on the astronomical work carried out there, but the effect on Egypt, which he also occupied, was substantial. Not only did the Greeks bring their geometry and theoretical skills to Egypt in Alexander's wake, but there were strong links established with Babylon which improved the Egyptians' arithmetical and observational skills.

Alexander spent the winter of 332–331 BC in Egypt where he was treated with enthusiasm by the local population, as he had released the country from Persian domination. He started reorganising the country and founded the city of Alexandria, that was to take over from the Phoenician city of Tyre as the centre of commerce between Europe and the East. Alexandria also became a centre of Greek learning, with a great museum and a library that was reputed to hold about 500,000 volumes.

## Eratosthenes

Aristarchus, who has been discussed above as one of the last Pythagoreans, was a great scholar in Alexandria. Eratosthenes of Cyrene (c. 276–195 BC), a contemporary of Archimedes,[34] a director of the Alexandrian library and

---

[34] Archimedes moved to Alexandria in about 250 BC, but left a few years later.

a tutor of Ptolemy III's son, was another. Although better known as a geographer, Eratosthenes determined the size of the earth by noting that the sun was exactly overhead at noon on midsummer's day in the town of Syene (now called Aswan) in southern Egypt.[35] At noon on midsummer's day the sun also produced a shadow in Alexandria that had an angle to the vertical of $\frac{1}{50}$ of a total circle. Since the distance from Syene to Alexandria was known to be about 5,000 stadia, he deduced that the earth's circumference was about $50 \times 5,000 = 250,000$ stadia, assuming that Alexandria is due north of Syene. He later updated his estimate by less than 1% to 252,000 stadia for some unknown reason. But how accurate was this estimate?

It is generally thought that the stade used by Eratosthenes was 157.5 m long. If this is correct, his estimate of size of the earth's circumference of 252,000 stadia converts to 39,690 km, which gives a diameter of 12,634 km. Although Eratosthenes did not know it, the earth is not absolutely spherical and he was estimating its polar dimensions. We now know that its polar diameter is 12,714 km, so his estimate was only 80 km or 0.6% too small.

This accuracy looks almost too good to be true, and in a way it was, because individual errors in his estimate had largely cancelled out. In particular, he measured a difference in latitude between Alexandria and Syene as 7° 12' (i.e. $\frac{1}{50} \times 360°$), whereas the true difference is 7° 7'. But Alexandria is 3° 14' west of Syene, not due north as he had assumed, so the length of the arc between these two towns is 7° 41', not 7° 12', giving an error of about 7% in his estimate. With too small an arc, his estimate of the circumference of the earth would have been too large, unless the linear distance was greater than he had assumed. In fact the distance must have been 5,412 stadia, or about 7.4% higher than he had estimated.[36] Nevertheless, in spite of the fact that he appears to have had two errors of about 7% that had almost cancelled each other out, his method was remarkably good for the time.

Eratosthenes also appears to have calculated the obliquity of the ecliptic by measuring the difference in the altitude of the sun at the winter and summer solstices to be $\frac{11}{83}$ of a total circle. This corresponds to an obliquity of 23° 51', which was only 8' from the true value at that time.

---

[35] This was not strictly true, as at that time the obliquity of the ecliptic was 23° 43', whereas the latitude of Syene was 24° 5', giving an angle at noon on midsummer's day of 0° 22' to the vertical. As a result, a thin vertical pole of 3 m height would have cast a 2 cm shadow.

[36] Assuming his estimate was $\frac{252,000}{250,000} \times 5,000$ stadia. That is assuming that he updated his 250,000 figure to 252,000 because of a slight underestimate in the 5,000 stadia distance.

## Hipparchus

Meton had proposed his calendar cycle (see Section 1.1) in about 430 BC, in which there were 235 synodic months in 19 tropical years or 6940 days, making a synodic month of 29.53192 days and a tropical year of 365.2632 days. About one hundred years later, Callippus modified this by reducing four Metonic cycles, that total 76 years, by one day, making a synodic month of 29.53085 days and a tropical year of exactly 365.25 days. We now know that both were slightly too large, but two hundred years later Hipparchus (c. 185–120 BC) improved the estimate even further.

Hipparchus, who was born in Nicaea in Asia Minor, is most celebrated for his star catalogue, which was completed to a previously unattained accuracy in 129 BC. In fact one of Hipparchus' key legacies to astronomy was his emphasis on the accuracy of observations, and his attempt to estimate the error of his results. His star catalogue is outside the scope of this present book, but he also carried out observations of the sun, moon and planets. These were undertaken mainly from an observatory on Rhodes, which at that time was, alongside Alexandria, a major centre of Greek culture. The observations allowed him to define the periods of the sun, moon and planets more accurately, and so to extend theoretical work on planetary orbits and the structure of the solar system. Although he wrote a number of books, only one survives, and so we are indebted to Ptolemy (c. 100–170 AD) who described his work.

In his *Syntaxis*, Claudius Ptolemaeus, better known as Ptolemy of Alexandria, describes how Hipparchus improved the 76 year cycle of Callippus. Hipparchus subtracted one day in four cycles, making 3,760 synodic months equivalent to 304 tropical years or 111,035 days. This makes a synodic month of 29.530585 days, or a tropical year of 365.2467 days, both closer to reality than Callippus' estimate of two centuries earlier. Hipparchus' estimate of the tropical year was still about six minutes too large, however, as was the slightly earlier Babylonian estimate for the sidereal year (see Section 1.2).

Hipparchus was well aware of Babylonian astronomy, and will have known that their estimate of the sidereal year (based on the stars) was longer than his estimate of the tropical year (based on the seasons). But his discovery of the precession of the equinoxes, which causes this difference, seems to have come via a different route. This precession of the equinoxes is now known, incidentally, to be due to the fact that the direction of the earth's axis is not fixed in space but is, instead, describing a cone at a rate of 50″.3 per year, thus completing one circuit in about 25,800 years.

Both the Babylonians and the Egyptians had been aware for some time that the sidereal and tropical years were not the same, because they had noticed the long-term drift of the stars compared with the solstices and other seasonal events. Nevertheless, Hipparchus was the first to quantify the difference in his treatise on the length of the year. In this, Hipparchus concluded that the equinoxes are moving relative to the sphere of the 'fixed stars' by at least 1° per hundred years, or at least 36″ per year.

In 294 and 283 BC, however, one of the earliest Alexandrian astronomers, Timocharis, had noted that the star Spica preceded the autumn equinox by 8°, whereas in constructing his star catalogue Hipparchus measured a value of 6° in 129 BC. The difference in dates between these observations was clear, because they were linked to known lunar eclipses. They gave values of 44″ and 47″ per year for the rate of precession, although these values were not quoted in any of Hipparchus' known work. There is evidence that he used a similar value in his work, however.

As mentioned earlier (Section 1.3), in about 330 BC Callippus had measured the length of the seasons as 94 (for spring), 92, 89, and 90 days, and in about 170 BC the Babylonians had devised a mathematical scheme (Section 1.2) that assumed two different solar velocities along the ecliptic. This produced seasonal lengths of 94.5, 92.7, 88.6 and 89.45 days, compared with Hipparchus' estimates of 94.5 for the spring and 92.5 days for the summer seasons. Whichever sets of numbers are used, however, it is obvious that the sun's apparent velocity is varying throughout the year, being fastest in the autumn and slowest in the spring. Callippus and the Babylonians had measured this variability, but left it there. Hipparchus, on the other hand, was driven to design an orbital scheme to explain it.

There were two basic alternatives for a geocentric universe based on circular orbits. Either the earth was at the centre of the sun's orbit, in which case the velocity of the sun had to vary in its orbit, or the earth was off-centre, and then the sun's velocity could be constant. So Hipparchus, in accepting Plato's and Aristotle's philosophy of uniform circular motion, had just one option and concluded that the earth was not at the centre of the sun's circular orbit (see Figure 1.5). He then showed that the seasonal lengths could be explained if the earth was offset from the centre of the sun's orbit by about $\frac{1}{24}$ of the orbit's radius, with an apogee in the direction of longitude 65° 30′. Although his estimate of the eccentricity was far too large, his direction of the apogee was in error by only 35′.

Aristarchus had estimated that the moon was about 50 earth radii away from the earth (Section 1.3), using his estimate of the size of the earth's shadow on the moon during a lunar eclipse. Hipparchus carried

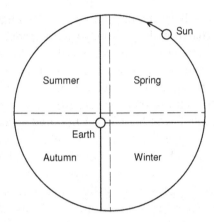

**Figure 1.5** Hipparchus' scheme to explain the different lengths of the earth's seasons. The sun orbits the earth in a circle at a uniform velocity, but the earth is offset from the centre.

out similar observations, and also tried to estimate the parallax of the sun and moon by observing total solar and lunar eclipses. For example, he was aware that a particular solar eclipse which had been total at the Hellespont was only an 80% eclipse at Alexandria, owing to parallax. As a result of this and other eclipse measurements, he estimated that the distance of the moon from the earth varies from 62 to $72\frac{2}{3}$ earth radii,[37] but he did not think he could detect any solar parallax. This is hardly surprising as the sun's parallax is so small, at about $9''$, that it can only be detected using telescopes, which were not to be invented for another 1,700 years or so.

## 1.5 Epicycles

As mentioned above (see Section 1.3), Heracleides had suggested in the fourth century BC that Venus and Mercury each describe a circular orbit around the sun, as it circles the earth. At about the same time, or possibly a little later, the concept of epicycles was introduced by an unknown Greek astronomer, in which a planet orbits a point in space, rather than the sun, as that point orbits the earth. The orbit of the planet around the empty point was called an epicycle, and that of the point around the earth was called a deferent. Both orbits are circular. The main advantage of this theory was that it could easily explain the temporary reversal in directions observed for the superior planets, as seen from the earth.

---

[37] We now know that the moon's distance from earth varies from about 55 to about 63 earth radii.

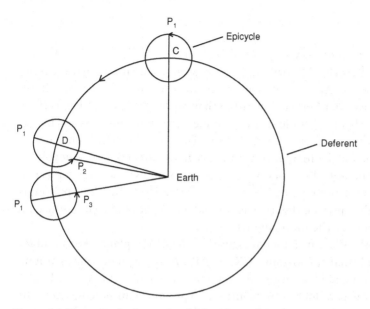

**Figure 1.6** The epicycle theory, in which a planet describes a circular epicycle orbit around a centre C which in turn describes a circular orbit around the earth. When the centre of the epicycle has moved to D, for example, the planet has moved in its epicycle from $P_1$ to $P_2$, see text for more details. (The velocity of the centre of the epicycle along the deferent has been exaggerated for clarity.)

In the basic epicycle theory a planet describes a uniform circular orbit, or epicycle, about a centre C, as shown in Figure 1.6. The centre C in turn describes a uniform circular orbit, or deferent, around the earth, with a slower velocity than that of the planet in the epicycle. When the planet is at point $P_1$ the net anticlockwise velocity of the planet around the earth, which is a composite of the epicycle and deferent velocities, is clearly a maximum. When the planet is at point $P_3$, however, because the velocity in the epicycle is greater than that in the deferent, the planet will have a net clockwise motion about the earth. At some intermediate point $P_2$, the anticlockwise and clockwise movements about the earth will be in balance, and the planet will appear to be temporarily stationary as seen from earth at one of its stationary points. After $P_3$ the net clockwise motion will gradually slow down until it falls to zero at the next stationary point, before the planet resumes its anticlockwise motion.

Not only did this epicycle theory explain the temporary reversals in motion of the superior planets, as observed from earth, but it also explained their apparent change in intensity as they moved to and from the earth, which Aristotle's theory, for example, was unable to do.

## Apollonius

The first person to write about the epicycle theory, as far as we know, was Apollonius of Perga (*c.* 265–190 BC), who examined the mathematical properties of epicycles. Apollonius, who spent most of his professional life in Alexandria, was a mathematician rather than an astronomer, and he is more famous for his theory of conic sections (circles, ellipses, parabolas and hyperbolas) than for his work on epicycles. His mathematical work is contained in eight books, seven of which have survived, whereas his work on epicycles is now lost. Fortunately, his work on epicycles was known to Hipparchus and Ptolemy, and Ptolemy recorded his work for posterity. Crucially Apollonius showed that the epicycle theory was equivalent to another system based on eccentric circles, as the following will explain.

Consider the system shown in Figure 1.7. When the planet is at $P_1$, mark a point $X_1$ such that the distance $CX_1 = AP_1$. As the planet rotates around its epicycle to the new point $P_2$, mark a point $X_2$ such that $CX_2 = BP_2$ and such that $CX_2$ is parallel to $BP_2$. Continue to produce similar constructions as the planet rotates around its epicycle. Then the points $X_1$, $X_2$, etc. will lie on a circle of the same diameter as the epicycle but centred on C. As the distances $P_1 X_1 = AC = BC = P_2 X_2 = r$, say, the planet can be thought of as rotating in a circle of radius $r$, but with a centre that moves around the central circle, which is the locus of $X_1$, $X_2$, etc.

In Figure 1.7 there is no linkage between the rate at which the planet goes around its epicycle, and the rate at which the centre of the epicycle goes around the deferent. If we now consider planetary motion starting at $P_2$ (i.e. ignoring $P_1$), and constrain the motion in the epicycle by insisting that not only does $CX_2 = BP_2$ and that $CX_2$ is parallel to $BP_2$, but the line $CX_2$ is fixed as the planet at $P_2$ goes around the epicycle and the centre of the epicycle goes around the deferent, then the planet will appear to orbit in a fixed circle centred on the so-called eccentric point $X_2$, offset from the earth at C. This was, of course, the type of orbit that Hipparchus assumed for the sun (see Figure 1.5 above) to explain the variation in length of the earth's seasons, by placing the earth off-centre at the eccentric point.

## Ptolemy

Ptolemy (*c.* 100–170 AD) was the main person to develop the epicycle theory some four hundred years after Apollonius (and three hundred years after Hipparchus). Whereas Apollonius had been mainly interested in the

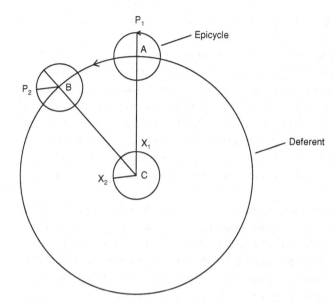

**Figure 1.7** The equivalence of the epicycle concept with motion in a circular orbit. When the planet is at $P_1$ the centre of the circular orbit is at $X_1$, and when at $P_2$ the centre has moved to $X_2$. So the planet can be thought of as rotating in a circle of radius $r = P_1 X_1 = P_2 X_2$, but with a centre that moves around a central circle with the same radius as the epicycle. (See text for more details.)

mathematical properties of epicycles, Ptolemy decided to use them to try to explain the detailed movements of the sun, moon and planets and so enable accurate predictions to be made of their future positions.

We know little about Ptolemy as a person, except that he carried out his astronomical observations from Alexandria between 127 and 141 AD, when it was under Roman rule. At that time Alexandria had the most complete library in the world, which Ptolemy undoubtedly used to write some of his numerous books, most of which have survived. His most famous book, now called the *Almagest*, is a veritable treasure house of astronomical information, covering both earlier Greek astronomy, as well as his own extensive contributions to the subject. It was to be the work of reference in astronomy for well over a thousand years. In addition he wrote works on astrology, music, optics and geography. In optics he studied refraction, and in geography he produced maps of Asia and parts of Africa which were to inspire Columbus over a thousand years later to attempt to reach India by sailing west across the Atlantic.

The structure of Ptolemy's universe, as defined in the *Almagest* and in a later work called *Planetary Hypotheses*, was a mixture of logic, calculation and guesswork. He discussed the arguments for and against a spinning earth and concluded, like Aristotle, that the earth does not spin on its axis. Then, having settled on an earth-centred universe like most of his predecessors, he considered the possible order of the sun, moon and planets in their distances from the earth. It was generally accepted at the time that, working inwards from the sphere containing the stars, the correct order was Saturn, Jupiter and Mars, in view of their apparent motions along the ecliptic (i.e. with Saturn being the slowest etc.) but, unfortunately, Ptolemy could find no way of determining the order of the remainder. In the end he decided to take the order taught by the Stoic philosopher Diogenes of Babylon in the second century BC which, again working inwards, was the sun, Venus, Mercury and the moon. But he acknowledged that he may be wrong, even though this scheme had the aesthetic advantage of having three planets outside the orbit of the sun and three inside, taking the moon as a planet.

It should be noted that, although some five hundred years earlier Heracleides had deduced that Venus and Mercury orbited the sun as it orbits the earth, Ptolemy assumed that the epicycles of Venus and Mercury, like those for the superior planets, orbited empty space. However, as both of the inferior planets appear almost symmetrically on either side of the sun at the longitudinal extremities of their orbits, he assumed that the centres of their epicycles were always on the line joining the earth and sun (see Figure 1.8). For the superior planets, on the other hand, he assumed that the lines linking their individual positions with the centres of their epicycles were always parallel to the earth–mean sun line, where the mean sun was taken to orbit the earth uniformly in one year.

To determine the size of the universe, Ptolemy assumed that there were no gaps between the most distant position of the moon in its epicycle and the closest position of Mercury, and between the most distant position of Mercury and the closest position of Venus, and so on up to the furthest position of Saturn, where he placed the fixed stars. He calculated that the maximum distance of the moon was 64 earth radii, and having calculated the sizes of the epicycles for the sun and planets, he concluded that the radius of the universe, as determined by the sphere of fixed stars was 19,865 times the earth's radius, or about 120 million kilometres.

Ptolemy made the centre of the moon's deferent move on a circle, whose centre was the earth, in order to improve the fit between the observed and calculated positions of the moon particularly around quadrature (i.e. half-moon). Interestingly, this caused the distance of the moon from the earth to vary from 33 to 64 earth radii. This should have caused the apparent size of the moon in the sky to vary also by a factor of two, which even the most casual observations indicate is not so. We do not know how concerned Ptolemy was with this discrepancy, as he does not mention it in any of his known works. In fact we do not even know whether Ptolemy was trying to

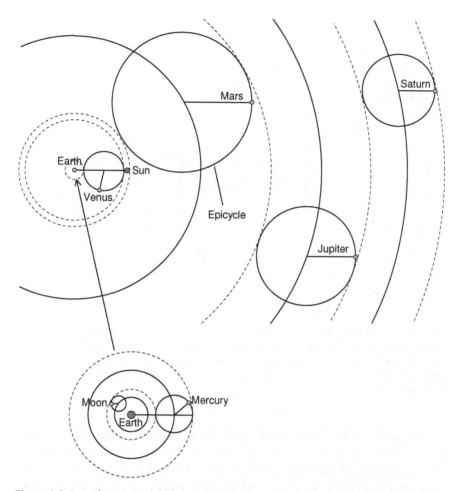

**Figure 1.8** In Ptolemy's model of the universe there were no gaps between the circle enclosing the furthest distance of one planet, and that just touching the epicycle of the next planet out from the earth. Unlike Heracleides, he assumed that the epicycles of Venus and Mercury were not centred on the Sun, but like the other planets on empty space.

produce an accurate physical representation of the universe, or whether he was simply trying to use geometrical constructions to produce reasonably accurate predictions for the positions of the sun, moon and planets. Although we now know that the physical structure of Ptolemy's universe was indeed wrong in almost every respect, nevertheless his estimate of its size was important as it showed, for the first time, how much larger it was than the earth and our local environment.

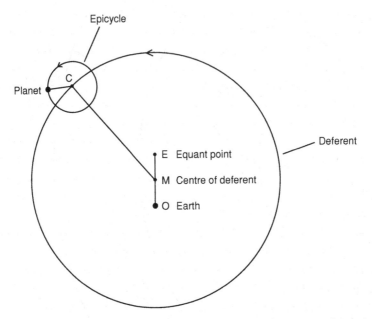

Epicycle

Planet

C

E  Equant point

M  Centre of deferent

O  Earth

Deferent

**Figure 1.9** Ptolemy modified the epicycle theory for the superior planets by moving the earth O from the centre M of the circular deferent, and by defining an equant point E such that the distance EM = MO. He then assumed that the angular velocity of C, the centre of the epicycle, is uniform about the equant point E, rather than about the centre M of the deferent.

In parallel with this general approach, Ptolemy modified the detailed working of the epicycle theory for the superior planets as the velocities along their deferents did not seem to be constant. To do this he moved the earth from the centre of the deferent, and defined a point, called the equant, that he placed at an equal distance to the earth on the opposite side of the centre (see Figure 1.9). He then assumed that the centre of the planet's epicycle appears to move at a constant velocity as seen from the equant, which of course meant that it no longer moved at a constant velocity along the circular deferent. This was a major change and caused much difficulty both during his lifetime and afterwards, as it violated the almost sacred principle that all celestial objects move at uniform velocity. Of course, the whole epicycle theory had previously disguised the fact that the planets do not orbit the earth in circular orbits, which was another fundamental principle to the Greeks.

There were further complications as Ptolemy had also to explain the deviations of the planets from the ecliptic. To do this he introduced, for the

Table 1.1 *A comparison between the orbital elements calculated by Ptolemy and those calculated for his era using modern theory*

|         | Ptolemy | | Correct | |
|---------|--------------|--------|--------------------------|--------|
|         | Eccentricity | Apogee | Eccentricity[a] | Apogee |
| Mars    | 0.200        | 116°   | 0.186                    | 121°   |
| Jupiter | 0.092        | 161°   | 0.096                    | 164°   |
| Saturn  | 0.114        | 233°   | 0.112                    | 239°   |

[a] This is using the old definition of eccentricity which gives values twice that today.

superior planets, small inclinations of the deferent to the ecliptic and of the epicycle to the deferent. For Venus and Mercury, however, he introduced small vertical circles to their deferents and epicycles to provide vertical oscillatory motion.

As he added more and more complications to overcome specific problems, Ptolemy must have wondered if there was not a simpler solution to planetary motions, but if there was no-one else had found it. Furthermore he could point to the accuracy of his model, with which he was able to make accurate position estimates for all but the moon and Mercury. To give some idea of the accuracy of his model, Table 1.1 compares his estimates of eccentricity and apogee longitude for the superior planets with those calculated today for his time. As can be seen, the agreement is very good.

In the case of Venus, Ptolemy calculated an eccentricity of $\frac{1}{24}$, which is the same as he deduced for the sun, and he also calculated an apogee longitude that was almost the same as for the sun. This indicated that the sun was probably at the centre of Venus' epicycle, but for some reason he did not draw such a conclusion in his extant work. The case of Mercury was more demanding, however, as not only was it difficult to observe Mercury accurately for most of its orbit, but its orbital eccentricity was too large for Ptolemy's model to handle satisfactorily.

Shortly after Ptolemy's period, the Roman empire began to collapse, and astronomy in Europe was to enter a 'dark' age, as far as astronomical progress was concerned, from which it would not surface for over one thousand years.

# Chapter 2 | COPERNICUS AND THE NEW COSMOLOGY

## 2.1 The 'Dark' Ages

It seems incredible that for well over a thousand years, between the work of Ptolemy in the second century AD and that of Copernicus in the early sixteenth century, the development of astronomical ideas was at a virtual standstill. The causes were many and various, but the lack of interest in science shown by the Romans,[1] and various wars, both religious and secular, did not help. Unlike the priests of the Babylonian era, the Christian priests showed little interest in developing their understanding of the heavens beyond that of their predecessors, and there were no secular alternative groups of intellectuals like those in the earlier Greek communities. However, that is not to say that the ideas of Plato, Aristotle and the other Greek philosophers were forgotten in this period, but they were not extended, as far as planetary astronomy was concerned.

The rise of Christianity in Europe was a particular problem for science. This was because many early Christians believed that all truth lay in the Bible, and so the structure of the cosmos could only be deduced with reference to Biblical texts. However, as Thomas Aquinas pointed out in his *Summa theologiae*, the highly influential St Augustine (354–430) had taught that where science and Scripture disagree, scientific ideas could be adopted over Scriptural interpretation, but only if they could be demonstrated to be unambiguously correct, which was rarely the case. So science did have a

---

[1] The Romans were much more interested in practical engineering than in the development of scientific principles, per se. They were also highly superstitious, and so astrology took root more in their thinking than astronomy.

role to play in interpreting the cosmos, but it was heavily constrained by Christian Scripture.

Meanwhile Mesopotamian astronomy had reached India in the late fifth century BC, followed by Greek astronomy subsequent to Alexander the Great's conquest of the western extremities of India (which are now in Pakistan). From time to time these contacts continued, then during the Gupta dynasty (c. 400–650) new Greek cosmological ideas reached India, possibly as a result of the persecution of Nestorian Christians[2] by the Byzantine Church. Whatever the route, Indian cosmology inherited an Aristotelian system of perfectly concentric spheres, but they were modified by the addition of epicycles. They had no equant, however, so their geocentric system owed more to Apollonius than Ptolemy. The sun and moon each had one epicycle, but the planets had two concentric epicycles each, arranged according to a complex set of rules. In some Indian systems the planetary epicycles were oval in shape, but in others this applied only to the epicycles of Mars and Venus.

In 628 an astronomer called Brahmagupta wrote a book in which he described Indian cosmology, and in 773 an Indian savant explained the mathematical techniques used in Indian astronomy to caliph Al Mansur in Baghdad. This prompted Al Mansur to order that Indian astronomical texts be translated into Arabic, which resulted, inter alia, in the adoption from India of the decimal system and the concept of zero. Translations of ancient Greek texts were also undertaken at second and third hand, until in the reign of Al-Ma'mûn (813–833) translations of the original Greek texts of Ptolemy's *Almagest* and other important works were produced.

Although the Arabs undertook some original work, their main importance to astronomy in this period was that they resurrected Greek astronomy, without the problems caused by Christian Scripture, using more complete copies of Greek texts than were available in 'Dark' Age Europe. Eventually Arab work started to percolate into Europe, mainly through Islamic Spain in the eleventh and twelfth centuries. In particular, Euclid's *Elements* were rediscovered by Christian Europe when Adelard of Bath came across an Arab translation in Cordoba in 1120, and shortly afterwards an Arab translation of Ptolemy's *Almagest* was also found. The route by which these Arab translations reached Europe was, unfortunately, severed in the thirteenth century with the overthrow of the Moors in Spain. It was during this period of the

---

[2] The Nestorians were a Christian sect in Asia Minor and Syria who refused to accept the condemnation of Nestorius by the councils of Ephesus and Chalcedon in the fifth century.

Islamic occupation of Spain that the influential *Toledo Planetary Tables* were produced in the eleventh century by Ibn al-Zarqâla (otherwise called Arzachel), followed in 1272 by the *Alfonsine Tables* produced under the patronage of Alfonso The Wise, King of Castile.

The most original Arab astronomer was probably Ibn al-Shātir (1304–1375), an astronomer working in Damascus who succeeded in getting rid of Ptolemy's equant, which was still a subject of controversy, by adding epicycles in a different way to the Indians. In al-Shātir's system, both the sun and moon had two epicycles riding one on another, whereas the planets each had three similar types of epicycles, which were not concentric like in the Indian system. His scheme had the big advantage, compared with the Ptolemaic system, of being able to explain the motion of the moon without the large apparent size variations implied by the latter. Strangely the movement of the sun in al-Shātir's geocentric scheme was similar to that of the earth in Copernicus' heliocentric scheme, and yet there is no evidence that Copernicus was aware of al-Shātir's work. A remarkable coincidence indeed.[3]

Meanwhile in Europe, universities had been founded in Bologna (in about 1080), Paris (*c.* 1160), Oxford (*c.* 1180), Padua (1222) and Naples (1224). The Church had become more organised and wanted a more accurate calendar to define the date of Easter, and the Franciscan and Dominican orders had been founded. Great thinkers now came to the fore like the Dominican Thomas Aquinas (1225–1274), who argued that reason is able to operate within faith, and who lectured on Aristotle at the University of Paris, and the Franciscan Roger Bacon (*c.* 1220–1292), who argued for and developed experimentation in science. Then in about 1310 Dante wrote his *Divine Comedy*, which combined theology and astrology with a modified Aristotelian description of the universe.

The University of Paris lost its dominant position on the continent during the Hundred Years War (1337–1453) between England and France. At about the same time, there was a great flowering of the arts and sciences in Italy and in German-speaking Europe, and new universities[4] were founded in Prague (1348), Vienna (1365) and Heidelberg (1386). It was, however, the

---

[3] Some people think that the similarities are too great to be a coincidence. They suggest that Copernicus must have known about al-Shātir's work although, as mentioned above, we have no evidence of this.

[4] These were probably the most important new universities of the period. By 1380 there were, in fact, a total of about thirty universities in Europe, although some of them were little more than schools.

invention of the printing press in the middle of the fifteenth century that was to have the most far-reaching consequences for astronomy. No more would astronomers have to rely on texts that had been copied successively by scribes, who often did not really understand what they were copying, and whose mistakes were generally not corrected at the next transcription, thus adding errors to errors.

By the fifteenth century the European religious calendar was in much need of reform, as the vernal equinox was by then ten days in advance of its correct date, and so the computation of Easter Day was compromised. This led the Pope to summon Johann Müller of Königsberg (1436–1476), otherwise known as Regiomontanus, to Rome in 1475 to seek his advice on what should be done. Regiomontanus advised the Pope that new observations were needed before an accurate system could be put in place but, unfortunately, Regiomontanus died the following year before they could be undertaken.[5]

Navigation also called for a wider understanding of astronomy at this time, when explorers were leaving the relative security of the Mediterranean, and the coastal waters of western Europe and Asia for the first time. For example, in 1420 Zarco and Teixeira discovered Madeira, and in 1487 Dias rounded the Cape of Good Hope. At these previously unexplored southern latitudes, navigators were able to observe new stellar constellations for the first time, particular the Southern Cross, which they erroneously took to be the constellation of four stars mentioned by Dante in his *Divine Comedy*. Not only were explorers travelling further south, however, but in 1492 Columbus crossed the Atlantic to try to find a western route to India. This added a new requirement on astronomy to help to measure longitude accurately, which was not to be solved for well over two hundred years.

These calendar and navigational interests helped to keep astronomy alive in the fifteenth and sixteenth centuries. Together with its theological significance, they helped to maintain its position as one of the seven core subjects[6] at European universities as the Renaissance got into full swing.

---

[5] Regiomontanus was not the first person to be invited to advise a Pope on calendar reform, nor would he be the last. John of Murs and Firmin de Bellaval had been invited to Rome to advise Pope Clement VI in the previous century, but it was not until 1582 that the calendar of the Roman Catholic Church was finally reformed. Interestingly, many years before, in 1074, Omar Khayyam (the author of the famous *Rubáiyát*) had also been asked by the sultan to help to reform the Islamic calendar.

[6] The seven core subjects taught at medieval universities were the arts of grammar, rhetoric, and logic (called the trivium), and the sciences of arithmetic, harmony, geometry and astronomy (the quadrivium).

## 2.2 Copernicus

Although there had been occasional attempts to improve Ptolemaic cosmology, it was still taught at universities in late fifteenth century Europe. There were, however, both philosophical and practical objections to it: the concept of the equant was generally disliked, and the apparent size variations of the moon, which was a consequence of the model, clearly did not exist. Whether these problems were the impetus behind Copernicus' attempt to produce a new cosmology, however, is unknown, although he was clearly aware of them.

Nicolaus Copernicus (Niklas Koppernigk) (1473–1543) was born in Torún on 19th February 1473, when it was under the suzerainty of the king of Poland. Ten years later he was adopted by his maternal uncle, Lucas Watzelrode, when his father died. At the age of eighteen he began to study mathematics and classics at the University of Cracow, where he became sufficiently acquainted with astronomy to buy his own copy of the *Alfonsine Tables*. Encouraged by his uncle, he travelled in 1496 to the University of Bologna in Italy, where he studied law and astronomy, and where his study of Greek enabled him to read ancient astronomical and philosophical texts. This was the era of Leonardo da Vinci and Michelangelo and, with the Renaissance in full swing, Italy must have been a wonderful and challenging place for a man of learning and ideas like Copernicus. In 1501 he moved to Padua, where he studied law and medicine, and then in 1506 returned to Poland as Canon of Frauenburg (Frombork) Cathedral.[7] There he also worked as a doctor and private secretary to his uncle, who was the bishop.

Copernicus had begun to make astronomical observations in 1497 when he was in Bologna, but it was not until his uncle's death in 1512 that he wrote the brief, anonymous text *Commentariolus*, which he distributed to his friends. In this he pointed out the deficiency, as he saw it, of the Ptolemaic system, in which the planets do not move at uniform speed either along their deferent or with respect to the centre of their orbit. Instead he outlined a system still based on epicycles, but using combinations of uniform circular motion without Ptolemy's equant. More importantly, the planets in his system orbit the sun (see Figure 2.1), rather than the earth in Ptolemy's system, and the earth spins once per day about its axis, so it was no longer necessary to assume that the whole universe rotated around the earth once per day.

---

[7] Copernicus was appointed canon in 1497 by his uncle, but he did not take up this position until his permanent return in 1506. Even then he spent most of the next six years at Heilsberg castle where his uncle lived.

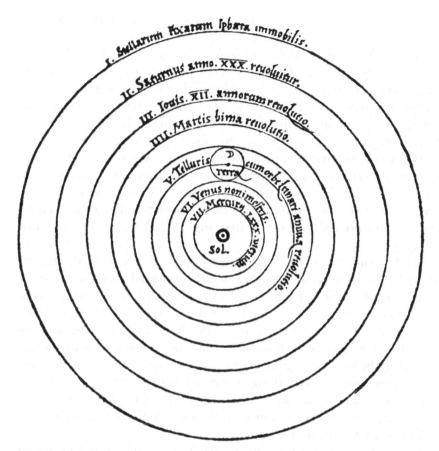

**Figure 2.1** The heliocentric universe as described by Copernicus in *De Revolutionibus*, with the planets orbiting the sun (Sol) and the moon orbiting the earth (Terra).

So the Copernican system is basically the same as that of Aristarchus of almost two thousand years earlier (see Section 1.3), but with the addition of epicycles. Copernicus then explained that the motion of the earth round the sun could not be detected by looking at the stars, as they were too far away for parallax to be observable. He also explained an important consequence of his model, in that the orbital motion of the earth and planets around the sun naturally produced retrograde loops for the outer planets.

Copernicus' cosmology was explained to Pope Clement VII and his advisors by the Austrian chancellor, Johann von Widmanstadt (1506–1557), in 1533. Interestingly, in the light of Galileo's subsequent problems with the Roman Catholic Church, no one seems to have raised any serious objections to it. In fact, three years later Nicholas Schönberg, the Cardinal Archbishop of Capua, and a member of the Roman Curia, suggested to Copernicus that he should have his work published. Schönberg even offered to send Theodoric of Rheden (Dietrich von Rheden) to Copernicus' house to make the arrangements, but Copernicus did not take up the offer. Then in 1539, Georg Joachim Rheticus (1514–1576), a teacher of mathematics at the University of Wittenberg, visited Copernicus and persuaded him to allow a *First Report* of his work to be published. When this also produced no serious adverse reaction, Copernicus finally agreed to allow his full work *De Revolutionibus Orbium Caelestium* (*On the Revolutions of the Heavenly Spheres*), or *De Revolutionibus*, as it is now generally known, to be published. This was done in Nuremberg in 1543, the year that Copernicus died. As a result he was to see the printed book for the first time on 24th May 1543, the day of his death.

*De Revolutionibus* was well received, possibly aided in theological circles by an anonymous Foreword, which explained that the model of the universe described in the book should not be taken literally as a true representation of the universe, but rather as a convenient mathematical interpretation of it. At the time it was generally thought that the Foreword had been written by Copernicus himself, but later research indicates that it was the work of the Lutheran theologian Andreas Osiander (1498–1552), who had been charged by Rheticus with supervising the book's publication.[8] In addition to the Foreword, the book starts with a copy of Cardinal Schönberg's letter urging Copernicus to publish his theory, together with a Preface and Dedication to Pope Paul III.

In *De Revolutionibus* Copernicus described his new cosmology, which had been outlined previously in *Commentariolus*, but the majority of the new book was an extensive numerical analysis which showed what his theory could achieve in practice.[9] Although these result were generally not much more accurate than those in Ptolemy's *Almagest*, the model was more intellectually satisfying in the way that it explained the complex motions of the outer planets, and showed that the orbits of Mercury and Venus were, in principle, no different from those of the outer planets. The reason why the orbits of these inner planets appeared to be different, as seen from earth, was simply due to the fact that they were inside

---

[8] The first time that Osiander was identified in print as the author of the Foreword was by Kepler in 1609 in his book on Mars.

[9] The planetary models in *De Revolutionibus* were also more complicated than those in *Commentariolus*, particularly for Mercury and Venus.

the orbit of the earth, rather than outside. Furthermore, no longer was it necessary to guess the order of the inner planets from the sun, as had Ptolemy, because they could now be deduced from the theory. Not only that, but the relative distances of all the planets from the sun could also be calculated.[10]

Copernicus was able to substantially improve the calculated orbit of the moon, compared with Ptolemy's, implying an apparent variation in the moon's diameter, as seen from earth, of only ±13%, compared with the ±30% of Ptolemy's model. Although this was still larger than observed, it was much closer to the truth. Unfortunately, however, the planetary latitudes in *De Revolutionibus* were still as badly in error as those in the *Almagest*, even though the longitude predictions in both were quite accurate.

It is often thought that Copernicus improved Ptolemy's theory by significantly reducing the number of circular motions used, but this was not so[11] for three reasons:

- As Copernicus eliminated the equant, he had to replace it with an extra epicycle for each planet.
- He made his theory unnecessarily complicated by insisting that the centre of the earth's orbit, rather than the sun, was at the centre of the planets' deferents. (This was also a philosophically unattractive idea, as it meant that the whole universe was orbiting the same empty point in space, rather than a solid body like the sun or earth.)
- He erroneously thought that both the period of precession of the earth's axis and the eccentricity of its orbit were changing markedly with time.[12] As a result, he added more circular motions to his model.

---

[10] The distances (in units of the earth's distance from the sun, now called the astronomical unit) were remarkably accurate. For example, the mean distances varied from 0.3763 (modern value 0.3871) for Mercury, to 5.219 (5.203) for Jupiter, and 9.174 (9.539) for Saturn.

[11] According to Arthur Koestler in *The Sleepwalkers*, (Penguin-Arkana, 1989), Copernicus used 48 circular motions in his model of the universe in *De Revolutionibus*, compared with 40 for Ptolemy.

[12] Ptolemy used Hipparchus' initial estimate for the rate of precession of the earth's axis of 1° in 100 years, whereas later measurements gave a figure of 1° in 66 years. Medieval astronomers erroneously took this to indicate a real variation in the rate of precession. Likewise, Ptolemy used Hipparchus' estimate of $\frac{1}{24}$ for the eccentricity of the sun's orbit around the earth, whereas more recent measurements had produced a figure of $\frac{1}{30}$, which was again erroneously taken to show a real change.

It is interesting at this stage, therefore, to ask why Copernicus' work is considered to be so important to the development of astronomy.

Copernicus was not the first person to suggest that the earth rotated on its axis, that had been proposed by Hicetas and Heracleides two thousand years earlier,[13] as Copernicus acknowledged in his Preface. Neither was he the first to propose a sun-centred universe, that was Aristarchus in the third century BC, as Copernicus also acknowledged in his book.[14] The detailed structure of the Copernican cosmos was also not very much simpler than that of Ptolemy, nor were his results much more accurate. It is true that his lunar orbit did not imply the large apparent variation in the size of the moon implied by Ptolemy's model, but that problem had also been solved by Ibn al-Shātir in the fourteenth century.

Nevertheless, Copernicus was important because he had the right ideas at the right time. Although in the fourteenth century Jean Buridan and Nicole Oresme had both seriously considered that the earth spun on its axis, they had rejected the concept and adopted the traditional Aristotelian idea of a non-spinning earth that was assumed by the other learned men of the time. Also, no-one had given serious thought to a heliocentric universe for two thousand years. So Copernicus *did* provide an invaluable service in resurrecting these concepts at a time when learned men were eager for new ideas at the height of the Renaissance.

## 2.3 Tycho Brahe

Eight years after Copernicus' death, Erasmus Reinhold (1511–1553) published a new set of tables, based on Copernican parameters, to replace the obsolete *Alfonsine Tables*, although he did not specifically refer to Copernicus' heliocentric universe in the book. These tables, called the *Prutenic* (Prussian) *Tables*, in honour of his patron Duke Albrecht of Prussia, were a significant improvement on the *Alfonsine Tables*. But in 1563 the young Tycho Brahe (1546–1601) found that neither the *Prutenic* nor the *Alfonsine Tables* accurately predicted the date of the conjunction of Jupiter and Saturn of that year. This was a crucial problem to someone like Tycho who was interested in astrology, where such conjunctions were

---

[13] A spinning earth was also a consequence of Philolaus' theory, but was not specifically proposed by him (see Section 1.3).

[14] This acknowledgement was in a handwritten draft that he deleted before publication.

very important, as well as astronomy. It encouraged him, therefore, to try to measure the positions of the heavenly bodies much more accurately than had previous observers, to enable more accurate predictions to be made.

Tycho Brahe was born on 14th December 1546 in Skåne, Denmark[15] of a wealthy noble family. He was brought up by his paternal uncle, who sent him to the Lutheran University of Copenhagen at the age of thirteen to study rhetoric and logic, as befitted a future nobleman. In 1560, however, Tycho observed a partial solar eclipse, which stimulated his interest in astronomy. Two years later he moved to the University of Leipzig, and in 1564 he started serious observing. Tycho then moved to the University of Rostock where he graduated in 1566, the year after his uncle died, moving from there to Basle and thence to Augsburg in 1569. There he designed a vertical wooden quadrant, 19 feet (6 metres) in radius, for two brothers Paul and Johann Hainzel who were members of the town council. Unfortunately, like many later instruments, although it was very accurate, it was somewhat unwieldy and so was only used intermittently.

No one had any major reason to doubt Aristotle's doctrine of the unchangeability of the universe beyond the moon (see Section 1.3) until November 1572, when a new bright object was observed in the constellation of Cassiopeia. Tycho first observed it by accident on 11th November when he was returning home from his laboratory, but later investigations showed that it had first been observed on 6th November by Francesco Maurolyco, a Sicilian mathematician, and Wolfgang Schüler at Wittenberg. On 7th November it had been independently observed by Paul Hainzel of Augsburg, Bernhard Lindauer of Winterthur (Switzerland) and Michael Mästlin (1550–1631), the future teacher of Kepler. Mästlin, Thomas Digges (c. 1546–1595), an English mathematician, and Tycho all independently proved that the star did not move against the other stars and so it was not a comet. Tycho could also find no parallax,[16] so the new object must be much further away than the moon, and it twinkled so it could not be a planet.[17] This led him to conclude

---

[15] Skåne is now in Sweden.

[16] The new object was so bright and near the pole star that it was also visible in daylight from Denmark. This enabled Tycho to measure its position 12 hours apart, when the rotation of the earth should have shown its parallax against the background stars, if it was very near to the earth. The lack of a measurable parallax led Tycho to conclude that the object must be at least ten times as far away as the moon.

[17] Planets do not twinkle like stars as they subtend a finite disc at the earth, with light from various parts of the disc smoothing out any possible twinkling due to the earth's atmosphere.

that it was a new star. He was reluctant to say that it was a 'brand new' star, however, because of the immutability of the stellar regions preached by Aristotle. So he suggested that the star had probably always been there, but that God had only just made it visible to humans.

This new star had a quite remarkable effect on both scholars and the ordinary people alike. Some thought that it was like the star of Bethlehem, this time announcing the second coming of Christ. Others were sure that it presaged disaster like a comet, but in the event it did neither. After November 1572 it gradually decreased in brightness and changed colour, and then in 1574 it finally disappeared.

The appearance of this new star, now called Tycho's star or supernova, persuaded Tycho Brahe to found an astronomical observatory, possibly in Basel, to observe and measure the heavens to unprecedented accuracy. In 1575 he set off on a tour of Germany looking for a potential site. But when his search came to the attention of King Frederick II of Denmark, the king decided to offer Tycho the lordship of the small island of Hven near Copenhagen to keep him in Denmark. The king also offered to provide money for an observatory. Not surprisingly, Tycho accepted, and in the following year he started to build his observatory of Uraniborg (meaning 'heavenly castle'), which he equipped with a range of state-of-the-art instruments.[18]

In 1577 Tycho demolished another of Aristotle's theories when he observed the comet of that year. Aristotle had taught that comets were in the earth's upper atmosphere, but Tycho tried to measure the parallax of the 1577 comet and found that it was so small that he could hardly detect it. This meant that the comet was appreciably further away than the moon, proving that comets were astronomical rather than atmospheric phenomena. Along with the new star of five years earlier, it provided convincing evidence that the universe was not immutable. His calculations showed that the comet, which was last seen on 26th January 1578, was in an almost circular orbit around the sun, outside the orbit of Venus.[19] This was consistent with his model of the solar system in which Mercury and Venus orbit the sun, as outlined at the end of this chapter.

---

[18] None of these instruments were telescopes, of course, as telescopes were not used for astronomical observations until early in the following century (see Section 3.2).

[19] Michael Mästlin also concluded that the comet had an orbit around the sun, outside the orbit of Venus, but he used an epicycle to explain its precise orbit. Tycho, on the other hand, suggested that its orbit could be oval. This was the first time that a non-circular orbit had been proposed for a comet.

In his drive for accurate observations, Tycho began to study the effect of refraction of light in the earth's atmosphere. This had first been discussed by both Ptolemy and Cleomedes, a near contemporary of Ptolemy, but Ibn Yûnus (d. 1009) was probably the first to clearly quantify its effect. He gave a deviation of 40′ for a ray of light at 0° altitude which Tycho refined to 34′.

Tycho Brahe started to examine the effects of refraction in detail when he noticed that the altitude of the earth's equatorial plane at his observatory, determined using the sun, was 4′ higher than its altitude deduced using the pole star. The sun-based measurement was half the sum of the sun's altitude at mid-day at the summer and winter solstices. As the latter was only 11° at the latitude of Tycho's observatory, he thought it reasonable to suppose that the error was due to $2 \times 4′ = 8′$ of refraction at the winter solstice. His detailed measurements of refraction at 11° altitude yielded a figure of 9′ which is, within measurement error, the same as he had previously deduced. With these relatively large values at 11° it was clear to Tycho that he would have to make detailed corrections if he was to produce accurate measurements of the positions of celestial objects. Accordingly, he produced a correction table for refraction that reduced from 34′ at 0°, to 10′ at 10°, and 10″ at 40°. Interestingly, he concluded that the refraction of starlight was less than that for light from the sun.[20] Then later he produced yet another set of atmospheric refraction estimates, this time for moonlight.

It had been known since at least the time of Hipparchus that the moon's orbit is inclined at about 5° to the ecliptic, that its line of nodes (where its orbit cuts the ecliptic) regresses with a period of 6,796 days (or about 18.6 years), and that its apogee precesses with a period of 3,232 days (or about 8.85 years). Both Hipparchus and Ptolemy were aware of these numerical values, which had been deduced from the timing of ancient eclipses, and Ptolemy had derived a model of the moon's orbit to fit. Ptolemy had gone one step further, however, as he had measured the position of the moon in its orbit around the earth, rather than just at the time of eclipses (i.e. at new or full moon). When he did this, he found that the moon deviated from the position predicted by his model by about 1° 20′ at the quadratures (half-moons), so he had to modify his model. This effect, now called the 'second anomaly' or 'evection', is caused by the variation in the sun's gravitational force on the moon, as the latter orbits the earth.

---

[20] Tycho concluded that refraction of sunlight was zero above 45°, but for stars it was zero above 20°.

Tycho Brahe found that even Ptolemy's modified model of the moon's orbit, that took account of evection, did not predict the position of the moon correctly at the octants, as the moon was 40′ 30″ ahead of its expected position 45° before full and new moon, and 40′ 30″ behind at 45° after full and new moon. It is now thought that this effect, now called 'variation', may have been discovered much earlier by Abu'l Wefa (Abul Wafa, 940–998), but if it was, Tycho was not aware of it. Later Tycho also found a further small inequality, now called the 'annual equation', as the moon in spring was found to be always behind its expected position by up to 11′, and in autumn it was always ahead by the same amount. Kepler was later to correctly conclude that this is caused by the varying effect of the sun on the moon's orbit as the sun–earth distance changes over the course of a year.

Finally, when Tycho analysed his observations of the comet of 1577 he noticed that the inclination of the moon's orbit was not exactly the 5° found by Hipparchus. Tycho's further analysis showed that the value oscillated between a minimum of 4° 58′ 30″ at full and new moons, and a maximum of 5° 17′ 30″ at the quarters, and that the rate of the retrograde motion of the nodes also varied, taking place more rapidly at the full and new moons and more slowly at the quarters.

Tycho made numerous observations of the planets, but most of the analysis of his data was carried out after his death by some of his assistants, of whom Kepler was by far the most renowned. In 1590, however, some eleven years before Tycho's death, Giovanni Antonio Magini of Bologna had written to him saying that he thought that the eccentricity of Mars' orbit was variable. Tycho replied to the effect that he thought that it was also, but he was going to make more observations to prove and quantify the effect. This was the work which Kepler eventually completed after Tycho's death, resulting in his discovery that the orbit of Mars was elliptical. But more of that in the next chapter.

When it came to cosmology, Tycho disagreed with both Ptolemy and Copernicus, on the one hand disliking the equant, and on the other finding the idea of a moving earth absurd. As many had pointed out before him, if an arrow is fired straight up in the air it returns to its starting point, whereas if the earth is moving beneath it, as Copernicus believed, it should not do so. Tycho was also unable to measure any stellar parallax, as the earth supposedly orbits the sun, even with his most accurate instrument. So this led him to conclude that the stars would have to be more than 700 times as far away as Saturn. Clearly this was ridiculous as, Tycho reasoned, God would not have wasted so much space.

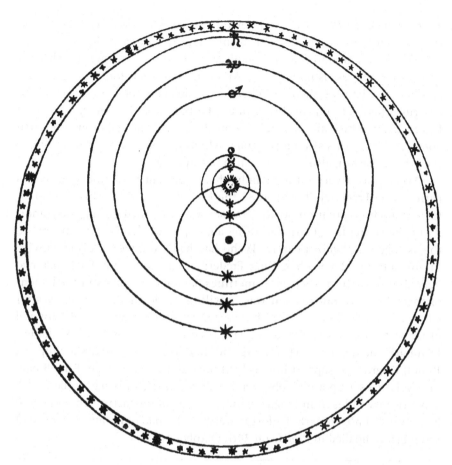

**Figure 2.2** The Tychonic system in which the sun and stars orbit the earth, whilst all the planets, except the earth, orbit the sun. In terms of relative motion it is the same as the Copernican system, as the daily orbits of the sun and stars around the earth are equivalent to the earth's axial spin.

By 1578, therefore, Tycho had decided to adopt the intermediate cosmological system advocated by Heracleides in the fourth century BC (see Section 1.3), and by Martianus Capella of Carthage (c. 365–440 AD), in which Mercury, Venus and, in Tycho's case, the comet of 1577 orbit the sun, whilst the sun, moon and the other three planets orbit the earth. Like Aristotle, Tycho thought that all these bodies were carried on real spheres as they orbited their parent body. Five years later, however, Tycho was beginning to have second thoughts, thinking it likely that all the planets orbit the sun, as the sun orbits the earth.[21] Unfortunately, this implied that the sphere that carried Mars around the sun would have to intercept the sphere that carried the sun around the earth. This was clearly impossible if both were crystalline spheres, as proposed by Aristotle, so he was forced to reject yet another of Aristotle's theories. Therefore in his book of 1588, in which he described his work on the 1577 comet, he proposed the system shown in Figure 2.2, in which all the planets orbit the sun, as the sun orbits the earth once per day. In this system the stars, which he concluded are just outside the orbit of Saturn, also orbit the earth once per day.

Tycho was a very difficult individual who had been tolerated by his patron King Frederick, but in 1588 the king died. Initially this had no effect on Tycho's situation, as the country was governed during the minority of the young king Christian IV by a group of noblemen sympathetic to the astronomer. Unfortunately, Tycho had had a number of disputes with his tenants on Hven when King Frederick was alive, but Frederick had helped to smooth things over, sometimes paying the tenants in settlement of their disputes. But his son was unwilling to do this and, shortly after Christian achieved his majority, he stopped Tycho's annual stipend. This and other problems caused Tycho to quarrel with the king, and in 1597 he left Denmark, moving first to Hamburg and then to Prague, where the Emperor Rudolf II agreed to support him as his Imperial Mathematician. Rudolf paid for Tycho to set up a new observatory in the castle of Benatky (Benatek), some twenty miles from Prague, using as many of his instruments as could be moved from Denmark. Unfortunately, Tycho had only one year's use of them before he died in Prague on 24th October 1601.

---

[21] Ursus (otherwise known as Nicolaus Reymarus, 1550–1599) and Helisaeus Roeslin had both independently arrived at the same solution at about the same time as Tycho. As a result, Tycho accused both men of plagiarism, although the charge was never proved.

# Chapter 3 | KEPLER AND GALILEO – THE FALL OF EPICYCLES AND THE START OF TELESCOPIC ASTRONOMY

## 3.1 Kepler

Johannes Kepler (1571–1630), or Keppler, was born into a modest family in Weil der Stadt near Stuttgart on 27th December 1571. Weil was mainly Roman Catholic, but the Keplers were Protestants, a fact that was to stand in Johannes' way a number of times in his career. Although the young Kepler was very bright, he suffered more than his fair share of childhood illnesses. This militated against him following his father into the army, and a Church career was envisaged instead. In 1584 he gained a scholarship to a school which was held in a confiscated monastery at Adelberg, and five years later he entered the philosophical faculty of the University of Tübingen, near Stuttgart, one of the great centres of Protestant learning.

Whilst at Tübingen, Kepler became acquainted with Copernicus' cosmology through the professor of mathematics and astronomy, Michael Mästlin. Kepler was much taken with this Copernican theory which, to him, seemed much more logical and beautiful than the various geocentric alternatives. At this stage, however, Kepler was still aiming to enter the Church, so after he took his masters degree in 1591 he embarked on a more detailed study of theology. Unfortunately for the furtherance of his chosen career, however, in 1594 a certain Georg Stadius died, and the Protestant Estates of Styria, in southeast Austria, asked the University of Tübingen to recommend a replacement as a lecturer in mathematics and astronomy at the Lutheran school in Graz. They recommended Kepler, who was then offered the position. Kepler was initially reluctant to leave his theological work and his established career for a job, as he saw it, of a lowly teacher in an unfamiliar subject. Eventually he did decide to take up the offer, however, as he had previously criticised some of his contemporaries for failing to take

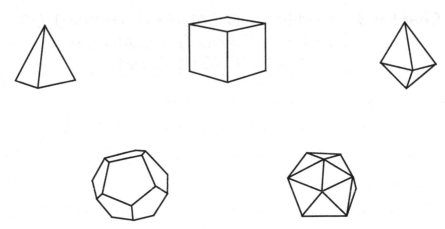

**Figure 3.1** The five regular solids of Euclid, from left to right, first row, a tetrahedron, cube, octahedron, second row, a dodecahedron and icosahedron.

up challenging appointments, and he thought that he ought now to take his own advice. Nevertheless, he decided to keep his options open by stipulating that he would accept the position only on condition that he could revert to his Church career if the new job proved unsatisfactory. It was whilst he was at Graz that his interest in astronomy and astrology came to the fore.

Early in 1595 Kepler started to extend Copernican cosmology by considering why there are six planets, and not five or seven, why their distances from the centre of their orbits (the centre of the earth's orbit) are what they are, and why those planets furthest from the centre move more slowly than those nearer in. It was logical that planets should move more slowly in angular velocity the further they are from the centre, as they have further to travel per orbit, but their linear velocity is also lower. Why was this?

During a lecture that he was giving in July 1595, Kepler suddenly saw a possible solution to the first and second questions, based on Euclid's geometry. Initially he tried two-dimensional schemes but, when these did not work, he was forced to consider the problem in three dimensions. Then the solution dawned on him. Euclid had taught that there are five and only five regular solids, that is solids bounded by a number of identical sides. They are the four-sided tetrahedron, the six-sided cube, the eight-sided octahedron, the twelve-sided dodecahedron, and the twenty-sided icosahedron (see Figure 3.1). Each of these solids had the interesting property that a sphere can be drawn thorough all its vertices, and another sphere can be drawn touching all its sides internally. So Kepler wondered if the distances between the spheres containing the orbits of each of the six planets could be such that each one of Euclid's regular solids could fit between each pair of adjacent planetary spheres. As there are five solids, they could cover five gaps between six planetary spheres. Hence the reason why there are six planets.

In Copernicus' cosmology the orbits of the planets are eccentric, rather than concentric circles. Therefore, instead of defining one (hypothetical) sphere per planet of infinitesimal thickness, Kepler defined spheres whose thickness was each just enough to contain these eccentric orbits. So his scheme was to have a different regular solid between the orbits of adjacent planets, such that the vertices of the solid defined the inner skin of the sphere containing the orbit of the outer planet, and its sides defined the outer skin of the sphere containing the orbit of the inner planet.[1] To match the observed distances of the planets, this meant that the ratio of the radii of these two spheres would have to be, for the gaps between the planets:

| | |
|---|---|
| Saturn–Jupiter | 1.57 |
| Jupiter–Mars | 3.00 |
| Mars–Earth | 1.32 |
| Earth–Venus | 1.26 |
| Venus–Mercury | 1.38 |

The question now was, which of the regular solids could best be fitted between the planetary spheres to give these ratios?

Geometry shows that the ratio of the radii of the outer and inner spheres containing an icosahedron or a dodecahedron is 1.26, for an octahedron or cube it is 1.73, and for a tetrahedron is 3.0. Therefore, looking at the measured values above, the best fit of theoretical and real distances is achieved if there is a tetrahedron between the spheres for Jupiter and Mars, an octahedron or cube between the spheres for Saturn and Jupiter, and Venus and Mercury, and an icosahedron or dodecahedron between the spheres for Mars and Earth, and Earth and Venus.

Although it does not matter in trying to compare actual and calculated distances, Kepler believed that there was a logic behind everything, and he wanted to know unambiguously which solid fitted where. Kepler then divided the regular solids into two categories based on a complicated set of arguments, not all of them scientific, namely primary solids (i.e. the dodecahedron, cube and tetrahedron) and secondary solids (i.e. the icosahedron and octahedron). It was obvious to him that God would have used the earth to separate these two types in space so, because there are two planets inside the

---

[1] Kepler did not believe that these spheres or regular solids were really present in space, of course. They were just geometrical devices to explain the dimensions of the system.

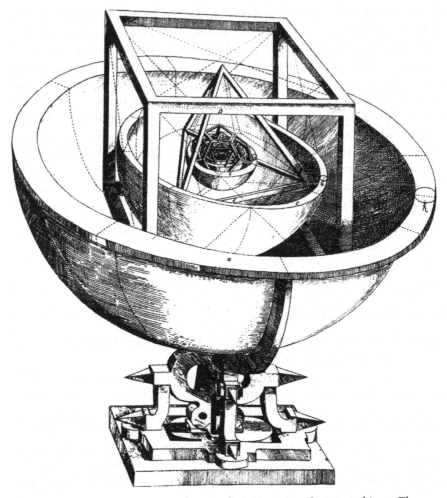

**Figure 3.2** A model of the universe from Kepler's *Mysterium Cosmographicum*. The outer sphere is that for Saturn, with a cube just inside, followed by a sphere for Jupiter, with a tetrahedron, etc.

earth's orbit and three outside, the two secondary solids would have to be inside the earth's orbit and the three primary solids outside. This produced the order from Saturn inwards of cube, tetrahedron, dodecahedron, icosahedron and octahedron (see Figure 3.2).

The match between the real values for the distances between the planetary spheres (given in the table above) and the calculated values, using these various solids, was good for Earth–Venus, Mars–Earth and

Jupiter–Mars, not so good for Saturn–Jupiter, and poor for Venus–Mercury (1.38 versus 1.73). Kepler pointed out that in the latter case the agreement would be much better if the radius of a circle just touching the sides of the square formed by the four median edges of the octahedron was used, instead of the sphere touching the sides of the octahedron. But this was not a very satisfactory modification to his theory. Kepler also wondered what would happen to his theory if, instead of using the centre of the earth's orbit as the centre of the planetary orbits, as Copernicus had done, he used the sun. His old teacher Mästlin agreed to undertake the necessary calculations, but the agreement between theory and reality was no better. Nevertheless, from this time onwards Kepler continued to assume that the sun is the centre of the solar system, which we now know to be correct, in his future work.

Kepler outlined much of his thinking and results in a book called *Mysterium Cosmographicum*, which was published in 1596. Also in this book, he considered the question of the velocities of the planets in their orbits. Kepler could simply have tried to get some sort of mathematical correlation between their distances from the sun and their orbital velocities. Instead he took the major step in trying to understand what physical effect caused the planets to orbit the sun, with the idea that once he had deduced that, the mathematics would follow naturally. This was a radical departure from the normal approach. The Babylonians had tried to understand the movement of the planets across the sky in strictly arithmetical terms. The Greeks, on the other hand, had tried to explain planetary orbital movements in geometrical terms. Many people had introduced various gods to explain some aspects of planetary motion, but Kepler was the first person to try to understand the solar system by considering physical forces.

Kepler considered two types of 'anima motrix' ('moving spirit'), as he called the planetary driving force. Either there is a separate anima motrix driving each of the planets which is weaker the more distant the planets are from the sun, or there is only one anima motrix which emanates from the sun and whose effect decreases with distance. After some thought he settled on the second explanation, and concluded that the magnitude of this force reduces with distance from the sun in the same way as does the intensity of light. The intensity of light is inversely proportional to the square of the distance from the source, but for some reason Kepler did not appear to have realised this at the time,[2] as he mistakenly considered light expanding over

---

[2] Kepler recognised the correct intensity/distance relationship for light in a book called *Optics* published in 1604.

Table 3.1 *Ratio of the planetary periods calculated using Kepler's theory, compared with their true values*

|  | Kepler | Actual |
|---|---|---|
| Jupiter:Saturn | 0.375 | 0.403 |
| Mars:Jupiter | 0.172 | 0.159 |
| Earth:Mars | 0.488 | 0.532 |
| Venus:Earth | 0.567 | 0.615 |
| Mercury:Venus | 0.365 | 0.391 |

two dimensions instead of three. As a result he considered the force from the sun acting on the circumference of progressively larger circles, instead of on the surface of progressively larger spheres. So, according to him, the force is inversely proportional to the distance from the sun, rather than to the distance squared, resulting in the linear velocity of the planets being inversely proportional to their distance from the sun also.

The distance travelled by a planet in completing one orbit is proportional to its distance from the sun, and if the linear velocity of the planets is inversely proportional to distance, as Kepler deduced, he should have concluded that the time taken for the planets to orbit the sun once, their so-called periods, is proportional to their distance squared. In fact he made an arithmetic error and concluded that 'the greater distance from the sun acts twice to increase the period'[3] giving him the relationship:

$$\frac{P_2 - P_1}{P_1} = \frac{2(r_2 - r_1)}{r_1} \quad \text{instead of} \quad \frac{P_2 - P_1}{P_1} = \frac{r_2{}^2 - r_1{}^2}{r_1{}^2} \quad \text{or} \quad \frac{P_2}{P_1} = \frac{r_2{}^2}{r_1{}^2},$$

where $P_1$ and $P_2$ are the periods of two planets 1 and 2, and where $r_1$ and $r_2$ are the radii of their orbits.

The ratio of the periods of the planets calculated using Kepler's theory and their known average distances from the sun, are compared with the actual ratio of their periods in Table 3.1. The correlation is very good, but by no means perfect. This correlation is fortuitous, however, as had Kepler

---

[3] Ch. Frisch, *Joannis Kepleri Astronomi Opera Omnia*, Vol. 1, p. 173, Frankfurt and Erlangen, 1858, as translated in Dreyer, *A History of Astronomy from Thales to Kepler*, p. 379.

not made the two errors mentioned above, the correlation would have been far worse.[4]

## Kepler and Tycho

Kepler thought that he could get a better correlation between his theory of planetary distances (using regular solids) and observed distances, and of planetary periods (using the mathematical relationship just described) and observed periods, if he could get more accurate planetary observations than those used by Copernicus. So he sent a copy of his book *Mysterium Cosmographicum*, which contained all the above, to the one man in the world who would be able to help, Tycho Brahe, asking if he could have access to his observational data.

Kepler's letter to Tycho, dated 13th December 1597, reached Tycho whilst he was at Wandsbeck castle near Hamburg in early 1598, just after he had finally left Denmark. Tycho's reply of April 1598 was both critical and encouraging. Whilst criticising his work, Tycho recognised in Kepler an up and coming young man with original ideas, and he invited Kepler to visit him to discuss his ideas in more detail. This reply, which did not reach Kepler until February 1599, somewhat offended Kepler, because of some of Tycho's critical comments. But it was out of the question for him to visit Tycho to discuss his work at that time, as the distance was too far, and things were becoming decidedly difficult in Styria, caused by Archduke Ferdinand's repression of the Protestants. Towards the end of that year, however, Kepler learned from his friend and patron Herwart von Hohenburg, the Bavarian chancellor, that Tycho had settled at Benatky, near Prague. This was not very far away from Graz, so Kepler decided to take up Tycho's offer to visit him.

Tycho received Kepler at Benatky on 4th February 1600 with great enthusiasm. Kepler stayed there for two months working, at Tycho's request, on the theory of Mars' orbit. At the end of his stay, Tycho offered him a job as one of his assistants, but Kepler could not make up his mind whether to accept, as the two men had recently had a violent disagreement over the terms of Kepler's continuing collaboration. Later, when Kepler eventually returned to Graz, he found that there was still a great deal of religious unrest there. He was then advised by the councillors responsible for carrying out

---

[4] The correct relationship between period and distance is neither of the two given above, however, as Kepler was to discover later when he found what we now know as Kepler's third law. The correct relationship is $P^2 \propto r^3$.

the so-called Counter Reformation that it would probably be best if he took up medicine in Graz, as this would be of much more use to the state than astronomy.

Kepler did not know what to do, and he thought of suggesting to the Archduke Ferdinand that he become his mathematician, in the same way that Tycho was mathematician to the Emperor. Whatever Kepler wanted to do, however, it was important that he keep in the good books of the Archduke. So he wrote him an astronomical essay, using the forthcoming solar eclipse of 10th July as an excuse. Interestingly, in this essay Kepler introduced a radical new concept in astronomy when he suggested that 'There is a force in the earth which causes the moon to move'. He also suggested that the force on the moon would be weaker when it is furthest from the earth, so that at that time the moon would move more slowly in its orbit. In effect this force 'in the earth' appeared to be similar to that in the sun controlling the planets.

The Archduke had other things on his mind at the time when he received Kepler's essay. Nevertheless, he thanked Kepler and sent him a gift in return. Then in July the Archduke decided to evict all Protestants in Graz who refused to convert to Roman Catholicism, including Kepler(!), thus making up Kepler's mind for him. Kepler then returned to Prague, arriving on 19th October 1600.

The relationship between Tycho and his new assistant was very much 'up and down', as both men had radically different ideas of what Kepler's new job was. Tycho really wanted an astronomical clerk to do his bidding, whereas Kepler wanted the job to gain access to Tycho's extensive sets of observations, which Tycho was very loath to release. Nevertheless, they both made the best of the situation, as they knew that they needed each other. Then in 1601, just before Tycho died, Kepler was made Tycho's senior assistant, and a few days after his death Kepler succeeded Tycho as Imperial Mathematician.

## The orbit of Mars

Kepler was to remain in Prague as Imperial Mathematician until the death of the Emperor Rudolf II in 1612. During the early part of this period, he continued with his analytical work on the orbit of Mars that he had begun at Benatky in 1600. As it happens, Mars was an ideal planet to choose, as its orbit is sufficiently non-circular for its ellipticity to be very clear. This meant that any errors in the theoretical orbit would be readily apparent. Mars also

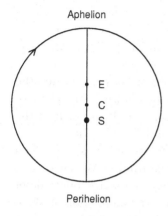

Aphelion

E

C

S

Perihelion

**Figure 3.3** The circular orbit of a planet according to Kepler's initial theory. The sun is at S, the centre of the circle at C, and the equant at E. They are all on the line of apsides joining perihelion and aphelion. The eccentricity $e$ equals the distance CS divided by the radius of the circle.

had the advantage over Mercury, whose orbit is the most eccentric of all the planets known to the ancients, in that it is far easier to observe as it is not as close to the sun, so the observational data on Mars was more comprehensive.

Kepler's analytical work on Mars was published in his book *Astronomia Nova* which was published in 1609. In this he assumed that all the planets, including the earth, orbit the sun, and not the centre of the earth's orbit as Copernicus had assumed. This did not mean that the sun was at the centre of the planetary orbits, but that the sun controlled the planetary movements in some way. Kepler also reintroduced the equant that Copernicus had been at pains to reject, as the point from which the planet's velocity appeared to be uniform.

In his initial orbital model, therefore, Kepler assumed that the planets describe circular orbits, as shown in Figure 3.3, about a centre C, sun S and equant E. In this model the sun replaces the earth of Ptolemy's model (see Figure 1.9). Now Ptolemy had assumed that the equant was the same distance on one side of the centre of the orbit, as the earth was on the other side. Kepler did not assume this for the sun and his equant, however, but left their positions to be determined by calculation based on observations. Nevertheless Kepler followed Ptolemy in assuming that all three points C, S and E were collinear.

Finally Kepler assumed that the planes containing the orbits of the planets all go through the sun, and have fixed inclinations to that of the earth's orbit around the sun (the ecliptic).

As a planet appears to move at constant velocity as seen from the equant (Figure 3.3), its angular velocity with respect to the sun is greatest when nearest the sun at perihelion, and smallest when furthest away at aphelion. The same applied to the moon with respect to the earth, as Kepler had suggested in his essay to Archduke Ferdinand in 1600. Kepler concluded that the reason that both the planets and the moon move faster when closest to their parent body was because the force of the parent body on the planet or moon is greatest at that point.

Kepler's first objective in working out the orbit of Mars was to find the heliocentric longitude[5] of the ascending node, which is where the plane of Mars' orbit cuts the ecliptic, as Mars travels from south to north of the ecliptic. He also wanted to find the angle between these two planes.

To find the longitude of the ascending node, Kepler looked through Tycho's records to find those occasions when the latitude of Mars was zero. Six of these observations produced a value of 46° 40' for the longitude of the ascending node. Following that, he determined the inclination of Mars' orbit to the ecliptic by three different methods, also using Tycho's data. In the first he looked for occasions when Mars was at 90° from the nodes of its orbit, and when it was equidistant from the earth and sun. At that time the observed latitude is equal to the orbital inclination. In the second, he looked for times when Mars was at quadrature (i.e. 90° away from the sun, as seen from earth), and when the sun and earth were both in the line of nodes of Mars' orbit. Again the latitude is equal to the inclination. Finally he used a method based on the latitudes at opposition. This work produced an inclination of 1° 50', and proved that the plane of Mars' orbit goes through the sun, and that its inclination is constant.

Kepler had thus established the orientation and inclination of Mars' orbital plane with respect to the ecliptic. He then set out to find the position of Mars' line of apsides (i.e. the line through its aphelion and perihelion), the centre C of its circular orbit, and the position of the sun S and of the equant E, as shown schematically in Figure 3.3. In order to do this he needed observations from four oppositions. So he used Tycho's data for 1587, 1591, 1593 and 1595, but the method that he was forced to adopt was by no means straightforward. Instead he had to make a number of assumptions to try to define the position of the circle that went through the four positions of Mars at opposition, with a centre on the line joining the sun S and the equant E.

The method Kepler used, in what he called his vicarious theory, was one of successive approximations, in which he had to progressively refine his assumptions to ensure that the centre of the circle through the four points lay on the line SE. In the end it took him over seventy iterations, before he finally found a configuration that worked. In that configuration the longitude

---

[5] The heliocentric longitude is measured relative to the 'First Point of Aries'. This is where the earth's equatorial plane cuts the ecliptic at the spring equinox.

of the aphelion was 148° 48′ 55″ (for 1587), with SC = 0.11332r and EC = 0.07232r, where r is the radius of the circle. So Kepler had been apparently correct to assume, contrary to Ptolemy, that SC ≠ EC. In the process Kepler had apparently also proved that the orbit was circular, in so far as he had found such a solution that matched the data of the four oppositions that he had used.

Kepler had observational data on twelve oppositions of Mars, ten from Tycho and two of his own. Before Kepler could check the accuracy of his new model against these observations, however, he need to know the annual movement of the longitude of Mars' aphelion relative to the first point of Aries. Fortunately he had observations from Ptolemy for the year 140, and from Tycho for 1587, which allowed him to calculate the annual movement which turned out to be 1′ 4″.[6] Using this value, and the parameters of his orbital model of Mars that he had just calculated, he was able to compare his calculated longitudes for each of the twelve oppositions of Mars with observational data. The results show that the maximum difference between his model and observations was only 2′ 12″. At first glance, this looks highly encouraging but, unfortunately, as he explained in his book, the results of this work, on which he had spent so much time and effort, were erroneous. Much to his surprise his new work that showed this, and which will now be described, indicated that Ptolemy had probably been correct in assuming that SC = EC.

The problem came to the fore when Kepler tested his theory by considering the *latitudes* of the oppositions of 1585 and 1593, when Mars was near to the limits of its greatest north and greatest south latitudes, and when it was also near its aphelion and perihelion positions respectively. His analysis of this data produced a value for SC of 0.08000r or 0.09943r, depending on the assumptions made. Either way, the value was lower than that of 0.11332r previously determined for SC. In fact, it was nearer to that of ½(SC + EC), i.e. of ½(0.11332 + 0.07232)r = 0.09282r. Kepler, noting this, decided to see what would happen if he assumed that SC = EC = 0.09282r. In this case, while the calculated positions at 90° from the apsides agreed reasonably well with observations, those 45° on either side differed by about 8′ which, although well within Ptolemy's measurement error, was clearly outside Tycho's.

When Kepler examined *longitudes* at two points outside the oppositions, but near the apsides, he found that the value of SC was between 0.08377r and 0.10106r, with an average of 0.09242r, which is again very close to the case of

---

[6] This 1′ 4″ consisted of 13″ due to the movement of the aphelion in celestial coordinates, and 51″ due to the precession of the earth's orbit.

SC = EC = 0.09282r. So there was not a unique pair of values of SC and EC for all parts of the orbit, and Kepler began to wonder if the orbit of Mars really was circular, or whether, if it was circular, there was no fixed equant point from which the motion appeared uniform. To get to the bottom of this, he decided to examine the orbit of the earth from which all these observations had been made. Previously it had been thought that the earth's orbit around the sun (or vice versa) was a simple eccentric circle, with the sun (or earth) off-centre. As it was thought that the earth traversed this orbit at a uniform speed, there was no need for an equant point. But Kepler now wondered whether the addition of an equant to the earth's orbit could explain and solve the problem with the orbit of Mars.

The question now arose as to how to determine the earth's orbit accurately, and to do this Kepler hit on a revolutionary idea. He would use Mars! The sidereal period of Mars was known to be 687 days, so every 687 days Mars would be back in the same place in its orbit. But because the earth's sidereal period is not an exact multiple of that of Mars, the earth would be at a different position in its orbit every 687 days. Three points on the circumference are required to uniquely define a circle, so Kepler needed to find at least three observations of Mars from Tycho's observations that were separated by multiples of 687 days. In fact, he found four, from which he was able to show that the best configuration of the earth's orbit was a circle with EC = SC = 0.0180r, which was one-half of the eccentricity calculated by Tycho Brahe.

Instead of returning to the problem of Mars, now that he had solved the problem of the earth's orbit, Kepler decided to turn his attention from numerical models of planetary orbits to the general dynamics of the solar system.[7] He wanted to try to understand what forces caused the planets to orbit the sun in the way that they did and their effect on planetary velocities.

Kepler had already concluded that, as the sun clearly controlled the solar system, the forces controlling the planets must come from the sun. As mentioned previously, Kepler had also assumed that the magnitude of the force from the sun is inversely proportional to distance, and that consequently the linear velocity of a planet in its orbit is also inversely proportional to its distance from the sun. As a result, he now concluded that the area swept out by a planet in its orbit around the sun in unit time is not dependent on its distance from the sun. This gives what we now know as Kepler's second

---

[7] It is not clear why Kepler changed tack at this time and did not complete his work on Mars, but he may well have wanted to have a rest from his extensive calculations that ran to hundreds of pages of his notebooks.

law, namely that 'the radius vector from the sun to a planet sweeps out equal areas in equal times'. It is interesting to note, however, that Kepler did not prove this relationship rigorously, but deduced it using somewhat flawed logic and then checked its applicability only for the earth and Mars, before stating its universal applicability to all planets. Nevertheless, as the errors in Kepler's assumptions fortunately cancelled each other out, his conclusion is valid.

Returning to the orbit of Mars, Kepler still encountered serious difficulties in getting its circular orbit to match the observational results, even with the corrected orbit for the earth. He could produce an accurate match at the apsides (i.e. at aphelion and perihelion), but outside those points the orbit seemed to be some sort of flattened circle. He first of all tried an egg-shaped oval, wider at one end than the other, based on an eccentric circle and an epicycle, but that did not work, no matter how he juggled with the parameters. When he had started this work, however, Kepler had written,[8] 'if only the shape [of the orbit] was a perfect ellipse', but for some reason he avoided trying an ellipse as a possible fit until he had no alternative. When he did do so in 1604, he quickly found that the orbit of Mars is an ellipse, with the sun at one of its foci. At last with this result the concept of circular orbits for all heavenly bodies had bitten the dust, never to return.[9] This law, which states that 'the planets describe elliptical orbits around the sun with the sun at one focus', is now called Kepler's first law.

Kepler next considered how the sun causes the planets to move, and concluded that the basic orbital motion was generated by vortices produced by a rotating sun.[10] These vortices would only carry the planets round the sun in circular orbits, however, so he needed another force that would cause the orbits to become non-circular. Kepler had been interested for some time in magnetism, and this interest had been much encouraged by his reading of William Gilbert's (1540–1603) important new book *On the Magnet* which had been published in 1600. In this Gilbert showed that the earth behaves as a magnet. This led Kepler to suggest that the sun and all the planets are also magnetic, and that the ellipticity of their orbits is caused by

---

[8] W. v.Dyck, and Max Casper (eds.), *Johannes Kepler, Gesammelte Werke*, Vol. XIV, p. 409, as translated in Koestler, *The Sleepwalkers*, Penguin-Arkana, 1989, p. 335.

[9] Except for one or two die-hards like Galileo, see Section 3.3, who still accepted Copernicus' cosmology, epicycles and all.

[10] The rotation of the sun was discovered by Galileo a few years later in 1612, but the rate of rotation was found to be almost a factor of ten slower than the three days deduced by Kepler.

forces on the magnetic poles of the planets which vary as the planets orbit the sun.

It took Kepler a considerable effort not only to write his book *Astronomia Nova* but also to get it published, as this followed protracted arguments with Tycho's heirs about who owned the copyright of Tycho's observational results. Interestingly, its eventual publication in 1609 caused no great stir, as not many people understood its complex arguments. Even his old tutor Mästlin, in answer to some questions from Kepler, had said, 'I must further confess that your questions are sometimes too subtle for my knowledge and gifts, which are not of the same stature'.[11] Although this *Astronomia Nova* was his seminal book, being the first to describe his elliptical theory of planetary motion, he had still only proved its validity for Mars. In addition, he had not yet found an exact relationship between the orbital periods of the planets and their mean distances from the sun, for which he had been seeking since his youth.

Two years after publication of *Astronomia Nova* disaster befell Kepler, when in February 1611 one of his children died of smallpox, whilst Prague was under attack by Leopold of Austria, and then in May his imperial patron Rudolf was forced to abdicate by his brother Matthias. There had been political unrest in Prague for some time, and Kepler had been looking for another job that would take him away from the area, hopefully back to Austria where his wife would feel much more at home. So he was naturally delighted when in June 1611 he was offered the job of Provincial Mathematician in Linz, the capital of Upper Austria. Further disaster befell him the next month, however, when his wife died.

In deference to the wishes of the deposed emperor, Kepler delayed his move to Linz, but on 20th January 1612 Rudolf died. Although the new Emperor Matthias confirmed Kepler in his old title of Imperial Mathematician, it was little more than an honorary appointment, as Matthias was not very interested in astronomy. Life was very unpleasant in Prague at that time, riven as it was by war, so in April 1612 Kepler left the city for Linz to take up his new appointment as Provincial Mathematician. His life was still not plain sailing, however, as his mother was accused of witchcraft and threatened with torture during proceedings which lasted from 1615 to 1621. It is nothing short of amazing, therefore, that Kepler was able to concentrate on his astronomy and continue to write major works during this traumatic period.

---

[11]  W. v.Dyck, and Max Casper (eds.), *Johannes Kepler, Gesammelte Werke*, Vol. XIV, p. 131, as translated in Koestler, *The Sleepwalkers*, Penguin-Arkana, 1989, p. 352.

## Venus, Mercury and the third law

In 1614 Kepler showed that the orbit of Venus was an ellipse with an eccentricity of 0.0060 (modern value 0.0068), and in the following year that that of Mercury was also an ellipse with an eccentricity of 0.210 (modern value 0.206). These results and others were published in his *Epitome Astronomiae Copernicanae*, now generally called the *Epitome*, which was published in three parts in 1618, 1620 and 1621. This book, which was written in the form of questions and answers, covered topics ranging from spherical astronomy, the shape and size of the earth, refraction, lunar and planetary orbits, the size of the sun, moon and planets, the distances of the stars, and so on. In fact, it was a veritable compendium of astronomy, written according to Kepler's new principle.[12] Slipped into it was a statement, which we now know as Kepler's third law, that the squares of the periods of the planets are proportional to the cubes of their mean distances from the sun.

Actually Kepler had already explained this relationship between the periods and distances of the planets in his book *Harmonice Mundi* (Harmonies of the World) which had been published in 1619. This book, much of which was more in keeping with the ancient Greeks than with Renaissance astronomy, looked at harmony in mathematics, music, astrology and astronomy. For example, Kepler defined perfect and imperfect polygons and tried to link them to concordant and discordant musical sounds. Later in the book he tried to link the ratios of the value of various planetary parameters to musical tones. In particular, he looked at the ratios of the periods of revolution of the planets, to see if they formed a harmonic series, and found that they did not. Then he looked at the ratio of the sizes of the planets, to see if they formed a harmonic series. They did not either. He had varying degrees of success with other planetary parameters, but on 15th May 1618, whilst he was still working on these ratios, he suddenly discovered the relationship now given by the third law.

Kepler's last great book, the *Rudophine Tables* published in 1627, listed the predicted positions of the heavenly bodies based on Tycho's and his own extensive set of observations. These tables were much more accurate than anything that had gone before, and in them he was the first person to predict the transits of Mercury and Venus across the sun's disc; the next transits

---

[12] In this book, Kepler, for the first time, recognised that the driving force from the sun is inversely proportional to the square of the distance, and not inversely proportional to the distance, as he had originally assumed.

being due on 7th November and 6th December 1631, respectively.[13] Unfortunately he died the year before these occurred, although the transit of Mercury was seen on the predicted day[14] by Pierre Gassendi in Paris, Remus Quietanus in Ruffach, and Johann Cysat in Ingolstadt, near Munich. Unfortunately the transit of Venus could not be seen in Europe as it occurred at night-time there.

We must now retrace our steps to discuss major developments on the observational side of astronomy that were occurring in parallel with Kepler's cosmological work. The key development here was the invention of the telescope.

## 3.2 Early telescopes

Convex lenses had been used in spectacles to correct for long-sightedness (hypermetropia) since the late thirteenth century, and concave lenses had been used to correct for short-sightedness (myopia) from about two centuries later. But the idea of putting two lenses together to form a telescope was much slower in coming. Roger Bacon is known to have experimented with lenses in the thirteenth century, and some think that he invented the telescope, although the written evidence for this is very ambiguous. Likewise, cases have been submitted for Leonard Digges, John Dee and William Bourne, all of whom were English, and Giambattista della Porta of Naples as the inventor of the telescope in the sixteenth century, but again the evidence in each case is ambiguous. In fact, the first clear reference to a telescope is in 1608 when Hans Lippershey, a spectacle maker from Middleburg in the Netherlands, submitted a petition dated 2nd October 1608 to the States-General, with a covering letter dated 25th September. At that time the States-General governed the Netherlands and Lippershey wanted patent protection (to use a modern phrase) for a telescope that he had made.

The States-General were very interested in his device, as the Netherlands was then still at war with Spain, and they decided to purchase such an instrument from him, although they turned down his request for patent protection. Subsequent to Lippershey's initial approach, James Metius (otherwise

---

[13] Kepler thought that he had observed an earlier transit of Mercury on 28th May 1607, but the black spot on the sun was later found to be a sunspot.

[14] The transit of Mercury occurred within 6 hours of the predicted time, corresponding to a longitude error in its position of 14'. This compares with an error of 5° in the Copernican tables.

known as Jacob Adriaanzoon) of Alkmaar in Holland, also sent a petition to the States-General claiming prior invention, although he provided no evidence. Claims were also made that Zacharias Janssen, another Middleburg spectacle maker, had also made a telescope before Lippershey. But, although his son said that his father had invented the telescope in 1590, there was no evidence of this. So the best that we can say is that Lippershey was the first known maker of a telescope, but we do not know who invented it.

News of the telescope spread quickly throughout Europe. According to Simon Marius a telescope was available for sale at the Frankfurt fair in the autumn of 1608, and by April 1609 telescopes were for sale in Paris. Galileo Galilei (1564–1642) first heard of these instruments when visiting Venice in May 1609,[15] and he made one with a magnification of about three diameters immediately on his return to Padua. Early telescopes were very crude instruments made with spectacle lenses, and consisting of a convex main lens and a concave eyepiece. They were mounted in a metal, cardboard or paper tube about one foot (30 cm) or so long, and about one inch (2.5 cm) in diameter. They had a very low magnification and produced an upright image.

Galileo experimented with the design of his telescope, and soon made a new instrument with a magnification of 8 diameters. He was so impressed with the results, and being well aware of its potential military and naval applications, he took it to Venice in August 1609 to show the senators. A few days after his demonstration he presented the telescope to the Doge, and then returned to Padua to try to improve his design yet again. He finally produced an instrument 49 inches (1.25 m) long and of 1 3/4 inches (4.5 cm) aperture that magnified some 30 diameters, using lenses that he ground himself. Unfortunately, the field of view was very small for these telescopes, being only about 7' for Galileo's largest instrument. In fact Kepler recognised that the reason for this was the concave eyepiece, and in his book *Dioptrice*, published in 1611, he proposed an improved design with a convex eyepiece. This produced a wider field of view, but at the expense of producing an inverted image which, whilst not a problem for astronomical work, was a problem for terrestrial use. Kepler was not a very practical man, and he never made such a telescope, as far as we know. The first such Keplerian instrument appears to have been made by Christoph Scheiner (1573–1650) at Ingolstadt in about 1617.

We must now consider the practical applications of these early instruments, with particular reference to the unique contribution of Galileo, who was both an outstanding observer and a theoretician.

---

[15] Some authorities (e.g. Stillman Drake in *Galileo at Work*, University of Chicago Press, 1978, republished by Dover, 1995, p. 138) give July 1609.

## 3.3 Galileo

Galileo Galilei was born at Pisa on 15th February 1564. His father Vincenzio was a talented musician, both composing and writing about music and playing the lute, whilst an earlier relative called Galileo Buonaiuti, from whom the family took its name, had been a well-respected doctor. In 1572 the family moved to Florence, leaving the young Galileo who stayed with a relative of his mother at Pisa. Two years later he rejoined his family.

Galileo's early education was at the Camaldolese monastery of Vallombrosa, about thirty kilometres east of Florence, where he studied grammar, logic and rhetoric. He was later moved to a monastery at Florence by his father, and then in 1581 Galileo returned to Pisa, where he enrolled as a medical student at the university. Although he started the course well enough, during his second year he started attending a series of lectures on Euclid, given by the mathematician Ostilio Ricci (1540–1603), who also introduced him to the mechanics of Archimedes. As time went on Galileo became more and more interested in mathematics and philosophy and less and less interested in medicine. Consequently, after an abortive attempt by his father to get him to concentrate on his medical studies, he persuaded his father to let him leave university in early 1585 without taking his medical degree. Instead he started teaching mathematics privately at Florence and Siena.

In 1586 Galileo began to write a book in the form of a dialogue on various problems of motion in which he challenged Aristotelian concepts. In this book, which he never finished, he considered the motion of balls of various masses in both air and water, in both free fall or in forced motion. At this early stage of his career his concepts of motion, although different from those of Aristotle, had not yet matured.

Whilst he continued with his private teaching, Galileo was looking for more permanent employment, and in 1587 he unsuccessfully applied for the chair of mathematics at the University of Bologna. Two years later, however, he had greater success when he was appointed professor of mathematics at the University of Pisa, teaching both Euclidean geometry and Ptolemaic astronomy. Unfortunately the salary of sixty florins per year was very modest, being just half of that of his predecessor.

Although there is a wealth of primary material on Galileo's career,[16] there is, surprisingly, considerable disagreement amongst experts on the

---

[16] Including the twenty volume *Le Opere di Galileo Galilei* of Antonio Favaro, published between 1890 and 1909 (afterwards referred to as *Le Opere di Galileo Galilei*).

timing of a number of events attributed to him. For example, Arthur Koestler, in his well-known book *The Sleepwalkers*,[17] says that Galileo discovered the isochronism of the pendulum (i.e. that a pendulum of a given length swings at the same rate, no matter how large the swing) in 1582. That is whilst Galileo was still a medical student at Pisa. Vincenzia Viviani, on the other hand, Galileo's first biographer, attributes a date of 1583 to this,[18] whereas the highly respected Stillman Drake in his *Galileo at Work*[19] deduces a date of 1588. Koestler[20] also mentions that Galileo invented the pulsilogium, a device for timing pulses, probably about 1582, whereas Drake[21] attributes this invention to a Venetian doctor called Santorre Santorio in 1603.

The most interesting disagreement, however, about Galileo's early career revolves about his famous experiment of dropping two balls of different weights from the Leaning Tower of Pisa. Aristotle had maintained that the ratios of the velocities of two different weights of the same material, falling through the same medium, would be the same as the ratio of their weights. In 1553 Giovanni Benedetti disagreed, theorising that the velocity of two different weights of the same material falling through the same medium should be the same. In his dialogue of 1586, Galileo came to the same conclusion as Benedetti. Galileo was also reputed to have shown this experimentally, whilst at Pisa between 1589 and 1592, by dropping two different weights of the same material from the Leaning Tower, and timing their descent. Viviani, in his biography of Galileo published in 1657, says that Galileo told him near the end of his life of this experiment, but there is no independent evidence that the experiment actually took place. Galileo never mentioned it in his extensive writings, for example. Some people go further in the opposite direction and maintain that, not only did the experiment take place, but that it was the first such one to show the effect surmised by Benedetti. This is clearly not so, however, as Simon Stevin had undertaken the experiment from a height of about 30 feet (10 metres) in 1586.

Just before joining the University of Pisa, Galileo had tried unsuccessfully to get appointed to the chair of mathematics at either the University of Padua or Florence. At that time, the University of Padua was the most prestigious Italian university, but Galileo had not given up hope of such an

---

[17] Koestler, *The Sleepwalkers*, p. 359.

[18] See Stillman Drake, *Galileo at Work*, University of Chicago Press, 1978, reprinted by Dover, 1995 (afterwards referred to as Drake, *Galileo at Work*), p. 20.

[19] Drake, *Galileo at Work*, p. 21.

[20] Koestler, *The Sleepwalkers*, p. 359.

[21] Drake, *Galileo at Work*, p. 21.

appointment in the future. His contract with the University of Pisa was for three years, and towards the end of it his father died, leaving Galileo financially responsible for the family. Sixty florins per year was hardly enough for one person to live on, let alone a family. In addition, Galileo had upset the authorities at Pisa by his criticism of Aristotle's theory of motion. So he resolved to have another attempt at the chair in Padua before the end of his three year contract at Pisa had expired, as the likelihood of it being extended was remote. Fortunately the chair at Padua was still vacant, and in 1592 he was given a four year contract there with a salary of 180 florins per year. This was excellent news because, not only was the University of Padua more prestigious than that of Pisa, it was far more tolerant of new ideas. Galileo was to stay at the University of Padua, which at that time was under the control of the Venetian Republic, for eighteen years, until he moved on once more in 1610.

The development of Galileo's views on the competitive cosmologies of Ptolemy, Tycho Brahe and Copernicus is not clear. In a text that is thought to date from the period 1586–1587, Galileo originally supported the earth-centred cosmologies of Aristotle and Ptolemy. He was certainly familiar with Ptolemy's *Almagest* during his tenure at Pisa from 1589 to 1592, but it is not clear whether he had read Copernicus' *De Revolutionibus* at that time. It seems that Galileo gradually moved from an earth-centred universe, to a composite model similar to that of Tycho Brahe, and finally to the Copernican structure. In about 1591 Galileo began to wonder whether the earth may rotate on its axis, and by 1595 he probably favoured the Copernican structure of the universe, although he still taught the Ptolemaic version during his lectures at university.

In 1597 Kepler sent a copy of his *Mysterium Cosmographicum* to Galileo. In his acknowledgement dated 4th August 1597 Galileo said,'...I adopted the teaching of Copernicus many years ago...I have written many arguments in support of him and in refutation of the opposite view–which, however, so far I have not dared to bring into the public light, frightened by the fate of Copernicus himself[22]...I would certainly dare to publish my reflections at once if more people like you existed; as they don't, I shall

---

[22] In view of Galileo's subsequent problems with the Catholic Church, it may be thought that he was referring to similar problems experienced by Copernicus. But this is not so, as Copernicus died peacefully in his bed, and his book was subsequently largely ignored by the Roman Catholic Church, certainly up to 1597 (the year of Galileo's letter). In fact what Galileo was most likely alluding to was potential ridicule from other academics.

refrain from doing so.'[23] In his reply Kepler asked Galileo if he would carry out some careful position measurements on two fixed stars to see if they exhibited any parallax. Galileo did not reply, and that was the end of their correspondence for thirteen years.

So, according to Galileo's letter to Kepler of 4th August 1597, Galileo had been converted to Copernicanism well before that date, yet he continued to teach Ptolemy's cosmology and repudiate Copernicus' in his university lectures for many more years. A document dated 1606 confirms this, and it was not until Galileo discovered the phases of Venus in October 1610 that he finally and unequivocally accepted the Copernican system.

Meanwhile in 1604 a new object had been observed in the heavens that was as bright as Jupiter for a time. Unlike Tycho's new star of 1572, however, this new object was seen before it reached maximum intensity, because astronomers and astrologers had been observing the conjunction of Mars and Jupiter that had been predicted to take place in October 1604 in this part of the sky. Surprisingly, the new object was first seen on 9th October, which was the day of closest approach of the two planets to each other. Astrologers saw this agreement in timing as an amazing vindication of their philosophy, as it seemed too exact to be a coincidence. If the new object was connected with the two planets, which seemed likely, then it appeared to indicate that it was in the solar system, and was not a new star.

The first people known to have observed this object were the astronomer Ilario Altobelli at Verona, and a doctor in Calabria, both of whom observed it on 9th October. On the following night a court official in Prague, named Brunowsky, observed it through broken clouds, and immediately notified Kepler. Because of inclement weather, however, Kepler did not see it until 17th October when, fortuitously, it reached maximum brightness. It then gradually reduced in intensity until in mid October of the following year when it became too dim to be seen.

The new object was first seen at Padua on 10th October by Simon Mayr and Baldessar Capra but, at that time, Galileo was not an observational astronomer,[24] and he apparently did not observe it until 28th. Both Galileo and Kepler instantly recognised that the object's sudden appearance, like that of Tycho's star in 1572, disproved Aristotle's doctrine of the unchangeability of the universe beyond the moon. Altobelli at Verona was in touch

---

[23] W. v.Dyck and Max Casper, (eds.), *Johannes Kepler, Gesammelte Werke*, Vol. XIII, p. 130, translated in Koestler, *The Sleepwalkers*, Penguin-Arkana, 1989, p. 361.

[24] There is no evidence that Galileo carried out any astronomical observations before 1604 when he was forty years old.

with Galileo, and neither he nor Galileo could detect any parallax before late November 1604. After then the star was too close to the sun in the sky to be seen, and observations had to be suspended.

It seemed possible that the new object was associated with the conjunction of Mars and Jupiter and, if that was the case, it must be in the solar system. Galileo hypothesised that the reason for its reduction in intensity since mid October was because it was moving away from the earth. If that was the case, its movement should be detectable unless, fortuitously, it was moving *directly* away from the earth. Even then, if the object was somewhere near Mars or Jupiter, and if Copernicus was correct and the earth did go round the sun, then this should be evident in the movement of the object against the background stars.

No parallax could be detected when the object became visible again in late December 1604, however, and none could be detected by the time that it finally disappeared from view in the following October. So either it was a distant star, or Copernicus was wrong. Interestingly, Galileo did not come out clearly in favour of either explanation, which may indicate that he was still unsure of the correctness of Copernican cosmology. Kepler, on the other hand, took the object's lack of parallax to show that it was a new star, like Tycho's new star of 1572.

## Early telescopic observations

As mentioned above (Section 3.2), Galileo made his first telescope in mid 1609. Then, shortly after his return from Venice in September of that year, he made another telescope giving a magnification of 20 diameters, which he used to observe the moon. Galileo was, in fact, not the first person to turn a telescope on the moon, as the Englishman Thomas Harriot (1560–1621) had used a 6 magnification instrument to produce the first telescope-based sketch of the moon in July 1609.

Galileo undertook a series of observations of the moon starting on 30th November 1609, producing a number of drawings in various phases as a result. He summarised his observations in a letter of 7th January 1610, pointing out that the terminator[25] was not a uniform curved line, as it should be if the moon was a smooth sphere, but was very irregular in shape with bright points in the dark area. From this he concluded that the moon has

---

[25] The terminator is the line dividing the illuminated from the non-illuminated part of the moon.

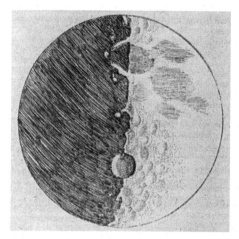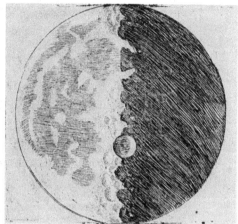

**Figure 3.4** Drawings of the moon from *Sidereus Nuncius* by Galileo, published in 1610. The large crater on the terminator, just below the centre, is thought to be Albategnius. The reason for doubt is that Galileo considerably exaggerated its size, possibly to show the effect of the side illumination more clearly.

mountains[26] and valleys, and is definitely not the pure spherical object of Aristotle's cosmology. Galileo gave more details in his little book *Sidereus Nuncius* (*The Sidereal Messenger*) published in Venice in March 1610, which included a number of drawings (two of which are shown in Figure 3.4). In this book he calculated that the highest lunar mountains are over 4 miles (6 km) in altitude, which he thought was much higher than terrestrial mountains.[27] He also concluded that the light seen on the dark part of the moon, when the moon in new, is caused by sunlight being reflected from the earth.[28]

Galileo seems to have begun using his largest telescope, that magnified 30 diameters, in early January 1610.[29] He first of all noticed that the

---

[26] Giordano Bruno, William Gilbert and Johannes Kepler all pre-dated Galileo in their belief that there are mountains on the moon, but Galileo was the first to prove it.

[27] Galileo erroneously thought that the highest mountains on the earth are no more than a mile (1.6 km) high. Even the Alps are much higher than this (Mont Blanc is 3 miles high).

[28] Galileo's later work on the moon is covered in the next chapter.

[29] This is according to Henry C. King in *The History of the Telescope*, Charles Griffin, 1955, republished by Dover, 1979, p. 36, and to Louis Bell in *The Telescope*, McGraw-Hill, 1922, republished by Dover, 1981, pp. 8–9. But Drake in his *Galileo at Work*, p. 148, concluded that Galileo continued to use his telescope that magnified 20 diameters at this time.

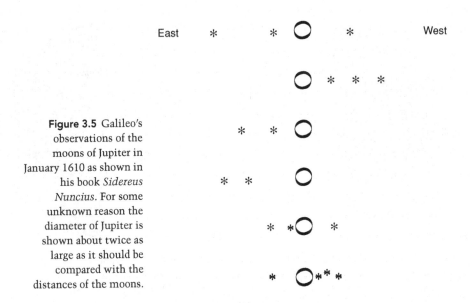

East    *    *   ◯   *     West

**Figure 3.5** Galileo's
observations of the
moons of Jupiter in
January 1610 as shown in
his book *Sidereus
Nuncius*. For some
unknown reason the
diameter of Jupiter is
shown about twice as
large as it should be
compared with the
distances of the moons.

planets showed small discs, whereas the stars did not, thus vindicating
Copernicus' view that the stars are much further away than the planets.
But probably Galileo's most significant discovery of the period was that
there were four moons around Jupiter.[30] In a draft letter of 7th January 1610
Galileo said that he was able to see many more fixed stars with the telescope
than with the naked eye, and that[31] 'only this evening I have seen Jupiter
accompanied by three fixed stars'[32] (see Figure 3.5). What struck Galileo
was that the three stars were in a straight line which was collinear with
Jupiter, but at this stage he still probably thought that he was observing
fixed stars.[33]

---

[30] Simon Mayr claimed in 1611, a *year* after Galileo had published his results, that he
had seen Jupiter's satellites in December 1609, but he provided no real evidence of this
(see Drake, *Galileo at Work*, p. 235).

[31] From *Galileo at Work; His Scientific Biography*, by Stillman Drake, University of
Chicago Press, 1978, p. 146.

[32] Galileo used the term 'fixed stars' to distinguish stars from planets, which were
usually called 'wandering stars'.

[33] Galileo says in *Sidereus Nuncius* (translated by Albert Van Helden, University of
Chicago Press, p. 64) that he thought that they were fixed stars when he saw them on
7th January. But in his draft letter of that same date he says, 'The planets are seen very
rotund, like little full moons...', see *Galileo at Work; His Scientific Biography*, by
Stillman Drake, University of Chicago Press, 1978, p. 146.

At this time Jupiter had just passed opposition (on 8th December 1609), and so it should be moving east to west compared with the fixed stars, before it returned to its normal west to east motion (see Figure 1.3). Galileo observed Jupiter again on 8th January and noticed that the planet appeared to have moved to the east of the three stars, not to the west, and that the three stars and Jupiter were still in a straight line (see Figure 3.5). For a time he wondered if 'contrary to the astronomical computations, his [Jupiter's] motion was direct [i.e. west to east] and that, by his proper motion, he had bypassed those stars',[34] although he also mentions that the three stars were closer to each other than on the previous night. He was reluctant to believe that the 'stars' had moved. Maybe his separation distance estimates of 7th January had been in error. To clarify matters, he set out to observe Jupiter on 9th January, but it was too cloudy. On 10th he saw just two 'stars', this time to the east of the planet (Figure 3.5), clearly showing that it was not Jupiter that was moving but the 'stars', with the third star behind Jupiter and thus hidden from view. As a result Galileo decided to make more careful observations of time and position in future to sort out the detailed movement of these 'stars'. Galileo's observations on 11th January led him to conclude that he was observing three moons of Jupiter, and when on 13th he saw four 'stars' for the first time near Jupiter, he concluded that Jupiter had four moons, not three.

Galileo's observations of Jupiter's moons up to and including 2nd March were contained in his *Sidereus Nuncius*, a bound copy of which was sent to Grand Duke Cosimo II of Tuscany on 19th March 1610. In this book, Galileo pointed out that the moons orbit Jupiter in 'unequal circles', and that the periods of the moons 'describing smaller circles around Jupiter are faster',[35] although he was unable to specify their periods accurately. He concluded that the outermost moon (number 4), now called Callisto, had a period of about half a month (modern value 16d 17h), whilst the innermost (number 1), now called Io, had a period of a day or so (modern value 1d 18h).[36] Although not proving the Copernican model of the universe, the discovery of these

---

[34] *Sidereus Nuncius*, as translated by Albert Van Helden, University of Chicago Press, 1989, (afterwards referred to as *Sidereus Nuncius*, translated by Van Helden), p. 65, © 1989 by The University of Chicago.

[35] *Sidereus Nuncius*, as translated by Van Helden, University of Chicago Press, 1989, p. 84, © 1989 by The University of Chicago.

[36] By March 1612 Galileo had calculated the periods of each of Jupiter's four satellites to an accuracy of a few minutes, see Section 4.2.

moons eliminated one objection that had been raised against it, namely that it seemed unreasonable to believe that the earth was the only planet to have a moon.[37]

For those who like puzzles, it may be tempting to try to work out which of Jupiter's moons was where from Galileo's observations shown in Figure 3.5. This is not as easy as it appears, however, as Galileo's estimates of their separation are not very accurate.[38] The easiest way to solve this problem observationally is, of course, to observe the moons every two or three hours, as the inner moons can be seen to move in this sort of timescale, but Galileo did not do this with his early observations. In 1962 Jean Meeus of the Kesselberg Observatory in Belgium computed, from their known positions and orbits now, where the moons would have been at the time of Galileo's observations.[39] Doing this he found that the 'three' moons seen by Galileo on 7th January were (from east to west), numbers 4, then 1 and 2 together, and 3.[40] Meeus found that numbers 1 and 2 were only 0.3 Jupiter radii apart that night about an hour after sunset, when Galileo made his observations. On the next night Galileo did not observe number 4, which was near its greatest eastern elongation, presumably because he was only looking for three 'stars', and 4 was outside his telescope's field of view. The others on that night, from the east, were 1, 2 and 3. So, although Galileo did not realise it, he had resolved moons 4 and 3 on 7th January, and 1, 2 and 3 on 8th, some five days before he saw all four together on 13th. On 10th January Galileo actually saw, from the east, 4, then 3 and 2 together, and no number 1, as it was either too close to or was behind Jupiter, depending on when exactly Galileo made his observations, Io having disappeared behind Jupiter just after 5 pm Universal Time (UT) that night according to Meeus. So Galileo may have been correct when he concluded that he couldn't see a third moon on 10th because it was behind Jupiter.

Galileo's book *Sidereus Nuncius*, which contained all the above observations and more, caused a great stir amongst the cognoscenti of Italy and elsewhere. Many thought that Galileo had not seen any moons of Jupiter at all, and that they were either optical illusions, or that his telescope was exhibiting some sort of flaw. The fact that he did not see the same number of

---

[37] This objection cannot be levelled at the Ptolemaic model, as in that the moon was just the closest of all the planets orbiting the earth (see Figure 1.8).

[38] This is not to criticise Galileo as it was very difficult to measure these distances with his telescope.

[39] See *Sky & Telescope*, September 1962, pp. 137–139.

[40] 1 is Io, 2 Europa, 3 Ganymede, and 4 Callisto.

moons each night did not help to persuade the doubters. On 24th–25th April 1610 Galileo stopped over in Bologna to show the astronomer Giovanni Magini and his colleagues Jupiter's moons, but the visit was a disaster as no one could see them with Galileo's telescope. Meanwhile news of Galileo's discovery had reached Kepler in mid March, and on 8th April he received a copy of *Sidereus Nuncius* with a request from Galileo for his opinion. Even though Galileo had not replied to Kepler's request made in 1597 for parallax measurements (of two stars), Kepler responded rapidly to Galileo's request with a pamphlet entitled *Conversation with the Sidereal Messenger* that was sent to Galileo on 19th April. This pamphlet, which was published the following month, was very supportive of Galileo, even though Kepler said that he didn't have a powerful enough telescope to observe the moons.

The support of Kepler helped to quieten some of the criticisms of *Sidereus Nuncius*, but many respected astronomers still doubted the observations, amongst them the influential Jesuit mathematician and astronomer Christopher Clavius of the Roman College. Kepler indirectly asked Galileo if he could let him have a good telescope, so that he could personally attest to Galileo's observations. Galileo made the excuse that he had sent the telescope that he had used to discover the moons of Jupiter to Grand Duke Cosimo, but said that he was going to make some new telescopes for his friends. In the event he did not provide Kepler with such an instrument, possibly because he did not want to create a rival observer. It was not until the Elector Ernest of Cologne, one of Kepler's patrons, lent him a telescope that he had received from Galileo, that Kepler was able to see the satellites[41] of Jupiter for himself in August 1610. This provided the first independent evidence of their existence.

The written communications between Kepler and Galileo clearly show their contrasting personalities. Kepler, trusting, full of enthusiasm, and wanting to strike up a professional relationship with Galileo. The latter, haughty, discourteous and secretive. In fact, Galileo only answered two of Kepler's numerous letters directly, in spite of the fact that he relied very much on his professional support against the chorus of critics following the publication of *Sidereus Nuncius*. Galileo didn't even have the courtesy to write a personal letter to Kepler when asking for his views on this book, but transmitted the request via Guiliano de Medici, the Tuscan Ambassador in Prague, who asked Kepler verbally. To add insult to injury, when Galileo

---

[41] Kepler was the first to call the moons of Jupiter 'satelles' or 'satellites', in a letter to Galileo on 25th October 1610.

did finally write to Kepler on 19th August 1610, he made no mention of his discovery on 25th July of what he thought were two moons of Saturn (see below).

The year 1610 was also important as far as Galileo's academic life was concerned, as he was offered a lifetime appointment by the Venetian Republic at the University of Padua, with a large increase in salary. But lectures and lessons at Padua took up valuable time, and Galileo resolved instead to try to get an appointment at Florence, where his ex-pupil Cosimo was now Grand Duke, that would not involve any teaching commitments.

As part of this plan, Galileo had dedicated the *Sidereus Nuncius* to Cosimo, had called the four moons of Jupiter the 'Medicean stars' in honour of Cosimo and his three brothers, and had sent the telescope that he had used to discover the moons to Cosimo as a present. On 7th May Galileo wrote to the Grand Duke's Secretary of State, Belisario Vinta, explaining that he wanted to take up scientific research full time, and asking if he could have a non-teaching appointment at Florence that would allow this. Vinta replied positively on 22nd May, and on 10th July Cosimo gave Galileo the lifetime appointment of 'Principal Mathematician of the University of Pisa, and Principal Mathematician and Philosopher to the Grand Duke of Tuscany', without the requirement to live in Pisa, and with no teaching commitments.

On 25th July 1610, shortly before he left Padua, Galileo was observing Saturn and noticed that it appeared to have two companions, one on each side, very close to the planet. Five days later he mentioned this discovery to Vinta in a letter, saying that 'Saturn is not alone but is a composite of three, which almost touch one another, nor do they move with respect to one another nor change, and they are located in a line along the length of the zodiac...'[42] A few days later, Galileo communicated his discovery to the Tuscan Ambassador in Prague, for onward transmission to Kepler, in the form of an anagram to establish priority of discovery. Kepler was unable to solve the anagram correctly,[43] and it was not until 13th November 1610 that

---

[42] From *Le Opere di Galileo Galilei*, Vol. 10, p. 410, as translated by G.V. Coyne, quoted in *Galileo for Copernicanism and for the Church*, by Annibale Fantoli, second edition, Vatican Observatory Publications, p. 119.

[43] Kepler thought that the anagram had announced the discovery of two moons around Mars, to match the one around the earth and the four around Jupiter. As Mars is between the earth and Jupiter in distance from the sun, this would give the sequence of moons as 1, 2 and 4, one number being double the number before. Interestingly, Mars does have two moons (discovered in 1877), but Jupiter has many more than four.

Galileo provided the solution, this time, not to Kepler, nor even to the Tuscan Ambassador, but to the Emperor Rudolf himself, who was intrigued by the puzzle. Galileo naturally thought that he had seen two moons of Saturn, but he had, in fact, seen its rings, although his telescope was not good enough to resolve them as such.

In October of this momentous year, Galileo also started to observe Venus, and by December he had managed to observe its changing phases. In the Ptolemaic system (see Figure 1.8) Venus could never be seen as more than about half illuminated, but in the Copernican or Tychonic systems it could exhibit a full range of phases, like the moon, up to a fully illuminated disc. The phases that Galileo observed in the last months of 1610 were inconsistent with the Ptolemaic theory, but fully consistent with those of Copernicus or Tycho. This finally settled, in Galileo's mind, the question of the structure of the universe in favour of the Copernican system, as he had previously rejected Tycho's model. Galileo announced his discovery of the phases of Venus on 11th December 1610 in another anagram sent, yet again, to the Tuscan Ambassador in Prague.[44]

A week earlier, on 5th December 1610, Benedetto Castelli (1578–1643), a former student of Galileo's, had written to him to point out that Venus should exhibit a full range of phases if the Copernican system was correct. As Castelli did not have a telescope, he asked Galileo if he had noticed these phases. Galileo confirmed this to Castelli on 30th December,[45] and two days later he sent a solution to his previous anagram to the Tuscan Ambassador in Prague. As it happened, the latter communication crossed with a plea from Kepler to put him out of his misery and send him the solution to the anagram.[46]

Finally the year 1610 ended on a very good note for Galileo, when he received a letter from Christopher Clavius in Rome dated 17th December,

---

[44] The discovery of the phases of Venus also disproved the idea, espoused by some astronomers including Kepler, that Venus is self-luminous.

[45] There have been some suggestions that Galileo was less than honest in claiming that he had discovered the phases of Venus before he had received Castelli's letter of 5th December. As Galileo had later claimed that he had started observing Venus in October, he would undoubtedly have observed its changing phases before he heard from Castelli. The pros and cons of this case are discussed in *Galileo for Copernicanism and for the Church*, by Annibale Fantoli, pp. 154–155, and in *The Great Copernicus Chase*, by Owen Gingerich, Sky Publishing and Cambridge University Press, 1992, pp. 98–104.

[46] Kepler thought that the anagram had announced the discovery of a red spot on Jupiter. Interestingly such a red spot exists, but it was not discovered until 1664 by Hooke.

which said that he and other Jesuits at the Roman College had finally seen the four moons of Jupiter. This was quite a change, as Clavius had previously expressed severe doubts as to their existence. Clavius was in a very influential position in the Church, and his confirmation of Galileo's discovery was a major boost to his credibility.

Whilst Galileo was convincing his fellow astronomers as to the validity of his observations, the theological establishment stirred. For example, at the end of 1610 the Aristotelian philosopher, Ludovico delle Colombe (1565–?) wrote a dissertation *Ludovico delle Colombe against the motion of the earth* which was distributed in manuscript form. In this he attacked Copernicanism from both a scientific and biblical point of view. Shortly afterwards Francesco Sizzi (1585–1618) published a book which also attacked, on biblical grounds, Galileo's interpretation of his observations in terms of Copernican cosmology.

Galileo, forever sensitive to attacks from any source, quickly realised that he needed the support of the Church against these attacks on Copernicanism, otherwise his scientific arguments would be eclipsed by theological ones. The obvious place to start was with Clavius and the Jesuits of the Roman College, and so on 23rd March 1611 Galileo left Florence for Rome, where he arrived about a week later. On the day after his arrival, Galileo met with Clavius and his colleagues Grienberger and Odo Van Maelcote. During his stay in Rome Galileo also visited other establishment figures, giving successful demonstrations with his telescope. In parallel, Cardinal Robert Bellarmine, a prominent theologian of the time, had written to Clavius asking for his views on Galileo's various discoveries, namely:

- Were there many stars shown by the telescope that could not be seen with the naked eye?
- Did Saturn have two companions?
- Did Venus have phases like the moon?
- Were there mountains on the moon?
- Did Jupiter have four moons in orbit around it?

To which Clavius and his colleagues replied positively on four out of five questions, whilst reserving their position on the question as to whether there were mountains on the moon.

Many years earlier, in 1584, the maverick philosopher Giordano Bruno (1548–1600) had written a series of six dialogues on various cosmological and moral matters in which, amongst other things, he supported Copernicus' cosmology, and argued that the universe was infinite with a limitless number of solar systems which supported intelligent life.[47] Bruno was eventually burnt at the stake for his religious beliefs in February 1600. But, interestingly, Bellarmine had played a significant part in Bruno's trial by the Roman Inquisition, so it was no accident that he took a keen interest in Galileo's work. Whatever theological reservations Bellarmine and his fellow cardinals may have had in 1611, however, Galileo was respected by them as a prominent scientist during his visit to Rome, and he was even received in audience by Pope Paul V. In fact, Galileo was also the guest of honour at an academic assembly held by the Roman College which was attended by a number of cardinals. Galileo returned to Florence about two weeks later on 4th June 1611.

At face value, Galileo's visit to Rome had been a great success, but he had not managed to still the minds of the Church authorities on the revolutionary aspects of Copernican cosmology. Galileo's scientific results seemed clear, but their interpretation was another matter. In fact, on 24th May 1611, whilst Galileo was still in Rome, Claudio Acquaviva (1543–1615), the Superior General of the Jesuit order, had sent a letter to all Jesuit professors recommending that they continue to teach and adhere to Aristotelian doctrine. In the meantime, however, a further threat to Aristotle's ideas was maturing.

Aristotle had argued in his book *De Caelo* that the sun was a perfect body. Since then, from time to time, Chinese, Arab and European astronomers had occasionally seen what appeared to be dark spots on the sun. At the time these spots had been thought to be bodies between the earth and sun, and even Kepler, who had seen a large sunspot in May 1607, took it to be the transit of the planet Mercury across the sun's disc. In the late sixteenth century, however, Giordano Bruno had put forward an alternative explanation that the dark spots were actually on the surface of the sun, and he even suggested that their movement probably indicated that the sun was rotating. Naturally, the idea of a rotating, spotted sun was completely unacceptable to the Aristotelian philosophers.

Galileo first observed sunspots with his telescope in mid-1610, but Thomas Harriot (1560–1621) of Oxford was probably the first to make telescopic observations of sunspots earlier that year.[48] The Dutchman Johann Fabricius (1587–1616) was the first to publish, in June 1611, his observations

---

[47] Nicholas of Cusa (1401–1464) had also argued, over a century earlier, for an infinite universe in which the stars may support life. In his case, however, he thought that the stars themselves may be inhabited.

[48] The exact month of Galileo's first observations of sunspots is not clear, but it was almost certainly after Harriot's first observations.

of sunspots, in which he concluded, like Bruno before him, that their movement probably indicated that the sun was rotating on its axis. Earlier in 1611 Galileo had shown sunspots to people during his visit to Rome and, at about the same time, a Jesuit astronomer called Father Christoph Scheiner (1573–1650) had entered the fray.

Scheiner and his assistant Cysat had been observing the sun at Ingolstadt in March or April 1611, when Cysat noticed several black spots on its disc. They investigated the phenomenon further and Scheiner concluded, later that year, that the sunspots were really small planets between the earth and sun. His results were contained in three letters sent by him in November and December 1611 to his friend Marcus Welser, a banker and businessman. Copies of Father Scheiner's letters, which he had written under the pseudonym 'Apelle', on the advice of his provincial Jesuit superior, were forwarded by Welser to Galileo in early 1612, asking for his opinion on their contents. In one of the letters Scheiner refers to the predicted transit of Venus of 11th–12th December 1611 which he had tried to observe. Scheiner was unsuccessful,[49] concluding that this was probably because Venus had passed behind rather than in front of the sun, showing that Venus must orbit the sun as Tycho[50] had proposed. His logic was, of course, heavily flawed, but Galileo was uncharacteristically cautious in his response to Scheiner's letters. He disagreed that the sunspots were small planets, and suggested, instead, that they were probably something similar to clouds on or near the surface of the sun.

Aware of the sensitivity of his conclusion to Aristotelian philosophers and the Church, Galileo sent the three letters from Apelle, together with his reply, to Cardinal Carlo Conti asking for a theologian's point of view. Conti replied in July 1612 that Scripture did not support Aristotle's philosophy of the incorruptibility of the heavens, and of the sun in particular, but that Copernicus' idea of a moving earth was more of a problem.

In 1612 Scheiner, still using his pseudonym, sent three more letters to Welser on sunspots, which were quickly published, repeating his view that sunspots were small planets between the earth and sun. Scheiner also claimed erroneously to have discovered a fifth satellite of Jupiter, as well as being the first to prove that Venus orbited the sun, which he was not. In addition, in these letters Scheiner mentioned that the Jesuit astronomers in Rome had been observing the phases of Venus at about the same time as

---

[49] No such transit took place. We now know that the first transit of Venus in the seventeenth century took place on 6th December 1631 (see Sections 3.1 and 4.3).

[50] Scheiner preferred Tycho's to Copernicus' cosmology.

Galileo's first observations of them, implying a challenge to Galileo's claim of priority. Whether these Jesuit observations had started at the same time as Galileo or not, Galileo had been the first to correctly conclude that the observed phases of Venus proved that the planet orbited the sun. Scheiner in these letters was trying to undermine Galileo on what were false grounds, and Galileo, quite naturally, did not like it.

Galileo had undertaken a new series of observations of sunspots in the spring and summer of 1612, with the able assistance of Benedetto Castelli, in between receiving Scheiner's two sets of letters. These observations showed that sunspots changed their shape with time, and their movement across the sun showed that the sun rotated about once per month. These results he discussed in a second letter to Welser, which was quickly followed by a third letter to Welser in answer to Scheiner's second set of three letters. In Galileo's third letter he strongly disagreed with Scheiner's proof that Venus orbited the sun, because of his inability to observe the transit, and also disagreed strongly with Scheiner's theory that sunspots were not on the sun.

In March 1613 Galileo had his three letters to Welser published as *A History and Demonstrations Concerning Sunspots* or *Letters on Sunspots*, as it is more popularly known The preface, written by Angelo de Filiis, secretary of the prestigious Accademia dei Lincei, of which Galileo was a member, was contentious in tone, claiming, amongst other things, Galileo's priority for the discovery of sunspots. Galileo was unhappy with the boldness of this preface and thought that it would cause trouble, but his friends persuaded him to let it remain. Galileo's instincts had been correct, however, as the preface upset the Jesuits of the Roman College, and also caused a distinct chill in the relationship with Scheiner.[51] Although the latter problem was eventually patched up, Scheiner and Galileo were to remain distrustful of each other for the rest of their lives.

## Galileo's dispute with the Church

This present book is not the proper place to go into the full details of Galileo's controversy with the Church and his subsequent trial, as many of the arguments were based on the interpretation of Scripture, not on science. Nevertheless, the dispute did cast a shadow on the relationship between the Catholic Church and science for well over a century, constraining

---

[51] Galileo had deduced in 1613 that Apelle was probably a Jesuit, but he did not know that Apelle was Scheiner until the following year.

astronomical research in Italy, in particular, to the advantage of Protestant Europe. So a short summary is called for.

In outline, the argument between Galileo and the Church was not about his observations, but about his interpretation of them. Unfortunately, instead of limiting himself to a scientific analysis of his observations, and drawing conclusions based purely on those, Galileo felt obliged to try to convince the Church authorities of their errors by using theological arguments. The Church naturally did not like this, regarding theological analysis as their preserve, but they made a similar error in trying to disprove Galileo's conclusions based on a scientific analysis. Galileo was basically an amateur theologian, and the Church authorities were not nearly as well versed in science as Galileo. A situation that was bound to lead to trouble, unless one party was willing to give way.

At this time the Church was quite content to let Galileo use Copernicus' cosmology as a working hypothesis, as long as he did not claim it to be the universal truth, as that would require the Church to reinterpret Scripture. Unfortunately, Galileo took the view that he had established a universal truth, so the Church would have to prove him wrong, even though he himself declined to prove his point scientifically, as he considered the Church hierarchy to be scientifically illiterate. The Church, on the other hand, in the guise of Cardinal Bellarmine, Master of Controversial Questions, said that if Galileo could prove scientifically that Copernican cosmology was not just a hypothesis but was in fact true, then the Church would have to reconsider its position.[52] As a matter of interest, however, Galileo was actually advocating Copernicus' cosmology, epicycles and all, which he would never have been able to prove as it was flawed in many ways.

Unfortunately, during the period 1613–1615 Galileo gradually painted himself into a corner, by ignoring Kepler's work on planetary motions, which Kepler was developing at this time, and by asserting more and more positively the correctness of Copernicus' cosmology, which was flawed.[53] Fame appeared to go to Galileo's head, and he began to become more and more arrogant, particularly when he was pushed by the Church authorities to

---

[52] This attitude of the Church is clearly stated in a letter, dated 4th April 1615, from Cardinal Bellarmine to Father Paulo Foscarini.

[53] Kepler had proved that the orbit of Mars was an ellipse in his *Astronomia Nova* in 1609 (see Section 3.1), which showed that the Copernican epicycles were not needed for Mars. This was proved more generally in his *Epitome* in 1618–1621. So Kepler's work would disprove Copernicus' model of the universe well before Galileo's argument with the Church came to a head at his trial in 1633.

provide unambiguous evidence that Copernicus was right. Unfortunately for Galileo, the absence of an observable stellar parallax made it impossible for him to prove that the earth orbited the sun, and, what was worse, it tended to suggest that the earth was at the centre of the universe, as in Tycho's system, for example. The phases of Venus could also be explained by Tycho's system, so Galileo had his work cut out to prove that Copernicus was correct.

At first Galileo tried to belittle his opponents, claiming that it was a waste of time presenting scientific evidence to them, as they would be incapable of understanding it. Then, when forced into a verbal confrontation from time to time, he tried to tie up his opponents in knots and make them appear ridiculous. When that eventually failed, he was forced to throw what he considered was his 'killer punch', namely his new theory of tides. But that was clearly scientifically wrong. Finally, in desperation, he asked Cardinal Alessandro Orsini to intercede directly with the Pope on his behalf. This Orsini did but without immediate success.

However, following Orsini's intervention, on 19th February 1616 the Pope asked the theological experts of the Holy Office to consider the matter. Four days later after much discussion they declared that the proposition that the sun is at the centre of the universe was heretical. But their findings were overruled by higher authority and were not published until seventeen years later. Instead, on 5th March 1616 the Congregation of the Index declared that the Copernican system was opposed to Holy Scripture, rather than being heretical.

As a result of the decree of 5th March, Copernicus' *De Revolutionibus* was suspended pending modification, which turned out, eventually, to be of a relatively minor nature. In fact, *De Revolutionibus* stayed on the Index of Forbidden Books for only four years, and Galileo's *Letters on Sunspots* was not put on the Index at all, as he had only referred to Copernican cosmology as an hypothesis in those letters, which was acceptable to the Church.

Galileo had been told in 1616 that he could continue to refer to Copernican cosmology, but only as an hypothesis, to which he agreed. Instead he decided to attack Aristotle's philosophy of nature. In the end, however, he could not resist attacking Aristotle and advocating Copernicus' cosmology in his *Dialogue on the two Great World Systems*, which was published in February 1632. This, as far as the Church was concerned, was the last straw, and his *Dialogue* was available for just six months, before it was confiscated by the Church. Galileo was then forced to appear before the Inquisition in Rome.

In the *Dialogue* Galileo described and analysed Aristotle's and Copernicus' ideas through three characters, Salviati, who propounded Galileo's views, Sagredo, the intelligent neutral, and the ultra-conservative Simplicio, who supported Aristotle and Ptolemy, and whose name Galileo chose carefully to imply that he was a simpleton. The result of the dialogue was, of course, a foregone conclusion, but not before Galileo had completely misrepresented the Copernican model. For example, he ignored the large number of epicycles that Copernicus had been forced to introduce, and he failed to point out that the centre of the earth's orbit, and not the sun, was at the centre of the Copernican universe.

Galileo's case before the Inquisition was not helped by the fact that Pope Urban VIII felt that he had been made a fool of by Galileo, as not only had Galileo duped the Chief Censor to get his *Dialogue* published, but Urban's favourite argument had been put into the mouth of the simpleton Simplicio. Galileo had also upset two of the senior Jesuits of the day, Fathers Scheiner and Grassi,[54] which did not help either. Nevertheless, he was treated with a great amount of respect by the Inquisition during his trial that took place in 1633.[55] Galileo, who was almost seventy at the time, realised that he had pushed the Church too far, and that he could not continue to play games with such powerful people as the Pope and senior Jesuits. As a result he willingly agreed to recant his views, as required by the Inquisition, after which he was placed under a very comfortable house arrest for the rest of his life. He died aged seventy-eight in 1642, the year that Newton was born.

---

[54] Galileo's dispute with Scheiner has been outlined above. In addition, Galileo also had a bitter argument with Grassi in 1619 about Tycho Brahe's theory of comets.

[55] Strictly speaking, the counts on which he was charged were incorrect. On the other hand, the Inquisition knew that much of Galileo's evidence was untrue, but they did not push him to tell the truth. During his trial he was not put in prison, but was allowed to stay at the Tuscan Embassy, as well as in a large flat overlooking the Vatican gardens, where he was allowed to have two servants.

# Chapter 4 | THE MID AND LATE SEVENTEENTH CENTURY

## 4.1 Saturn

It is evident, from the description of Galileo's work above, that planetary astronomy had taken on a new dimension with the invention of the telescope. No longer were the planets just points of light wandering across the sky, but they were now observed as discs with individual shapes and, in the case of Jupiter and Saturn, as planets with their own moons.

Unfortunately, when Galileo observed Saturn in November 1612 the two 'moons' on either side of Saturn had disappeared. He put this down to changes in the sun–earth–Saturn geometry affecting the illumination of the moons and their visibility from earth. As a result, he predicted that they would be visible again for about two months around the summer solstice of 1613, and for a few months in the winter of 1614–15. Galileo noticed their reappearance, as predicted, in mid-1613, and so did Giovanni Agucchi of the Pope's office. Unfortunately for Galileo's theory, however, the 'moons' stayed visible for a number more years. Then in mid-1616 Galileo noticed that Saturn's 'two companions are no longer two small perfectly round globes as they were before, but are at present much larger bodies and no longer round in shape, but as seen . . . [as] two half-ellipses.'[1]

Over the next forty years or so a series of astronomers including Scheiner, Biancani, Fontana, Gassendi, Hevelius, Riccioli, Grimaldi, Divini and Sir Christopher Wren,[2] observed Saturn and its strange attendants called

---

[1] From a letter written by Galileo to Federico Cesi, translated by Stillman Drake in his *Galileo at Work; His Scientific Biography*, University of Chicago Press, 1978, p. 259.

[2] Sir Christopher Wren was a scientist and astronomer before he became an architect.

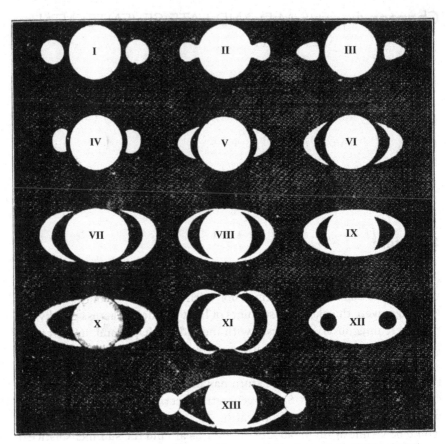

**Figure 4.1** Drawings of Saturn in Huygens' *Systema Saturnium*. I is by Galileo (1610), II Schneider (1614), III Riccioli (*c.* 1641), IV–VII Hevelius (theoretical), VIII & IX Riccioli (*c.* 1648), X Divini (*c.* 1646), XI Fontana (1636), XII Biancani (1616), and XIII Fontana (1644).

ansae,[3] noticing their varying shapes over the years (see Figure 4.1). These observations produced a great deal of disagreement over the size and shape of both the planet and its ansae, principally because of the poor quality of the telescopes then available. According to Giovanni Riccioli of Bologna, his former pupil and fellow Jesuit, Francesco Grimaldi (1618–1663), was in 1650 the first to draw attention to Saturn's polar flattening. But it was left

---

[3] The attendants were called 'ansae' after the Latin for 'handles', indicating their appearance in rudimentary early telescopes.

to Christiaan Huygens (1629–1695) to correctly interpret the ansae as a thin flat ring surrounding Saturn.

Christiaan Huygens was born on 14th April 1629 into a prominent Dutch family. His father, who was a diplomat, sent Christiaan to the University of Leiden in 1645 to study mathematics and law, followed by a further two years at Breda, just south of Rotterdam, where he specialised in law. This was intended as a prelude to a diplomatic career, but Christiaan decided to become a gentleman of leisure, undertaking scientific and mathematical research, funded by his father. As a young man he met René Descartes (1596–1650), who was an occasional visitor to his parent's house, and who naturally had a great influence on the young Christiaan. Later in life Huygens visited London and met and discussed his gravitational theory with Isaac Newton (1642–1727). He also knew both Blaise Pascal and Gottfried Leibnitz, with whom he communicated regularly, so Huygens was very much at the centre of events. Although he was clever, however, Huygens did not quite have the brilliancy of these great men.

Huygens, like Galileo before him and Newton afterwards, undertook research into a vast range of scientific and mathematical problems. In mechanical engineering he invented the pendulum clock, in optics he proposed the wave theory of light, and he invented a telescope eyepiece (now named after him) that significantly reduced spherical aberration in telescopes. He was also a first class mathematician, but in astronomy he is best known for his work on Saturn.

The telescopes available at that time were rather small and of relatively poor quality, so Christiaan and his brother Constantyn decided to make a much larger instrument. Their first large telescope, which they finished in early 1655, was 12 feet (3.6 m) long and 2 inches (5 cm) diameter, producing a magnification of 50. Soon after it was completed, on 25th March 1655 Christiaan discovered the first moon of Saturn, which we now call Titan. He observed its movement over the next few days, confirming that it was in orbit around the planet, and by the end of the year he had determined its orbital period to be about 16 days 4 hours.[4] At about the same time he and his brother had completed an even larger telescope of 23 feet (7.0 m) length and 2.3 inches (5.9 cm) diameter, with a magnification of about 100, which Christiaan also used to study Saturn.

---

[4] After many years of observations, Huygens corrected this to 15d 22h 41m 11s, which is within a few seconds of the correct value accepted today.

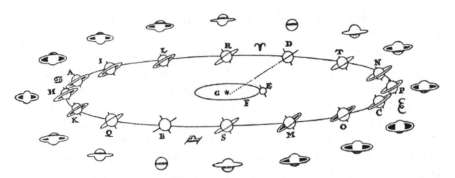

**Figure 4.2** Huygens drawing in his *Systema Saturnium* to explain the appearance of Saturn's rings. They are edgeways on, as seen from the earth, at B and D about 15 years apart, and could not then be seen in the telescopes of the time as the rings are too thin.

Encouraged by his friend Jean Chapelain, in March 1656 Huygens announced his discovery of Titan in a short publication entitled *De Saturni lunâ observatio nova*. Other astronomers quickly confirmed the discovery, whilst Johannes Hevelius (1611–1687) and Christopher Wren (1631–1723) both said that they had seen Titan earlier but had taken it to be a star. In *De Saturni* Huygens also said that he had solved the riddle of Saturn's ansae, which had disappeared again in January 1656. He predicted that they would reappear at the end of April 1656, and he invited anyone who thought that they had found a solution to the ansae riddle to send him their proposal. Huygens then gave his solution in the form of an anagram to cover his claim to precedence, and show any respondents who may reply that he had not copied their solution. Unfortunately for Huygens, the ansae did not reappear before Saturn was lost to daylight in June 1656, but when he observed the planet on 13th October 1656 he found that they were clearly visible again.

Huygens finally disclosed his theory of the ansae in his *Systema Saturnium*, published in July 1659, in which he commented on the various drawings of Saturn produced by various astronomers starting with Galileo (see Figure 4.1). He then presented the theories of Hevelius, Hodierna and Roberval, sent in response to his request of three years earlier, and dismissed each one in turn. Finally he gave the solution to his anagram, explaining that Saturn is surrounded by a thin, flat, solid ring, inclined at an angle of about 31° to the ecliptic (the true figure is 28°). So the configuration of the ring and planet varies cyclically as Saturn orbits the sun about every 29½ years (see Figure 4.2). In making this proposal, Huygens was crucially able to see the shadow of the ring on the planet (see Figure 4.3(a)), enabling him to know which part of the ring was in front of the planet and which

(a)

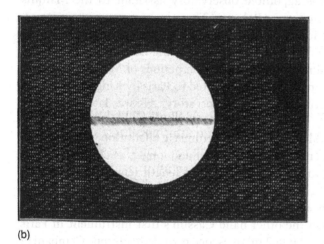

(b)

**Figure 4.3** (a) & (b) Huygens' drawings of Saturn in his *Systema Saturnium*, clearly showing the shadow of the ring on the planet.

was behind. Even when the rings were edge on to the earth, he was able to see a dusky band across the middle of the planet, which was the shadow of the ring (see Figure 4.3(b)). Anticipating questions as to how the ring could stay in the same place *with respect to Saturn* over time, he explained that it rotated around the centre of the planet, like Titan did, thus providing it with stability.

Some astronomers, including Wren, accepted Huygens theory of the ansae as a brilliant insight, but others attacked it for various reasons. For example, Ismael Boulliau objected that the ring should not completely disappear when it was seen edge on, but he did not appreciate that the ring was too thin to be seen with contemporary telescopes.[5] Others objected that it was unreasonable to propose a solution to the ansae problem that was unique to one heavenly body.[6] But the main opposition to Huygens' theory of Saturn and its ring came from Eustachio Divini and Honoré Fabri, who wrote two pamphlets attacking it with many and varied reasons, most of them erroneous. For example, they claimed that the dusky band across Saturn, when the ansae disappeared, did not exist. But Huygens was able to quote independent evidence of its existence from the English observer William Ball, who had seen it when the ansae were not present between January and June 1656.

The second great Saturn observer of the second half of the seventeenth century, and the greatest planetary observer of that period, was Jean Dominique Cassini (1625–1712), who had been known as Gian (or Giovanni) Domenico Cassini before his move to France. Born in Perinaldo near San Remo, in north-west Italy, Cassini studied in Genoa and, at the age of nineteen, he was appointed observatory assistant to the Marquis Mavasia at his observatory near Bologna. There he met two prominent astronomers, Giovanni Riccoli (1598–1671) and Francesco Grimaldi. Six years later, Cassini became professor of astronomy at the University of Bologna, where he observed and measured the rotation periods of Mars, Jupiter, and Jupiter's moons. Then in 1669 he was invited to Paris by King Louis XIV to supervise the construction of its new observatory.[7] He was later to become its first director.

Because of the problems caused by chromatic aberration, telescopes of the mid seventeenth century became longer and longer, as chromatic aberration was less of a problem with long focal length lenses. For example, Huygens eventually used a 210 ft (64 m) long instrument with an 8½ inch (21½ cm) diameter objective lens, whilst Hevelius used a 150 ft (45 m) long telescope at Danzig. On the other hand Cassini's first instrument in Paris was a relatively modest 17 ft (5.2 m) telescope made by Guiseppe Campani of Rome, who was, with Eustachio Divini, probably the best maker of long focal

---

[5] The true thinness of Saturn's rings was not fully appreciated until the Voyager spacecraft fly-bys of the late twentieth century, when they were seen to be less than 1 km thick.

[6] Since then, of course, rings have been discovered in the late twentieth century around Jupiter, Uranus and Neptune, although they are not nearly as impressive as those of Saturn.

[7] The observatory was designed more for its appearance than for its use as an astronomical observatory. It was half built when Cassini arrived and, although some modifications were made at his request, it was not a very practical building.

length lenses of the period. With this instrument Cassini discovered Saturn's moon Iapetus in October 1671, the second of Saturn's moons to be found, followed by the discovery of Rhea in December 1672 using a 34 ft (10.4 m) Campani instrument. Then in March 1684 Cassini discovered Tethys and Dione with Campani telescopes of 100 and 136 ft (30.5 m and 41.1 m) focal lengths. These were the last two of Saturn's moons to be found by Cassini.

Cassini observed Iapetus from 25th October until 6th November 1671, and noted that it reached its maximum lateral distance from Saturn of about three times the distance of Titan (or the 'Ordinary Satellit' as he called it) on 1st November. Then using Kepler's period/distance relationship he concluded that its period was about five times that of Titan (which is correct to the first decimal place).[8] Cassini recorded one proviso, however, in his discovery announcement, that the new satellite, unlike Titan, seemed to be below the ring plane when Iapetus was at its furthest point from Saturn. As a result, he speculated that either Iapetus' orbit was inclined to the ring plane, or that it could be a planet of the sun, rather than a satellite of Saturn. However, further observations over the next few years proved that it was a satellite of Saturn, with an orbital inclination of about 15°.

Cassini, in his report of the later discovery of Rhea, noted that he had been unable to see Iapetus after its discovery period in 1671, until he saw it again in December 1672, when it disappeared again until February of the following year. He could only see it when it was to the west of Saturn or near conjunction. As a result, Cassini concluded that Iapteus had two hemispheres of greatly different reflectivity, and that its axial rotation period was the same as its orbital period (as in the case of our moon), so it always has the same side facing Saturn in what is termed synchronous rotation. In September 1705, however, Iapetus began to be visible on the eastern side of Saturn, so in 1707 Cassini dropped his theory of synchronised rotation (which is a pity as it was correct).

Huygens had envisaged Saturn's ring as a solid, opaque body, but in 1675 Cassini observed that it was divided in two by a dark line going all the way around the planet. Cassini observed this line as a permanent feature dividing a brighter inner ring (now called the B ring) from the outer ring (A ring). In his first published drawing of the dark demarcation line, now called the Cassini division, he also showed that the B ring was broader than the A one.[9] Interestingly, it appears that Cassini also thought that the rings may consist

---

[8] Iapetus is, on average, 2.91 times the distance of Titan from the centre of Saturn, with a period of about 4.97 times that of Titan.

[9] The B ring is now known to be about 75% broader than the A ring.

of swarms of small satellites, whereas his contemporaries still thought of them as solid. Cassini's letter of August 1676,[10] in which he describes the discovery of the two rings, also mentions 'a dusky zone [on Saturn], a little further south than the centre, similar to the zones of Jupiter'. This appears to refer to Saturn's south equatorial belt and, as such, marks its first known observation.

## 4.2 Jupiter's satellites

One method of trying to sort out which of Jupiter's four satellites is which, and obtain an approximate estimate of their periods, is to try to find repeat patterns in their positions. On 10th December 1610 Galileo had found that the pattern of the three inner satellites was the same as it had been some seven days less one hour earlier, whilst the outer satellite, number 4, had moved from nearly greatest western elongation to greatest eastern elongation. Further observations enabled Galileo to conclude, early in 1611, that the periods of the four satellites, now called Io, Europa, Ganymede and Callisto, are approximately in the ratio of 1:2:4:8, with satellite number 1 having a period of a little less than 2 days and satellite 4 of a little more than 16 days.

A planetary satellite moves very slowly relative to the planet near greatest elongation, but very much faster near occultation[11] or during transit. So Galileo tried to refine his estimates of the periods of Jupiter's satellites by noting their times at occultation and transit. Unfortunately, it is not easy to see the satellites at these times because Jupiter is very much brighter than its satellites, and they are often lost in its glare. Nevertheless, Galileo persevered, only to find that his results were inconsistent. He then realised that he had forgotten to take into account the fact that the movement of the earth around the sun changes our angle of view at Jupiter by up to $\pm 11\frac{1}{2}°$ (see Figure 4.4). So Galileo introduced a correction called prosthaperesis into his analysis, to take this into account, before he presented his results in his *Discourse on Bodies in Water*, published in mid-1612. As a result, he was able to deduce the periods of the four satellites to within a few minutes of their modern values.

---

[10] *Phil. Trans.*, **11**, p. 689.

[11] A satellite is said to be occulted when it is hidden behind the planet as seen from earth.

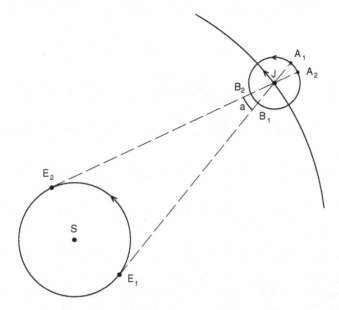

**Figure 4.4** The varying aspect angles of satellites orbiting Jupiter (J) are shown as seen from earth. When the earth is at $E_1$ a satellite at $A_1$ will be occulted by Jupiter, whereas when the earth is at $E_2$ a satellite at $A_2$ will be occulted. The angle a is about 23°. The difference in time for positions $A_1$ and $A_2$ is a little under 3 hours for satellite 1, and just over 1 day for satellite 4. (Drawing not to scale).

Finally, Galileo had been puzzled for a little while because he had noticed that the satellites did not always seem to reappear from behind Jupiter when they were supposed to. In particular, on 18th March 1612 satellite number 4 was not visible until six hours after its occultation by Jupiter should have finished. Galileo concluded correctly that it was still being eclipsed by Jupiter when its occultation had finished. This is because a satellite will enter or exit an eclipse by the planet at a different time compared with its occultation, if the sun, earth and Jupiter are not collinear (see Figure 4.5). As a result Galileo modified his calculations in late 1612 to take this effect into account.

In the same year it occurred to Galileo that the timing of eclipses or occultations of Jupiter's satellites could be used as an absolute clock to assist in the determination of longitude at sea.[12] If the time of local noon, deduced from observations of the sun, was compared with the absolute time, deduced from satellite eclipses or occultations, then the longitude could be determined. But in order to do this, it was essential to have both an accurate clock and an accurate set of predictions for satellite eclipses and occultations. In

---

[12] Nicholas de Peiresc (1580–1637) came up with the same idea at about the same time, and even sent an observer to compare the timings of Jupiter's satellites in Marseilles, Cyprus, Malta and Tripoli. The results were inconclusive, however, so he did not take the idea further.

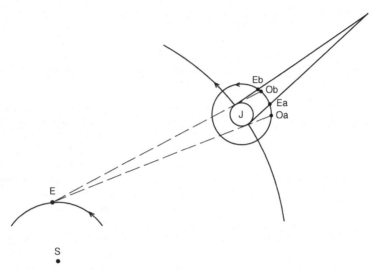

**Figure 4.5** The difference in times of occultation and eclipse for one of Jupiter's satellites is shown for the sun at S, the earth at E, and Jupiter at J. A satellite is occulted, as seen from earth, between Oa and Ob, whereas it passes through Jupiter's shadow and is in eclipse between Ea and Eb. So we first see the satellite emerging from its eclipse at Eb, after the end of the occultation at Ob. If the earth is on the other side of the sun, however, the eclipse starts before occultation, and ends whilst the satellite is still being occulted. (Drawing not to scale.)

1612 Galileo sent a proposal based on this principle to the Spanish king, and in 1636 he sent a similar proposal to the States General of the Netherlands, but nothing came of the idea. The main practical problem was trying to accurately observe the ingress and egress of the moons behind the bright planet with a telescope on the deck of a heaving ship. In addition, Jupiter is not always observable, either because of its position in the sky, or because of cloudy weather. But, above all, the proposed method would only work if the predictions of the eclipses and occultations were accurate enough, and that was not the case with Galileo's predictions.

Gian Domenico Cassini was the first person to publish accurate tables for the four satellites of Jupiter in March 1668, under the title *Ephemerides Bononienses*.[13] He had taken up the challenge in 1650, just after he had been appointed to his post of professor of astronomy at Bologna. First of all he reviewed Galileo's and Gassendi's observations, which in toto covered three of Jupiter's orbits of the sun, and then in 1652 he started observations of his

---

[13] This is an abbreviation of the full title, which was *Ephemerides Bononienses Mediceorum syderum ex hypothesibus et tabulis Io. Dominici Cassini.*

own. In 1664, with the aid of a new telescope, he was also able, for the first time, to observe the shadows of the satellites as they moved across Jupiter.

In producing these tables Cassini assumed that:

- The satellites all move uniformly in circular orbits. (This is almost correct, as the maximum eccentricity of any of the four orbits is 0.01.)
- The angle of inclination of each of the satellite's orbits is double the inclination of Jupiter's orbit to the ecliptic. (Again, this is almost true, although the inclinations of the satellites' orbits are slightly larger than this.[14])

Paris now became the leading centre for planetary astronomy, as not only did Cassini move there in 1669 to supervise the construction of the new observatory, but Huygens had also moved to Paris three years earlier. It was essential that, when this new observatory was in full operation, there should also be available as complete a set of observational data as possible, including the invaluable data of Tycho Brahe. But to use this data, it was necessary to know what correction to apply because of the difference in longitude of Paris and of Tycho's observatory at Uraniborg.

Cassini decided the best way of measuring the difference of longitude between Paris and Uraniborg was to use the method proposed by Galileo some time earlier, based on the accurate timing of eclipses and occultations of Jupiter's satellites. Jean Picard (1620–1682) made the measurements in Denmark during 1671–72, assisted by Ole Römer (1644–1710), whilst Cassini made them in Paris. In the process of analysing the results, however, it began to appear that Cassini's tables predicting satellite events were in error by several minutes.

The subsequent story is a little confused.[15] There is some evidence that Cassini put forward the correct reason for this timing error in 1674, but retracted his suggestion later. However, most historians of astronomy attribute the correct reason to Römer, namely that light has a finite velocity and so when we see an event is dependent on how far away it is from us. For example, in the *Journal des Sçavans*, published in August 1676, Cassini

---

[14] The inclinations vary from $0°.04$ for satellite number 1, to $0°.47$ for number 2, relative to Jupiter's equatorial plane, which is inclined at $3°.12$ to that of its orbit around the sun. This orbital plane is inclined at $1°.3$ to the ecliptic.

[15] For a discussion of this see *Planetary Astronomy from the Renaissance to the Rise of Astrophysics. Part A: Tycho Brahe to Newton*, by R. Taton and C. Wilson (eds.), Cambridge University Press, 1989, pp. 152–4.

predicted the eclipse of satellite 1 for 9th November 1676 at 5 h 27 min in the evening, whereas in September 1676 Römer forecast that eclipses in November would occur 10 minutes later than predicted by Cassini. This was because Cassini's timings were based on Jupiter being at its average distance from us, but in mid November 1676 it would be near conjunction, and so its light would have to travel the extra distance of the earth's orbital radius.

Picard recorded the actual eclipse on 9th November at 5 h 37 min 49 s and Römer at 5 h 35 min 45 s, thus vindicating Römer's prediction. Cassini, whilst accepting the time difference of about 10 minutes from his predictions, refused to accept Römer's reason, pointing out that the error should be the same for all four satellites at a given date. But the data on the other three satellites was inconclusive,[16] although it was accurate enough to rule out an alternative theory of Cassini's based on modifying the orbits of all four satellites. So, although Römer's theory, which was supported by Huygens, was not completely accepted at the time, it was the best available.

## 4.3 New planetary characteristics

Galileo had discovered the phases of Venus in 1610 (see Section 3.3), but his telescopes were not good enough to show the phases of Mercury, which were known should exist because its orbit, like that of Venus, is inside that of the earth. These phases of Mercury were apparently first recorded by the Italian Jesuit, Giovanni Zupus (or Zupo), in 1639, and then by Hevelius in 1644. Francesco Fontana (1580–1656) also showed them in his book *Novae Coelestium Terrestriumque Rerum Observationes*,[17] published in 1646, in which he recorded drawings of other planets.

Kepler had predicted that there would be a transit of Venus in 1631 (see Section 3.1) but, unfortunately, it was not observed in Europe, as the transit occurred when it was night-time there. A little later a young English astronomer, Jeremiah Horrocks (or Horrox) (1618–1641), an avid follower of Kepler, made some slight changes to Kepler's orbital parameters for Venus

---

[16] Later work was to show that mutual interactions between the satellites was the basic cause of perturbations which affected these results. Once this cause had been taken into account, each of the four satellites showed the same time error.

[17] Or in English *New Observations of Celestial and Terrestrial Objects.*

and the earth in his *Venus in Sole Visa*, and was able to predict that the next transit of Venus would take place on 24th November 1639 (OS).[18,19] Horrocks was keen to observe this transit, particularly as he wanted to obtain an accurate measurement of the angular diameter of Venus. In the event, both Horrocks and his friend William Crabtree were able to observe part of the transit, enabling Horrocks to measure Venus' diameter as $76'' \pm 4''$ (modern value about $60''$), by projecting an image of the sun using a camera obscura. This compared with an angle of $20''$ for Mercury (modern value about $11''$), which had been measured by Gassendi during its 1631 transit. Horrocks then concluded that both planets subtend an angle of $28''$ at the sun, which confirmed his earlier belief that the diameters of the planets are proportional to their distances from the sun.[20] As a result, he concluded that the earth subtended this same angle of $28''$ at the sun, giving a sun to earth distance of about 15,000 earth radii or 95 million kilometres.

In 1643 Riccioli found that the unilluminated part of Venus appeared to glow with a bluish-green light, which is now called the 'ashen light'. (This glow in the unilluminated part of the planet has been reported by a number of observers since then, but its existence is still in dispute.[21]) Then in 1646 Fontana mentioned in his book *Novae Coelestium* that Venus' terminator sometimes appears to have an irregular outline, which he erroneously took to be due to mountains. During the period 1666–67 Gian Domenico Cassini also managed to observe a few spots on Venus, which enabled him to deduce a rotation period of a little less than 24 hours. Later analysis of Cassini's

---

[18] OS stands for 'Old Style', and is used to signify dates according to the unreformed calendar, which was still in use in England at that time. It was ten days behind the Gregorian calendar, which had been introduced into Catholic Europe over the period 1582–84, but the ten days became eleven days in 1700, when the Gregorian calendar added a leap year. Britain only introduced this new calendar in 1752, with other Protestant countries and Eastern Orthodox countries making the change in other years.

[19] Kepler did not realise that transits of Venus occur in pairs about eight years apart, and so he failed to predict this transit of 1639.

[20] Remus Quietanus, a correspondent of Kepler, held the same opinion, and in the 1620s he had concluded that all the planets subtended an angle of about $34''$ at the sun. In fact, the true values range from $36''$ for Jupiter, through $23''$ for Venus, about $17''$ for Mercury, Earth and Saturn, to just $6''$ for Mars.

[21] The ashen light is only observed occasionally, and during controlled observing sessions some observers see it whilst others do not (see 'The enigmatic Ashen Light of Venus', by R.M. Baum, *J. Br. Astron. Assoc.*, **110**, 2000, pp. 325–9). So the suspicion exists that it is either a trick of the eye, or that it is caused by the earth's atmosphere.

observations by his son Jacques yielded a period of 23 h 20 min. Again this was completely wrong, but the error was not discovered until the second half of the twentieth century.

Fontana also noted a marking on Mars in his *Novae Coelestium*, which he took to be a permanent feature. From his description it is thought that the observation was spurious, but it may have been a sighting of the region later known as Syrtis Major, which had first been clearly observed by Christiaan Huygens in 1659. Huygens used this feature to deduce that the rotation period of Mars is about 24 hours, although he did not publish this result as he was not certain of it. Then in 1666 Cassini deduced a rotation period for Mars of 24 h 40 min, which is remarkably close to the currently accepted figure of 24 h 37 min 23 s. In the same year Cassini also discovered the bright Martian polar caps, which may have been observed by Huygens in 1656, but Huygens' drawing is very inconclusive. Nevertheless, Huygens definitely observed the south polar cap in 1672.

Galileo's pupil Evangelista Torricelli discovered the main belts of Jupiter in 1630, which were later described in Fontana's *Novae Coelestium* of 1646 and in Huygens' *Systema Saturnium* of 1659. Both Fontana and the English astronomer Robert Hooke (1635–1703) suspected that Jupiter's changes in appearance showed that it was rotating, but Cassini was the first to prove that was the case in 1663. Cassini deduced a rotation period of 9 h 56 min, based on the movement of small irregularities in Jupiter's equatorial bands.[22] This is a remarkably rapid rate considering that Jupiter is the largest planet in the solar system. As a result, Jupiter shows a noticeable polar flattening which Cassini measured in 1691 as $\frac{15}{16}$ (polar to equatorial diameter). Then in 1690 he observed that the rotation rate of the higher latitudes of Jupiter is about 5 minutes longer than for the equatorial regions.

In 1664 Hooke observed a small spot in what is now believed to be the North Equatorial Belt.[23] Then in the following year Cassini recorded a particularly impressive spot, which he called 'the Eye of Jupiter', in what is now called Jupiter's South Tropical Zone.[24] In fact Cassini's spot was to remain visible until 1713 when it was last recorded by Giacomo Maraldi. It

---

[22] The modern value of the rotation period for the equatorial regions is 9 h 50 min 30 s.

[23] Hooke simply said that the spot was in 'the biggest of the 3 obscurer belts', which has been interpreted as the North Equatorial Belt. See 'The Discovery of Jupiter's Red Spot', by Clark R. Chapman, in *Sky and Telescope*, **35**, pp. 276–8.

[24] Hooke's and Cassini's spots are often referred to as one and the same, but it now appears that they were not, as they were probably seen in different belts, and their sizes were also probably different. See Chapman in *Sky & Telescope*, **35**, pp. 276–8.

is now thought that it may have been an early appearance of the Great Red Spot which was not seen again until 1878.

## 4.4 Solar parallax

If one could observe the sun against the background stars, its position would be seen to change over the course of a day, owing to the varying displacement of the observer from the line linking the centre of the sun to the centre of the earth.[25] On the equator, at the equinoxes, for example, the sun is overhead at noon and on the horizon at 6 pm (ignoring the effect of atmospheric re-fraction), and the observer's position has moved during this time from being on the sun–earth centre line to being one earth radius to one side of it. By midnight the observer is back on the centre line again. This diurnal parallax is a maximum at the equator, reducing at higher latitudes, as only at the equator will the observer move by plus and minus one earth radius, to give what is called the solar parallax. Measuring this change in position of the sun against the background stars at the equator, and knowing the radius of the earth, clearly enables the distance of the earth from the sun, the so-called astronomical unit, to be determined.

There are two basic problems with this approach, however. Firstly the effect of atmospheric refraction has to be taken into account, and secondly the sun can only be seen against the background stars during a total solar eclipse.[26] A surrogate object is thus called for, to replace the sun, which can be seen against background stars and which is close enough to exhibit a clear parallax effect. Mars is an obvious candidate,[27] as it is usually seen at night against a fully dark sky, when faint stars can be seen. Mars is at maximum altitude, due south in the sky, when it is at opposition, and in this case the parallax, and hence the distance of Mars, can be determined by measuring how its position has changed, relative to the stars, over a six hour period. The sun's parallax can then be deduced, as well as the sun–earth distance, knowing the geometry of the solar system.

The orbit of Mars is relatively eccentric, and so the distance between Mars and the earth at opposition can vary by almost a factor of two. The best

---

[25] This is in addition to the gradual movement of the sun along the ecliptic, of course, which it circuits once per year.

[26] This is no longer the case with modern instruments, however, as an occulting disc can be used to create an artificial eclipse in them. We can also now observe the sun from space, where it is considerably easier to see the background stars near to the sun.

[27] The idea of using Venus during a transit came later.

oppositions, from a parallax point of view, are obviously those in which Mars is closest to earth, and one of these was due to occur in 1672. So the French Authorities decided to carry out an expedition to Cayenne in South America to allow simultaneous measurements of the parallax of Mars in Cayenne and Paris. Cayenne[28] was chosen as it is near to the equator and is very far from the latitude of Paris. So the combination of atmospheric refraction, which is dependent on the angle of observation above the horizon, and the diurnal parallax of Mars would be quite different for the two sites. Jean Richer (1630–1696) carried out the observations in Cayenne, and Cassini and Jean Picard carried them out simultaneously in Paris.

The Richer results were excellent, showing, amongst other things, that the obliquity of the ecliptic was 23° 29′, rather than the 23° 31½′ deduced by Tycho Brahe, or the 23° 28′ estimated by Copernicus. It was also clear that the gravitational force measured by the pendulum clock at Cayenne, which had been used for accurate time-keeping, was different from that measured in Paris, indicating that the earth has an equatorial bulge. Finally, the sun's parallax was shown to be less than the 12″ maximum value previously deduced by Cassini (see Section 5.2). Cassini finally settled on a figure of 9″.5 as a result of the Cayenne expedition, which gave a value for the sun–earth distance of about 140 million kilometres. This distance is only about 7% below its true value, but the accuracy was more apparent than real, as Cassini had based his estimate on a highly dubious analysis. Nevertheless, it was similar to the value of 10″ that had also been deduced from Mars' diurnal parallax in 1672 by the respected English astronomer, and future Astronomer Royal, John Flamsteed (1646–1719). As a result, these solar parallax values of Cassini and Flamsteed were considered to be very close to the truth by contemporary astronomers.

## 4.5 **The moon**

The moon to most astronomers today is no longer of any great interest. A grey, unchanging world, with no atmosphere, it is more of a nuisance to astronomers than anything else, blocking out earth-based, optical observations of deep sky objects for a number of days per month because it is so bright.[29]

---

[28] Incidentally, the launch site for the European Ariane satellite launcher is now in Cayenne.

[29] The pedant could take exception to this description of the moon, pointing out that the moon is not completely grey, that there have been observations of minor changes

But to seventeenth century astronomers with their newly developed telescopes, this was not the case. Here was another world, they reasoned, that could be observed and mapped at leisure, and it may be inhabited.

Galileo had been first to seriously study the moon (see Section 3.3), concluding that it is a body like the earth, having mountains and plains. He deduced that the dark areas on the illuminated face are mostly flat and lower than the bright areas, which are generally mountainous. He also thought that the moon has an atmosphere, but he attributed the faint light seen in the dark part of the new moon[30] to sunlight reflected off the earth,[31] not to the moon's atmosphere. Galileo also deduced that there was probably very little water on the moon, as there were no clouds, and that any life that may exist on its surface would therefore be very different from that on earth. Not only would life have to survive with little water, but it would also have to be used to 15 days of heat followed by 15 days of cold and darkness. Galileo also concluded that such life would experience only limited seasonal effects, as he thought that the moon's inclination changes by only about $\pm 5°$ relative to the ecliptic,[32] rather than the $\pm 23\frac{1}{2}°$ for the earth.[33]

On 14th December 1611 Thomas Harriot found that the Sinus Roris and Mare Frigoris were nearer the northern edge of the moon than he had observed previously, but whether he took this as a real effect, or put it down to an inaccuracy in observing is unknown. In fact he had found the so-called 'libration in latitude', which has a period of one month, and which is caused by the fact that the moon's spin axis is not perpendicular to the plane of its orbit around the earth. The angle between the spin axis and the orbit normal is, in fact, about $6\frac{1}{2}°$.

---

taking place on its surface, and it has a very tenuous atmosphere. But the observations of change are disputed, and the atmospheric pressure is only about $10^{-8}$ times that at the surface of the earth, which is less than that of a man-made vacuum.

[30] Often referred to as 'the old moon in the new moon's arms'.

[31] This is now called 'earth-shine'.

[32] Galileo knew that the moon's orbit is inclined at about 5° to the ecliptic, and he assumed that the moon's spin axis is perpendicular to its orbit, which it is not. The angle between the moon's spin axis and the normal to the ecliptic is now known to be only $1\frac{1}{2}°$, not 5°, so the seasonal effect described by Galileo is even less than he had assumed.

[33] Huygens in his book *Cosmotheoros*, which was published posthumously in 1698, also speculated on possible life-forms on other planets, suggesting that as 'it is certain that the earth and Jupiter have clouds and water, it can hardly be doubted that they [clouds and water] are found on the surface of the other planets'. As a result, he said, there was probably vegetation on the planets, and possibly animals and intelligent beings also.

Galileo had initially theorised that the line joining the centres of the earth and moon passed through a fixed point on the surface of the moon. As a result, he predicted that we would observe a slightly different aspect of the moon as the earth spins on its axis, to provide a libration that moves back and forth with a period of one day. This would result in an east–west libration of the Moon, compared with the north–south nodding libration observed by Harriot. Galileo erroneously announced the discovery of this diurnal, east–west libration in 1632.[34] Then in 1637, in a letter to Fulgenzio Micanzio he withdrew his theory that the earth–moon centreline goes through a fixed point on the moon's surface, although he still maintained that he had detected a diurnal libration in longitude. In fact the libration in longitude which he did detect (of 7° 45′) is caused by the eccentricity of the moon's orbit, and its period is monthly, not daily.

In his letter to Micanzio of 1637, Galileo also announced that he had observed a libration in latitude, with a period of one month. This was the libration observed by Harriot. Galileo also told Micanzio that he had found a third libration, which is seen as a clockwise–anticlockwise oscillation of the lunar face, with a period of one year. There is some doubt, however, about whether Galileo could have detected this relatively subtle effect, which is caused by the 1½° tilt of the moon's spin axis to the ecliptic.

In the early seventeenth century, numerous astronomers had sketched the surface of the moon using telescopes. They included Harriot (1609), Galileo (1609), Scheiner (1610), Malapert (1619), Biancani (1620), Borri (1627) and Fontana (1629), but the best early lunar map was produced by Claud Mellan (1598–1688), as part of the Gassendi–Peiresc–Mellan programme.

Nicholas de Peiresc (1580–1637) had had the idea in about 1611 of using the timing of Jupiter's satellites as an absolute clock, to help to determine differences in longitude on the earth, but he had given up the idea shortly afterwards owing to inconclusive results. Then some years later he decided to use eclipses of the moon for the same purpose. On 20th January 1628 Peiresc, assisted by Joseph Gaultier and Pierre Gassendi, observed an eclipse of the moon from Aix en Provence, whilst Marin Mersenne and Claude Mydorge observed it from Paris. Similarly the lunar eclipse of 28th August 1635 was observed from Cairo, Naples, Aix and Paris, amongst other places, and the estimated length of the Mediterranean was reduced from 60° to 41° as a result.

---

[34] A diurnal libration does exist but its magnitude was too small to be detected by Galileo.

Gassendi and Peiresc realised, from their experience with lunar eclipses, that it was essential to have a reliable lunar map, if accurate longitudinal differences were to be determined during lunar eclipses. This was to enable the unambiguous timing of the occultation and reappearance of numerous small lunar landmarks to be made at the various observing sites. Claud Mellan produced the first three map engravings for the project in 1636, but Peiresc died the following year and the project was abandoned. At about the same time, Michael Van Langren (otherwise known as Langranus) (1600–1675) decided to use the timing of the monthly sunrise or sunset of various lunar peaks and craters to determine longitude on earth. He produced many sketches to aid the production of his lunar map, which was published in 1645. Unfortunately, however, his timing scheme proved impracticable as sunrise and sunset of lunar features occurred far too slowly to allow accurate timings to be made.

In 1647 Johannes Hevelius (or Hewelcke) (1611–1687) published a book called *Selenographia*, which was to be the standard work on lunar cartography for well over a hundred years. Unlike previous works by other astronomers, his full moon maps (see, for example, Figure 4.6) show the effect of libration, which enables us to see a total of 59% of the lunar surface over time. Then a few years later Riccioli, in his *Almagestum Novum*, produced a modified version of Hevelius' map, drawn by Grimaldi, whilst adding the names of features devised by himself. These names, which were completely different from those proposed previously by either Van Langren or Hevelius, are those generally in use today.

Although these and later maps of the moon have been useful reference tools for astronomers over the years, what is of more interest to the modern astronomer and geologist is how the various types of lunar feature were formed. In fact, Robert Hooke was the first to speculate on this in a scientific way in his *Micrographia*, published in 1665. In this he suggested that the 'substance of the moon be very much like that of the earth, that is, [it] may consist of an earthy, sandy or rocky substance...' He also undertook laboratory-like experiments and noted that, if round objects are dropped into a viscous mixture of clay and water, features are temporarily produced that resemble lunar craters. But as he could not think of a source of any such bodies in space that could impact the moon, he dismissed this idea as a possible source of craters. Instead, he found that he could produce crater-like features if he boiled dry alabaster powder in a container, leading him to favour a volcanic-like origin of lunar craters.

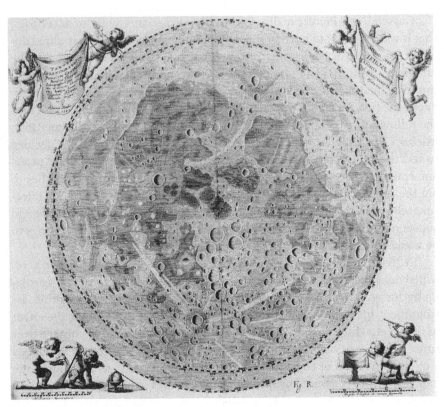

**Figure 4.6** One of Hevelius' maps of 1647 from his *Selenographia*. The effect of lunar libration is indicated by the two dotted circles.

## 4.6 Early concepts of gravity and Descartes' vortex theory

Copernicus had concluded in the early sixteenth century that the mutual attraction of the earth's constituent material had resulted in its spherical shape, but neither he nor anyone else of that period understood gravity, in the way that we understand it today, as pervading all of space. It was seen as a purely local effect limited to the earth. Some years later, Kepler had come nearest to understanding the all-pervading nature of gravity in his *Astronomia Nova* of 1609, where he attributed the tides on earth to the attraction of the moon. Then in his *Somnium seu Astronomia Lunari* of 1630 he attributed the tides on the earth to the combined attractions of the moon and sun. But, surprisingly to us, he ignored this attraction between the sun and the earth when he was trying to understand the force that keeps the earth in orbit around the sun.

Kepler had reasoned in his *Mysterium Cosmographicum* of 1596 that there is an anima motrix (moving spirit) in the sun, which drives the planets in their orbits, and whose magnitude reduces with distance. This force is *tangential* to the planetary orbits, and not radial (like gravity). In his letter to Archduke Ferdinand in 1600, Kepler also proposed that there was a similar force in the earth controlling the movement of the moon. Later, in his *Astronomia Nova*, he suggested that the driving force controlling the planets was produced by magnetic vortices that were produced by a rotating sun. Then, over the period 1629–1633, Descartes developed a more detailed, alternative vortex theory, which he published in his book *Principia Philosophiae*.

René Descartes (1596–1650), a radical French philosopher of the first order, enumerated three laws of nature and a number of rules concerning the impact of bodies, before developing his vortex theory in full. Unfortunately, Descartes' rules of impact were largely incorrect, but his second law of nature, that each particle of matter tends to move along straight lines, if subjected to no external influences, was the first accurate statement of the law of inertia.

In his *Principia*, Descartes recognised two endeavours[35] or tendencies for a body moving in a circle. That is, it wants to move:

(i) at a tangent to the circle. So, for example, in the case of a mass being whirled around in a circle at the end of a piece of string, if the string is released, the mass flies off at a tangent.

(ii) away from the centre of the circle. In the case of the mass being whirled around on a string, this movement away from the centre is prevented from happening by the tension in the string.

Descartes now used these principles to examine the motion of the planets around the sun.

Descartes developed his theory assuming that the universe is filled with a thin, fluid ether,[36] mainly composed of very small globules in rapid motion, which rotates in vortices about the sun and other stars. These vortices carry the planets around the sun, and smaller vortices around the earth, Jupiter and Saturn carry their moons around these planets. In Descartes' scheme, a body floating in ether can be moved by the smallest force and, once moving, it would see no resistance to its motion from the ether. Ingeniously, he explained, for the benefit of the Church authorities, who thought, at the time, that the earth was at rest at the centre of the universe, that the earth

---

[35] Descartes used the word 'conatus' which is usually translated as 'endeavour'.
[36] Sometimes spelt aether.

is at rest in this fluid ether. He likened the situation to that of a ship, carried along by a current of water, the ship being at rest relative to the current.[37]

Descartes' vortices were very different from those of Kepler, as Kepler saw his vortices as pushing the planets around the sun, As a result, he had no need to propose a radial, restraining force from the sun to keep them in orbit. On the other hand, Descartes in considering his two endeavours clearly said that, unless there was some restraining influence, the planets would fly off into space. So, what stopped that from happening? According to Descartes, it was the pressure in the vortex itself. To Newton, who started to study Descartes' theory in 1664, it was the effect of some particles of ether moving sunward at a great speed.

## 4.7 Isaac Newton

Isaac Newton was born at Woolsthorpe, Lincolnshire, on 25th December 1642 (OS),[38] three months after his father's death. His mother remarried when Isaac was three to the rector of north Witham, which required her moving out of the immediate locality. So Isaac was brought up at Woolsthorpe by his grandmother and educated locally, before entering Grantham grammar school at the age of twelve. There he studied Latin and Greek, and possibly mathematics.

After his mother was widowed again, she moved back to Woolsthorpe, and later withdrew Isaac from the grammar school to run the family business. His mother quickly realised her mistake, however, as her son was not interested in the slightest in running the business. So he was briefly allowed to return to the grammar school in 1660 to study for university. In the following year he was accepted at Trinity College, Cambridge, as a 'sizar' or servant to one of the lecturers, probably Dr Humphrey Babington, a Fellow of Trinity College. In fact Babington has often been identified as being mainly responsible for Isaac going to the university in the first place, so such an arrangement was probably convenient to both parties.

As with other European universities of the time, the curriculum at Cambridge still revolved around Aristotle's philosophy, just as if Copernicus, Kepler and Galileo had not existed. But the undergraduates obtained a general

---

[37] This assumes that the only force acting on the ship is from the current. So there is no wind, and the ship is not restrained by an anchor. (Ships did not have engines at that time, of course.)

[38] This was 4th January 1643 according to the Gregorian calendar.

view of these 'new' developments through word of mouth and private study. Then in 1664 Newton read and started to study Descartes' natural philosophy and mathematics. At about the same time he also discovered the work of Pierre Gassendi on the atomic basis of matter, of Robert Boyle the brilliant chemist, and of Henry More who believed in alchemy and magic. All these men were to have a profound influence on Newton and his ideas.

In early 1665 Newton received his Bachelor of Arts degree, shortly before the plague was to close the university for eighteen months.[39] During this period, working in isolation at home, Newton laid the foundations for his subsequent extensive work on calculus, optics and dynamics. Then in 1667 he returned to Cambridge where, two years later, at the tender age of twenty-six, he became Lucasian Professor of Mathematics. With some regret I will now have to ignore Newton's substantial contributions to science in all but the areas with which we are directly concerned, namely celestial mechanics and dynamics. But, before outlining his astronomically oriented work, we need to return to the work of Christiaan Huygens.

## Motion in a circle

Huygens had invented the pendulum clock in 1657 but, as Section 4.1 indicates, he was very much more than just a practical engineer. During subsequent years, Huygens developed the theory of the simple pendulum and considered motion in a circle, publishing his results in 1673. In particular he found that the force acting radially on a body moving in a circle is proportional to $\frac{mv^2}{r}$ (where $m$ is its mass, $v$ its velocity, and $r$ the radius of the circle). He included this particular result, without proof, in an appendix to his book *Horologium Oscillatorium* published in 1673; the appendix being written in 1659.

In late 1664 Newton also turned his attention to the subject of motion in a circle, and in the following year he derived the same result as Huygens for the radial force. However, Newton was unaware that Huygens had come to the same conclusion some years earlier, as Huygens was not to publish his result until 1673. Unlike Huygens, however, Newton had proved his result from first principles.

Newton now understood that, for a body moving in a circle, the outward, radial force (the centrifugal force), which is proportional $\frac{mv^2}{r}$, must

---

[39] The university was closed from August 1665 to March 1666, and from June 1666 to April 1667.

be balanced by a centripetal force of the same magnitude directed towards the centre of the circle. At this stage in the seventeenth century, gravity was known to operate on objects on the surface of the earth, to produce the trajectories of projectiles deduced by Galileo, for example. But it was not obvious that gravity may be the centripetal force keeping the moon orbiting the earth and the planets orbiting the sun. Descartes, for example, certainly did not see it like that (see Section 4.6).

In the late 1660s Newton investigated theoretically how the force acting on two planets *in circular orbits* would vary with their distances from the sun, as follows.

Consider two planets of unit mass in circular orbits of radii $r_1$ and $r_2$, with velocities $v_1$ and $v_2$. Assume that the centrifugal forces acting on them are $P_1$ and $P_2$.

Then according to Newton:

$$\frac{P_1}{P_2} = \frac{v_1^2}{v_2^2} \times \frac{r_2}{r_1}, \tag{1}$$

but according to Kepler's third law (see Section 3.1):

$$\frac{T_2^2}{T_1^2} = \frac{r_2^3}{r_1^3}, \tag{2}$$

where $T_2$ and $T_1$ are the periods of the planets. Now

$$T = \frac{r}{v},$$

hence (2) becomes

$$\frac{r_2^2}{v_2^2} \times \frac{v_1^2}{r_1^2} = \frac{r_2^3}{r_1^3} \quad \text{or} \quad \frac{v_1^2}{v_2^2} = \frac{r_2}{r_1},$$

and (1) implies

$$\frac{P_1}{P_2} = \frac{r_2^2}{r_1^2}.$$

So the force on a planet varies as the inverse square of the distance from the centre of its orbit, for a planet *in a circular orbit*. So far, no assumptions have been made about the source of the centripetal force which balances this centrifugal force, thus keeping the planet in orbit.

## The moon test

Newton outlined the above calculation in a document written in the late 1660s, in which he also tried to compare the gravitational force at the surface of the earth with the force experienced by the moon in its orbit, in his so-called 'moon test'. Years later he claimed that he did this to test the inverse square law of gravity. This may be true, but nowhere in his original document does he make this clear.

For his 'moon test' Newton measured the gravitational force at the earth's surface, using a simple pendulum, and he compared this with the centrifugal force at the earth's equator as determined by his $\frac{mv^2}{r}$ formula. This showed that the gravitational force was about 350 times the centrifugal force at the equator, thus explaining why objects are not spun off the earth as it rotates. Then, comparing the moon's period of revolution around the earth with the rotation period of the earth, and using his $\frac{mv^2}{r}$ formula, he concluded that the centrifugal force at the earth's surface is about $12\frac{1}{2}$ times greater than the centrifugal force acting on the moon.[40] Hence the gravitational force at the earth's surface is about $12.5 \times 350 = 4{,}325$ times the centrifugal (and centripetal) force acting on the moon or, as Newton put it, the force of gravity 'is 4,000 times greater than the tendency of the moon to recede from the earth, and more'. Unfortunately these figures of 4,325 and >4,000 are significantly greater than the square of the ratio of the distance of the moon to the radius of the earth, or the $60^2 = 3{,}600$ figure that Newton used. So gravity did not appear to be the force keeping the moon in its orbit. The unsatisfactory result of these 'moon test' calculations caused Newton to stop work on gravity and go on to other matters. In fact it appears, according to William Whiston, who was Newton's successor as Lucasian professor at Cambridge, that Newton thought, at the time, that the moon was probably kept in its orbit by a mixture of forces due to gravity and the movement of the ether through the vortex around the earth. So maybe Newton was not too surprised at the result of his calculations.

Interestingly, it is now clear that the reason for the discrepancy between the 4,325 and 3,600 figures was Newton's adoption of a value of 3,500 miles for the earth's radius which was taken from Galileo's *Dialogue*. If he had used the correct value, which was known at the time, the discrepancy would have disappeared.

---

40 $\frac{(27.3 \text{ days})^2}{(1.0 \text{ day})^2} \times \frac{1}{60} \approx 12\frac{1}{2}$, assuming that the moon is about 60 earth radii from the centre of the earth.

Newton had shown, as outlined above, that the centrifugal force on a planet varies as the inverse square of its distance from the sun for a planet *in a circular orbit*. This had, of course, to be balanced by an equal and opposite centripetal force. In fact, Newton concluded that, like with the moon, the centripetal force was somehow connected with the movement of the ether through the vortex,[41] even if that was not the sole source. Interestingly, Newton believed, like some of his predecessors, that the centripetal and centrifugal forces acting on the planets were not balanced along their orbits because their orbits were elliptical. In particular, he thought that planets moved from aphelion to perihelion because there was a net inward force acting on them because of the ether, and that from perihelion to aphelion there was a net outward force. In fact, in 1666 Giovanni Alfonso Borelli (1608–1679) had published a detailed theory based on such a premise of out-of-balance forces.[42]

## Theory of gravity

The above discussion may be puzzling to some who understood that Newton produced his universal theory of gravitation in 1666, or thereabouts, following the famous incident with the apple. However, that story, like many others in astronomy, is a myth. In fact, the catalyst that was to cause Newton to produce his universal theory of gravitation was a series of letters between Robert Hooke and Newton, exchanged over a few months in late 1679 and early 1680, and a visit by Edmond Halley to Newton in 1684.

Robert Hooke started this correspondence with Newton in November 1679,[43] as Hooke was the secretary of the Royal Society at the time, and he had been asked to maintain contact with members who did not live in London. In his reply, Newton outlined a possible method of proving that the earth rotated on its axis, knowing that Hooke was interested in such things. In essence Newton suggested that, if a heavy body is stationary above

---

[41] It was thought that this vortex was produced by the spinning sun, whereas that affecting the moon was produced by the spinning earth.

[42] In his *Theoricae Mediceorum Planetarum ex Causius Physicis Deductae*, or *The Theory of the Medicean Planets deduced from Physical Causes*.

[43] This was not the first time that Hooke and Newton had communicated with each other, but on both previous occasions, in 1672 and 1675, it had ended in a quarrel. This is hardly surprising as Hooke had been very dismissive of Newton's ideas on the nature of light in 1672, and in 1675 Hooke had even accused Newton of stealing his ideas, which he had expounded in his (Hooke's) book *Micrographia* published in 1674.

a fixed point on earth, and it is then allowed to fall, it will not hit the surface of the earth directly underneath its original position, but to one side of it, assuming no air resistance. In particular, it will fall to the east of its original footprint,[44] not to the west as most people might think, because its original lateral velocity was higher than that of the footprint. This is because the body had the same angular velocity as the footprint, but it was further from the centre of the earth. Unfortunately for Newton, he went on to describe the imaginary trajectory of the body after it had passed through the surface of the earth, without specifying what resistance, if any, he assumed it would experience. In a letter of 8th December 1679 Hooke challenged Newton's trajectory, which Newton was forced to revoke, and an interesting exchange of letters then ensued. In the process a key question arose as to what gravitational force the body would experience as it got closer to the centre of the earth.

Hooke assumed in his analysis that the gravitational force would be uniform as the body approached the centre of the earth, although he actually believed that the force would reduce the closer the body got to the centre. Unfortunately, Hooke had no way of working out the consequences of the latter, as he did not have Newton's developing calculus to help him, hence his assumption. Newton was unsure at this stage how the gravitational force would vary, and speculated that it may actually increase as the body got closer to the centre. In fact, Hooke was correct, as Newton himself was to prove mathematically five years later. The gravitational force exerted by a solid sphere, on a particle inside it, is proportional to the mass of the sphere whose radius is defined by the distance of the particle from the centre; the mass of the sphere outside that radius can be ignored.

In 1680 Hooke also concluded that the gravitational force exerted by a solid sphere on a distant body can be calculated assuming that the whole of the mass of the sphere is at its centre. Five years later, Newton was to show that that is also true for a body anywhere outside of, or even on the surface of the sphere.[45] This exchange of letters with Hooke also encouraged Newton to prove in 1680 that, assuming an inverse square law of attraction, planets or moons will orbit a central body in an ellipse, with the central body at one focus. So Newton had shown that Kepler's first law, that 'the planets describe elliptical orbits around the sun, with the sun at one focus', was consistent with an inverse square law of attraction. Newton did not send his

---

[44] The footprint is the point directly underneath its original position.

[45] Newton had assumed this in his earlier 'moon test', although he did not specifically recognise it as an assumption at the time.

proof to Hooke, however, even though it was Hooke who had asked him to prove the proposition in his letter of 17th January 1680. Instead Newton sat on his proof until in August 1684 Halley asked Newton if he could prove the same thing that had previously been requested by Hooke in 1680. Newton assured Halley that he had already done so some years earlier but, on checking, Newton found that he had to reconstruct the proof, as he had lost it in the meantime.

Newton had, apparently, in 1680 also shown that Kepler's second law (or areal rule, as it is sometimes called) that 'the radius vector from the sun to a planet sweeps out equal areas in equal times' is also a consequence of an orbit based on the inverse square law of attraction. In addition, by reversing his analysis on circular orbits described above, he knew that, at least for circular orbits, Kepler's third law, that 'the period of a planet squared is proportional to its orbital radius cubed', could be proved assuming an inverse square law of attraction.

So in 1680 Newton had proved all three of Kepler's laws of planetary motion, by assuming that the force on a planet varies as the inverse square of its distance from the sun, with the exception that he had only proved the third law for circular orbits. Somewhat surprisingly, however, rather than publishing this major piece of work, Newton had sat on it for a number of years. The reason for this is unclear. His possible disappointment in not proving the inverse square law for the moon in his 'moon test' may be part of the reason, but it is not clear that he expected gravity to be the sole attractive force between the earth and the moon at that time, because of his belief in aetherial vortices. Then in 1684, for some unknown reason, he gave up the idea of aetherial vortices, and began to construct a mathematical model of the universe, based on a gravitational force which varied as the inverse square of the distance.

## The comet of 1680

A bright comet had appeared in 1680, whilst Newton was trying to understand the forces controlling orbital motion in the solar system. Clearly any theory that was to explain the motion of the planets should also be able to explain that of comets. Kepler had managed to show (see Section 3.1) that the orbits of planets were ellipses, with the sun at one focus, but were the orbits of comets similar? Initial indications suggested that this was not so.

Kepler and most of his contemporaries in the first half of the seventeenth century believed that comets followed rectilinear orbits, but two observers of the comet of 1664, Johannes Hevelius and Giovanni Borelli, began to have doubts. In his *Prodromus Cometicus* of 1665, Hevelius concluded that the orbit of the 1664 comet was a conic section,[46] with the sun at one focus, but in his later *Cometographia* of 1668 he modified this to a parabola or hyperbola, with the sun no longer at the focus. Borelli also suggested, in a letter to Duke Leopold of Medici in May 1665, that the orbit of the comet of 1664 was a curved line resembling a parabola. Nevertheless, most astronomers still believed in rectilinear cometary orbits when the comet of 1680 appeared.

On 4th November 1680 (OS)[47] Gottfried Kirch (1639–1710) of Coburg, near Nuremberg, rose early in the morning to make observations of the moon and Mars, and saw something that he was not expecting – a comet.[48] A little later, Georg Dörffel (1643–1688) who lived in Plauen, not far away from Coburg, also observed the comet. In total Dörffel observed the comet for three nights in November, and issued a short report on his observations. Not only did he anticipate the comet's reappearance after conjunction, which was to occur on 9th December 1680, but he continued to observe the comet with the naked eye until it became too dim to be seen in February 1681.[49] Then in his second publication on his observations, entitled *Astronomische Betrachtung des Grotten Cometen*,[50] Dörffel calculated an orbit for the comet, concluding that it was a parabola with the sun at its focus (see Figure 4.7). This work was remarkable, being undertaken, as it was, by a part-time astronomer who made observations and positional measurements without a telescope. It was also at a time when even Newton thought cometary orbits were rectilinear.

John Flamsteed, the first Astronomer Royal, took a particular interest in the comet of 1680 but, because of poor weather, he had been unable to observe it in November. Nevertheless, he, like Dörffel, predicted that it would reappear after conjunction, as he believed that comets described closed orbits.

---

[46] Conic sections are circles (a conic section with an eccentricity of 0), ellipses (eccentricity of < 1), parabolas (eccentricity 1) or hyperbolas (eccentricity > 1).

[47] This and all subsequent dates in this discussion of the 1680 comet are Old Style dates.

[48] This was the first discovery of a comet made with the aid of a telescope.

[49] The last sighting before conjunction was made in the Philippines at noon on 8th December, when the comet's head was less than 2° away from the sun. Flamsteed first detected its tail after opposition in the evening sky at Greenwich on 10th December, with its head being observable two days later. The last observation of the comet was made by Isaac Newton on 9th March 1681.

[50] *Astronomical Observation of the Great Comet.*

Figure 4.7 Dörffel's method of determining cometary orbits consisted of measuring the angle between the sun and the comet at strictly regular intervals. This gave the angles x, y and z, but the position of the comet in space could be anywhere along the lines aA, bB and cC. Dörffel then found, by trial and error, that he was able to draw a parabola, such that the comet's speed between B and C was greater than between A and B. The exact parabola was then chosen such that the comet's speed increased regularly on its inward course, and decreased regularly on its outward course.

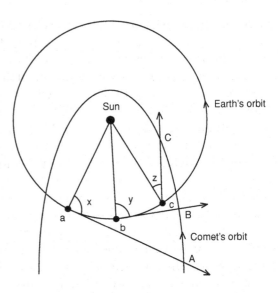

Even after it reappeared on 10th December, however, there was much disagreement amongst astronomers as to whether the pre 9th December comet (hereafter called comet I) was the same as the post 9th December comet (comet II). In a letter to Halley of 17th February 1681, Flamsteed suggested that the comet had changed its direction of travel before it had reached the sun (see Figure 4.8), thus passing in front of the sun, and not behind it. Flamsteed explained this orbital movement as being caused by a combination of magnetic forces, originating in a magnetic sun, and motion produced by the solar vortex. He supposed that the comet was initially attracted to the sun by the sun's magnetic field, but as it got closer to the sun the solar vortex caused the comet to move sideways, and so avoid falling in to the sun. As it passed in front of the sun, the orientation of the magnetic dipole of the comet's nucleus changed with respect to the sun, with the result that the sun now repelled the comet, forcing it away.

Newton was sent a copy of Flamsteed's letter, to which he replied on 28th February with a series of objections to Flamsteed's conclusions. Newton maintained that:

(i) The solar vortex would, if anything, cause the comet to move counterclockwise around the sun, as seen from the north pole of the ecliptic, in the direction of the planets, and not clockwise in front of the sun. As a result, Newton suggested that the comet could have changed direction behind the sun (see Figure 4.9), instead of doing so in front of it, although Newton himself did not subscribe to this idea. Instead, he

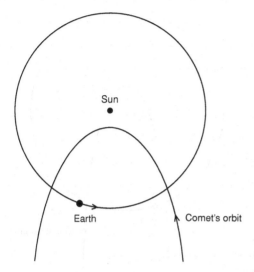

**Figure 4.8** Flamsteed's proposed orbit for the comet of 1680 showed the comet turning back before it reached the sun.

thought that there were two comets, not one, travelling in rectilinear orbits (see (iv) below).

(ii)   The sun was too hot to be magnetic, as it was well-known that a lodestone loses its magnetic properties when it is heated to red heat.

(iii)   Even if the sun was magnetic, once the sun had attracted the comet, it would never repel it, as the same magnetic pole of the comet would always point towards the sun.

(iv)   Comet I could not be the same as comet II, as the data provided by Flamsteed indicated that one comet would have to accelerate and decelerate too many times, whatever the orbit. So Newton retained his previous view that comets follow rectilinear orbits and, in this case, the 1680 comet must be two comets, as one rectilinear orbit could not explain both the pre- and post-opposition data.

The last objection was based on errors in Flamsteed's original communication, which were caused by him failing to transform all of the Gregorian dates to Julian (i.e. Old Style) dates.[51] Flamsteed put this right in his reply of 7th March, in which he continued to argue that the sun was magnetic, but he was no longer as adamant about the comet's orbit passing in front

---

[51] Some of the observations used by Flamsteed were made by Father Jean Charles Gallet in France, who naturally used the Gregorian calendar then in use in Catholic Europe. Unfortunately, Flamsteed, by accident, only transformed four of Gallet's six observations to the Julian calendar. Hence the confusion.

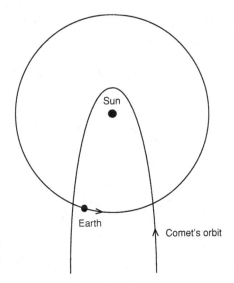

**Figure 4.9** Newton's modification of
Flamsteed's orbit for the 1680 comet
showed the comet orbiting around the sun.

of the sun. Nevertheless, he was convinced that the 1680 comet was a single comet because, as he pointed out to Newton, Marco Antonio Collio in Rome had observed that the comet had moved 26° 30′ in the six days before first perigee on 17th November, and Flamsteed had measured a movement of 25° 36′ in the six days after second perigee on 30th December. Furthermore, for the nine days after first perigee the comet had moved 41° 15′, compared with 42° 30′ for the nine days before the second perigee. The symmetry of the results clearly indicated to Flamsteed that this was one comet and not two. Newton, in his reply of 16th April, pointed out that Flamsteed was quoting the comets' positions in geocentric longitudes, instead of giving their true positions in space. Newton continued to object to Flamsteed's magnetic hypothesis but, although he still believed that there had been two comets, he appears to have softened his position a little on this.

Newton then dropped the subject of cometary orbits, but the spark of curiosity had been lit by Flamsteed. Eventually Newton returned to the subject, and in 1686 he concluded that cometary orbits are highly elliptical or parabolic in shape. He then assumed that the eccentricities of the elliptical orbits are so large that they could be assumed to be parabolic, to a first approximation. Newton then went on to show how it was possible to work out the parameters of such a parabolic orbit for any comet using just three positional observations. Applying this method to the 1680 comet he found that comets I and II had the same perihelia, and that their orbital velocities were the same at the same distance from the sun. So, he concluded, it had been one comet after all.

## The *Principia*

Newton was now able to complete book 3 of his *Principia*,[52] which was published in 1687, and which had been held up pending a solution to the problem of cometary orbits. In this *Principia* he fitted a parabola through Flamsteed's measured positions of the comet on 21st December 1680, and 5th and 25th January 1681, and calculated that the perihelion had occurred at 0 h 4 min on 8th December, just one day before the comet's opposition. He found that the maximum differences between observed and calculated positions of the comet, post-perihelion, were just 2' in longitude and 10' in latitude, for sample dates of 12th and 29th December, 5th February and 5th March. The pre-perihelion discrepancies were larger, however, but Newton put this down, with some justification, to less accurate observational measurements.

There were two catalysts that had caused Newton to write *De Motu Corporum in Gyrum*,[53] a forerunner of the *Principia*, in November 1684. The first was the correspondence with Hooke in 1679–1680, which had eventually led Newton to analyse the gravitational force outside of, and within a spherical body. The second was Halley's visit to see Newton in August 1684, when he had asked Newton what type of orbit a planet would traverse around the sun, if the force of attraction varied as the square of the distance. Newton told Halley, in answer to his question, that the orbit would be an ellipse, and that he had proved it mathematically some time ago. Halley, Wren and Hooke had been struggling to prove this for some time, but to no avail. So Halley suggested to Newton that it would be of great benefit if he published his proof, along with his other work in the same area.

The *De Motu* which Newton sent to the Royal Society in November 1684 was much smaller, at twenty-two pages, than the subsequent *Principia*, which ran to over 500 pages in its first edition. Nevertheless *De Motu* included three definitions, four hypotheses, four theorems, and seven solved problems, together with various additional analyses. In it he produced, amongst other things, his earlier proofs of Kepler's first and second laws and, importantly, proved Kepler's third law for an elliptical as well as for a circular orbit. Then, over the winter of 1684–85, Newton gradually expanded and

---

[52] The full title was *Philosophiae Naturalis Principia Mathematica*, or *Mathematical Principles of Natural Philosophy*.

[53] Translated as *On the Motion of Bodies in an Orbit*.

improved on his first *De Motu*, issuing a second version only a month or so after the first. It was in the process of undertaking this work that Newton began to conceive and develop the idea of universal gravitation for the first time.

Hooke was furious when he read book 1 of the *Principia*, which Halley had brought to the Royal Society in May 1686, as Hooke claimed that he had been the first to deduce the inverse square law of force between the sun and planets, and not Newton. Hooke went even further, suggesting that he had provided Newton with the idea in 1679. Newton dismissed these claims in a letter to Halley of 20th June 1686, in which he says that he had already concluded, sometime before 1673, that the force of attraction between the sun and planets followed an inverse square law, and that Wren had come to the same conclusion in 1677. Whatever the rights or wrongs of the case, whereas Hooke had come up with the idea intuitively, Newton had eventually been able to prove the relationship mathematically.

Newton was so annoyed with Hooke's attitude that he expunged any reference to Hooke's work from his published texts and, at one stage, he even threatened to withdraw book 3 of the *Principia* from publication. Fortunately Halley was able to calm Newton down, however, and Newton withdrew his threat. Halley then continued to supervise the production of the *Principia*, even meeting the costs of printing it out of his own pocket. In fact, we are greatly indebted to Halley as, had it not been for Halley's powers of persuasion, Newton would never had written the book in the first place.

In the event, Newton's *Principia* consisted of three books. Book 1 developed the mathematical theory of motion in free space, book 2 considered motion against resistance, and book 3 applied the mathematics of book 1 to celestial problems. So it is in book 1 that Newton proved Kepler's second and third laws from first principles. This was for a body in an elliptical orbit, where the force on the body is proportional to the inverse square of the distance from the focus to which the force is directed. The application of Newton's theories to the orbit of comets, which has just been described, is in book 3, together with his work on planetary and lunar orbits.

Newton applied his new-found theories to the planets to determine their masses and densities. His analysis is somewhat complicated, and so will only be summarised here. It was, in effect, based on two ideas:

- That a force applied to a free moving body produces an acceleration, in the direction of the force, whose magnitude is given by the relationship

$$\text{Force} = \text{Mass} \times \text{Acceleration}. \tag{1}$$

This is what we now call Newton's second law.

- That the gravitational attraction $P$ between two bodies of mass $M$ and $m$, separated by a distance $d$, is given by

$$P = G\frac{M \times m}{d^2},\qquad(2)$$

where $G$ is the universal gravitational constant.

The above two equations can be used to give

$$\frac{M_P}{M_S} = \frac{R_{MP}^3}{R_{PS}^3} \times \frac{T_{PS}^2}{T_{MP}^2},\qquad(3)$$

where $M_P$ and $M_S$ are the masses of a planet and the sun, respectively, and where $R_{MP}$ and $T_{MP}$ are the radius and period for a moon in orbit around the planet, and where $R_{PS}$ and $T_{PS}$ are the radius and period for the planetary orbit around the sun. Essentially the moon is used to 'weigh' the planet, and the planet is used to 'weigh' the sun. Clearly this equation cannot be used to calculate the mass of a planet if it does not have a moon.

The most accurately known planetary orbit in Newton's time was that of Venus,[54] so Newton used that to 'weigh' the sun. Venus' mean orbital radius was estimated to be 0.724 AU (astronomical units) and its period was calculated as 224.67 days. Newton then used Callisto, with an estimated orbital period of 16.75 days, to 'weigh' Jupiter. So, all Newton now needed, to calculate the mass of Jupiter relative to that of the sun using equation (3) above, was an estimate of Callisto's orbital radius.

The greatest heliocentric elongation of Callisto, at Jupiter's mean orbital distance from the sun of 5.211 AU, had been measured as $8'\,13''$. This gave $5.211 \times \tan 8'\,13'' = 0.01245$ AU as the radius of Callisto's orbit, which was known to be virtually circular. Hence, using equation (3), the relative mass of Jupiter to that of the sun was given by

$$\frac{M_{Jupiter}}{M_{Sun}} = \frac{0.01245^3}{0.724^3} \times \frac{224.67^2}{16.75^2} \approx \frac{1}{1,100}.$$

This was Newton's value for the relative mass of Jupiter in the first edition of his *Principia*. By the third and last edition he had modified it to $\frac{1}{1,067}$, against a true value which is now known to be $\frac{1}{1,047}$.

Newton calculated the mass of Saturn, using the parameters of Titan's (and Venus') orbit, as $\frac{1}{2,360}$, relative to that of the sun, in the first edition. By

---

[54] Strictly speaking, the orbit of the earth was more accurately known than that of Venus, as the mean radius of the earth's orbit was 1 AU by definition, and its mean period was very accurately known. But the case of the earth caused particular problems in Newton's work, as explained below.

the third edition he had modified this to $\frac{1}{3,021}$, compared with a true value of $\frac{1}{3,498}$.

The mass of the earth could not be determined in exactly the same way as for Jupiter and Saturn, as the earth and moon could not be seen from afar, and so the elongation of the moon's orbit could not be observed directly. In the case of the earth, Newton modified equation (3) above by adding the radius of the earth $R_E$ cubed to the top and bottom of the first fraction on the right hand side. This produced, for the planet earth (i.e. also changing 'P' to 'E' in the equation)

$$\frac{M_E}{M_S} = \frac{R_E^3}{R_{ES}^3} \times \frac{R_{ME}^3}{R_E^3} \times \frac{T_{ES}^2}{T_{ME}^2}. \tag{4}$$

Now $\frac{R_E}{R_{ES}}$ = the solar parallax and $\frac{R_{ME}}{R_E}$ = the inverse of the lunar parallax.

In the first edition of the *Principia* Newton took the solar parallax to be 20″, and the inverse lunar parallax to be 60, giving a mass of the earth of about $\frac{1}{28,700}$ times the mass of the sun. However, Newton soon realised that the solar parallax was about 10″, not 20″, and this would reduce his estimated relative mass of the earth by a factor of about 8, because the relative mass is dependent on the cube of the solar parallax in equation (4). In fact in the third edition of the *Principia* Newton assumed a solar parallax of 10″.50 which produced a relative mass of the earth of $\frac{1}{169,282}$.[55] We now know that this value of the solar parallax was still too high, however, and a currently accepted value of 8″.7936 gives a relative mass for the earth of $\frac{1}{332,946}$.

Newton now turned to the subject of planetary densities relative to that of the sun. Given the above relative planetary masses, these relative planetary densities could easily be calculated, knowing the relative diameters of the sun and planets. In the third edition of the *Principia* he deduced densities of 94.5 for Jupiter and 67 for Saturn, compared with a nominal 100 for the sun.[56]

Newton realised that the solar parallax values that he had assumed were somewhat unreliable, and so he used a different method for determining the relative density of the earth that did not rely on that parallax. As the density $\rho$ is mass per unit volume, for a uniform solid sphere it is proportional to the mass and inversely proportional to the radius cubed. As a result, the ratio of

---

[55] The value of $\frac{1}{169,282}$ for the relative mass of the earth given in the *Principia* was in error, due to a transcription error in Newton's analysis. The change from 20″ to 10″.50 in the solar parallax should have produced a figure of $\frac{1}{198,337}$.

[56] The true values known today are 94.4 for Jupiter and 50 for Saturn, again assuming 100 for the sun.

Table 4.1 *A comparison of Newton's results, in the third edition of his Principia, with modern values*

|  | Mass | | Density | |
|---|---|---|---|---|
|  | Principia | Modern value | Principia | Modern value |
| Sun[a] | 1 | 1 | 100 | 100 |
| Earth | $\frac{1}{169,282}$ | $\frac{1}{332,946}$ | 400 | 392 |
| Jupiter | $\frac{1}{1,067}$ | $\frac{1}{1,047}$ | 94.5 | 94.4 |
| Saturn | $\frac{1}{3,021}$ | $\frac{1}{3,498}$ | 67 | 50 |

[a] Values in this line are by definition, see text.

$\rho_E$, the density of the earth, to $\rho_S$, the density of the sun, is given by

$$\frac{\rho_E}{\rho_S} = \frac{R_S^3}{R_E^3} \times \frac{M_E}{M_S},$$

where $R_S$ is the radius of the sun.

Therefore, using equation (4) above gives

$$\frac{\rho_E}{\rho_S} = \frac{R_S^3}{R_E^3} \times \frac{R_E^3}{R_{ES}^3} \times \frac{R_{ME}^3}{R_E^3} \times \frac{T_{ES}^2}{T_{ME}^2}$$

or
$$\frac{\rho_E}{\rho_S} = \frac{R_S^3}{R_{ES}^3} \times \frac{R_{ME}^3}{R_E^3} \times \frac{T_{ES}^2}{T_{ME}^2}.$$

So the relative density of the earth no longer depends on the solar parallax $\frac{R_E}{R_{ES}}$, but on the radius of the sun as seen from the earth, and on the lunar parallax. As a result, Newton calculated in the third edition of the *Principia* that the density of the earth was about 400, compared with 100 for the sun. The agreement with the modern value of the earth's density (see Table 4.1) is very good.

Unfortunately it was not possible to estimate the mass or density of Mercury, Venus or Mars, as in Newton's time they were not known to possess moons.[57] Nevertheless, Newton observed that the densities of the earth, Jupiter and Saturn reduced with increasing distance from the sun, and concluded that the unknown densities of Mercury, Venus and Mars would

---

[57] Although two moons of Mars were discovered in the nineteenth century, Mercury and Venus still appear to have no moons, as far as we know. If they have any moons, they must be tiny to have escaped detection by the spacecraft that have now visited these planets.

probably fit into this sequence. This was because the higher temperature of the planets nearer the sun would cause them to lose their lighter constituents, compared with those planets further away.

Newton realised, in developing his theory of universal gravitation, that every object in the solar system would gravitationally attract every other object. So the planets do not orbit the sun, as stated by Kepler's first law, but the centre of mass of the solar system, which is not quite the same thing. In addition, those planets with moons have a more complicated orbit, as the centre of mass of each planetary/moon system orbits the centre of mass of the solar system.

Newton recognised that it is impossible to analyse all the mutual interactions between the sun, planets and moons to get a precise definition of their orbits at any particular time. These orbits must be continually changing, but some effects may be so small as to be insignificant. For example, the gravitational attraction of any of Jupiter's moons on Mercury can be ignored, but does the gravitational attraction of the sun distort the orbits of Jupiter's moons around the centre of mass of Jupiter's system? It was to answer this question that Newton asked Flamsteed in 1684 for information on the orbital sizes and periods of Jupiter's moons. This data, to the order of accuracy possible at the time, showed that Jupiter's moons followed Kepler's third law. So clearly the effect of the sun on these moons could be ignored, which made their orbital analysis a good deal easier. But what about the gravitational effect of Jupiter on Saturn, for example?

Flamsteed explained that Kepler's laws did not seem to fit exactly for Saturn, which indicated to Newton that the gravitational attraction of Jupiter was having some effect on the orbit of its sister planet. This delighted him as it indicated that the gravitational force really was universal, and was not just limited to the effect of the sun on its planets, or the planets on their moons. Using the inverse square law he calculated that, at its closest approach to Saturn, Jupiter would have about $\frac{1}{217}$ the gravitational attraction of the sun. Unfortunately, although Newton had correctly identified one cause of the perturbation of Saturn's orbit, it was only one of a number of possible causes. So he was unable to fit his analysis to the observational results, although it did not stop him writing his *Principia* as if he had done so.[58]

---

[58] This is particularly so with the second and third editions of the *Principia*, see *Isaac Newton; The Principia*, by I.B. Cohen and A. Whitman, University of Chicago Press, p. 206 *et seq*. Analytical solutions to the complex problem of the gravitational interaction of Jupiter and Saturn, and the resulting changes in their orbital motions, were developed successively by Euler (in his essay 'Recherches sur la question des inégalités du

The sun, Jupiter and Saturn is not the only three-body system of significance in the solar system; the sun, earth and moon is another but, in this case, their mutual interactions are more complicated. This is due to the proximity of the moon to the earth, and the effect of the moon and sun on the earth's equatorial bulge, which Newton realised would have an effect on the orientation and movement of the earth's axis relative to the ecliptic.[59]

During the attempts to measure the solar parallax in 1672, Richer, Cassini and Picard had found evidence that the earth had an equatorial bulge (see Section 4.4), as the gravitational force at Cayenne, near the equator, was clearly different from that measured in Paris. As a result, Richer's pendulum clock in Cayenne had lost 2 min 28 s per day compared with a similar pendulum clock in Paris.[60] A little later, Huygens developed his theory of the cause of gravity based on vorticular motion and, as a consequence of the earth's rotation, he calculated that it had a polar flattening or oblateness[61] of $\frac{1}{578}$. This he published in a supplement to his discourse on the cause of gravity in 1690. But in the first edition of his *Principia*, published three years earlier, Newton had used his own theory of gravity to calculate an oblateness of $\frac{1}{230}$, which compares with the currently accepted value of $\frac{1}{298}$.

In the first edition of the *Principia*, Newton also calculated that the ratio of Jupiter's equatorial diameter to its polar diameter is about 1.107, or $10\frac{1}{3}$ to $9\frac{1}{3}$, as Newton put it. This shows that Jupiter has a much greater equatorial bulge than that of the earth, because of Jupiter's lower density and faster rotation rate. Newton pointed out that the ratio of $10\frac{1}{3}$ to $9\frac{1}{3}$ that he had calculated was assuming that Jupiter's density was uniform throughout its volume. In the third edition of his *Principia*, published in 1726, he suggested that, if Jupiter is more dense in its equatorial plane than in its polar regions,

---

movement de Saturne et de Jupiter' of 1748), Lagrange (in Vol. 3 of 'Miscellanea Taurinensia' of 1763) and Laplace (in his memoir 'Sur les inégalités séculaires des planètes' of 1785, and his 'Théorie de Jupiter et de Saturne' of 1786). The mathematics of this work is too long and complicated to give here. It is probably best summarised, however, in *Planetary Astronomy from the Renaissance to the Rise of Astrophysics, Part B, The Eighteenth and Nineteenth Centuries*, by R. Taton and C. Wilson (eds.), Cambridge University Press, 1995.

[59] The sun also has an effect on the moon's orbit around the earth, which is considered below.

[60] Part of this timing difference is due to the non-spherical earth, and part due to the difference in centrifugal force acting on the pendulum at the two sites due to the earth's rotation.

[61] The oblateness is the difference in the equatorial and polar radii divided by the former.

the ratio of the equatorial to the polar diameter could be 12 to 11 (i.e. 1.091) or as low as 14 to 13 (1.077). Newton then pointed out that in 1691 Cassini had *observed* that the ratio of Jupiter's equatorial to its polar diameter was about 16 to 15 or about 1.067, and that the English astronomer James Pound had produced a series of measurements in 1719 giving ratios of from 1.091 down to 1.073. So, Newton concluded, his theoretical predictions were in agreement with observation.[62]

One result of the non-spherical shape of the earth is that the length on the surface of the earth of 1° at the meridian will vary from the equator to the poles. In fact, as the equatorial diameter is greater than the polar diameter, the length of 1°, as it is termed, on the surface of the earth, will increase from the equator to the poles, and Newton calculated the magnitude of this. Unfortunately, Cassini and his colleagues measured smaller rather than higher values at higher latitudes, leading them to dispute Newton's analysis, and conclude erroneously that the earth has a polar bulge.[63]

Newton now found that the gravitational attraction of the sun and moon on a non-spherical, spinning earth would have a fundamental effect on its dynamics. If the earth were spherical, the gravitational attraction of the sun and moon could be calculated as if the whole mass of the earth were concentrated at its centre. But the oblate earth has its equatorial plane inclined at an angle of about $23\frac{1}{2}°$ to the ecliptic, and the moon's orbit is inclined at only about 5° to the ecliptic. So both the sun and moon will exert a gravitational attraction on the equatorial bulge of the non-spherical earth at an appreciable angle to the earth's equatorial plane. In addition, as the moon is only about 60 earth radii away from the earth, it will exert a significantly different pull on the near side bulge to that on the far side. As a result, both the sun and moon will exert a turning force, or couple, on the earth in a plane perpendicular to the earth's equator. Such a force on a spinning earth, Newton reasoned, will cause the earth's spin axis to precess.

Newton calculated that the sun's gravitational attraction acting on the non-spherical earth will cause it to precess at about 9".12 per annum. The lunar effect he calculated to be about 4.482 times larger, giving a lunar contribution of 40".88 to the earth's precession. So the total effect of the moon and sun would be to cause the earth's spin axis to precess at a rate of

---

[62] The true value, known today, is 1.076.

[63] Measurements in the early eighteenth century by Bouguer and La Condamine in Peru, and by Maupertuis and Clairaut in Lapland showed that Newton was correct, however.

about 50″.00 per annum, which is very close to the currently accepted value of 50″.3.

With this work Newton was able to explain, for the first time, the physical cause of the precession of the equinoxes. But, although he got the basic cause and net result correct, his detailed analysis was, as it turned out, somewhat flawed. As a result, he overestimated the contribution of the moon and underestimated the contribution of the sun; the two errors just balancing.

As mentioned earlier, the sun, earth and moon is a dynamical, three-body system producing complex effects in the moon's orbit. For example, Hipparchus had deduced that the line of nodes of the moon's orbit regresses with a period of about 18.6 years (see Section 2.3), and that its line of apsides precesses with a period of about 8.85 years. Later Ptolemy had found that the moon was about 1° 20′ away from its expected position at the quadratures (half-moons), in an effect now known as the 'second anomaly' or 'evection'. Then Tycho Brahe had found that the moon was about 40′ 30″ ahead of its expected position 45° before full and new moon, and about 40′ 30″ behind some 45° after full and new moon. This effect is now called the 'variation'. Tycho also found that the moon was up to 11′ behind its expected position in the spring, and up to 11′ ahead in the autumn.

Newton developed his theory of the moon's orbit, which is variously described in the *Principia* and in his *Theory of the Moon's Motion*, first published in 1702, to try to explain the above effects. The subject was so complicated, however, that even the brilliant Newton was forced to admit that it 'made his head ache', and his theory was not always able to produce satisfactory results.

It was evident that, as the moon orbits the earth, the moon is subjected to a varying disturbance from the sun. This is dependent on the varying sun–moon–earth angle and the varying sun–moon distance over the moon's orbit. As the moon's orbit is inclined at about 5° to the ecliptic, and as the earth's orbit around the sun is non-circular, these effects are very complicated to analyse. Nevertheless, Newton was able, using his gravitational theory, to calculate that the line of nodes of the moon's orbit should regress at the rate of 19° 18′ 1″ per year, compared with an observed value of 19° 21′ 21″.[64] In addition, he calculated that the average maximum value of the 'variation' of the moon was 35′ 10″, compared with Tycho's observational value of 40′ 30″,

---

[64] Newton did not give his calculations in any of his books to show that the line of apsides precesses by about 40° per year. As he says in the first edition of his *Principia*, the calculations seem 'too complicated and encumbered by approximations', so he did not include them in his published work.

and the currently accepted value of about 39′ 30″. Newton also concluded that the maximum value of the variation varied from 37′ 11″ to 33′ 14″, as the earth's distance from the sun changed. Finally he gave a value of 11′ 49″ as the maximum annual inequality of the moon's mean motion, which compares with a currently accepted value of about 11′ 12″. Unfortunately we do not know how he calculated his figure of 11′ 49″. This is particularly annoying as using the data in his published work produces a value of closer to 13′.

Newton's lasting contribution to astronomy[65] was in quantitatively explaining the observed motions of solar system bodies according to physical principles, based mainly on his theory of universal gravitation. In some cases he managed to explain these motions unequivocally, but in others, such as in the precession of the equinoxes, he produced the correct result by accident. Sometimes we are not sure how he got his results at all, but we do know that he was not averse to introducing 'fudge factors' from time to time to get his theory to match observations. Although his legitimate methods did not generally produce better predictions for future positions of solar system bodies than the numerical methods then in use, nevertheless they did provide an invaluable physical basis on which future theories could be developed.

---

[65] Ignoring his invention of the calculus, which has been of inestimable value to all branches of physical science.

# Chapter 5 | CONSOLIDATION

## 5.1 Halley's comet

Newton had assumed in his *Principia* of 1687 that the orbits of comets are ellipses of such large eccentricities that they can be assumed, to a first approximation, to be parabolas.[1] He had then proceeded to show how these orbits can be determined from the observed cometary positions in celestial coordinates, and had successfully applied this system to the orbit of the 1680 comet, to which he managed to fit a parabola.

In 1687 Halley suggested to Newton that his method of orbital determination be applied to other comets than that of 1680, but Newton's mind was on other things, and it was left to Halley himself to undertake this analysis, which he duly did. As a result, Halley concluded in two letters sent to Newton in September 1695 that the orbit of the 1683 comet seemed to be a parabola, but that of the 1680 comet appeared to be an ellipse. In the second of these letters, Halley also mentioned that he thought that the comets of 1531, 1607 and 1682 were actually successive appearances of the same comet, and so their orbit must also be an ellipse. He was concerned, however, that the time differences between the successive appearances were not exactly the same, and wondered if this could be caused by the gravitational effects of Jupiter and Saturn.

In 1705 Halley published a treatise called *Synopsis of the Astronomy of Comets*, in which he gave the orbital elements for twenty-four comets which had been observed between 1337 and 1698. As a first approximation,

---

[1] As mentioned earlier (see Section 4.7), ellipses have an eccentricity of less than 1, parabolas have an eccentricity of exactly 1, whilst hyperbolas have an eccentricity of greater than 1.

he assumed, like Newton, that their orbits were parabolic, but he then modified their eccentricities to achieve a better fit with observations. In the process he showed that none of the twenty-four orbits were hyperbolic, so these comets were clearly permanent members of the solar system.[2] Halley also concluded that the comets of 1531, 1607 and 1682 were successive appearances of the same comet, as their orbital elements were very similar, and the time intervals between appearances seemed to be about the same.

In his *Synopsis* Halley estimated that the 1531 comet had passed perihelion on about 24th August, the 1607 comet on 16th October, and the 1682 comet on 4th September, so the time intervals between successive appearances were actually 76 years 2 months and 74 years 11 months. He attributed this change in period to the gravitational effect of Jupiter, and further noted that the comet of 1456 may have been an earlier appearance of the same comet. As a result, Halley predicted the return of the 1682 comet for 1758, which he modified in 1717 to be late 1758 or early 1759, after taking the perturbing effects of Jupiter into account. This modified prediction was not known until 1749, however, when his *Astronomical Tables* were published posthumously, seven years after his death.

Halley was not the only person to predict the return of the 1531/1607/1682 comet that we now call Halley's comet. Jean Philippe Loys de Chéseaux (1718–1751) concluded in 1744 that, because of the unequal time intervals, we were, in fact, seeing two comets travelling in identical orbits, such that when one comet was at perihelion the other was near aphelion. He then concluded that, as the time between successive perihelia of the 1531 comet was 151 years 11 days, the 1607 comet should pass perihelion on 27th October 1758 in the Julian Calendar, or 7th November in the Gregorian.

In 1758 Thomas Stevenson, who lived on Barbados, and may not have known of Chéseaux's work, also proposed a two comet theory, assuming that the bright comet of 1305 was a previous appearance of Halley's comet. Although this proved not to be correct,[3] as Halley's comet had reached perihelion in 1301, his predicted perihelion date for Halley's comet of 3rd February 1759 was, fortuitously, only six weeks out.

The most thorough work on the predicted perihelion date of Halley's comet in 1759 was undertaken by Alexis Claude Clairaut (1713–1765), assisted by Joseph Jérôme le François de Lalande (1732–1807),

---

[2] If a comet is in a hyperbolic orbit, it will only be a temporary member of the solar system, as it will visit the sun only once and never return.

[3] Stevenson was not alone in believing that the 1305 comet was a previous appearance of Halley's comet, as Halley had thought the same at one time.

and Mme Nicole Lepaute in 1757 and 1758. Previously Euler, Clairaut and D'Alembert had published an approximate solution to the three-body problem (see Section 4.7), to enable the moon's orbit to be analysed more accurately, so Clairaut used a modified version of this three-body analysis to refine his predictions for Halley's comet's return.

In the beginning, Clairaut was only intending to take into account the disturbing effect of Jupiter on the comet's orbit, but this effect proved to be so large that he decided to examine the effect of Saturn also. In the event he concluded that the interval between the 1682 and 1759 perihelia would be 618 days longer than the interval from 1607 to 1682; 518 days being due to Jupiter and 100 days due to Saturn. He also calculated the effect of Jupiter and Saturn on the interval between the 1531 and 1607 appearances, and came up with a figure which was within 27 days of that observed, giving him confidence in his approach. As a result, he concluded that Halley's comet would reach perihelion on 15th April 1759 ± 1 month. He later modified the perihelion date to 31st March 1759, which proved to be just 18 days later than observed.[4]

Clairaut naturally wanted his initial prediction of the date of Halley's perihelion to be known before the comet was sighted, so he and his team worked day and night to finish the calculations. In the event Clairaut presented his result to the Academy of Sciences in Paris on 14th November 1758, just six weeks before Halley's comet was first observed.

As so often happens in science, what is supposed to happen doesn't do so, which is what helps to make science so interesting. In the case of Halley's comet, it was expected that Charles Messier (1730–1817) would be the first to observe it on its reappearance, as he had begun to search for the comet in mid 1757, looking for it on every clear night from the new French Naval Observatory at the Hotel de Cluny. In the event, it was first observed by a German amateur Johann Georg Palitzsch (1723–1788) on 24th December 1758. Palitzsch recognised the object as a comet, and communicated his observations to Christian Hoffman, who observed the comet himself on 27th December, but neither Palitzsch nor Hoffman realised that it was Halley's comet.

The first published announcement of the observation of Halley's comet was made in an anonymous paper published in Leipzig on 24th January 1759.

---

[4] In 1985 Peter Broughton concluded that six days of this eighteen day difference were due to Uranus and Neptune, which had not then been discovered, six days were due to Mercury, Venus, the Earth and Mars, and four days were due to errors in the masses of Jupiter and Saturn.

The author had apparently heard of Palitzsch's observations and, realising that the comet was Halley's comet, decided to calculate its orbit. He observed it for the first time on 18th January, to check his calculations, and in his paper predicted that the comet would reach perihelion on 14th March, which was just one day after it did so. It is not certain who wrote this paper, but the author is believed to be Gottfried Heinsius, who was professor of mathematics at the University of Leipzig.

Meanwhile Messier was being hindered by poor weather in Paris at the end of 1758, and it was not until 21st January 1759 that he first saw the comet. He continued to observe it on six further nights in January, and five nights in February, until it got too close to the sun to be seen. For some inexplicable reason, Messier was told by his superior, Joseph Nicolas Delisle (1688–1768), to keep the discovery strictly to himself, until Delisle would decide when a formal announcement should be made. In the event, Delisle agree to the publication of the discovery when Messier first saw the comet after perihelion on 1st April. Unfortunately for Messier, this was the very day that news of the prior German discovery reached Paris.

In Halley's treatise of 1705, he had not only identified the comet of 1531/1607/1682 as a recurrent comet, but he had also suggested that the comet of 1661 was probably the same as that of 1532, as it appeared to have a similar orbit. He was unsure, however, as Halley thought Peter Apian's observations of the 1532 comet were too inaccurate for any clear conclusion to be drawn. In fact we now know that these two comets were not the same. Likewise in the third edition of Newton's *Principia*, published in 1726, Newton mentions that Halley thought that the comets of 44 BC, 531, 1106 and 1680 were also probably one and the same, but again this turned out not to be so.

There is no doubt that the successful prediction of the return of Halley's comet by Halley, Clairaut and others was a major triumph, not only for the individuals concerned, but also for Newtonian dynamics. We should note, however, that, alongside this successful prediction of the return of Halley's comet, there were also these erroneous cometary identifications. Such erroneous identifications were inevitable, however, considering the inaccuracies of some of the early observations on which they were based.

## 5.2 Atmospheric refraction

It became evident as the seventeenth century progressed, and as the measurement of celestial coordinates became more accurate, that a much more accurate correction was required for atmospheric refraction. In the late sixteenth century Tycho Brahe had produced separate tables of atmospheric refraction for the sun, moon and stars (see Section 2.3), assuming that refraction of sunlight and starlight was zero above 45° and 20° altitude, respectively.

Tycho, in producing the refraction table for sunlight, had assumed a horizontal solar parallax of 3′, which was far too high, and he had used this value to correct his measurements. Although he did not realise it at the time, this was part of the reason why his atmospheric refraction values for sunlight and starlight were different. We now know that these, and the values for moonlight, should have been the same, but Tycho was unsure of the cause of atmospheric refraction. He thought that it was caused by vapours in the lower atmosphere, and consequently that it may vary with season, location or weather on the earth. Tycho was not very confident in his refraction tables, however, and so he used them with caution.

Kepler in his *Astronomiae pars Optica* of 1604 found a single law of refraction for light from all heavenly bodies. This enabled him to produce a table of corrections which went all the way up to the zenith, and which gave good results for altitudes above about 10°. However, in his *Rudophine Tables* of 1627, he still included Tycho's refraction tables.

Some years later, Thomas Streete (1622–1689), a London Excise Office clerk, realised that Tycho's correction for the horizontal solar parallax was too large, so he adopted Tycho's stellar refraction values as being valid for the sun, moon and planets also. Although this approach made good physical sense, it meant that he necessarily assumed that atmospheric refraction was zero above 20° altitude, which we now know is not the case. In fact the true value of refraction at 20° is about 3′, which is still very significant.

One of Cassini's first research projects, after being appointed professor of astronomy at Bologna in 1650, was in what used to be called solar theory, which involved measuring the apparent movement of the sun across the sky throughout the year. Shortly after starting this work, however, he found, like Tycho before him, that the latitude of his observatory deduced by using half the sum of the noon altitudes of the sun at the two solstices, was different from that calculated using the pole star. Cassini put this down to faulty refraction and solar parallax corrections. In this work he had used a parallax correction of 1′ which, although better than the 3′ used by Tycho, still required him to produce different refraction tables for winter, summer and the equinox periods. This seemed inherently unsatisfactory to him, but he found that only by assuming a solar parallax value of ≤12″ could he reduce the three tables to one. Unfortunately, adopting such a small value for the solar parallax implied that the sun was, what seemed to him, an incredible distance away. Clearly a new measurement of the solar parallax was required.

The solar parallax problem was solved, to a first approximation, by the Cayenne expedition of 1671–73 (see Section 4.4), which was mounted by the French Authorities following strong recommendations by the French Academy of Sciences and Cassini, who had by now become head of the Paris Observatory. The results of the expedition led Cassini to adopt a value of 9″.5 for the horizontal solar parallax, allowing him to use a single atmospheric refraction table for all of the year. This was eventually published in 1684 in his *Les Élémens de l'Astronomie Verifiez*.

## 5.3 Aberration of light

As astronomers strove, in the second half of the seventeenth century, for greater and greater accuracy in their measurement of the positions of the planets, it became apparent that another effect, previously undetected, was causing problems. At that time, the positions of the planets were measured with reference to the fixed stars, but it became apparent in about 1670 that the positions of the fixed stars were not fixed. Jean Picard, in particular, noticed that the declinations of stars sometimes showed deviations of >10″. Flamsteed thought that this was due to stellar parallax, which is caused by the orbital motion of the earth, but others disagreed, as the detailed movement of the stars did not seem to be consistent with such a cause.[5] The measurements were very difficult to make with high accuracy, however, because of atmospheric refraction. As a result, Robert Hooke decided to measure the movement of a star that would appear near the zenith of his observing site, where the effect of refraction would be very small. He chose the star γ Draconis because it was bright, and so presumably was near to the earth. If that was the case it should show a large parallax.

Hooke's results were generally inconclusive, and so Samuel Molyneux (1689–1728) asked George Graham, a well-known instrument maker, to build him a telescope that would measure the effect with maximum accuracy. The instrument had to be rigidly mounted, but only a very limited

---

[5] As the earth orbits the sun, so the nearest stars should appear to move relative to those more distant, owing to the effect of parallax. The orbits should be circular for those near the pole of the ecliptic, and linear for those in the ecliptic plane. This did not seem to be so in the case in Flamsteed's observations.

movement was required. To achieve this, a bizarre solution was adopted with the top of the telescope being attached to the chimney of Molyneux's house on his estate at Kew. The telescope then passed through the roof to an observer some 20 ft (7 m) below. Molyneux decided to measure γ Draconis, the same star as Hooke, and he and his friend James Bradley (1692–1762), who was Professor of Astronomy at Oxford, started observing on 3rd December 1725. Unfortunately, shortly afterwards Molyneux was required for duty by the Admiralty, and he had to pass the task on to Bradley to complete.

Bradley found that the position of γ Draconis varied with a period of one year, as he had expected, but, much to his surprise, it reached its maximum southerly position in March and not December, as it should have done if the effect was due to parallax. The total amplitude of the movement was about 40″, which was also very much larger than he had expected. Bradley thought that the effect could be caused by the earth's axis nutating, but he needed to measure the movement of other stars to check on this.

Bradley could not use Molyneux's telescope to undertake his new measurement programme, as he wanted to measure stars in different parts of the sky. So he had a new telescope built. With this, starting in 1727, he observed about 200 stars, and found that their movement could neither be explained by stellar parallax nor by a nutating earth.

Then in 1728 Bradley found an explanation. What he was observing was the apparent deflection of the star from its true position by the fact that the velocity of the earth, in its orbit around the sun, is a significant fraction (albeit a very small one) of the velocity of light. This effect, now known as the 'aberration of light', allowed the ratio of the velocity of light to the velocity of the earth in its orbit to be determined and, knowing the size of the earth's orbit around the sun, it enabled the velocity of light to be determined, and vice versa. The aberration that Bradley measured, of 20″.2 half-angle, enabled him to deduce that it takes light about 8 min 12 s to reach us from the sun,[6] which was, within error, the same time as estimated using the eclipses of Jupiter's satellites (see Section 4.2). In fact, this work of Bradley finally proved that Römer had been correct in 1676 in attributing the time differences in the eclipses of Jupiter's satellites to the finite velocity of light.

Bradley continued observing γ Draconis for a number of years, and found that its apparent movement was actually a mixture of two effects. One, with

---

[6] Strictly speaking this is the time taken for light to travel 1 astronomical unit; that is it ignores the finite size of the sun. It is given by $\frac{\tan 20''.2 \times 365.25 \times 24 \times 3600}{2\pi}$ seconds. Bradley also calculated the ratio of the velocity of light to that of the earth in its orbit as $\frac{1}{\tan 20''.2} = 10{,}210$.

a maximum amplitude of about ±20″ and a period of 1 year, was due to the aberration of light, but the other had an amplitude of ±9″ and a period of about 18 years. The latter effect was identical for all stars, whereas that due to stellar parallax varied depending on the angular distance of the star from the pole of the ecliptic. The period of regression of the line of nodes of the moon's orbit was known to be 18.6 years (see Section 2.3), so Bradley suggested in 1748 that this second effect was due to the nutation of the earth's axis, caused by the moon.[7]

As a result of this work by Cassini and Bradley in particular, astronomers were able, in the second half of the eighteenth century, to make corrections to their observations for atmospheric refraction, the aberration of light, and the nutation of the earth's axis. This made a considerable difference to the absolute accuracy of their results derived from their observations.

## 5.4 **The 1761 and 1769 transits of Venus**

In 1663 James Gregory had suggested, in his *Optica Promota*, that observations of a transit of Mercury across the face of the sun could be used to determine the solar parallax, and hence allow the distance of the sun from the earth (the astronomical unit) to be calculated. The method consisted essentially of timing the ingress and egress of Mercury across the solar disc from two different places on the earth of known separation. Knowing the relative size of the orbits of Mercury and the earth, the relative parallax between the sun and Mercury could be determined, followed by the distance of the sun from the earth.

The young Edmond Halley was aware of James Gregory's proposal when he sailed for St Helena in November 1676 with a commission from the British Admiralty to accurately measure the positions of the southern stars. Whilst Halley was on St Helena he was fortunate enough, in spite of the poor weather, to observe a transit of Mercury across the sun. As just mentioned, estimates of the solar parallax required observations of the transit from more than one place on earth, but on Halley's return to England, he was disappointed to find that Jean Charles Gallet in Avignon was the only other person to have observed the transit. Unfortunately, there were far too many uncertainties involved in comparing Halley's and Gallet's measurements, so Halley was not too happy with the resulting solar parallax of 45″.

---

[7] D'Alembert and Euler showed analytically that this was so a year or so later.

As Mercury approaches the sun, just before a transit, the separation between the sun and the planet becomes smaller and smaller, as seen from the earth, until the leading edge of the planet appears to just touch the sun at the first external contact. Mercury then starts to cross the edge of the sun until it is completely 'on' the sun. As the planet continues in its transit, a continuous, very thin, filament of sunlight is seen between the trailing edge of the planet and the edge of the sun. When this continuous filament is first detected, this is called the first internal contact. When the transit has virtually finished, these internal and external contacts are seen in reverse. Halley, during his observations of the transit of Mercury from St Helena, found that it was easier to unambiguously time the internal contacts, rather than the external ones. As a result he recommended that the timings of internal contacts be adopted in future transit observations.

Not only had Halley estimated the value of the solar parallax from the transit of Mercury in 1677, but he had also assisted Flamsteed in trying to estimate the solar parallax using Mars (see Section 4.4). However, Halley was not convinced that either method would ever yield accurate results, so in his *Catalogue Stellarum Australium*[8] of 1678 he proposed that transits of Venus should be used instead. In particular he pointed out that as Venus is much further from the sun, and closer to the earth, than Mercury, the observation of a transit of Venus should yield much more accurate results for the solar parallax than using Mercury. The problem was, as he pointed out in his *Catalogue*, that the transits of Venus are few and far between, and the next one was not due until 26th May 1761 (Julian date, or 6th June Gregorian date), when none of the then current astronomers would be still alive.

Over the next ten years or so Halley refined his analysis of the orbit of Venus (and Mercury), giving his results in a paper that was published in the *Philosophical Transactions* in 1691. There he gave accurate values for the inclination of Venus' orbit to the ecliptic, the location of its ascending node, and the orbital period of Venus. These new values then enabled him to list seventeen transits of Venus between 918 and 2004, including those of 1518 and 1526, 1631 and 1639, and 1761 and 1769. Although Halley's predictions were more accurate than those of Kepler,[9] his predictions inevitably became less accurate the further away they were in time from his own period. Nevertheless, his predictions for the next two transits of 1761 and 1769 were accurate enough, for the moment.

---

[8] In English *Catalogue of Southern Stars*.

[9] For example, Kepler had failed to predict the transits of 1639 and 1769.

Halley died in 1642, but not before he had written a second detailed paper, which was published in the *Philosophical Transactions* in 1716, giving more information on the methodology to be followed for observing the forthcoming transits of 1761 and 1769. Unfortunately, Halley's requirement that the times of both ingress and egress of Venus be observed restricted the places on earth from which observations could be made. For example, for the transit of 1761 such observations could only be made from Asia and the north polar regions. In addition, weather constraints may also restrict observation at any of these locations to just ingress or egress. Accordingly, in 1723 Delisle suggested a slight modification of Halley's method which required a measurement of just ingress or egress, provided that the longitude of the observing site could be accurately determined.

It would be over one hundred years before the next pair of Venus transits could be observed, so it was essential that as many practical details as possible be cleared up before those of 1761 and 1769. Nicolas-Louis de Lacaille (1713–1762) pointed out that problems with atmospheric refraction at low altitudes would make it impossible to obtain accurate timings of ingress or egress, so such observing sites should be avoided, if possible. It was also thought best to measure the timings of internal contact of Venus on the sun, rather than of external contact, following Halley's experience with the timing of Mercury's transit of 1677. Even so, there were uncertainties of up to 40 seconds reported in the timing of internal contacts during the Mercury transit of 1743. This augured badly for the Venus transit, as the Mercury transit should be easier to time than that of Venus, as Mercury moves across the face of the sun that much faster.

It was obvious that to get the most accurate results for the Venus transit, all of the observers should use the same procedure and have a trial run if possible. Fortunately, there was a transit of Mercury predicted for 1753, and for this Delisle revised Halley's data, and distributed diagrams, instructions, and an updated set of timing predictions to as many observers as possible to aid their observations. Much to his chagrin, however, the results of the Mercury transit were disappointing in so far as any reasonably accurate estimate of the solar parallax was concerned. Nevertheless, Delisle set about preparing the astronomical community for the Venus transit of 1761.

The main task, which was to occupy Delisle for the next few years, was to produce predictions for the timing of the 1761 transit to aid observations, and enable decisions to be made by the national authorities as to where to send their observers. Many transits of Mercury had been observed by then, and yet Delisle's prediction for the 1753 Mercury transit had been in error by seventeen minutes. In the case of Venus, however, only one transit had been observed (in 1639), so large timing errors were anticipated for the 1761 event.

Delisle's updated predictions were published in May 1760 by the French Academy of Sciences, which left barely enough time for the expeditions to the various observation sites to be organised. In England the situation was even worse, as it was not until the following month that the Royal Society started to consider their options, after discussing Delisle's paper.

In the event, the French sent Alexandre-Gui Pingré (1711–1796) to the island of Rodrigues in the Indian Ocean, Chappe d'Auteroche to Tobolsk in Siberia, Guillaume Le Gentil (1725–1792) to Pondicherry in India, and César-François Cassini de Thury (1714–1784) to Vienna, whilst the British sent Nevil Maskelyne (1732–1811) to St Helena, Charles Mason (1728–1786) and Jeremiah Dixon (1733–1779) to Benkulen on Sumatra (Indonesia), whilst Maximilian Hell went from Vienna to Vardö, in northern Norway, and John Winthrop (1714–1779) was sent by the government of Massachusetts to St Johns, Newfoundland.

Unfortunately, Britain and France were engaged in the Seven Years War at this time and, although the governments of both countries had instructed their navies not to intercept ships carrying observers for the transit of Venus, it was impossible to be sure who was on which ship. As a result, a French frigate shot at, and severely damaged, the ship carrying Mason and Dixon to Benkulen, killing eleven of the crew. As the damage to the ship was extensive, it had to return to England for repairs. Unfortunately, when it set sail again, it was too late for Mason and Dixon to reach their original destination, so when they reached the Cape of Good Hope they decided to make their observations from there, instead. Similarly, when Le Gentil finally arrived at Pondicherry aboard a French warship, he found that it was now under British control. As a result, it was decided to sail to Mauritius, obliging Le Gentil to make his observations from the deck of the ship.

The best observations were made by Chappe in Siberia, and by Mason and Dixon at the Cape of Good Hope, all of whom were able to observe the entire transit, and by Winthrop in Newfoundland, who measured egress and five positions of Venus on the sun. But the timing measurements proved to be much more difficult to make than anyone had anticipated, owing particularly to the so-called 'black drop effect'. In this the planet does not separate from the edge of the sun cleanly, as it starts to cross the sun's disc after what should have been the first internal contact. Instead there seemed to be a thick, dark

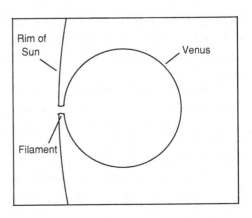

**Figure 5.1** The black drop effect, indicated in this diagram, made exact timing of the start of Venus' transit difficult. The spherical dark planet appeared to be connected to the rim of the sun by a thick, dark filament, whose appearance varied from observer to observer.

thread (see Figure 5.1), which took some time to 'break' to leave a clear gap between the edge of the sun and the planet. Unfortunately the exact nature of the effect varied from observer to observer, and from one telescope to another, making precise timing of the first internal contact impossible.

In addition to the problems caused by the black drop effect, several observers saw a luminous ring around Venus that also affected accurate timing. This ring was interpreted by Mikhail Lomonsov (1711–1765) as showing that Venus has an atmosphere. He was correct, but to the observers of the time the luminous ring was simply a nuisance, making exact timing even more difficult. Finally, there were problems in making exact longitude measurements of some of the observing sites. These problems conspired together to produce estimates of the solar parallax ranging from $8''.3$ to $10''.6$ which, although better than the previous range of $9''$ to $15''$ using the parallax of Mars, was still disappointingly imprecise, particularly considering the amount of effort that had gone into the various expeditions. Nevertheless, the lessons learnt from these observations could be put into good effect for observing the next transit on 3rd June 1769.

Fortunately, by 1769 the Seven Years War had been brought to a close by the Peace of Paris, and Le Gentil, who had moved to Manila to await this second transit, was advised to return to Pondicherry. The two days preceding the transit in Pondicherry were sunny, but on transit day the weather intervened, with clouds obscuring the sun. Although the unfortunate Le Gentil did not realise it at the time, his colleagues in Manila had had a clear view. Then twice on his return journey back to France, Le Gentil was shipwrecked, and when he finally reached home, he found that he had been given up for dead.

Worse was to befall a four man Franco-Spanish expedition to Baja California, however, that was led by Chappe d'Auteroche. Leaving Cadiz in October 1768, the expedition arrived at Vera Cruz in March of the following year. It then took two months of arduous trekking across Mexico before they were able to cross the sea to the Californian peninsula, where they arrived three weeks before the transit. Unfortunately, although they had a good view of the event, three of the party died shortly afterwards, leaving Vincente de Doz to make the hazardous journey across Mexico alone with the observational results.

In all over 150 astronomers set out to view the transit of 1769 from over 70 sites, stretching from Siberia, the North Cape, and Hudson Bay, to Peking, Madras, California, and Tahiti. One of the most famous of the expeditions was Captain Cook's first voyage of 1768–1771, during which three groups of observers studied the transit in a cloudless sky at the newly discovered island complex of Tahiti.[10] However, this was not before the main quadrant, which was to be used during the transit, had been stolen by one of the natives. Fortunately, it was soon recovered in good condition in time for the observations.

It took some time for all the expeditions to return to civilisation and analyse and publish their results. In 1771 the Frenchmen Pingré and Lalande deduced values for the solar parallax of 8″.8 and 8″.6, respectively. Maskelyne settled on 8″.8, the Swede Anders Planmann on 8″.5, and Anders Johan Lexell (1740–1784) at the St Petersburg Academy on 8″.6. These results produced a general consensus among astronomers that the most likely value was about 8″.6; this best estimate lasting well into the next century.

There was a problem in the eighteenth century on how to establish the most likely value of a parameter, and its probable error range, when it had been observed by a large number of scientists of different skill and reliability. So the most likely value of the solar parallax deduced in the 1770s was based more on consensus than on a statistical analysis. However, in 1809 Carl Friedrich Gauss (1777–1855) led the way to a statistical solution when he published his method of least squares.[11] Then in 1835 Johann Encke (1791–1865) used Gauss' method to re-analyse the 1769 transit data, and so produce a best estimate of 8″.571 ± 0″.037 for the solar parallax, which is equivalent to an astronomical unit of 153.5 ±0.7 × $10^6$ kilometres.

---

[10] Tahiti had been discovered only twelve years earlier by Captain Wallis.

[11] Although Gauss published his method of least squares in 1809, he had initially developed it over ten years earlier when he was a student at the Collegium Carolinium in Brunswick.

## 5.5 Secular acceleration of the moon

Halley discovered in 1693 that the moon's position in the sky observed at that time was not consistent with ancient eclipse records that went back for about 2,500 years. The moon had, apparently, accelerated in its orbit over that period, so that it was now about 2° in advance of where it should be, were there to have been no acceleration. This so-called 'secular[12] acceleration' of the moon was confirmed in 1749 by Richard Dunthorne, who calculated its rate from eclipse data as being about $10''/\text{century}^2$. Then Tobias Mayer published a figure of $7''/\text{century}^2$ in his lunar tables of 1753, which he revised to $9''/\text{century}^2$ in 1770. The natural consequence of such an acceleration was that the moon's orbit was getting smaller and, if it continued into the future, the moon would eventually collide with the earth.

At that time, the cause of this secular acceleration was a complete mystery. As a result, in 1770 and 1772 the Paris Academy offered a prize for a more complete development of lunar theory, based on universal gravitation. In particular, the successful candidate had to examine whether this theory could account for the secular acceleration of the moon. Euler, in his prize essay of 1770, concluded that the moon's acceleration cannot be produced by gravitational forces, and in his essay of 1772 he went on to conclude that the effect is caused by the resistance of the medium in which the planets move.[13]

Laplace, in an essay of 1773 on universal gravitation, considered an alternative theory that could explain why the moon appeared to be ahead of its predicted position. Instead of the moon accelerating in its orbit, maybe the earth's rotation rate was gradually slowing down, which would produce the same observed effect. He examined whether the friction of the trade winds or of the interplanetary ether could have the desired effect, but concluded that it could not. The problem continued to nag away at his mind and on 19th December 1787 he read a draft paper to the Paris Academy outlining his novel solution. He had previously shown that planetary perturbations were

---

[12] Perturbations in the motions of celestial bodies were found to result in either periodic or secular changes in their orbital parameters. The word 'secular' was used when the change was always in the same direction over time, and when there was no indication that the effect was periodic. As indicated below, however, it turned out that the use of the word secular to describe the above effect was a misnomer, as the effect was eventually found to be periodic, albeit with a very long period.

[13] It may seem strange to propose that a resistance can cause an acceleration, but the acceleration concerned is an angular and not a linear acceleration.

causing the eccentricity of the earth's orbit to have been reducing for many thousands of years. He now pointed out that this would result in the mean distance of the earth from the sun increasing slightly, causing the effect of the sun on the moon's orbit to be marginally reduced, thus allowing the moon to orbit the earth slightly faster in a slightly smaller orbit. He showed that, although this effect is cyclical, its period is several hundred thousand years, so over recorded time the lunar acceleration has been constant. Laplace calculated the magnitude of the effect to be $10''.2/\text{century}^2$, in good agreement with observations. So yet again Newton's universal gravitational theory had managed to explain a most obscure effect, giving astronomers confidence that they would eventually be able to explain all the detailed, intricate motions of the solar system using gravitational theory.

Following on from this success, Laplace went on to publish an extensive theoretical analysis of the motion of the moon and planets in his five volume masterpiece *Traité de Méchanique Céleste*,[14] which was progressively published over the period 1799–1825. In it he derived the perturbations of eccentricity, inclination, apsides and node for the orbits of the moon and planets over hundreds of thousands of years. In the process he showed that the solar system was deterministic. That is, once it has been set in motion, it will continue to move in a completely predictable way, like an enormous clock. Furthermore, he showed that, although the magnitudes of the orbital parameters of the bodies of the solar system were always changing, there was no net effect and the system was stable. Laplace then went on to derive a figure of $\frac{1}{305}$ for the flattening of the earth and $8''.6$ for the solar parallax, from the effect of these parameters on the orbit of the moon. As a result he commented that he found it amazing that an astronomer could accurately calculate the shape of the earth, and the value of the solar parallax, without leaving the comparative comfort of his observatory, when previously this had required 'long and troublesome voyages in both hemispheres'.[15] This was a further remarkable confirmation of the theory of universal gravitation.

---

[14] In English *Treatise on Celestial Mechanics*.

[15] From Laplace's semi-popular *Exposition du Système du Monde*, Vol. IV, Ch. 5, which was published in six editions between 1796 and 1835.

# Chapter 6 | THE SOLAR SYSTEM EXPANDS

## 6.1 William Herschel and the discovery of Uranus

Friederich Wilhelm Herschel (1738–1822), now better known as William Herschel, was born on 15th November 1738 in Hanover to Anna and Isaac Herschel. The fourth of ten children, William followed his father and elder brother Jacob into the band of the Hanoverian Foot-Guards when he was fourteen years old, where, like his father, he played the oboe. At that time the English king, George II, was also the elector of Hanover, and in 1756 all three Herschels went with their regiment to England because of the threat of a French invasion. Whilst in England, William learned English, and Isaac and his two sons mixed with the local musical fraternity. Later in the year William and his father returned with the regiment to Hanover, but a disastrous campaign against the French saw William return to England in 1757 after leaving the army.[1]

William Herschel spent the majority of his first ten years in England as an itinerant musician, living mostly in the north of England, teaching and composing music and playing the organ and other instruments. At one of the concerts, the orchestra included the Duke of York, the brother of King George III,[2] who played the violoncello. During this period William not only perfected his English, but also learned Latin and Italian. Then in 1767

---

[1] There is some confusion as to the exact circumstances of the conclusion of William's army career, with some people claiming that he had deserted. Strictly speaking this was not so, however, as William had been too young when he joined the regiment to enlist formally. Nevertheless, to regularise matters his father persuaded the army to issue formal discharge papers which were signed by General von Spörcken in 1762.

[2] George III succeeded George II in 1760.

he was appointed organist of the Octagon Chapel in Bath, which at that time was a booming and fashionable health resort. Five years later he was joined in Bath by his sister Caroline, ostensibly to start a singing career, but she was soon to become William's housekeeper and musical assistant. Later she was also to play a crucial rôle in her brother's astronomical research, acting as his assistant during both his observational and telescopic manufacturing activities.

Although William Herschel's father had been interested in astronomy, we have no indication of William's interest in the subject until 1766, when we know that he made observations of Venus and of an eclipse of the moon. He made only casual astronomical observations over the next few years, however, but his interest in astronomy was rekindled by reading Robert Smith's *Compleat System of Opticks* and Ferguson's *Astronomy* during the winter of 1772–3. As a result, in the following spring and summer he made a series of refracting telescopes using bought lenses. But he found the results disappointing, and the telescope too long and unwieldy. So in September 1773 he hired a reflecting telescope of 2 foot (60 cm) length, but he found that to be too small. Herschel then enquired about the price of a 5 or 6 foot reflector, but found that it was more than he was willing to pay.

It was now clear to him that the only solution to his requirement for large, high quality telescopes was to make them himself. So, using Smith's *Opticks* as a guide, he started on a long and distinguished career as a designer and builder of reflecting telescopes, grinding and polishing the mirrors himself. By 1776 he had built a series of such reflectors, ending with a 20 foot (6 m) long instrument, with a 19 inch (48 cm) diameter mirror, which he used in his back garden. But it was not just the size of his telescopes that was impressive, as the quality of his telescopes turned out to be even better than those at the Royal Observatory at Greenwich.[3]

Early on in his astronomical career Herschel made measurements of the rotation periods of Mars and Jupiter and of the height of the lunar mountains. In fact, one night in December 1779, Herschel had moved his 7 foot (2.1 m) long, 6.2 inch (15.7 cm) diameter telescope into the street in front of his house to measure the lunar mountains, when a passing stranger asked if he could look through it. Ever willing to be helpful, Herschel obliged. The stranger

---

[3] This was proved by a side-by-side comparison that was carried out in June 1782, following Herschel's discovery of Uranus. This comparison was undertaken because many astronomers thought that Herschel may not have been telling the truth when he claimed that he had first seen Uranus with a 7 foot long reflector, with a mirror of just 6.2″ (15.7 cm) diameter.

turned out to be Dr (later Sir) William Watson, whose father was physician to the king; both father and son being fellows of the Royal Society. So started the lifelong friendship with a man who was to be a considerable influence on Herschel's early career.

Ever the methodical scientist, Herschel had already by 1779 made a complete survey of the sky with a small reflecting telescope, with which he catalogued all the stars visible from his house down to a magnitude of 4.[4] Then in August 1779 he had started a survey of stars down to magnitude 8 using his 7 foot reflector, looking for double stars.[5] It was, in fact, during this survey that he made his most important planetary discovery. As his notebook says for Tuesday 13th March 1781, 'in the quartile near ε Tauri the lowest of the two is a curious either nebulous star or perhaps a comet. A small star follows the comet at $\frac{2}{3}$ of the field's distance.' Then on 17th March, when he next appears to have observed the object, he says, 'I looked for the comet or nebulous star and found that it is a comet, for it has changed its place.'

In Herschel's discovery paper, which William Watson read to the Royal Society on 26th April, Herschel noted that stars stayed bright and distinct as he increased the magnification of his telescope, whereas this new object became less distinct with greater magnification. In addition, he knew that the apparent diameters of planets seen in his telescope increased linearly with magnification, whereas those of stars did not. As the diameter of this new object had also increased linearly with magnification, Herschel was convinced that it was not a star but a comet. It apparently never occurred to him that it may be a new planet. Herschel went on to present measurements, in his discovery paper, of the object's diameter between 17th March and 18th April, showing that it was increasing steadily from 2″ 53‴ to 5″ 2‴, apparently because the comet was getting closer to earth. But he was concerned because the object's path did not seem to resemble that of a comet,

---

[4] The intensities of stars, as seen from earth, are given in so-called magnitudes. These are on a logarithmic scale, such that adjacent magnitudes have an intensity difference of about 2.5. So a magnitude 1 star, for example, is about 2.5 times as bright as a magnitude 2, and so on. The dimmest stars visible to the naked eye, under good observing conditions, are of about magnitude 6.

[5] Herschel wanted to observe the angular separation of double stars over time to see if he could detect the effect of parallax caused by the earth's annual orbit of the sun. He reasoned that if one star is very close to us, and the other further away, he should be able to detect the apparent movement of the nearer star compared to the further one due to parallax.

although he explained this as being due to observational uncertainties. As it turned out, however, it was his diameter measurements that were in error, any gradual increase being due either to observational error or wishful thinking.

Shortly after his initial observations, Herschel had written to the Astronomer Royal, Nevil Maskelyne, via Dr Watson, notifying him of his discovery. Maskelyne wrote to Watson on 4th April, after observing for three nights, that 'it is a comet or a new planet, but very different from any comet I ever read any description of or saw.' By the time he replied to Herschel on 23rd April, Maskelyne was still unsure whether it was a planet or comet, but commented that he had 'not yet seen any coma or tail to it'.

The first orbits were deduced separately by Pierre Méchain, a French mathematician and comet hunter, and Anders Lexell, a Finnish[6] professor at St Petersburg, who happened to be in England at the time. Both orbits were calculated on the assumption that the object was a comet, but it became evident a little later that Uranus' orbit was essentially circular, and so Uranus was a planet. It is not clear who was the first to realise this, but Jean Baptiste de Saron, Pierre Simon de Laplace, and Anders Lexell all seem to have come to the same conclusion independently in May 1781.

It is difficult to exaggerate the effect that the discovery of the first planet since ancient times had on the astronomical community of that period. At a stroke Herschel had doubled the diameter of the known solar system, and encouraged those astronomers like Lexell who thought that there may be yet more planets. Lexell estimated the radius of Uranus' orbit to be 18.93 AU, which was within 1.4% of its true value of 19.19 AU, and Lalande and Méchain estimated its orbital inclination to the ecliptic as only 46', which is exactly correct. Then in 1788 Herschel estimated the diameter of Uranus as about 34,000 miles (54,700 km), which is not far from its correct value of about 31,400 miles (50,500 km).[7]

One of the first questions was what to call this new planet? Although names had had to be found for the moons of Jupiter and Saturn, a new planet was quite a different matter and there was no precedent. Herschel initially ignored a request from Sir Joseph Banks, president of the Royal Society, to suggest a name. Lalande generously suggested calling it *Herschel*, whilst Johann Bode, editor of the *Berliner Astronomisches Jahrbuch* proposed *Uranus*, as Uranus was the father of Saturn, who in turn was the father of Jupiter. Eventually, in mid-1782 Herschel suggested calling it

---

[6] At that time Finland was part of Sweden, so he is sometimes referred to as Swedish.

[7] Uranus is now known to be slightly flattened in shape. This is its average diameter.

*Georgium Sidus* (George's Star), after George III who had just offered Herschel an annual stipend so that he could continue his astronomical work full time.[8] Naturally the idea of naming the planet after a British king did not go down well abroad. In fact the new planet was called *Uranus* in Germany, and first *Herschel* and then *Uranus* in France, but it was not until the 1820s that its name was generally changed from the *Georgian Planet*[9] to *Uranus* in Britain.[10]

After the discovery, searches were made to see if Uranus had been observed before, and much to everyone's surprise it was soon found that it had been seen, and its position measured, no fewer than eleven times before Herschel's discovery. But its character as a planet had never been realised. The earliest recorded observation had been made by John Flamsteed, the first Astronomer Royal, in 1690. Tobias Mayer had seen it once in 1756, and Pierre Le Monnier (1715–1799) had observed it no less than nine times in all between 1764 and 1769, four of these on consecutive nights in January 1769. Unfortunately for Le Monnier, in January 1769 the planet was near the stationary point in its orbit, as seen from earth, and so its movement had been slight and hence he had missed it.[11]

---

[8] How much calling the planet after King George was because of Herschel's gratitude to the king for providing him with financial support, is unclear, but it was definitely part of the reason. As Herschel explained '...the name of *Georgium Sidus* presents itself to me, as an appellation which will conveniently convey the information of the time and country where and when it was brought to view. But as a subject of the best of Kings, who is the liberal protector of every art and science;– as a native of the country from whence this Illustrious Family was called to the British throne;...and, last of all, as a person now more immediately under the protection of this excellent Monarch, and owing everything to His unlimited bounty...'

[9] The name of *George's Star* or the *Georgian Star* did not catch on in Britain, as it seemed perverse to call a planet a star, although such had been the practice in ancient times. So, at first, it was generally called the *Georgian Planet* in Britain.

[10] The Nautical Almanac did not change the name from the *Georgian Planet* to *Uranus* until the 1850s, however, following a letter from John Couch Adams, the 'co-discoverer' of Neptune in 1846.

[11] Eventually it was found that Flamsteed or his assistant had observed Uranus another six times between 1712 and 1715, James Bradley had seen it three times between 1748 and 1753, and Le Monnier had observed it another three times, twice in 1750 and once in 1771. (The details of these pre-discovery observations are given in *The Planet Uranus* by Alexander, Faber and Faber, 1965, as modified by Rawlins in *Publications of the Astronomical Society of the Pacific*, **80**, 1968, p. 217–19, and by Rawlins in *Astronomy* magazine, Sept. 1981.)

On 2nd August 1782, following the award of his annual stipend from the king, William and Caroline Herschel left Bath and their musical friends to take up residence in Datchet, near Windsor, to become full time astronomers. This move was to satisfy a condition of his stipend, that he should live close enough to Windsor Castle to show the king and his family the heavens from time to time. Unfortunately, the land surrounding his property at Datchet suffered from flooding. So after suffering a severe attack of ague, brought on by the damp, the Herschels moved to Slough, near Windsor, where William resumed his astronomical observations on 3rd April 1786.

Ever since it had been established that Uranus was a planet, William Herschel had observed Uranus from time to time, hoping to discover satellites revolving around it. Initially he was unsuccessful. Then on 12th January 1787, using his 20 foot (6 m) telescope,[12] he noticed that two star-like objects, that he had observed near Uranus on the previous evening, were missing. He thought that they were two satellites but, to be sure, he observed the area near Uranus for a number of nights in January and early February to try to detect their movement around the planet, and so to prove their identity. By 5th February he was sure that he had found one satellite, and possibly a second one closer to the planet. He observed the outer satellite for about nine hours on the night of 7th/8th February to be sure that it was moving relative to the planet. Then, having satisfied himself that it was, he proceeded to concentrate on the possible inner satellite, which he found more difficult to observe, as it was often lost in the glare of the planet. On 9th February Herschel noticed that the inner satellite had apparently moved compared with its position of two days earlier, but he was unable to detect any clear movement over the $3^{1}/_{4}$ hours of his observations on the 9th. Finally, on the night of 10th/11th February he was able to detect the movement of this satellite over a period of five hours, thus confirming that it was also a satellite of Uranus.

In his brief discovery paper of February 1787, Herschel concluded that the inner satellite, which is now called Titania, has a synodic period of about $8^{3}/_{3}$ days, and the outer satellite, now called Oberon,[13] has a period of about $13^{1}/_{2}$ days. Importantly, he also noted that their orbits make a considerable

---

[12] Herschel had just started using this telescope in a so-called 'front view' mode, in which he looked straight down the tube from the open end, without the angled flat mirror of the Newtonian design. He found that this improved the intensity of the image, although there was some loss in resolution.

[13] The names Titania and Oberon, taken from Shakespeare's *A Midsummer Night's Dream*, were suggested many years later by William Herschel's son John.

angle to the ecliptic. By 1788 Herschel had determined their synodic periods as 8d 17h 1m 19.3s and 13d 11h 5m 1.5s, which, although they did not have the accuracy implied by quoting their periods to tenths of a second, proved to be within only a few minutes of their true values.

Herschel soon realised that the orbits of both satellites, although almost circular, were inclined at about 90° to the ecliptic. However, in 1787 he was unable to deduce when the satellites were on the far side of Uranus and when they were on the near side, as seen from earth, as the orientation of their orbits was such that neither satellite was eclipsed by the planet at that time. As a result, he produced two alternative dates when eclipses should be seen, either in 1799 or 1818, depending on whether the inclination of the orbits was more or less than 90°, respectively. In the event, the first eclipses occurred in early 1798, allowing Herschel to announce that the angle of the satellites' orbits to the ecliptic was greater than 90°. As he pointed out at the time, this meant that these were the first satellites of the solar system known to have retrograde orbits, with an inclination to the orbit of Uranus[14] which he later calculated to be about 101° 2', compared with the currently accepted value of 97° 52'.

Herschel realised that it was now possible, with the discovery of Titania and Oberon, to 'weigh' Uranus using Newton's laws (see Section 4.7). To get the mass of Uranus, in terms of the earth's mass, he used the relationship:

$$\frac{M}{m} = \frac{D^3}{d^3} \times \frac{t^2}{T^2},$$

where $M$ is the mass of Uranus, $m$ that of the earth, $D$ and $T$ are the diameter and period of Oberon's orbit around Uranus, and $d$ and $t$ are the diameter and the period of the moon's orbit around the earth. Oberon was chosen, rather than Titania, as it was thought that Oberon's orbit was known to a higher accuracy.

This relationship gave Herschel a value of 17.740612 for the relative mass of Uranus, which he produced, following the practice of the time, by quoting results to a larger number of significant figures than warranted by the accuracy of his measurements. In fact, the relative mass of Uranus to that of the earth is now known to be 14.531. Herschel's relative mass for Uranus implied a value of 18,567 for the *reciprocal* mass of Uranus in terms of the mass of the sun, which compares with the true value of 22,869 known

---

[14] For simplicity, Herschel assumed that the orbit of Uranus was in the plane of the ecliptic. In fact the angle between the two planes is very small, being just 46', as mentioned earlier.

today. Although Herschel's value for the relative mass of Uranus was some-what inaccurate, nevertheless, he had shown that Uranus is intermediate in mass between Jupiter and Saturn, on the one hand, and the earth on the other.[15]

In his 1798 paper on Uranus, Herschel concluded that the planet was not spherical but flattened at the pole, concluding that it 'has a rotation on its axis of a considerable degree of velocity'. In other words, Uranus must be rotating fast for it to produce the observed polar flattening due to centrifugal force. Herschel says that he first suspected Uranus' polar flattening in 1783, and that he had observed in 1792, and again in 1794, that the planet appeared to be 'a little lengthened out in the direction of the satellites'.

At the time of Herschel's discovery of Titania and Oberon in 1787, Jupiter was known to have four satellites and Saturn five. Then in 1789 Herschel discovered two more satellites of Saturn (see Section 6.2, below). This con-vinced him that Uranus, which he also considered to be a substantial planet like Jupiter and Saturn, must have more than two satellites. As a result, he continued his search for more Uranian satellites, after his discovery of Titania and Oberon, eventually concluding in 1798 that he had found four more. But this turned out not to be so, although it is thought that one of the four observations of his supposed 'interior satellite' may have been an early observation of Ariel, and some other observations of one of his four 'satellites' may be of Umbriel, but neither attribution is certain.[16]

## 6.2 Saturn

Jean Dominique Cassini had suggested in 1705 that the rings of Saturn may consist of swarms of small satellites. Ten years later his son Jacques came to the same conclusion, and yet in the eighteenth century the vast majority of astronomers still considered them as solid. The problem was that neither theory could be proved at the time. Then in 1785 Laplace published his

---

[15] Using Herschel's values for the relative mass of Uranus, and Newton's values for the relative masses of Jupiter and Saturn (see Section 4.7), we obtain for the masses of Jupiter and Saturn to that of Uranus as about 17.4 and 6.1, respectively, whilst the mass of Uranus to that of the earth is about 17.7. The correct values are now known to be 21.8, 6.5 and 14.5, respectively.

[16] Unfortunately, Herschel was observing at the limit of his telescope. In fact, only two other astronomers of the period, namely Johann Schröter (1745–1816) and Karl Harding (1765–1834), had even managed to see Titania and Oberon.

memoir *Théorie des attractions des sphéroids et de la figure des planètes* in which he analysed the stability of Saturn's rings mathematically.

At that time it was unclear as to whether the Cassini division was a clear division between two rings, or whether it was a dark marking on one ring. Either way Laplace showed that the one or two ring system could not be stable if the ring or rings were solid, as the ring(s) had to rotate around Saturn to retain their stability, and the particles nearer to Saturn would normally have a higher angular velocity than those further away (if the rings were not rigid), and this tendency would cause rigid rings to fragment. As a result, Laplace concluded that Saturn has two rings which must each consist of many narrower, concentric rings, separated radially by gaps that are similar in nature to the Cassini division, but much narrower. He further showed that each narrow ring could not be uniform, if they were solid, as the perturbing forces of Saturn's satellites, Jupiter and the sun would cause uniform rings to become unstable. This meant that each solid, narrow ring had to have density and thickness variations that were just enough for them to retain their stability. Although this seemed highly unlikely, Laplace's theory of Saturn's rings lead the field in the late eighteenth and early nineteenth centuries until Maxwell produced his theory in 1857 of particulate rings like those envisaged by Cassini.

In the meantime, it was obviously important to try to see observationally whether Saturn's two rings were uniform, or whether they consisted of many narrower rings. But, first of all, it was important to find out if the Cassini division was a real division in the ring, or whether it was just a dark marking on its surface.

Herschel had started to observe Saturn in early 1774, just as the earth was passing through the ring plane. By 1780, when the ring was wide open, he was able to see the Cassini division on the northern face of the ring, apparently separating it into two. The inner and outer edges of the division appeared to be two concentric circles, as the division gradually decreased in breadth from the ansae, to be a minimum when it was directly in front of the planet.

In 1789 the earth was due to pass through the ring plane once more. So Herschel eagerly awaited a view of the southern face of the ring after this date, to see if the Cassini division was the same width, and at the same distance from the centre of Saturn, as it was on the northern face. By 1791 the ring was sufficiently tilted towards the earth for Herschel to conclude that this was so and, as the division also appeared to be as dark as the background sky, he concluded that the division was a true gap dividing the one ring into two.

From his observations, Herschel calculated the width of the Cassini division to be 4,050 km,[17] which compares with a modern value of 4,700 km. This is an underestimate of just 14%, assuming that the division has not changed in size since his time. Similarly his estimate of the width of the B ring was about 11% too large, but his estimate of the width of the A ring appears to be about 30% too low. However, the total width of the ring system, from the outside of the A ring to the inside of the B ring was only about 6% lower than with the current measurement. How much of the difference between Herschel's estimates, and those of today, is due to inaccuracies in his measurements, and how much is due to real changes in dimensions, is impossible to say. It seems improbable, however, that so diligent an observer as Herschel could have made an error as large as 30% in the width of the A ring.

Herschel now reviewed the observational evidence for possible subdivisions of the A and B rings into many rings. Herschel himself had never seen such subdivisions, although he had recorded a possible thin division of the B ring near to its inner edge. However, he had seen it on only four occasions in June 1780 and, even then, it was only evident on the B ring on one side of the planet. The only person he could find who had apparently observed that the two rings were divided into multiple, narrow rings was the London optician and instrument-maker James Short (1710–1768). Short had told Lalande of his observations, but he does not seem to have recorded them, so the observational evidence backing up Laplace's theory seemed to be sparse.

Although Herschel could not find any convincing evidence of the multiple, narrow rings predicted by Laplace, he did find evidence that Saturn's rings rotated around the planet. He did this by measuring the rotation period of a bright spot that he thought was on the outer edge of the A ring. The resulting period of 10 h 32 min 15.4 s has since caused something of a problem, however, as we now know that that is the rotation period of the outer edge of the B ring. So either the spot was associated with the B ring, or Herschel made an error in measuring its rotation period, or, as P.H. Hepburn[18] suggested in 1926, the spot may have been an optical illusion.[19]

On 11th November 1793 Herschel observed a narrow, very dark band on the planet, just above the inner edge of B ring. Although Herschel believed that this was the shadow of the rings on the planet, the band was too broad

---

[17] This and the subsequent measurements in kilometres are derived from his estimates that were in miles, of course.

[18] P.H. Hepburn was the director of the Saturn section of the British Astronomical Association.

[19] Such an optical illusion has since been observed by other astronomers.

for this, assuming that his drawing is accurate. It is now generally thought that this was an early observation of the so-called crepe, crape, or C ring, that was discovered over 50 years later by W.C. and G.P. Bond and C.W. Tuttle in the USA and independently by W.R. Dawes in England (see Section 8.2). It has even been suggested that Campani in 1664 or Hooke in 1666 were the first to observe this crepe ring, but the evidence for this is substantially flimsier than for the case of Herschel.

Tethys and Dione, the fourth and fifth of Saturn's satellites to be discovered, had been first seen by Cassini in 1684 (see Section 4.1), but no more satellites of Saturn had been found for over one hundred years. Then on 19th August 1787 Herschel saw what he took to be a sixth satellite with his 20 foot (6 m) telescope. He was very busy at the time, and was unable to make follow-up, confirmatory observations. But on 28th August 1789 he observed Saturn with his new 40 foot (12 m) reflector, and saw what he took to be the sixth satellite again, in line with the rings and the four inner satellites of Tethys, Dione, Rhea and Titan. The rings were virtually edge on at that time, and so were less bright than normal, making it easier to see dim satellites close to the planet. Fortunately, at that time Saturn was moving quite rapidly against the background stars, and so it only took Herschel $2\frac{1}{2}$ hours to be sure that the new satellite, now called Enceladus, was moving with Saturn. Then on 8th and 14th September Herschel detected an even dimmer object, which he confirmed as Saturn's seventh known satellite, now called Mimas, on 17th September.

Cassini had earlier concluded that Iapetus had a dark and a light hemisphere (see Section 4.1), and that its axial rotation period was synchronised with its orbital period, but he had later abandoned this idea of synchronous rotation. Herschel now observed Iapetus and found, like Cassini, that it was always far brighter around western than eastern elongation. In fact Herschel observed Iapetus for over ten orbits and found that the intensity change was periodic, with a period equal to its orbital period, as far as he could determine. As his data was consistent with Cassini's observations of about one hundred years earlier, Herschel was convinced that Iapetus was exhibiting synchronous rotation which has, in fact, turned out to be the case.

Saturn is almost twice as far away from us as Jupiter, and Saturn's surface features are much more bland, so it is difficult to determine its rotation rate. The first clear indication of Saturn's rapid rotation was shown by its polar flattening, which Herschel measured in 1789 as $\frac{22.81''}{20.61''} = 1.107$.[20]

---

[20] Although this estimate of polar flattening is almost exactly correct, Herschel's measurements of both diameters of Saturn were somewhat on the high side.

This is considerable, being even larger than that of Jupiter (see Section 4.7). The orientation of the faint belts on Saturn, of its rings, and of the orbital planes of most of its satellites are almost identical. This had already indicated to Herschel that Saturn rotates on an axis perpendicular to its ring plane, and his observations of the orientation of the oblate planet confirmed this.

Encouraged by his measurement of the polar flattening, Herschel tried to estimate Saturn's speed of rotation by making observations between November 1793 and January 1794. The problem was that there were no clear spots on Saturn, and its bands were almost, but not quite, uniform in shape and density, so it was difficult to detect any movement. Nevertheless, he found pairs of observations that showed an apparently identical planet, and noted the difference in times between these pairs of observations. He then assumed that Saturn's period of rotation was, to a first approximation, similar to the 10 h 32 min period of the ring. On this basis Herschel deduced an average period of 10 h 16 min 0.4 s, with a possible error of ±2 min, for Saturn itself.[21] This first estimate of the rotation period of Saturn is remarkably accurate, considering the difficulties involved. Today the equatorial rotation rate of the visible surface is generally accepted to be 10 h 13 min 59 s, which is only about 2 minutes less than Herschel's estimate.

## 6.3 Origin of the solar system

The eighteenth century saw a flurry of activity amongst a small number of philosophers who tried to explain the origin of the universe in general, and the origin of the solar system in particular. Isaac Newton, Richard Bentley, David Gregory and Thomas Wright all tried, independently, to understand the structure of the universe as a whole, and its likely evolution under the action of gravity. But it was the young German philosopher Immanuel Kant (1724–1804) who focused on the problem of the origin of the solar system. In his *Universal Natural History and Theory of the Heavens*, published in 1755, Kant suggested that the solar system had condensed out of a nebulous mass of gas, which had developed into a flat rotating disc as it contracted. This disc rotated faster and faster as it continued to contract, and, eventually, it started throwing off masses of gas as it became unstable. These masses cooled to form the planets, and the core of the disc condensed to form the sun. A little later, the French physicist Georges-Louis Leclerc, Compte de Buffon

---

[21] A modern astronomer would not quote a result to within the first place of decimals in seconds, whilst quoting a possible error of ±2 min.

(1707–1788) proposed an alternative scheme in which matter was dragged out of the sun by a passing comet. Later this matter condensed to form the planets.

Although Laplace was aware of Buffon's idea, he was probably unaware of Kant's theory when he wrote his *Exposition du Système du Monde*, which was published in 1796. In this Laplace noted that the solar system had a number of characteristics which seemed unlikely to have occurred accidentally, and which thus indicated a common origin. They were:

- The seven planets and fourteen satellites known at that time orbit their primaries in the same direction.[22]
- The sun and those planets with a known spin direction all spin on their axes in the same direction.
- The orbital planes of the planets and their satellites are approximately co-planar,[23] and the equatorial planes of the sun and planets, in so far as they were known, make small angles to this general plane.
- The eccentricities of the planetary and satellite orbits are generally small.

Laplace's theory of the origin of the solar system, which attempted to explain these characteristics, was similar to Kant's theory in outline. However, Laplace's theory was more detailed, and it did not contravene the law of conservation of momentum which Kant's theory did, as Kant had assumed that his nebula only started rotating when it began to contract, whereas Laplace's nebula was already rotating when its started to cool and contract. Laplace also pointed out that Buffon's theory was deficient in a number of ways. In particular it failed to explain the low orbital eccentricity of the planets, and failed completely to explain the formation of the planetary satellites.

In Laplace's theory, the spinning, primeval nebula started to cool and, when its angular velocity became too fast for the nebula to remain stable, the matter near the outer edge separated to form a ring, which eventually coalesced to form the outermost planet, Uranus. This process continued, with the nebula throwing off ring after ring as it contracted and speeded up, thus producing rings that eventually produced the other planets. Laplace suggested that the planetary satellites formed in an analogous way, from condensing rings of material around each of the protoplanets that were thrown off as they contracted and speeded up. Saturn's rings were the remains of the

---

[22] This is not true for the satellites of Uranus, but their orbits were still unclear when Laplace wrote his book.

[23] Again, with the exception of the satellites of Uranus (see previous footnote).

original Saturn nebula that did not condense to form a satellite as it was too close to the planet.

Laplace expanded his theory in later editions of his book, encouraged by William Herschel's detailed observations of distant nebulae, that Herschel thought were stars at various stages in the process of formation. Some of the nebulae observed by Herschel were almost structureless masses of very tenuous material, whereas others showed clear signs of a central condensation, and yet others looked like stars surrounded by a very tenuous envelope. Laplace seized on these observations and suggested that the Pleiades star cluster, in particular, was an excellent example of star formation from a nebula.

## 6.4 The first asteroids

In the preface to his *Mysterium Cosmographicum* of 1596, Kepler mentioned that he thought that there may be a planet between Mercury and Venus, and another between Mars and Jupiter, to fill in what he saw as gaps in the solar system. He came to this conclusion when he was trying to find a pattern in the distances of the planets from the sun, by looking at the ratios of the distances of adjacent planets. Eventually he gave up this purely arithmetic approach for his geometrical approach, which is described in Section 3.1 above and which required no such extra planets.

Although later astronomers were not concerned with the gap between the orbits of Mercury and Venus, as that seemed relatively small (see Figure 6.1), that between Mars and Jupiter seemed exceptionally large. Interestingly, Newton was not concerned with this latter gap, as he thought that the Creator had deliberately placed the largest planets Jupiter and Saturn some distance away from the other planets as, if they were any closer, their gravitational attraction would disturb the whole planetary system.

Galileo's discovery of the four large satellites of Jupiter, and the subsequent discovery of satellites of Saturn, had shown astronomers that there were more bodies in the solar system than known to the ancients, so a few astronomers reasoned that maybe there was a planet between Mars and Jupiter that had so far gone unnoticed. Certainly the Scottish mathematician Colin Maclaurin thought so in the early eighteenth century. On the other hand, others, such as Thomas Wright of Durham (1711–1786) and Johann Heinrich Lambert (1728–1777) speculated that there may once have been a planet between Mars and Jupiter, but that it had been broken up by a collision with a comet, according to Wright, or had left the solar system,

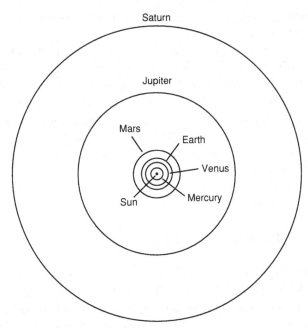

**Figure 6.1** This drawing, based on one published in 1702 by David Gregory, shows the orbits of the planets known at that time as idealised circles. The gaps between the orbits gradually increases, going outwards from the sun, but that between Mars and Jupiter appears to be too large for the sequence.

according to Lambert. Lambert's suggestion was contained in the first of his *Cosmologische Briefe* of 1761, whereas Wright's was in an unpublished letter. So Wright's suggestion could not have been known to Olbers who came up with the same idea in 1802, after the discovery of the asteroids Ceres and Pallas (see below).

In 1766 Johann Daniel Titius of Wittenberg (1729–1796) produced a German translation of the book *Contemplation de la Nature*, that had been written by Charles Bonnet two years earlier. In his first edition, Titius actually modified Bonnet's text to point out that the relative distances of the planets from the sun are almost consistent with the following series.[24] For Mercury a distance of 4 units, for Venus $4 + 3$ units, the earth $4 + 3 \times 2$, Mars $4 + 3 \times 2^2$, Jupiter $4 + 3 \times 2^4$, and Saturn $4 + 3 \times 2^5$. This series gives the distance values of 4, 7, 10, 16, 52, and 100 units compared with the actual average distances of the planets from the sun of 4, 7, 10, 15, 52 and 95 units, to the nearest whole number, with 10 units equivalent to 1 astronomical unit (AU). Now Titius pointed out that there was no 'planet or satellite' corresponding to a distance of $4 + 3 \times 2^3 = 28$ units, and suggested that 'this space belongs

---

[24] In the second edition Titius more correctly put the distance relationship that he had deduced into a footnote.

without doubt to the as yet undiscovered satellites of Mars'. This was clearly nonsense, as any satellite of Mars would have an average distance from the sun the same as that of Mars itself. Nevertheless, Johann Elert Bode (1747–1826) found the mathematical relationship interesting, and mentioned it in a footnote to the second edition of his book *Anleitung zur Kenntniss des gestirnten Himmels*, published in 1772, convinced that there was an undiscovered planet at 28 units or 2.8 AU in the gap between Mars and Jupiter. Then in 1781 Uranus was discovered at a distance of about 18.9 AU, which was very close to the expected value of $4 + 3 \times 2^6 = 196$ units or 19.6 AU for the next planet after Saturn, according to what is now called the Titius–Bode series or 'law'.[25]

The Hungarian Baron Franz Xaver von Zach (1754–1832), court astronomer to Duke Ernst of Saxe-Gotha, was so impressed by the success of the Titius–Bode series in anticipating the distance of Uranus that in 1787 he began a search for the missing planet between Mars and Jupiter. He soon realised, however, that the search was too much for one person and, in the following year, he convened a conference at Gotha to discuss the matter, which both Bode and Lalande attended. Lalande suggested that the search should be a cooperative effort, with parts of the sky allocated to different astronomers, but nothing came of the idea until in 1800 it was resurrected. In September 1800, whilst Europe was at war, Harding, Olbers, von Ende and Gildemeister assembled with von Zach and Schröter at the latter's observatory at Lilienthal, and agreed to found the Vereinigte Astronomische Gesellschaft (the United Astronomical Society). This group, that came to be called the 'Celestial Police', agreed to undertake a systematic search for the missing planet, and invited William Herschel, Giuseppe Piazzi, Bode and others to participate. But before this coordinated search could get underway chance intervened.

On 1st January 1801 Giuseppe Piazzi (1746–1826), the director of the Palermo Observatory in Sicily, was in the process of observing stars in the constellation of Taurus for his new star catalogue, when he observed what he thought at the time was a new eighth magnitude star. On the following night, however, he found that the object had moved by about 5', and on 3rd and 4th January he found, to his satisfaction, that it had moved by a similar amount. At this stage he thought that it was a comet, although it had no nebulosity or tail. The next few days were cloudy, however, but he was able to observe it again on 10th and 11th January. In all Piazzi observed the object

---

[25] The true average distance of Uranus was eventually corrected to 19.2 AU, which was even closer to the 19.6 AU expected.

on a total of twenty-four nights between 1st January and 11th February. After that, poor weather and a severe illness brought Piazzi's observations to a halt. The object then passed too close to the sun in the sky to be seen until late summer.

Piazzi announced his discovery on 24th January in separate letters to Barnaba Oriani of Milan, Bode and Lalande, confiding to his friend Oriani, that he thought that it might be a planet. However, in his letters to the other two astronomers he was more cautious, describing the new object as a comet, although he made it clear that it had no nebulosity or tail. Piazzi was determined to secure his priority of discovery,[26] so in his initial announcement he released only two of his observations, namely those made on 1st and 23rd January.[27] When Piazzi was no longer able to continue his observations after 11th February, he tried to calculate its orbit. At first he tried to fit a parabola, in case it was a comet, but found that a parabola did not fit. Then he tried circular orbits with more success, although he did not have enough results to produce a definitive orbit. By April he had done as much as he could in analysing the observations so he released his complete set to Oriani, Bode and Lalande.

Throughout these early months, Piazzi had a nagging feeling that the object may be a comet, rather than a new planet, as after 23rd January he thought that it was progressively reducing in size at such a rate that it must be a comet. This suspicion was reinforced when no-one, not even William Herschel, was able to observe it in July, when it should have moved sufficiently far away from the sun's glare to be observable once more.

It is at this stage that the brilliant young German mathematician Carl Friedrich Gauss (1777–1855) enters the story. Gauss had previously set himself the task of calculating celestial orbits using data spread over a very limited time period, and this new object was a perfect case on which to try his analytical methods, as it had only been observed over a geocentric arc of about 3°. Gauss produced his orbit in November, and on 31st December his orbit was accurate enough to allow von Zach at Gotha to recover the object about 1/2° from its predicted position.[28] On the following evening it was

---

[26] He was assisted in this by the time that his letters took to reach their destination during this troubled period in Europe's history. Bode's did not arrive until 20th March, and Oriani's until 5th April.

[27] Piazzi observed that on 14th January the object had changed from retrograde to direct motion.

[28] Von Zach thought that he had caught a glimpse of the object on 7th December, but was prevented by bad weather from following this up.

independently observed by Olbers at Bremen. The new object was clearly a planet at a distance from the sun of 2.767 AU, according to Gauss, which was almost exactly the 2.80 AU predicted by the Titius–Bode series. Piazzi proposed naming it Ceres Ferdinandea after Ceres, the patron goddess of Sicily, and Ferdinand IV, Piazzi's royal patron, although the name was soon shortened to Ceres. Its orbital inclination was found to be quite high, being about 10°, compared with 7° for Mercury, which had the most inclined orbit of all the known planets.

The solar system now seemed complete, at least as far out as Uranus, and the Titius–Bode series seemed to have been thoroughly vindicated, following the discoveries of Uranus and Ceres. But soon doubts began to set in. As early as mid February 1802, only six weeks after its recovery, Herschel announced to the Royal Society that the diameter of Ceres was less than five-eighths the diameter of the moon, making its diameter less than about 2,200 km. Then on 28th March, Heinrich Wilhelm Olbers (1758–1840) found another new object which moved over the course of just a few hours. A month later Gauss had calculated its mean distance from the sun as 2.670 AU, which was very similar to that of Ceres. However, the orbital inclination of the new object, which Olbers named Pallas, was a massive 34°. This was more reminiscent of comets than planets, but like Ceres it had no nebulosity or tail.

Herschel carried on trying to measure the diameters of Ceres and Pallas, which were clearly very small. As a result, in May 1802 he estimated the diameter of Ceres to be about 162 miles (261 km) and of Pallas to be about 147 miles (237 km).[29] Obviously Ceres was much smaller than he had previously thought, and Ceres and Pallas were nothing like the size of the previously known planets. But they were not comets, as they had no tail and did not have the parabolic or near parabolic orbits of comets. So they were like very small planets, and Herschel suggested calling these objects 'asteroids', to distinguish them from the planets themselves. This name is still used, although they are sometimes referred to as 'minor planets'.

It seemed to Olbers more than a coincidence that these two asteroids should be at about the same distance from the sun. So he suggested that they were really fragments of a full-sized planet that had exploded at some time in the past, either by the action of internal forces, or due the collision

---

[29] Modern diameter estimates are substantially larger than Herschel's figures, being about 1,000 km for Ceres and about 500 km for Pallas, but these are still very much less than the diameter of the moon (3,476 km).

of a comet. In making this suggestion he was thus unknowingly resurrecting Thomas Wright's unpublished theory of the previous century. Such an explosion could explain the high inclinations and relatively high eccentricities of the orbits of Ceres and Pallas. Olbers then went on to suggest that, if the original planet had exploded, there may be other fragments whose orbits should all intersect at the place where the explosion had taken place, as well as in a place on the opposite side of the sun. As a result he suggested looking for more asteroids in orbits that went through the points of intersection of the orbits of Ceres and Pallas.

On 1st September 1804, Schröter's assistant Karl Harding (1765–1834) discovered the third asteroid, Juno. It was found to have an orbit consistent with Olbers' theory, and so for a time it looked as though his theory was correct. Unfortunately, the orbit of the fourth asteroid Vesta, which was discovered by Olbers himself in 1807, did not appear to be consistent with the explosion theory, leading Olbers to abandon it. However, as Simon Newcomb pointed out some years later, Olbers appears not to have realised that the orbits of the asteroids would have been drastically changed with time by the gravitational attraction of other planets, particularly Jupiter, and so it was hardly surprising that their orbits did not all intersect in the same two places. As a result, Olbers had rejected his theory prematurely, although later discoveries were to prove that his theory was not valid.

## 6.5 The discovery of Neptune

Alexis Bouvard (1767–1843), a shepherd boy from Chamonix in the Alps, left home at the age of eighteen and went to Paris to study science. There he attended free lectures and, because of his proven ability in mathematics, became an assistant to Laplace. Under Laplace's tutelage he won a prize for his lunar theory, became a member of the Academy, and in 1808 produced new tables for the future motions of Jupiter and Saturn. In 1821 he modified these tables and added similar ones for Uranus.

Uranus has a period of 84 years and it had only been discovered in 1781, so it had traversed less than half of its orbit when Bouvard was trying to produce his tables. As a result, it was clearly important to have as much data from any pre-discovery observations as possible, in order to produce an accurate orbit. Some pre-discovery observations had already been unearthed at the time, dating back to Flamsteed's observation of 1690, and Bouvard was able to find nine more when he searched through

Le Monnier's fifteen volumes of observations, covering the period from 1736 to 1780.[30]

Bouvard was now able to calculate an orbit for Uranus by supplementing the extensive post-discovery observations with these earlier ones going back to 1690. Unfortunately, when he did this he found that the pre-discovery observations did not appear to be consistent with the newer ones. Believing that the pre-discovery sightings were the most likely to be in error, he decided to calculate an orbit using the post-discovery observations only. Bouvard clearly felt uneasy in doing this, as it implied that some of the earlier observations were in error by up to 65″, which was highly unlikely. Nevertheless, he could think of no alternative. As a result, he said in his 1821 tables that he left it 'to the future to determine whether the difficulty in reconciling the two sets of observations [i.e. before and after discovery] is a result of the inaccuracy of the old observations, or whether it depends on some extraneous and unknown influence which has acted upon the planet'.

Unfortunately, it took only four years for the first evidence to appear, from observations made at Kremsmünster, near Linz, that Bouvard's calculated orbit was in error. By 1828 the discrepancy between observations and his prediction was 12″ longitude and 13″ latitude, according to Slavinsky of the Vilna Observatory, which increased to 23″ long., 14″ lat. by the following year. Similar errors were measured at the Vienna and Cambridge observatories, and by 1845 the longitude discrepancy had reached about 2′, so there was something definitely wrong with Bouvard's orbit.

Various ideas were put forward to solve the problem. Maybe some modification was required to Newton's laws at such a great distance from the sun, or maybe Uranus was being affected by the ether, or its orbit had been changed by a collision with a comet, or the problem was caused by a large satellite of Uranus. There was a general reluctance to believe that there was something wrong with Newton's laws which had, so far, passed every test thrown at them with flying colours. If the ether was at fault, why, it was argued, did it have no effect on any other planet? A collision with a comet could possibly explain why the pre-discovery and post-discovery observations were not consistent, but why were the latest observations not consistent with Bouvard's orbit that had been produced using only the post-discovery observations? Finally, a large satellite of Uranus could produce an oscillation about a mean ellipse, but that was not what was observed. So what was the cause of the problem?

---

[30] Three of Le Monnier's pre-discovery observations had been found previously, so producing a total for him of twelve pre-discovery observations.

As mentioned above, Bouvard had suggested the cause could be some external influence, and in 1834 the Rev. T.J. Hussey of Hayes in Kent was more specific in suggesting that the external influence could be 'some disturbing body beyond Uranus', clearly a reference to a possible trans-Uranian planet. This suggestion was made in a letter to George Biddell Airy (1801–1892), the Plumian Professor of Astronomy at Cambridge, in which Hussey says he had already discussed the idea with Bouvard. Bouvard had replied that the same idea had already occurred to him and to the Danish astronomer Peter Hansen. Bouvard also told Hussey that he would try to calculate the likely position of the disturbing planet, but, in the event, he did not do this. In addition, Airy poured cold water on Hussey's idea of trying to look for the new planet, so Hussey gave up on the idea.

In 1835 J.E.B. (Benjamin) Valz, the director of the Marseilles Observatory, suggested that slight discrepancies in the orbit of Halley's comet were probably caused by an 'invisible planet, located beyond Uranus'.[31] He also suggested that this could be the cause of the discrepancies in Uranus' orbit. In the following year Friedrich Nicolai, the director of the Mannheim Observatory, also made a similar proposal, further suggesting that the new planet would be about 38 AU from the sun, as predicted by the Titius–Bode 'law'. Four years later, the German astronomer and mathematician Friedrich Wilhelm Bessel (1784–1846) mentioned that he and his pupil Friedrich Flemming would calculate the mass and orbit of this unknown planet, but Flemming died shortly after starting the calculations, and Bessel died shortly afterwards. As a result the work never got very far. So the baton now passed to England and a young mathematician called John Couch Adams.

John Couch Adams (1819–1892) was born on 5th June 1819 on a farm near Laneast in Cornwall, where his father was a tenant farmer. The family was of limited means, and John's first school was in a farmhouse at Laneast. Then at the age of eight he was put in the care of a Mr Sleep, a tutor who prided himself in his mathematical ability. It was at this stage that John's exceptional ability at mathematics became obvious. Four years later, he was moved to a school at Devonport but, sadly for John, it was weak in mathematics, so he decided to teach himself the subject from books kept at the Devonport Mechanics Institute. As a result, in 1839 he won a scholarship to St John's College, Cambridge, where he graduated as senior wrangler (i.e. with the highest marks in mathematics) in 1843.

---

[31] The cause of the discrepancy in Halley's comet's trajectory, to which he was referring, was later shown to be mostly due to the effect of evaporation of the comet. The solar wind and radiation pressure also contributed to the discrepancy.

Although John Couch Adams' first love was mathematics, he also showed an interest in astronomy whilst he was an undergraduate at Cambridge. For example, in 1840 he and two friends walked the 60 miles (100 km) to London to, amongst others things, visit the Royal Observatory at Greenwich. Then in the following year he visited the observatory at Cambridge, which had the excellent 11¾ inch (30 cm) Northumberland equatorial refractor. This was to figure later in the discovery of Neptune, in which Adams played a not insignificant part.

On 26th June 1841 Adams discovered, by chance, a copy of Airy's 1832 report on the problem of Uranus' orbit. Such a problem was a natural magnet for the young mathematician, and on 3rd July Adams wrote on a scrap of paper in his notebook, 'Formed a design, in the beginning of this week, of investigating, as soon as possible after taking my degree, the irregularities in the motion of Uranus, wh[ich] are yet unaccounted for; in order to find whether they may be attributed to the action of an undiscovered planet beyond it; and if possible thence to determine the elements of its orbit, etc. approximately, wh[ich] w[oul]d probably lead to its discovery.' This was two years before he took his finals.

Initially Airy thought that the reason for Uranus' inexplicable orbital behaviour was that Newton's theory of gravity did not apply universally,[32] needing some modifications at the distance of Uranus from the sun. He conceded that there may be other explanations, however. In particular, in response to Hussey's 1834 letter, Airy said that the deviation between theory and observation could probably be removed by making small modifications to Bouvard's orbit. In any case, Airy doubted that the cause was a trans-Uranian planet, but even if it was he wrote to Hussey 'I give it as my opinion, without hesitation, that it is not yet in such a state as to give the smallest hope of making out the nature of any external action on the planet [Uranus]...But if it were certain that there were any extraneous action, I doubt much the possibility of determining the place of a planet that produced it. I am sure it could not be done till the nature of the irregularity was well determined from several successive revolutions'.[33] This would take several hundred years.

---

[32] Airy was by no means the only person to doubt the universality of Newton's gravitational theory at that time. Earlier in his career, Bessel had gone one step further, thinking that Newton's concept of gravity was probably oversimplified. In particular, he had suggested that the magnitude of the gravitational force may depend on the composition of the bodies concerned, and not just their mass.

[33] *Monthly Notices of the Royal Astronomical Society, 7*, 1846, p. 124.

In 1835 Airy became Astronomer Royal. Then two years later, in reply to an enquiry from Eugène Bouvard (Alexis' nephew), Airy expressed the view that, if the effect on Uranus was that of an unseen body, 'it will be nearly impossible ever to find out its place.' Later, in a letter to Airy of 21st May 1844, Eugène said that, even after modifying his uncle's orbit, there were still residual errors that he could not explain. In response Airy referred to the 'absolute necessity of seeking some external cause of disturbance', which clearly indicates a change of mind on his part. Part of the reason for this was probably the work of Adams.

Adams started work in 1843 by correcting errors in Alexis Bouvard's tables,[34] and improving the calculation of the perturbations due to the known planets. He then had to make some assumptions about the likely orbit of the unknown planet. So far the Titius–Bode 'law' had held remarkably well for planetary distances up to and including Uranus, so he assumed that the unknown planet would be about 38 AU away from the sun, in accordance with this law for the next planet after Uranus. Adams further assumed that the orbit of the unknown planet was in the ecliptic, which also seemed reasonable, as the orbits of Jupiter, Saturn and Uranus were almost in the ecliptic also.[35] For simplicity he also assumed that the orbit of the unknown planet was circular.

Initially Adams used only the post-discovery observations of Uranus, and by October 1843 he had shown that a planet in an orbit beyond Uranus could explain the discrepancies. The worst fit, using his approximate orbit for the new planet, was for the years 1818–1826. So he asked James Challis (1803–1882), Airy's successor as Plumian Professor of Astronomy, if he could help him to obtain additional data from Greenwich for those years. Challis wrote to Airy in February 1844, explaining the situation and asking for this data. Airy replied by return giving not only the information requested, but enclosing all the Greenwich observations from 1753 to 1830.

By September 1845 Adams had calculated the orbital elements of the new planet, and its expected position in the sky, using both pre- and post-discovery observations. In this calculation he had retained the assumption that the mean distance of the unknown planet from the sun was 38.4 AU, but he had now abandoned his first approximation of a circular orbit in favour of an elliptical orbit. He found that the best fit with observations was for a planet with a reciprocal mass of 6,040, in an orbit with an eccentricity

---

[34] Eugène Bouvard's tables, which were much more accurate than those of his uncle, were never published.

[35] Their orbits made angles of only 1.3°, 2.5° and 0.8° to the ecliptic, respectively.

of 0.1610, and a period of 238 years. He gave his orbital elements and position estimates to Challis, hoping that he would start a search. But Challis was too busy and, instead, suggested that Adams should send his results to the Astronomer Royal. Adams was soon to visit his family in Cornwall and proposed to deliver his results to Airy en route. So on 22nd September Challis wrote a letter of introduction to Airy for Adams.

This is not the place to discuss this and subsequent details of the English side of the story, however, as this detail is outside the scope of this book, and it has been well covered elsewhere.[36] Suffice it to say that, for various reasons, Adams did not actually see Airy, but left a letter giving the orbital elements and the longitude of Neptune for 1st October 1845, so Airy was aware of his results, even though he had not seen Adams face to face. Meanwhile, in France a parallel investigation had just started into the problems of Uranus' orbit by Urbain Le Verrier (or Leverrier). This investigation, which was to achieve dramatic results, was, at this stage, unknown to the English astronomers.

Urbain Jean Joseph Le Verrier (1811–1877) was born in Saint-Lô, Normandy, on 11th March 1811. Although he showed an aptitude for mathematics at his school in Caen, he was turned down by the prestigious École Polytechnique in Paris. Determined to give Urbain every chance to succeed, however, his father then sold the family home in 1830 to pay for him to go to the Collège de Saint Louis in Paris to receive further tuition. As a result, in the following year Urbain was successful on his second application to the École Polytechnique, even winning the mathematics prize.

After graduating top of his year at the Polytechnique, Le Verrier started working under the outstanding chemist Gay-Lussac at the Administration des Tabacs. But in 1837 Le Verrier returned to the École Polytechnique as an astronomy assistant, where his first major piece of work was into the stability of the solar system. Then in 1840 Dominique François Jean Arago (1786–1853), director of the Paris Observatory, suggested that Le Verrier examine the motion of Mercury, as it seemed to be anomalous. This he did, predicting the start of the 1845 transit on 8th May to within 16 seconds of that observed, which was a clear improvement over the accuracy of previous predictions. A month after the transit, however, Arago suggested to Le Verrier that he look into the problem of Uranus' orbit. At this stage, neither Arago nor Le Verrier were aware of Adams' work.

---

[36] See, for example, Morton Grosser, *The Discovery of Neptune*, Harvard University Press, 1962, which is the standard work. Or, for a summary, see Richard Baum and William Sheehan, *In Search of Planet Vulcan*, Plenum, 1997.

Le Verrier's first paper on Uranus was presented to the Paris Academy of Sciences on 10th November 1845. In this he corrected Alexis Bouvard's tables, but he still found it impossible to reconcile the old and modern observations of Uranus. He then went on to calculate possible orbits for a perturbing planet, but found that none of his possible orbits satisfied the 1690 or 1750 observations. Then, after a short break from his calculations, Le Verrier applied himself to the problem with renewed vigour, and found a way out of his earlier impasse.

Le Verrier's work, which was published in the *Comptes rendus* of 1st June 1846, proved that the discrepancies in Uranus' orbit could be explained by the existence of another planet, which would be at a longitude of 325° on 1st January 1847. Astonishingly, no one offered to try to find the planet. Unknown to Le Verrier and the other French astronomers, Le Verrier's predicted position was only about 1° from that previously predicted by Adams.

Airy received a copy of Le Verrier's paper on or about 23rd June, and immediately realised the similarity between the position of the unknown planet as predicted by Le Verrier and that predicted by Adams. In his reply to Le Verrier, Airy congratulated him on his analysis, but omitted to mention Adams' work at all. Likewise Airy failed to tell Challis or Adams of Le Verrier's work, although Challis was to hear of Le Verrier's results at a meeting of the Board of Visitors of the Royal Observatory one week later. Eventually, after the prompting of George Peacock, Airy's old professor at Cambridge, Airy asked Challis to start a search for the planet, which he finally did on 29th July.

Even now problems were not over for the English search. Although Le Verrier's and Adams' position estimates were almost indentical, and although Adams predicted that the new planet would be no fainter than 9th magnitude, and have an observable disc, Challis was not going to take any risks of missing it. So, in accordance with Airy's suggestion, Challis decided to scan an area of sky some 30° by 10° in extent, and record the position of all 3,000 stars in it down to the 11th magnitude. Unfortunately, Challis had no reliable sky map down to the 11th magnitude, so he had no alternative but to scan his chosen area at least twice, to see if any of the stars moved. In fact he planned to scan the area three times. This was a massive task, which he anticipated would take several months using every clear night.

Meanwhile Le Verrier was still working on his calculations, presenting his third and final paper on Uranus to the Paris Academy of Sciences on 31st August 1846. In this he produced a slightly modified position estimate for the new planet, together with full orbital elements and an estimate of its mass. Le Verrier also predicted that the planet would be large enough to show a disc in a large telescope, but again no one offered to undertake a search. Airy had not offered to start a search in England when he had received Le Verrier's earlier paper, so Le Verrier thought that it would be a waste of time to ask him to start a search now. Then Le Verrier remembered that about a year previously he had received a doctoral thesis from a certain Johann Gottfried Galle (1812–1910), who was an assistant at the Berlin Observatory. So on 18th September Le Verrier finally thanked Galle for the copy of his thesis, and enclosed his predicted position for the new planet, suggesting that he might undertake a search.

Le Verrier's letter arrived at Berlin Observatory on 23rd September, and Galle and his assistant Heinrich Louis d'Arrest asked permission from the observatory director Johann Encke (1791–1865) to start the search that very night. With some reluctance, Encke agreed, so Galle and d'Arrest set to work as soon as it was dark. Their approach was very different to that of Challis, however, starting to look for the planet at Le Verrier's predicted position and then working outwards. They also decided to use a star chart to see if they could pick up an uncharted object.

Fortunately for Galle and d'Arrest, there was at the Berlin Observatory an unpublished chart of the search area that had recently been produced by Carl Bremiker, a member of the observatory staff, to assist in the search for asteroids. It was far better than any chart available to Challis. Using this Berlin chart, Galle and d'Arrest found the 8th magnitude planet within an hour of starting the search on their first night of observing! They quickly told Encke, who immediately went to the observatory with Galle and d'Arrest to confirm the new planet's position. It turned out to be less than 1° from that predicted by Le Verrier (see Figure 6.2). However, it was not until the following night that they could be sure that they had discovered the planet, when they found that it had moved. Encke then announced its discovery, crediting Galle and himself, but ignoring the young d'Arrest.

## Post-discovery events

Challis received his copy of Le Verrier's third paper on 29th September, but by then it was too late, as the new planet had already been found. In England the announcement of the discovery of Le Verrier's planet by Galle was made in a letter to *The Times* newspaper, published on 1st October, by John Russell Hind (1823–1895). Hind also mentioned that he had seen the new planet himself on 30th September with a 7 inch (18 cm) telescope at George Bishop's private observatory in Regent's Park in London.

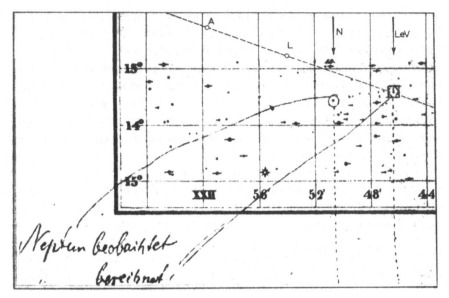

**Figure 6.2** A section of the chart used by Galle and d'Arrest to find Neptune. Le Verrier's calculated position is indicated by the arrow marked 'LeV', Adams' calculated position by 'A', and the observed position of Neptune by the arrow marked 'N'. The dotted line indicates the ecliptic. The axis along the bottom is in hours and minutes of right ascension, where 4′ (minutes) is equivalent to 1°. (From *Vistas in Astronomy*, vol. 3, 1960, p. 44.)

Galle in his letter to Le Verrier of 25th September, notifying him of the Berlin discovery of the planet, suggested calling the new planet Janus. Le Verrier, in his reply of 1st October, said that the Bureau des Longitudes had already named the planet Neptune, which, although not a formal decision, made the point that the planet would be named in France. Le Verrier then had second thoughts about the name Neptune, and persuaded Arago to call it 'Le Verrier', although Arago only agreed to do this provided Uranus was called Herschel, and the asteroid Juno was called Olbers. The Paris Academy of Sciences accepted Arago's decision, but it was emphatically rejected by the international astronomical community, which stuck with the name Neptune.

The discovery of Neptune was followed by an almighty row between members of the English and French scientific establishments as to the priority of its discovery. Although Adams had clearly been the first to predict its position, he had not published anything, and so the English had trouble claiming priority. But Adams, to his credit, did not enter into the argument, giving Le Verrier credit for its discovery, whilst retaining for himself the credit of prior prediction. However, more serious questions were being asked in England of Airy and Challis, as to why they had consistently dragged their feet. The Royal Astronomical Society even carried out an investigation into what appeared to many people to be a scandal. Eventually, the arguments and recriminations died down, but every now and again they resurface. For example, at the centenary celebrations at the Royal Astronomical Society, W.M. Smart heavily criticised Airy's treatment of the young Adams, whilst the then Astronomer Royal, Sir Harold Spencer Jones (1890–1960), defended Airy.

Le Verrier received many honours for his work that lead to the discovery of Neptune, including being awarded the Royal Society's prestigious Copely medal for 1846. Adams received the Copely medal two years later, and was also offered a knighthood, which he refused. In 1861 Adams was appointed to replace Challis as director of the Cambridge Observatory, and in 1887 he was offered the post of Astronomer Royal, which he declined. In addition to Le Verrier and Adams, another beneficiary of the discovery of Neptune was undoubtedly Newtonian mechanics, which had been used to good effect to find the new planet. As far as the general public was concerned, however, and even to some astronomers, it seemed miraculous that the position of an unknown planet could actually be predicted by mathematical calculations. It seemed more like magic than physics.

When Challis had heard of Galle's discovery, he had looked back through his own records, which he had not analysed as he went along, and found that he had actually seen Neptune on 30th July, and again on 4th and 12th August 1846. Using these observations, along with those of the Berlin Observatory, Adams was able to produce the first orbital analysis for the new planet, finding, much to his surprise, that its distance from the sun was only about 30 AU. This was much less the mean distances calculated just before discovery by Le Verrier and Adams of 36.15 and 37.25 AU, respectively, or that of about 38 AU expected according to the Titius–Bode law. So the discovery of Neptune finally resulted in the latter being abandoned.

Challis continued to observe Neptune until January 1847, when it was lost in the sun's glare, allowing an even better estimate of its orbit to be made. Meanwhile, in the United States, Sears Cooke Walker, of the US Coast Survey, who had been planning to search for the new planet when its discovery was announced, now looked for pre-discovery observations. Early in his search, he found that there was an 8th magnitude star recorded in Lalande's *Histoire Céleste Français*, published in 1801, at the position

Table 6.1 *A comparison between the principal parameters of Neptune, as predicted by Le Verrier and Adams, deduced by Walker after its discovery, and modern values*

|  | Le Verrier | Adams | Walker | Current values |
|---|---|---|---|---|
| Mean distance from sun (AU) | 36.15 | 37.25 | 30.25 | 30.06 |
| Orbital eccentricity | 0.1076 | 0.1206 | 0.0088 | 0.0087 |
| Sidereal period (yrs) | 217.4 | 227.3 | 166.4 | 164.8 |
| Reciprocal mass | 9,300 | 6,700 | 15,000 | 19,400 |

expected for Neptune at that time. Walker tried to observe that star in 1847 and found, as he expected, that it was not there. Then F.V. Mauvais consulted Lalande's original manuscript, and noticed that the 'star' had actually been observed by Michel de Lalande, J.J. de Lalande's nephew, on both 8th and 10th May 1795. In the manuscript, the observation of 8th May had been rejected completely, and the validity of the observation for 10th May had been questioned, as the two 'stars' appeared to be in slightly different positions. This clearly confirmed that it was a pre-discovery observation of Neptune.

In 1847 Walker used Michel de Lalande's observations of Neptune to produce its first accurate orbit. He found that, as intimated by Adams the previous year, Neptune was in quite a different orbit from that proposed by either Le Verrier or Adams (see Table 6.1).

Because of the discrepancies between the predicted and actual orbits of Neptune, Benjamin Peirce (1809–1880) of Harvard claimed in 1847 that the planet discovered by Galle was not the planet predicted by Le Verrier, describing its discovery near its calculated position as 'a happy accident'. Jacques Babinet, a member of the Académie des Sciences, went even further in 1848, suggesting that there were, in fact, two perturbing planets, namely Neptune plus another further away from the sun, which he called Hypèrion. But no-one took his theory seriously as its details lacked technical credibility.

Later it transpired that Sir John Herschel (1792–1871), the son of William Herschel, had observed Neptune on 14th July 1830, and John Lamont had seen it on 25th October 1845 and on 7th and 11th September 1846. Amazingly, in 1980 Charles T. Kowal and Stillman Drake found, on looking through Galileo's notebooks, that he had apparently observed Neptune on both 28th December 1612 and 28th January 1613, when it was near to Jupiter (see Figure 6.3).

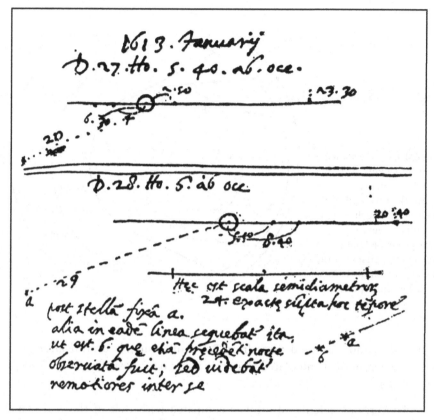

**Figure 6.3** Part of Galileo's notebook for January 1613. Jupiter is shown on 28th January in the centre of this extract, with its satellites along the solid horizontal line. To the bottom left is a dotted line with a star marked 'a' some 29 units from Jupiter. This line is shown continued at the lower right with another star marked 'b' shown. Star 'a' has been identified using a modern atlas, but there is no star 'b' today. Galileo remarked that 'a' and 'b' seemed further apart on the previous night. Star 'b' has since been identified as Neptune. (Courtesy Charles Kowal.)

There are a number of technical questions that arise from the discovery of Neptune:

(i) How can the planet in the orbit predicted by Le Verrier (or Adams) produce essentially the same disturbing effect on Uranus as Neptune does, considering the difference in orbits between Le Verrier's planet and Neptune?

(ii) Why was Alexis Bouvard unable to reconcile the pre-discovery (i.e. pre-1781) observations of Uranus with those post-discovery?

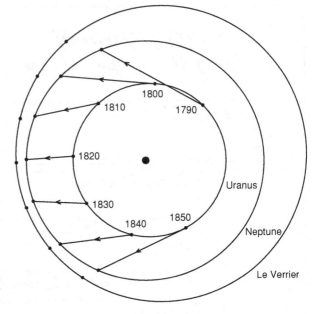

**Figure 6.4** Diagram showing the orbits of Uranus, Neptune and that deduced by Le Verrier for the planet effecting Uranus' orbit. The directions of pull of Neptune and Le Verrier's planet on Uranus are seen to be similar over the crucial period from 1790 to 1850 when Neptune and Uranus were close together.

(iii) Why did Uranus deviate from Bouvard's calculated orbit after 1821, when he had calculated the orbit using only post-discovery observations?

Le Verrier's calculated orbit is compared with that of the real planet Neptune, which was closest to Uranus in 1822, in Figure 6.4. It is clear from this figure that the direction of pull of Le Verrier's planet, and of Neptune, on Uranus were very similar over the crucial period from about 1790 to 1850, when Neptune and Uranus were closest together, and when Neptune's gravitational attraction on Uranus was therefore the largest. Le Verrier had also calculated a higher mass for his hypothetical planet than Neptune proved to have. It so happens that this higher mass and the greater distance between Le Verrier's planet and Uranus, compared with that between Neptune and Uranus shown in the figure, essentially balanced each other out. So the effect of Le Verrier's planet and of Neptune on Uranus were virtually identical over the crucial period from 1790 to 1850. This answers (i) above.

Figure 6.5 shows schematically the motion of Uranus relative to a fixed Neptune, indicated by the letter N. The orbital period of Uranus is about 84 years, and that of Neptune is about double that at 165 years. As a result, after 84 years, when Uranus has orbited the sun once, Neptune is about halfway round its orbit, and their separation is greatest. In fact the synodic period

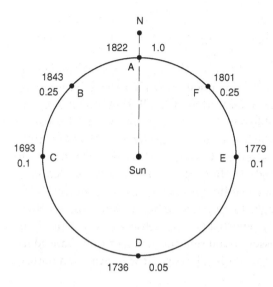

**Figure 6.5** The approximate orbit of Uranus is shown relative to a stationary Neptune at N. The two planets were closest together in 1822 and furthest apart in 1736. Figures adjacent to each location of Uranus indicate the approximate gravitational attraction of Neptune on Uranus (and vice versa) relative to a value of unity at closest approach. (Adapted from W.M. Smart's diagram.)

of Neptune with respect to Uranus is about 172 years, so as the two planets were closest together in 1822, they would have been furthest apart in 1736, at point D in the figure. Likewise, in 1693 and 1779, at points C and E, the heliocentric longitudes of the two planets differed by about 90°. As a result, in 1693 the gravitational attraction of Neptune on Uranus was only about $\frac{1}{10}$ of that in 1822. It decreased to about $\frac{1}{20}$ of the 1822 figure in 1736, before increasing to $\frac{1}{10}$ in 1779, and $\frac{1}{4}$ in 1801. It then went through a maximum in 1822 before reducing again to $\frac{1}{4}$ of this 1822 value in 1843.

The first pre-discovery observation of Uranus was by Flamsteed in 1690, whilst Herschel had discovered Uranus in 1781. During this period from 1690 to 1781 the gravitational attraction of Neptune on Uranus was slight (see Figure 6.5). But between 1781 and 1821, when Alexis Bouvard produced his Uranus tables, the gravitational attraction of Neptune was increasing rapidly, with Neptune causing Uranus to accelerate in its orbit, as can be seen by the direction of the arrows in Figure 6.4. This explains why Bouvard had trouble in trying to reconcile the pre-discovery (i.e. pre-1781) observations with those between 1781 and 1821.

After 1822 the gravitational attraction of Neptune on Uranus gradually decreased, but because Uranus had overtaken Neptune, that attraction now tended to slow Uranus down in its orbit. So before 1822 Neptune had accelerated Uranus in its orbit, but after that date it had slowed it down. This explains why Uranus' orbit that had been produced by Bouvard using the 1781–1821 observations, could not explain Uranus' behaviour after 1822.

In short, Alexis Bouvard was unfortunate in that he had tried to produce an orbit for Uranus in 1821, which, as it turned out, was just one year before the closest approach of Neptune and Uranus.

News of the discovery of Neptune had reached London on 30th September 1846, and on the following day Sir John Herschel wrote to the distinguished amateur astronomer William Lassell (1799–1880), suggesting that he look for any possible satellites of Neptune 'with all possible expedition'. At least if the English had missed discovering Neptune, they could be the first to observe a satellite. Lassell started his observations on 2nd October, and on that very night he thought that he saw a ring of Neptune. He had the same impression on the following night, and on 12th October[37] he wrote of his possible sightings of a ring to *The Times*. John Russell Hind and James Challis subsequently reported possible observations of the ring, but on 15th December 1852 Lassell found that the apparent ring changed its position when he rotated the tube of his telescope. It was clearly an optical illusion.[38]

On 10th October 1846 Lassell observed a 'star' very close to Neptune, which he thought may be a satellite, but a combination of the nearness of Neptune to the sun's glare, and poor weather prevented Lassell from confirming his suspicions until the following July. Subsequent observations allowed Lassell to deduce a period for the satellite, now called Triton,[39] of about 5 days 21 hours, with a greatest elongation of about 18″. Although he was correct in these estimates,[40] he also concluded that Triton, like Iapetus, has one face much darker than the other, but this is not so.

---

[37] His letter was published on 14th October.

[38] Neptune's ring arcs, discovered in 1984, could not possibly have been seen in Lassell's telescope, as it was far too small.

[39] The name of Triton, Neptune's son, was apparently suggested by Flammarion. See *Popular Astronomy*, by Flammarion, translated by Gore, 1907, p. 470.

[40] Triton's period is now known to be 5d 21h 3m.

## Chapter 7 | THE INNER SOLAR SYSTEM IN THE NINETEENTH CENTURY

### 7.1 Vulcan

Urbain Le Verrier had had only limited success in 1845 in refining the theory of Mercury's orbit and predicting the transit of that year (see Section 6.5). However, his subsequent work in explaining the discrepancies in the orbit of Uranus convinced him that Newton's gravitational theory was universally valid. So he felt ready to make another attempt to explain the discrepancies between theory and observation in the case of Mercury.

Before starting work, Le Verrier reanalysed the motions of all the inner planets and of the moon, using the latest observations. As a result, in 1858 he reduced the estimated value of the astronomical unit (i.e. the sun–earth distance) from $153.5 \times 10^6$ kilometres, as estimated by Johann Encke in 1835 (see Section 5.4), to $147.0 \times 10^6$ kilometres.[1] In addition he increased the mass estimate of the earth by 10%, and reduced that of Mars by 10%. Le Verrier then tackled the problem of Mercury's orbit using these modified values.

The most accurate data on the orbit of Mercury available to Le Verrier was that obtained during the transits of the planet across the face of the sun. Such transits can occur only within a few days either side of 10th November or 8th May, as that is when the heliocentric longitudes of the earth and Mercury coincide. During the November transits, Mercury is near perihelion, and the transit limit along the ecliptic, on either side of the ascending

---

[1] In the *Annales de l'Observatoire de Paris*, **iv**, p. 101, Le Verrier deduced a parallax of $8''.95$, which is equivalent to an astronomical unit of $147.0 \times 10^6$ km. There were two small numerical errors in the paper, however, which E.J. Stone of the Greenwich Observatory corrected to give values of $8''.85$ or $148.8 \times 10^6$ km.

node, is about $4° 45'$, whereas that during the May transits is about $2° 40'$. As a result there are about twice as many November as May transits. As Mercury moves faster in its orbit around perihelion, however, the maximum possible duration of a November transit is less than that in May.

Le Verrier used data from nine November transits from 1697 to 1848 and five May transits from 1753 to 1845, and found that the disagreement between the theoretical and observed orbits of Mercury was getting progressively larger with time. The only possible explanation appeared to be that the perihelion of Mercury's orbit was precessing at about $565''$/century. Unfortunately, when Le Verrier calculated the sum of the perturbations caused by the other planets he could only account for a precession rate of $527''$/century.[2] So the perihelion of Mercury's orbit was precessing at a rate of about $38''$/century too much. How could this be?

An increase in the mass of Venus by about 10% could explain the difference, but that would have an effect, which does not exist, on the obliquity of the ecliptic. Le Verrier then thought that the effect could be due to an unknown planet near the orbit of Mercury. First he showed that the new planet would have to have an orbit inside that of Mercury, as otherwise it would have observable effects on the orbits of Venus and the earth, which clearly did not exist. Then he deduced a relationship between the mass of the planet and its orbital radius. He found, for example, that if this new planet had a mass equal to that of Mercury, its orbital radius would be 0.17 AU.[3] But that would take the new planet about $10°$ from the sun at its greatest elongation, when it would have been easily visible to the naked eye. The nearer the new planet was to the sun, the larger and brighter it would have had to be, and although the planet would be smaller if its orbit were further from the sun, it would still have been easily visible at greatest elongation. So a single planet looked an unlikely explanation. As a result, on 12th September 1859 Le Verrier suggested, in a letter to Hervé Faye (1814–1902), the secretary of the Paris Académie des Sciences, that there may be a ring of small planets between the sun and Mercury, of such a size that they had not, so far, been observed.

In his letter to Faye, which was published in the *Comptes Rendus*,[4] Le Verrier suggested that the small intramercurial planets could be detected

---

[2] The largest contributions by far were $281''$ from Mercury's neighbouring planet Venus, and $153''$ from the largest planet Jupiter.

[3] This compares with Mercury's perihelion distance from the sun of 0.31 AU, for example.

[4] *Comptes Rendus*, **59**, 1859, p. 379.

as they transited the sun, and he asked astronomers to carefully study the movement of any spot across the sun to see if it may be such an object. Faye also suggested looking for such objects close to the sun during a total solar eclipse, the next one being visible from Spain on 16th July 1860, possibly using photography. Fast moving spots[5] had been seen on the sun since at least 1762, but their occurrences were rare, the observers were generally unknown amateurs, and some of the observations were clearly spurious. A more scientific observation campaign was evidently called for.

Le Verrier was not the first astronomer to think that there may be a planet (or planets) inside the orbit of Mercury. For example, in 1826 Karl Harding of Göttingen had suggested to Heinrich Schwabe, a Dessau amateur, that regular observing of the sun may reveal an intramercurial planet, but thirty years of observing had yielded nothing.[6] However, observations of fast moving spots on the sun, that may be intramercurial planets, had continued to be announced by various other astronomers, but considering the number of astronomers regularly observing the sun, the number of such observations was small. Nevertheless, if the intramercurial planets were very small, as Le Verrier had suggested, they could easily have generally gone undetected.

Shortly after Le Verrier inferred the possible existence of intramercurial planets, he received a letter, dated 22nd December 1859, announcing the discovery of such a planet. The author was a doctor and amateur astronomer called Edmond Modeste Lescarbault who claimed to have seen a small perfectly circular object crossing the face of the sun on 26th March 1859. Lescarbault, who had been looking for an intramercurial planet from time to time since 1853, was apparently a good observer. He estimated the position of the object on the sun during its transit using an eyepiece with crosswires and a circular scale, that he had made himself, estimating the transit time as 1 h 17 min 9 s.

Lescarbault's letter was delivered by hand to Le Verrier by a local colleague of Lescarbault, a M. Vallée, on 30th December 1859. The respected but intolerant Le Verrier was instantly suspicious of such a discovery by an unknown amateur, and decided then and there to descend on him to establish the truth face to face. In spite of Lescarbault's retiring and deferential manner, however, he convinced Le Verrier, during his visit, of the validity of his observations. Le Verrier then lost no time in presenting them to the

---

[5] i.e. not sunspots.

[6] Although Schwabe did not find an intramercurial planet, in searching for it he discovered the sunspot cycle instead.

Académie des Sciences. This he did on 2nd January, and on 25th January the quiet country doctor received the Legion d'Honneur, in absentia, on Le Verrier's recommendation.

The haughty Le Verrier and the timid Lescarbault were the toast of Paris. This time a planet had been discovered by Le Verrier and another Frenchman, unlike the case of Neptune where German astronomers had been involved. Congratulations poured in from many foreign countries, but a number of people began to wonder how such an object could have been discovered by an unknown amateur astronomer, when there were so many professional astronomers around the world with more time and better facilities. However, amateur astronomers had previously made important astronomical discoveries, so such a discovery was by no means out of the question. Then, shortly after the planet's discovery, Emmanuel Liais, a French astronomer employed by the Brazilian Coastal Survey, announced that he had been observing the sun at the same time as Lescarbault, and had not seen the planet. But Liais was known to have an intense dislike of Le Verrier and so his observations were treated with reserve. Nevertheless, it made a number of astronomers uncertain of the new planet's existence.

Immediately following its apparent observation by Lescarbault, Le Verrier calculated the orbit of the new planet, now called Vulcan, after the Roman god of fire. Assuming that the orbit was circular, he deduced an orbital radius of 0.147 AU, a period of 19 days 17 hours, and an orbital inclination of 12° 10',[7] with transits occurring around 3rd April and 6th October each year. In addition, using Lescarbault's estimate of the size of the disc on the sun, Le Verrier deduced a mass of about 6% that of Mercury, assuming the same density as Mercury. So something like twenty Vulcan-sized planets would be required to explain Mercury's motion. It seemed doubtful that these could have escaped detection, although Le Verrier speculated that this could be because the other intramercurial planets are much smaller than Vulcan.

As with Uranus and Neptune, as soon as Vulcan had been discovered, astronomers quickly looked to see whether it had been seen before. Rudolf Wolf of Zurich, a keen observer of sunspots, had already produced a list of fast-moving spots on the sun, that could be observations of intramercurial planets. From this list he suggested that three could be prior observations of Vulcan. J.C.R. Radau of Königsberg used two of these three observations, along with those of Lescarbault, to deduce an orbit for Vulcan

---

[7] As a reference, the values of these parameters for Mercury are 0.387 AU (average), 87 days 23 hours and 7° 0'.

that was completely different from that of Le Verrier. Radau's orbit, with an inclination of about 1°, implied that transits of Vulcan would occur on 29th March and on 2nd and 7th April 1860, but observations by numerous astronomers on these days yielded nothing. Likewise observations during the total solar eclipse of July 1860 yielded nothing either, so more astronomers began to doubt the existence of Vulcan.

If there really were no intramercurial planets, however, what could be the cause of the extra precession of Mercury's orbit? One suggestion was that it could be caused by a cloud of dust in orbit around the sun. It was proposed that the zodiacal light, which is seen either just before dawn or just after sunset, could be caused by such a dust cloud. But the dust cloud needed to explain Mercury's extra precession would have to be symmetrical about Mercury's orbit, otherwise it would cause a change in the node of the orbit, which it does not. Mercury's orbit is inclined at about 7° to the ecliptic, but the zodiacal light appeared to be symmetrical about the ecliptic. So this theory seemed to be untenable.

In about 1880 Simon Newcomb (1835–1909), of the American Nautical Almanac Office, re-examined the rate of precession of Mercury's orbit, which resulted in increased calculated and observed magnitudes. However, the difference between them of $41'' \pm 2''$/century, was only slightly larger than Le Verrier's estimate of $38''$/century. Then in 1894 Asaph Hall (1829–1907), of the US Naval Observatory, suggested that if the 2 in the inverse square law of gravity was increased to 2.0000001574 the problem of Mercury's orbital precession could be solved. Unfortunately, this change also modified the calculated orbit of the moon, for example, making it incompatible with observations. As a result the idea was generally dismissed.

So, as the nineteenth century drew to a close, the only solution to the problem of Mercury's orbit seemed to be Le Verrier's idea of a number of intramercurial planets. In fact, every now and again, some astronomer would report seeing Vulcan, or something like Vulcan, transiting the sun, but many of the observations were later attributed to sunspots. So doubts still prevailed as to whether there were any intramercurial planets or not, until in 1915 Albert Einstein showed, using his new general theory of relativity, that he could explain the precession of Mercury's perihelion without recourse to such planets. His calculated value of the precession of $43''$/century was, within error, the same as the unexplained amount calculated by Le Verrier over fifty years earlier, and Newcomb about twenty years later. So where Newton's theory had, in the past, been supreme, now there was at least one case where it needed supplementing by the new theory of relativity.

## 7.2 Asteroids

We left the asteroids or minor planets in 1807 (see Section 6.4) with the discovery in that year of the fourth asteroid Vesta. Over the next few years a number of astronomers, particularly members of the Celestial Police, looked for further asteroids, but none were found. Then in 1816 Olbers, the last active member of the Celestial Police, gave up his search. No more serious work was done in this area for a number of years, but in 1830 Karl Ludwig Hencke, an amateur astronomer from the Prussian town of Driessen, resumed the search.

Remarkably Hencke persevered for fifteen years, and on 8th December 1845 he was finally successful when he discovered the fifth asteroid, to be called Astrea, followed on 1st July 1847 by his discovery of Hebe. This was quickly followed by John Russell Hind's discovery of Isis on 13th August 1847, followed by his discovery of Flora on 18th October. One asteroid was found in 1848, one in 1849, and then the discovery rate increased dramatically with the total of 13 known at the end of 1850, increasing to 62 at the end of 1860, 219 at the end of 1880, and nearly 500 at the end of 1900. Max Wolf of the Heidelberg Observatory was the first to use photography to discover an asteroid, namely Brucia (No. 323)[8] on 22nd December 1891. The use of this new medium of photography then led to a veritable avalanche of discoveries.

The rapid increase in the discovery of asteroids in the years immediately after 1845 had taken almost everyone by surprise,[9] and it was not long before astronomers tried to make some sense of what they were finding. For example, in 1866, and later in 1876, the American astronomer Daniel Kirkwood (1814–1895) pointed out that there were no asteroids with periods of $\frac{1}{3}$, $\frac{2}{5}$ and $\frac{2}{7}$ of Jupiter's period. Later on Kirkwood discovered gaps at $\frac{1}{2}$, $\frac{3}{5}$ (see Figure 7.1) and other simple fractions of Jupiter's period. He attributed these gaps to the effect of a resonance between the asteroids and Jupiter, which caused asteroids with these periods to be forced out of their orbits by Jupiter. In their new orbits they may have collided with other asteroids causing them to break up, or they may have coalesced with them. Some years later De Freycinet suggested that Jupiter, rather than clearing asteroids from their 'resonance' orbits, may have prevented them from forming in those orbits in the first place.

---

[8] i.e. the 323rd asteroid to be discovered, called Brucia.

[9] The Report of the Council of the Royal Astronomical Society for the year ending February 1851 said, after reporting the discovery of three more asteroids, that the 'rate of increase ... can hardly be expected to continue for very long'.

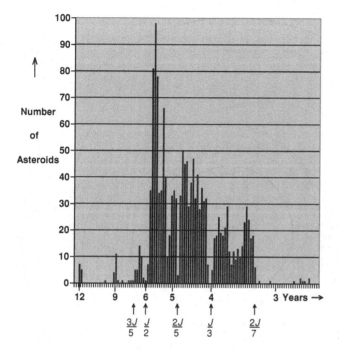

**Figure 7.1** This histogram showing the number of asteroids with various periods is based on one published in the early twentieth century. It clearly shows that there are some periods where there are few or no asteroids. These are the Kirkwood gaps, the main ones of which are identified by arrows, with periods which are simple fractions of Jupiter's period, $J$.

In 1855 the asteroid Fides (37) was discovered by Luther in an orbit of period 4.29 years, eccentricity 0.177 and inclination 3°.07. There was nothing unusual in that, but six years later the asteroid Maia (66) was found by Tuttle in almost exactly the same orbit. Then in 1881 Lespiault suggested that eventually the two asteroids may either coalesce or combine to form a binary asteroid system, although at that time no such binary asteroids had been found.

The asteroid Aethra (132), which was discovered in 1873 by James Watson at the Ann Arbor Observatory in the United States, had perihelion and aphelion distances of 1.61 and 2.61 AU, compared with Mars' aphelion distance of 1.67 AU. In fact Aethra was the only asteroid known, until late in the nineteenth century, to have at least part of its orbit inside that of Mars. Up to that time all the asteroids had orbits that were wholly between those of Mars and Jupiter (see Figure 7.2). Then on 14th August 1898 Eros (433) was photographed by Gustav Witt (1866–1946) at Berlin, and by A. Charlois at Nice Observatory.[10] Over the next few months, old plates were examined

---

[10] Witt developed his plate immediately, and noticed the new asteroid. But 14th August was a Sunday, and the following day was a holiday in France, so Charlois did not examine his plate until 16th. As a result he missed the opportunity of being credited as the co-discoverer of Eros.

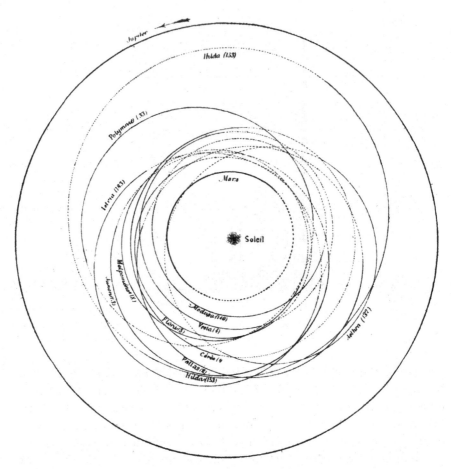

**Figure 7.2** The orbits of a selection of asteroids are shown in this plot published by
M. Niesten in 1881. At that time the only asteroid whose orbit was known to intercept
that of Mars was Aethra 132. (From *The Sun: Its Planets and their Satellites*, by Edmund
Ledger, 1882, p. 290.)

at various observatories, and it was found that Eros had appeared on about
twenty photographs taken at the Harvard Observatory between October 1893
and June 1896.

It soon became apparent that Eros had a unique type of orbit with a perihelion distance of 1.13 AU, and an orbital inclination of about 10°.8, which
meant that, not only was its orbit well inside that of Mars, it could approach
very close to the earth from time to time (see Figures 7.3 and 7.4). Because
of this, it could be used to provide an accurate estimate of the astronomical
unit at its closest approach.

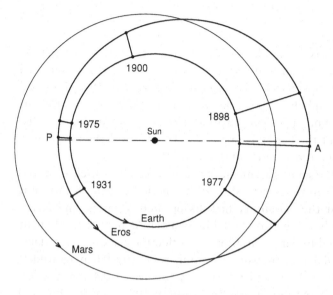

**Figure 7.3** Eros' highly eccentric orbit is seen to cross the orbit of Mars, and come very close to that of the earth near Eros' perihelion, P. The positions of Eros and the earth are shown at various dated oppositions, with that of 1975 being exceptionally close (see also Figure 7.4), and those around aphelion, A, naturally being the most distant. The dotted line shows the line of nodes of Eros' orbit (i.e. where its plane intercepts that of the ecliptic).

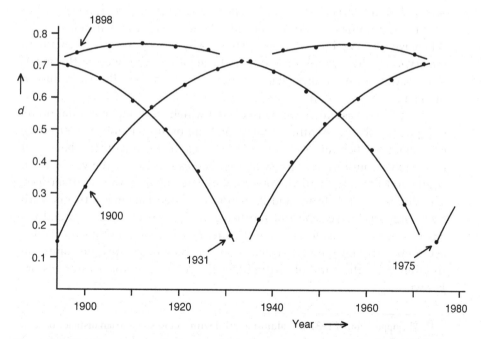

**Figure 7.4** The closest approach distances ($d$) for Eros around opposition, as indicated by dots, are at a minimum about every forty years. The closest approach distances repeat in a cyclical fashion, as indicated by the solid lines, which are the loci of such distances.

The perihelion of Eros' orbit is at a heliocentric longitude of 122°, which the earth passes on 22nd January of each year, so its most favourable oppositions are on about that date. The closest approach to earth is not exactly on that date, however, because the orbits of Eros and the earth are inclined to each other. As it happens, the day after its discovery in 1898, Eros had a very unfavourable aphelic opposition, with a closest approach of about 0.74 AU. The following opposition in 1900 was more favourable at a distance of 0.32 AU, but the perihelic opposition of 1975 was the best one in the twentieth century at a distance of 0.151 AU.[11]

In February 1901 Egon von Oppolzer found that Eros varied in intensity by about 1.5 magnitudes (or 75%), with a period of 2 h 38 min. But by May of that year the variability had disappeared. Charles André suggested that this was because Eros was a binary asteroid, but others thought that it may be caused by large variations in reflectivity across its surface. At opposition in 1903 the intensity variation was only 0.8 magnitudes, and at opposition in 1917 it had fallen even further to 0.4 magnitudes. The mystery was getting worse. But in February 1931 van den Bos and Finsen at Johannesburg, South Africa, solved the problem, when they observed that Eros was elongated in shape. It was estimated to be about 22 km long by 6 km broad, and to be rotating about its short axis in 5 h 16 min. When we observe Eros along the short axis, we see no variation in intensity, but when we observe it at right angles to this we see the maximum variation. At intermediate orientations we see smaller variations in intensity.

In 1906 two asteroids were discovered which travelled in similar orbits to Jupiter, with one (Achilles, 588) 60° in front of Jupiter and one (Patroclus, 617) about 60° behind. Now, in 1772 Lagrange had shown that a body will be in stable equilibrium with two other bodies if it lies at the vertex of an equilateral triangle, with the other two bodies at the other two vertices. The discovery of Achilles and Patroclus validated Lagrange's theory with each asteroid, Jupiter and the sun being at the vertices of equilateral triangles. These two asteroids were the first of what we now call the Trojan asteroids, orbiting the sun at about 60° in front or 60° behind Jupiter in its orbit.[12] At the time of writing (2002) over 1,200 Trojan asteroids are known.

---

[11] The opposition of 1931 was almost as good with a closest approach distance of 0.174 AU.

[12] The Trojans do not stay exactly at these positions, however, owing to the perturbations of Saturn.

Of the planets known in the second half of the nineteenth century, Mercury has the highest orbital eccentricity of 0.21, as well as the highest orbital inclination to the ecliptic of 7°.0. The orbital eccentricities and orbital inclinations of the asteroids are somewhat larger than those of the planets, however. For example, of the first 100 asteroids to be discovered, Polyhymnia (33) has the largest orbital eccentricity of 0.34, and Pallas has the highest inclination of 34°.8. Nevertheless, both of these figures are significantly less than those for the comets, as some comets have hyperbolic orbits (i.e. their eccentricities are >1.0) and some orbit the sun in the opposite direction to the planets and asteroids (i.e. their orbital inclinations are >90°).

As far as asteroid sizes are concerned, in the early nineteenth century William Herschel had deduced a diameter of about 260 km for Ceres and about 240 km for Pallas (see Section 6.4). In 1853, Le Verrier determined, from a detailed analysis of Mars orbit, that the total mass of all the asteroids could not exceed $1/4$ of the mass of the earth. However, by August 1880, when 216 asteroids were known, Niesten estimated that their total volume was only about $\frac{1}{4,000}$ that of the earth.

Late in the nineteenth century, the American astronomer Edward C. Pickering (1846–1919) deduced, from their photometric intensities, diameters of 510 km for Vesta, 270 km for Pallas and 150 km for Juno, and about 20 km for the smallest of the asteroids known at the time. This was assuming that they had similar albedos to that of Mars. But in the 1890s Edward Emerson Barnard (1857–1923) was able, for the first time, to measure the diameters of the largest asteroids directly, using the 40 inch (102 cm) Yerkes refractor. His values were 770 km for Ceres, 490 km for Pallas, 390m for Vesta and 190 km for Juno. Previously Vesta had been thought to be the largest asteroid, as it is the brightest, but Barnard found that the reason for this brightness was not its size, but its exceptionally high albedo of about 0.74. This compared with 0.18 for Ceres, 0.23 for Pallas and 0.45 for Juno.

A number of astronomers in the late nineteenth century thought that the asteroids may have complex surface features, together with a tenuous, but possibly extended atmosphere. In 1882, in his Gresham lectures, Edmund Ledger mentioned possible 'nebulous atmospheres' or 'atmospheric vapours' around some asteroids.[13] Then in 1897 Robert Ball hypothesised that the asteroids could be 'globes like our earth in miniature, diversified by

---

[13] Edmund Ledger, *The Sun: Its Planets and Their Satellites*, Stanford, 1882, pp. 277–278.

continents and oceans' and that their atmospheres may be 'in an extremely rarefied condition, though possibly of enormous volume.'[14]

Barnard's large albedo for Vesta created a problem for astronomers and geologists, as it was too high to be due to any known rock. It was also thought that it could not be due to snow, as Vesta was too small to retain water vapour in a possibly very tenuous atmosphere. In fact, this problem was not resolved until the 1970s when, for the first time since Barnard's measurements, better estimates were made of the diameters and albedos of the largest asteroids. It then transpired that Barnard's estimated diameter for Vesta was too low, and so his albedo estimate was too high. With its new albedo value, it was now possible that its surface could be composed of igneous rocks.[15]

## 7.3 Comets

### Orbits

Newton had shown in 1686 that the orbits of comets were parabolas or ellipses of large eccentricity with the sun at a focus (see Section 4.7), and Halley had confirmed this for the twenty-four comets that he considered in his treatise of 1705 (see Section 5.1). The most famous confirmation of the elliptical orbit hypothesis was, of course, the prediction that Halley's comet would return and pass perihelion in 1759. In the process, the rôle of Saturn, and particularly that of Jupiter, in modifying the orbit of Halley's comet became clear.

On 14th June 1770 Charles Messier discovered a comet (now called Lexell's comet) which was moving rapidly towards the earth. A short while later, on 1st July, it passed within just 0.015 AU of the earth, before being lost in the sun's glare a few days later. At closest approach its head or coma was seen to be a massive 2° 23′ in diameter. The comet was subsequently recovered on 4th August and observed until 2nd October.

The first attempts to fit a parabolic orbit to Lexell's comet failed, but Anders Lexell at St Petersburg was able to fit an elliptical orbit with a period of just 5.6 years and a perihelion distance of 0.7 AU. This short period orbit presented a problem, however, as the comet should then have been seen on previous orbits, but it had not been so. Lexell then pointed out that the

---

[14] Robert S. Ball, *The Story of the Heavens*, Cassell, revised edn., 1897, pp. 200–202.

[15] The best estimates today of the diameters and albedos of these large asteroids are 940 km and 0.11 for Ceres, 530 km and 0.16 for Pallas, 510 km and 0.42 for Vesta, and 260 km and 0.23 for Juno.

comet had passed very close to Jupiter in 1767, and that Jupiter had radically changed its orbit during that very close approach. Prior to then the comet had had a very different orbit, with a perihelion distance of 2.9 AU. That is why it had not been seen before.

Lexell's comet was not observed during its next perihelion passage in 1776, because of its unfavourable position in the sky relative to the sun. Lexell then suggested that another very close approach to Jupiter in 1779 could cause the following perihelion passage to be delayed beyond 1781. In fact, despite many attempts to recover the comet in 1781 and 1782 it was never seen again.

On 26th November 1818 Jean Louis Pons (1761–1831) of Marseilles discovered a comet, now called Encke's comet, that remained visible for about seven weeks. Olbers suspected that it was the reappearance of a comet first seen by Pierre Méchain in 1786, Caroline Herschel in 1795, and by Pons, Huth and Bouvard independently in 1805. But it was the German astronomer and mathematician Johann Encke who proved, in 1819, that these were all appearances of the same comet in an orbit with the very short period of 3.3 years. This took it to inside the orbit of Mercury at perihelion.[16] Encke predicted the comet's return to perihelion on or about 24th May 1822 but, unfortunately, the return would only be visible in the southern hemisphere. It was, nevertheless, observed for three weeks in June 1822 by Karl Rümker at Sir Thomas Brisbane's observatory at Paramatta in New South Wales, very close to its predicted position. This was the first confirmed prediction for the return of a periodic comet (see Figure 7.5) since that of Halley's comet over half a century earlier.

Interestingly, Encke noticed that the comet's return in 1822 was a few hours early, and in 1823 he suggested that this was because it was being affected by a resisting medium[17] whose density varied as the inverse square of

---

[16] The close approach of Encke's comet to Mercury, from time to time, enabled the first accurate estimate to be made of Mercury's mass. In 1835, for example, Encke calculated a reciprocal mass of about 4,700,000 for Mercury, using perturbations in the orbit of the comet named after him. In fact, this method of using Encke's comet was still in use in the 1960s, when Makower and Bochan in the USSR deduced a reciprocal mass for Mercury of 5,880,000 ± 200,000.

[17] The idea of a resisting medium was by no means new in the nineteenth century. Aristotle's concept of the ether had been developed in many ways in the seventeenth century, with some astronomers, like Descartes, postulating that it carried objects in their orbit (see Section 4.6), whereas others, like Hevelius, saw it as a resisting medium. Then in 1746 Euler had used the concept of a resisting ether to explain the delayed

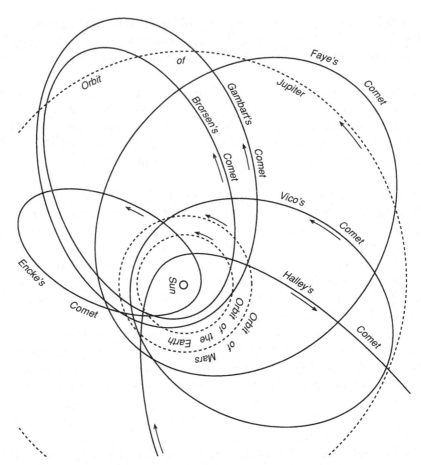

**Figure 7.5** The orbits of those periodic comets known in about 1870, showing the large variety of their eccentricities and orientations. (From *The Heavens*, by Amédée Guillemin, 1871, p. 247.)

its distance from the sun.[18] He reasoned that such a resisting medium would not affect the planet Mercury to nearly the same extent, however, as Mercury is much denser than a comet.[19] Encke found that the comet's orbital period

---

appearance of Halley's comet in 1682, and in 1772 he used the same concept to explain the secular acceleration of the moon (see Section 5.5).

[18] It may appear, at first sight, contradictory that a resisting medium should cause a comet to return earlier. This is because the comet's orbit would become smaller when subject to resistance, and so its angular velocity would increase.

[19] Such a medium was later proposed to try to solve the riddle of the precession of Mercury's perihelion, however (see Section 7.1).

was getting shorter at the rate of about $2\frac{1}{2}$ hours per orbit, and correctly predicted the times for return to perihelion from 1825 to 1858. Because of this success, contemporary astronomers tended to accept Encke's hypothesis of a resisting medium, as either an extension of the solar atmosphere, or as the material that causes the zodiacal light. One significant astronomer, namely Friedrich Wilhelm Bessel, did not agree, however.

Bessel had observed the central region of Halley's comet during its 1835 appearance, and had noticed a rocket-like jet apparently being ejected from the nucleus. Bessel then calculated what the effect would be on the comet's orbit of such a jet. Assuming that it acted for about three weeks before perihelion, and that there was a mass loss from the nucleus of 0.1% of its mass per day, the resulting change in the orbital period was about 3 years. Although this figure is now known to be far too large, the correct change in period due to this effect being about 3 or 4 days, it nevertheless showed that such an effect can be significant.

In 1878 Friedrich Emil von Asten (1842–1878) published his analysis of the orbital changes of Encke's comet over the period 1818 to 1875, and concluded that Encke's resisting medium was the most likely cause of these changes up to 1868. But because there appeared to be no such effect with other periodic comets of larger period, he concluded that the resisting medium has no measurable effect outside the orbit of Mercury. There was a problem, however, as the 1871 return of Encke's comet did not show the expected time delay. Von Asten noticed that the comet had passed reasonably close to the asteroid Diana in May 1869, and wondered if this could be the cause of the problem. But his calculations showed that only a collision with an asteroid in June 1869 could have caused such an effect.

In a series of papers between 1884 and 1910, the Swedish astronomer Jöns Oskar Backlund (1846–1916) analysed the observed orbital behaviour of Encke's comet, and eventually concluded that the decrease in its orbital period was not uniform, with discrete changes occurring in 1858, 1868, 1895 and possibly in 1904. He put this down to a variation in the density of the resisting medium with time, which he assumed to be a ring of meteoric material. As an alternative, Backlund suggested the effect could be caused by changes in the sun's surface activity.

The jury was still out on Encke's resisting medium when in 1882 the Great September Comet (otherwise known as comet 1882 II) was observed, both before and after its very close approach of about 500,000 km to the surface of the sun. Its orbit showed no measurable change during this close perihelion passage, effectively demonstrating that the resisting medium does not exist.[20] This was confirmed in 1933 when Michael Kamienski (1880–1973) found that the return of Wolf's comet was late, rather than early (as for Encke's comet). If the resisting medium did not exist, however, what did cause the changes to the orbits of Encke's and Wolf's comets?

The problem was finally solved in 1950 by Fred Whipple (b. 1906) when he returned to Bessel's solution of a jet-like emission from the cometary nucleus, although Whipple's model of the so-called 'non-gravitational forces' was somewhat more sophisticated. In Whipple's icy-conglomerate model, or dirty snowball theory, as it was more popularly known, the cometary nucleus was made of dirty ice which partially vaporised as the comet got closer to the sun. But the essential difference with Bessel's model is that Whipple assumed that the nucleus was rotating, so the effect of the jet was not simply along the comet–sun line. This allowed the force of the jet to either slow down or speed up the comet in its orbit, rather than just push the comet further from the sun around perihelion, as in Bessell's model.

## Density

Attempts were made in the nineteenth century to estimate the density profile of the heads of comets by observing the intensity and position of faint stars as comets passed in front of them. In 1832, for example, Sir John Herschel was able to see the stars of a star cluster through the head of Biela's comet, which was estimated to be at least 80,000 km thick. Three years later, stars were seen through the head of Halley's comet, independently, by Struve and Glaisher, and in 1847 Dawes saw a tenth magnitude star through the head of a comet. In fact, quite dim stars could be seen through the heads of comets, indicating that they were generally very tenuous objects.

## Illumination

On 3rd July 1819 Arago observed the tail of Tralles' comet (1819 II) with his newly developed 'polariscope' or 'polarimeter', as we would now call it, and found that the light was partially polarised, indicating that at least part of the light was reflected sunlight. Subsequently Arago also found that the light from Halley's comet was partially polarised. Similarly, although the observations differed in detail, the light from Donati's comet (1858 VI),

---

[20] Incidentally, this also showed that the zodiacal light and the solar corona must also be very tenuous.

Tebbutt's comet (1861 II), and Swift–Tuttle (1862 III) was also found to be partially polarised, so at least some, if not all of the light from these comets also came from the sun. In fact, in the case of Donati's and Tebbutt's comets, the plane of polarisation of the light was found to pass through the sun, so giving added credibility to the idea that some of the light was reflected sunlight.

Astronomy benefited considerably from developments in the new discipline of spectroscopy in the second half of the nineteenth century. The German physicist and optician Joseph Fraunhofer (1787–1826) had studied the absorption lines, now named after him, in the solar spectrum in the early part of the century. But the main work of developing spectroscopy, and linking spectral lines to specific elements, was begun by Gustav Kirchoff (1824–1887) and Robert Bunsen (1811–1899) in Heidelberg in 1859.

The first successful observation of a cometary spectrum was made in 1864 by Giovanni Donati (1826–1873), the director of the Florence Observatory. At that time it was thought that comets generally shone by reflected sunlight, but when Donati observed the spectrum of Tempel's comet (1864 II), when it was near to the sun, he found three faint luminous bands, indicating that it was self-luminous. Four years later, William Huggins (1824–1910) identified these bands with those emitted by various hydrocarbon compounds when they are made luminous in the laboratory by electrical discharges. Meteorites, when heated in experiments, also gave off hydrocarbons, and so it appeared that comets and meteorites were made of similar elements.

The second comet to be examined spectroscopically was Tempel–Tuttle (1866 I), which was observed in 1866 independently by William Huggins in England and Angelo Secchi in Italy. Huggins found that the coma or head of the comet produced a broad continuous spectrum, like the sun, with just one bright band, whereas Secchi found the three bright bands previously discovered by Donati for comet 1864 II. These observations clearly showed that this comet was both self-luminous and shining by reflected sunlight. Similar effects were found for the vast majority of comets observed over the next few years when they were relatively close to the sun, but when they were some distance away only a continuous spectrum was observed. Evidently they were just reflecting sunlight at such relatively large distances, not being hot enough to become self-luminous. Closer to the sun, however, the heat caused them to start emitting their own light.

Tebbutt's comet (1881 III) was the first comet to have its spectrum recorded photographically, when the spectrum of its head was photographed on 24th June 1881 by William Huggins during a one-hour exposure. He

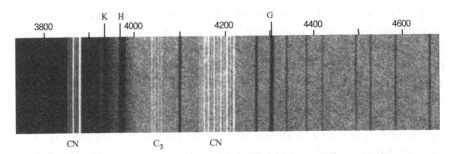

**Figure 7.6** An engraving showing the spectrum of Tebbutt's comet (1881 III) based on Huggin's photograph of 24th June 1881, which was the first successful photograph of a comet's spectrum. The Fraunhofer absorption lines K and H are due to calcium, whilst that at G is due to the CH radical. The emission bands with heads near 3880 and 4220 Å are due to cyanogen (CN), and that around 4050 Å is due to the carbon molecule $C_3$. (From *Comets: A Chronological History of Observation, Science, Myth and Folklore*, by Donald Yeomans, 1991, p. 212. Reprinted by permission of John Wiley & Sons, Inc.)

interpreted the photograph as showing a continuous solar spectrum, with its Fraunhofer absorption lines, plus three hydrocarbon emission bands produced by the comet. In fact, Huggins' drawing from this photograph still exists (see Figure 7.6), clearly showing that the cometary emission bands are actually due to cyanogen, CN, and the carbon molecule $C_3$.

So far none of the comets that had been observed spectroscopically had passed very close to the sun. But in 1882 the spectrum of Wells' comet (1882 I) was observed as it approached its perihelion distance of just 0.061 AU from the sun on 11th June. During the early part of April its spectrum was continuous, with weak hydrocarbon emission lines, but these died out as the comet approached closer to the sun. Then on 28th May, Ralf Copeland and J.G. Lohse, who were working at Dun Echt Observatory in Ireland, first identified the sodium line in the comet's spectrum. By 1st June the sodium line was brighter than any other part of the spectrum, and on 6th June, five days before perihelion, the sodium line was seen to be double.

Shortly afterwards, the spectrum of the Great September Comet (1882 II) was observed after it had passed even closer to the sun. In this case, Copeland and Lohse observed that, one day after perihelion, the double sodium line was also accompanied by several lines of iron. All these lines were slightly displaced in wavelength due to the Doppler shift, which was caused by the relative velocity of the comet and earth. In fact, the magnitude of the shift indicated that the comet was receding from earth at about 70 km/s. Later that day, the velocity was found by Louis Thollon and A. Gouy at Nice Observatory to have increased to over 100 km/s. It was then found that as the comet receded further from the sun, the sodium and iron lines faded and the usual hydrocarbon bands reappeared.

## Tails

The tail structure of comets is very variable from one comet to another, and over time with any particular comet. It appears that ancient Chinese astronomers were the first to note that the main tail streams away from the cometary nucleus on the opposite side to the sun (i.e. it is antisolar), no matter in which direction the comet is travelling,[21] a fact discovered in the West by Peter Apian (or Apianus, 1495–1552) in 1531. The ancient Chinese also seem to have discovered that occasionally some comets have a short, so-called 'antitail', which appears to be directed towards the sun.

In 1625 Kepler outlined his idea of cometary tails as being an emission 'from the head, expelled through the rays of the sun into the opposite zone'.[22] This is interpreted to mean that the tail is formed from material emitted from the head, which then streams in an antisolar direction due to solar radiation pressure.[23] Kepler also continued, perceptively, to say that the loss of material into the tail would eventually lead to the death of a comet. A little later, Claud Comiers, in his thesis on comets, described his theory of cometary tails which is less ambiguous than Kepler's. In particular, Comiers suggested that comets reflect sunlight from their atmosphere, and that their tails form when the sun's heat causes the material of the coma to become more rarefied. The sun's rays then push this rarefied material away from the sun.

Newton in his theory of comet's tails suggested that the heated ether carries the material in the tail away from the sun. Newton was somewhat inconsistent in making this suggestion, however, as he had already rejected the idea of the ether as there was no evidence that it was slowing down comets in their orbits.[24] Nevertheless, Newton did put forward one idea about cometary tails that returned in a somewhat different way in the twentieth century. He suggested that planetary surfaces would, over time, pick up and accumulate material emitted in cometary tails. In the earth's case, in

---

[21] So the tail is not trailing behind the comet in its orbit.

[22] In his *Tychonis Brahei Dani Hyperaspistes*.

[23] Some astronomers dispute that Kepler envisaged solar radiation pressure as the mechanism, suggesting instead that he just thought of the sun's rays pushing the material away from the sun. However, the latter is really just a simplistic way of explaining solar radiation pressure.

[24] This was before the discovery of the slowing down of Encke's comet, of course.

particular, Newton envisaged this material helping to replenish our atmosphere and stop it disappearing with time.

The Great Comet of 1811 was observed and subjected to a detailed analysis by many astronomers, of which the analysis of Sir William Herschel and Heinrich Olbers was the most interesting. When the comet came near to earth on 15th October, its yellowish tail covered about 23° of the sky, corresponding to a length of about 1.5 million km, and its bluish-green head was about 200,000 km in diameter. In the head Herschel found a reddish nucleus that he measured to be about 690 km in diameter, and which he deduced was shining by its own light, rather than by reflected sunlight, as it showed no phases. Olbers concluded that the tail of this comet consisted of very small particles, driven away by a solar repulsive force that was somehow based on electrical repulsion. In his memoir on this comet, Olbers also referred back to the comet of 1807, which had both a long, straight tail and a shorter, curved tail, reasoning that 'the longer and straight tail must consist of particles repelled more strongly by the sun than the matter forming a curved tail'.[25]

In 1836 Bessel developed this theory of Olbers, spurred on by the reappearance of Halley's comet the previous year, and calculated that the repulsive force of the sun acting on cometary particles was about twice that of the sun's gravitational attraction. A few years later, there was a dramatic demonstration of the very high velocity of particles in the tail of a comet, when the tail of the Great March Comet of 1843 (1843 I) was observed to be always pointing away from the sun during its perihelion passage. This was even though the tail was many millions of kilometres long (see Figure 7.7), and around perihelion the comet had moved 180° around the sun in just a shade over two hours, with the tail moving through a similar angle.[26] The velocity of the tail particles must have been enormous for the tail to remain antisolar during such a rapid perihelion passage.

The Russian physicist Fëdor Alexandrovich Bredikhin (1831–1904) analysed the tails of seventy-six comets over the period from 1877 to 1882, and concluded that, in addition to antitails, they were of three types. There were long straight tails, called Type I, for which he calculated a solar repulsive force about 11 to 18 times that of the solar gravitational attraction. Then there were the curved, scimitar-type tails, called Type II, that were subjected

---

[25] In his *Ueber den Schweif des grossen Cometan von 1811* of January 1812.

[26] The tail of this comet was quite phenomenal. At noon on 28th February the comet was observed when it was only about 1° from the sun, with a 5° long tail. By 3rd March the tail was seen to be about 25° long, and on 17th March it was some 40° long around sunset. At one stage the tail's length was estimated to be well over 500 million km.

**Figure 7.7** An engraving showing the long, bright tail of the Great March Comet of 1843, as seen over Paris. (From *Comets: A Chronological History of Observation, Science, Myth and Folklore*, by Donald Yeomans, 1991, p. 179. Reprinted by permission of John Wiley & Sons, Inc.)

to a solar repulsive force about 0.7 to 2.2 times that of the solar gravitational attraction. And finally there were the short, stubby, highly curved Type III tails, where the ratio of the solar repulsive force to gravitational attraction was about 0.1 to 0.3. Type I tails were attributed by Bredikhin to hydrogen, type II to hydrocarbons and light metals, and type III to heavy elements like iron or chlorine. In his view the repulsion was caused in each case by the sun's electric charge and the like polarity of charge of the cometary gas molecules. At the time, hydrocarbons and iron had been detected in the spectra of comets, but hydrogen had not.

## Origin

A key question that exercised the minds of astronomers in the nineteenth century was are the non-periodic comets part of the solar system? In the early nineteenth century both William Herschel and Laplace theorised that non-periodic comets come from interstellar space, whilst in 1783 William Herschel had shown that the sun was moving in space towards the star λ

Herculis. Therefore, in 1860 Carrington and Mohn independently suggested that if non-periodic comets originated in interstellar space, and so were not participating in the general movement of the solar system, the sun should sweep up more of them in its forward direction, and those comets should, on average, have a higher velocity relative to the sun, than those behind it. No such effect could be found, however, in the case of more than 100 comets considered, and so it appeared as though non-periodic comets must be an integral part of the solar system, even if their aphelia are some distance away from the sun.

Over the period 1878–1893, Hubert Newton of Yale examined the effect of planetary perturbations on the orbits of comets. In particular he examined the efficiency with which a comet, in a parabolic orbit of any inclination, would have its orbit perturbed enough by Jupiter to change it to one of short period. Somewhat surprisingly, the mechanism turned out to be remarkably inefficient. Only about one comet in a million that were originally in parabolic orbits that came closer to the sun than Jupiter, had their periods reduced to less than that of Jupiter (i.e. about 12 years). Of these modified orbits, most were direct. Newton also showed that those comets in direct orbits would be far more likely than those in retrograde orbits to have their orbital periods reduced at subsequent returns to the vicinity of Jupiter.

## 7.4 Meteor showers

### Origin

Showers of meteors, or shooting stars, as they are often called, had been known since antiquity. In the seventeenth century they were thought to be of terrestrial origin, but in 1714 Edmond Halley had suggested that they may be caused when the earth, in its orbit around the sun, meets matter formed in the ether. This, one of the first suggestions of their cosmic origin, was not accepted at that time, as the vast majority of scientists were still convinced that they were purely terrestrial in origin.

In 1794 the German physicist Ernst Chladni (1756–1827) thought that meteors were produced when very small particles in space are attracted to the earth, heating up in its atmosphere and so becoming visible. He also explained that iron meteorites could not have been produced by natural processes on the earth, so they must be of cosmic origin. Furthermore, fireball material was debris left over from the formation of the planets. As a result, meteors, meteorites and fireballs were all of cosmic origin.

Chladni then encouraged Johann Benzenberg (1777–1846) and Heinrich Brandes (1777–1834), who were students at Göttingen University at the time, to study meteors in more detail. Accordingly, they simultaneously observed twenty-two meteors between 11th September and 4th November 1798 from two locations a few kilometres apart, to try to establish the heights of the meteors above the ground. They found that the heights varied from about 10 to 210 km, clearly indicating that, whatever their cause, when observed they were in the upper reaches of the atmosphere.

## Orbits

On the morning of 12th November 1799, Friedrich Alexander von Humboldt (1769–1859) and others observed a spectacular display of shooting stars (see Figure 7.8), which Humbolt noted appeared to come from one point in the sky. Thirty-four years later, on the night of 12th–13th November 1833, another spectacular display was seen, and, again, a number of observers noted that the meteors seemed to originate from one point, now called the 'radiant', in the constellation of Leo (see Figure 7.9). One of the observers, Denison Olmsted (1791–1859) of Yale College in America, analysed the various observations of the 1833 November meteors and published a report in the following year.

Olmsted, in his report, noted the similarity between the 1833 meteor shower, and that which had been observed one day earlier in 1799, and with the less intense shower that had taken place on the same day in 1832. He showed that the apparent origin of the meteors from a fixed radiant was due to a perspective effect. The relative velocity between the earth and the source of the meteors was in the direction of the radiant, and the paths of the meteors were all parallel, ending in the radiant as their vanishing point. Olmsted suggested that the meteors were material from a comet-like body being heated, and hence made visible, by friction in the earth's atmosphere. He estimated the velocity of the meteors in the atmosphere to be about 6 km/s for the 1833 event, and concluded that the orbit of the comet-like source of these meteors around the sun had a period of 182 days, with an aphelion near the earth's orbit. So the source orbited the sun twice a year, meeting the earth on 12th November every year near the source's aphelion. Alexander Twining (1801–1884), a civil engineer at West Point, came to similar conclusions a little later, although he estimated the velocity of the meteors in the atmosphere as at least 23 km/s.

**Figure 7.8** A woodcut showing the spectacular meteor shower of 12th November 1799. (From *The Midnight Sky*, by Edwin Dunkin, 1869.)

The display of the November meteors in 1834 and subsequent years was less intense than in 1833. In fact, in 1837 Olbers noted not only the extreme variability in the intensity of the display from year to year, but suggested that the most intense showers occurred 34 years apart. As a result he predicted that the next intense display would occur in 1867.

It had been known for some time that there was often a meteor shower in August. So, with the success in observing in 1833 that the November meteors or Leonids,[27] as they became known, had a fixed radiant, it was natural that attention should now be focused on the August meteors. Thus it was that

---

[27] So named because their radiant is in the constellation of Leo.

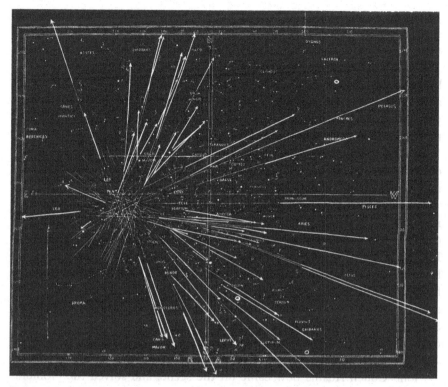

**Figure 7.9** Paths of the Leonid meteors showing their apparent origin from a common radiant point, which Olmsted correctly attributed to parallax. (From *Popular Astronomy*, by Simon Newcomb, 1898, p. 403.)

John Locke (1792–1856), of the Ohio Medical College, observed that the meteors in the shower of 8th August 1834 seemed to originate from a point in the constellation of Perseus. Then in 1836 Lambert Quetelet (1796–1874) of Brussels showed, by analysing historical records, that the August meteors returned annually, and he correctly predicted another shower on or about 9th August 1838.

The discovery of the annual nature of these so-called Perseid meteors proved fatal to Olmsted's theory, however. It was acceptable to envisage that the comet-like body producing the Leonids has an orbit with a period of exactly half a year, and an aphelion near the earth's orbit. But it was pushing probability too far to suggest that a second annual shower, i.e. the Perseids, was caused by a similar body in such a closely prescribed orbit. Then in 1839 Adolf Erman of Berlin suggested a way out of the dilemma. He proposed that meteoric material does not orbit the sun in a comet-like cloud, but rather in a ring, thus eliminating any requirement to have orbital periods that are a simple fraction of that of the earth. The showers simply occurred when the earth passed through the ring on an annual basis.

In 1864 Hubert Newton (1830–1896) of Yale College, assisted by the brilliant Yale student Josiah Gibbs, collected and analysed data on the Leonid and Perseid meteors, and showed that they returned in periods of one sidereal year rather than one tropical year. As a result, the point of intersection of these meteor streams with the earth's orbit was changing at the rate of about 1° or 1 day every 70 years. Newton also showed that major Leonid showers occurred on average every 33.25 years, leading him to predict that the next major shower would occur on 13th–14th November 1866.

Newton found that the node of the Leonids' orbit was advancing at the rate of 1′ 42″.66/year. Some 50″.26/year of this figure was due to the precession of the earth's equinoxes, so the advance of the orbital node of the Leonids' orbit in space was 52″.4/year or 29′ in 33.25 years. Analysing the historical data, Newton showed that there were five possible orbital periods for the Leonid meteor stream of 0.49, 0.51, 0.97, 1.03 or 33.25 years consistent with the data. Newton himself favoured the 0.97 year figure, but an analysis was required of the perturbations experienced by particles in each of these orbits, to see which of them would show a nodal advance of 29′ in 33.25 years, in order to come to a definitive conclusion.

John Couch Adams now came on the scene, and solved the problem when he found that only a particle in a 33.25 year orbit would have the observed nodal precession of 29′. In fact, Adams found that in that orbit 20′ would be due to Jupiter, 7′ to Saturn and 1′ to Uranus, with the other planets having an insignificant effect. So the reason why the Leonid meteor showers peaked on average every 33.25 years, was because that was the period of the meteor particles in their orbit around the sun. Although there was one main cloud of particles, however, which the earth intercepted on average every 33.25 years, there were other particles strung out in the same orbit. So the Leonids were still observed annually, although the number of meteors observed on those occasions was considerably less than in the 33 year storms.

## Links with comets

As already mentioned, Olmsted had suggested in 1834 that the Leonids may be caused by particles from a comet-like body in solar orbit, interacting with the earth's upper atmosphere. However, when his proposed half-year orbit seemed unlikely, following the discovery of the Perseids, Erman had suggested that the particles could be distributed in an orbiting ring around the sun which the earth intercepted on an annual basis.

Olmsted was not the only astronomer of the period to link meteors and comets, however, as W.B. Clarke had also suggested in 1834 that the previous year's Leonid and Perseid meteor showers may be caused by cometary fragments. Nine years later, Sears Cooke Walker (1805–1853) likened meteor orbits to cometary orbits. Then in 1861 Daniel Kirkwood suggested that periodic meteors were caused by the debris of old comets whose matter had become distributed around their orbits.

Giovanni Virginio Schiaparelli (1835–1910), director of the Brera Observatory in Milan, outlined his meteor theory in 1866 in a series of letters to his fellow Italian Angelo Secchi (1818–1878). First Schiaparelli showed that the orbital velocities of meteor streams were similar to those of long-period comets. Then he hypothesised that both meteors and comets had been captured by the sun from outside of the solar system. Finally, assuming that the Perseid meteors move in a parabolic orbit, he calculated their orbital elements. This showed that the Perseids' orbit was, within error, the same as that of the comet Swift–Tuttle (1862 III) as calculated in 1862 by Theodor von Oppolzer (1841–1886). This was the first identification of a meteor stream with a specific comet. So there was a ring of particles around the sun, in the same 120 year orbit as the comet, and the Perseid meteors were seen every year when the earth passed through that ring (see Figure 7.10).

Hubert Newton had predicted in 1864 that the next intensive shower of Leonid meteors after that date should occur on 13th–14th November 1866. In the event this was correct for Europe, although in the USA the shower of the following year was better. The position of the radiant was accurately measured during the intense shower of 1866, and this enabled Schiaparelli and Le Verrier to independently calculate the orbital elements for the Leonids (see Figure 7.10).

Schiaparelli's orbit for the Leonids was too inaccurate to allow the parent comet to be defined, but Le Verrier's orbit, which he described in a presentation to the French Academy of Sciences on 21st January 1867, was far better. Unknown to Le Verrier, however, Von Oppolzer had analysed the orbit of comet Tempel–Tuttle (1866 I), publishing his results on 7th January 1867. Towards the end of January, Carl Peters (1844–1894), whose father was the editor of the *Astronomische Nachrichten*, which had published Von Oppolzer's orbit for comet Tempel–Tuttle, recognised the striking similarity between that cometary orbit and that calculated by Le Verrier for the Leonid meteor

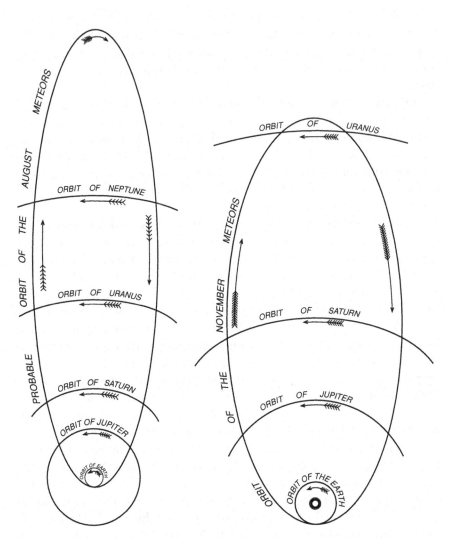

**Figure 7.10** The orbits of the August Meteors or Perseids and the November Meteors or Leonids, as deduced in the 1860s. (From *Popular Astronomy*, by Simon Newcomb, 1898, pp. 407 & 408.)

stream (see Table 7.1). A few days later, Schiaparelli also recognised the similarity of an improved orbital estimate that he had made for the Leonids and Oppolzer's orbit for comet Tempel–Tuttle. The orbits were identical within observational error.

Le Verrier was not aware of this link between comet Tempel–Tuttle and the Leonid meteors when he gave his presentation to the French Academy

Table 7.1 *A comparison of the orbits of comet Tempel–Tuttle and the Leonid meteors*

|  | Comet Tempel–Tuttle (Von Oppolzer) | Leonid meteors (Le Verrier) |
| --- | --- | --- |
| Period | 33.18 years | 33.25 years |
| Eccentricity | 0.905 | 0.904 |
| Perihelion distance | 0.977 AU | 0.989 AU |
| Orbital inclination | 163° | 165° |
| Longitude of descending node | 51° | 51° |
| Longitude of perihelion | 42° | 51° |

of Sciences in January 1867 on the origin of meteors. In that presentation Le Verrier explained that planetary perturbations, acting on a cloud of meteor particles, should eventually spread the particles uniformly around their orbit. However, in the case of the Leonids, this had clearly not yet happened, as there was a much heavier Leonid shower every 33 years or so. Le Verrier concluded that this was because the stream was relatively young, and so the cloud had not yet had time to completely disperse. To check on this, he traced the orbit of the main cloud back through successive earth encounters, to see if he could determine what had caused it to start to disperse. He was, therefore, delighted to find that in AD 126 the cloud of particles had gone relatively close to Uranus. This led him to suggest that the precursor to the Leonids had originally been following a much more elliptical orbit before 126, but that the orbit had been radically changed by the close encounter with Uranus (see Figure 7.11). A few days after this presentation, the link was discovered between the Leonid meteor stream and comet Tempel–Tuttle.

There were many other associations found between comets and meteor showers since the discoveries of the association between the Perseids and comet Swift–Tuttle, and between the Leonids and comet Tempel–Tuttle. One of the most interesting of these associations was that connected with Biela's comet.

The French amateur astronomer Jacques Leibax Montaigne of Limoges, in France, discovered a faint comet on 8th March 1772 which was observed about three weeks later by Charles Messier. It then disappeared from view in the following month. No-one suspected at the time that it was a periodic comet, but on 10th November 1805 another comet was discovered by Jean Louis Pons of Marseilles, which passed within about 6 million km (0.04 AU) of the earth. Bessel and Gauss independently calculated the orbit of this

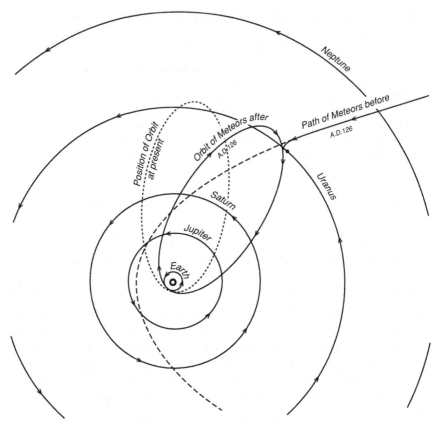

**Figure 7.11** Le Verrier's suggested path of the Leonid meteor stream showing it being affected in AD 126 by its close encounter with Uranus. (From *The Story of the Heavens*, by Robert Ball, 1897, p. 338.)

comet, and suggested that it was the same comet as had been seen in 1772. The comet was not seen again until 28th February 1826, however, when it was observed by the Austrian infantry captain, Wilhelm von Biela (1782–1856), who calculated its period to be about 6 years 8 months.[28] Biela's comet was next seen at its 1832 approach, but it was too poorly placed at its visit in 1839 to be observed. The comet was due to create quite a stir during its return of 1845–1846, however.

When Biela's comet was recovered on 26th November 1845 by Francesco Di Vico (1805–1848) in Rome, it looked perfectly normal. On 19th December

---

[28] It is not clear if the discovery was an accident, or whether Biela was deliberately looking for the return of the comet of 1805.

it appeared to be pear-shaped, but on 29th December the Americans Edward Herrick and Francis Bradley found that there were now two comets. These two comets gradually separated so that they were about 250,000 km apart when they disappeared from view in April 1846. Later analysis indicated that Biela's comet may have been broken up during a close encounter with Jupiter in 1841. On their next appearance in 1852 the two comets were about 2 million km apart, but they were never to be seen again on subsequent orbits. In 1867 Edmund Weiss (1837–1917) of Vienna found that the earth passed within about 0.018 AU of the orbit of Biela's comet on 28th November each year. As a result he predicted a Bielid meteor shower, in place of the comet, on about 28th November 1872 or 1879. He was proved correct, as on 27th November 1872 the earth was treated to an impressive display of meteors, and on 27th November 1885, about two orbits later, there was an even more spectacular display.

## 7.5 Kreutz sungrazing comets

The tail of the Great Southern Comet of 1880 (1880 I) was first observed at Cordoba in Argentina just after sunset on 31st January 1880 very close to the south-western horizon. But it was not until 4th February that its head could be seen very close to the sun. By the following day its tail, which was virtually straight, was about 50° long, reminding Janisch, who observed it from St Helena, of the Great March Comet of 1843 (1843 I). The tail was initially not as bright as that of comet 1843 I, however, and it rapidly lost intensity over the next week or so.

It was clear to the most casual observer that comet 1880 I was moving very fast, very close to the sun, just like comet 1843 I, and the first orbital calculations of Gould, Hind and Copeland independently showed that the orbits were virtually identical near the sun. The problem was that Hubbard of Washington had calculated an orbital period of 533 years for comet 1843 I, so how could it have returned just 37 years later? It did not appear to have passed close enough to any planet to have had its orbit changed so dramatically, so what had happened?

Wilhelm Meyer pointed out that, as both comets had been observed for only a very small part of their orbit, small errors in observation could have produced large errors in their orbital periods. Using this argument, he managed to convince himself, but very few others, that the comets 1843 I and 1880 I were reappearances of the same comet in an orbit with a period 37 years. If that was so, however, the comet should have been observed at regular intervals of 37 years before 1843, but there was no evidence of this. So his theory looked suspect.

A superficially more attractive theory was based on the effect of the resisting medium close to the sun, as proposed by Encke. The centre of comet 1843 I had passed only about 125,000 km above the surface of the sun, so if there was such a resisting medium, it would certainly have slowed down this comet and reduced its orbital period dramatically. Von Oppolzer showed that if 1843 I had approached the sun in a parabolic orbit, for example, and if the sun was surrounded by a resisting medium whose density varied inversely as the square of its distance from the sun, as Encke had proposed, then its orbital period after its first perihelion passage would be about 24 years. This looked a very promising result but, unfortunately, the observations of both comets 1843 I and 1880 I showed that such a rapid change in orbit had not taken place.

Wilhelm Klinkerfues (1827–1884) proposed that the comets of 371 BC, 1668, 1843 I and 1880 I were one and the same comet that had undergone a progressive reduction in velocity at perihelion. Gould, on the other hand, suggested that the comets 1843 I and 1880 I were the same as those of 1688 and 1702. In fact all sorts of theories were put forward linking 1843 I and 1880 I with various other comets, but all the theories that explained how 1843 I and 1880 I could be one and the same comet were seriously flawed in one way or another. As a result, the only reasonable conclusion appeared to be that these two comets were the remnants of a previous comet that had broken up like Biela's comet, these two remnants being in essentially the same long period orbit some 37 years apart.

Then in 1882 another sungrazing comet appeared. The Great September Comet (1882 II) was first seen from Auckland, New Zealand, on 3rd September, and by Finlay of the Cape Observatory, South Africa, five days later. It approached to within about 500,000 km of the surface of the sun on 17th September, but on 30th September its nucleus was seen to have broken into two by observers both at the Cape of Good Hope and in the United States. It fractured into three components in mid October, and by 27th January 1883 it had five separate nuclei all orbiting the sun in line looking like 'pearls on a string', to use a contemporary description.

Towards the end of the nineteenth century Heinrich Kreutz (1854–1907) analysed the orbits of many sungrazing comets, finding that 1843 I, 1880 I and 1882 II had similar orbital elements. As a result, he concluded that they were all parts of the same original comet that had been broken up by the sun at perihelion many years previously. This family were now travelling in similar orbits of large eccentricity with periods ranging from 500 to 800 years. Today this group of so-called Kreutz sungrazers has, according to Brian Marsden, four members observed before 1900, namely 1843 I, 1880 I, 1882 II and 1887 I, plus possibly the comets of 1668 and 1702 and the one seen during the solar eclipse of May 1882.

## 7.6 Mercury

There was considerable uncertainty in the nineteenth century about the mass of Mercury, as it had no satellite to enable an accurate mass estimate to be made. However, Isaac Newton had pointed out (see Section 4.7) that the densities of the earth, Jupiter and Saturn reduced with increasing distance from the sun, and suggested that the unknown densities of Mercury, Venus and Mars would probably fit into this sequence. On this basis Joseph Lagrange had deduced a reciprocal mass for Mercury of about 2,026,000.

The discovery of Encke's comet in the early nineteenth century (see Section 7.3) enabled the first accurate mass estimate to be made of Mercury, by observing the deviations caused in the comet's orbit during one of its closest approaches to the planet. As a result, in 1835 Johann Encke calculated a reciprocal mass for Mercury of about 4,700,000. Von Asten later deduced a value of about 7,600,000, using Encke's closest approaches from 1814 to 1848, whilst in 1894 Backlund deduced a value of about 9,600,000, using similar data from 1871 to 1891. This compares with a modern value of 6,023,600.

A luminous ring had been seen around Mercury by John Flamsteed when observing the planet's transit across the sun on 5th May 1707. As a result, he concluded that the planet was surrounded by a thick atmosphere. A similar luminous ring was also seen by De Plantade from Montpellier during the transit of 1736, and again by Prosperin and Flaugergues during the transits of 1786 and 1789. Other observers of the period did not see the ring, however.

In April 1792 Johann Schröter concluded from his observations of the gradual reduction of light on Mercury's partially illuminated disc that the planet has an atmosphere. He was more confident of this when he, and a number of other observers, saw the luminous ring around the planet during the transit of 1799. However, William Herschel could find no evidence of the ring during the transit of 1802. Nevertheless, during the nineteenth century the ring continued to be reported from time to time. For example Moll of Utrecht reported in 1832 that the ring appeared to have a violet tinge. Huggins and Stone observed the ring in 1868, as did Christie and Dunkin from Greenwich in 1878, but a larger number of equally reliable observers could not see it. As a result, by the end of the century most, but not all astronomers had concluded that the ring was an optical illusion.

An apparently obvious way of detecting Mercury's atmosphere, if it has one, is to detect its spectral lines, but that is more difficult than it appears. This is because the light coming from Mercury is scattered sunlight, so Mercury's spectrum is covered in Fraunhofer solar lines. In addition the light has to come through the earth's atmosphere, which adds its own so-called telluric lines to the spectrum. As a result, it is difficult to disentangle from Mercury's spectrum which lines are Fraunhofer lines, which are telluric lines, and which lines are due to Mercury itself. Nevertheless, in 1871 the respected astronomer and spectroscopist Hermann Vogel (1841–1907) concluded that the water vapour lines in Mercury's light seemed to be slightly more intense than those in the telluric spectrum, indicating that Mercury has an atmosphere containing water vapour. In the same year Angelo Secchi concluded from his observations that not only does Mercury have a dense atmosphere, but there are also clouds in it.

Notwithstanding the above, nineteenth century photometric observations tended to suggest that Mercury has, at most, a thin, mainly transparent atmosphere. Friedrich Zöllner (1834–1882) had found that the change in intensity of light with phase[29] is different for a smooth sphere compared with a rough one. In 1874 he measured this variation of intensity with phase for Mercury and concluded that it was very similar to that of the moon. He also measured Mercury's albedo as about 0.13, which again is almost the same as the moon's. These phase and albedo results implied that the surface of Mercury was rocky like the moon. They also indicated that Zöllner was measuring the surface of Mercury and not its atmosphere, so if there is an atmosphere on the planet, it must be thin and transparent. Gustav Müller (1851–1925) at the Potsdam Astrophysical Observatory made similar, more extensive photometric measurements between 1885 and 1893, with similar results.

In March 1880 Johann Schröter observed that the southern horn of Mercury's crescent was clearly blunted, with a small illuminated point just inside the unilluminated part. He attributed this to a mountain of about 55,000 ft (18,000 m) height on the limb. This is an enormous height for a planet that has a diameter of less than 40% that of the earth, so his mountain idea was never generally accepted. Schröter also noted that the terminator had an indented edge, indicating that there were other large mountains on

---

[29] That is going from a crescent phase to a fully illuminated phase (or vice versa) of the disc.

the planet. His study of the illumination of his assumed mountain with time led him to conclude that Mercury's rotation period is 24 h 4 min. In the following year Schröter and his assistant Karl Harding also recorded a central streak and patches on Mercury whose appearance also changed over time. From these observations Bessel concluded that Mercury rotates about an axis inclined at about 70° to its orbital plane, with a period of 24 h 0 min 53 s. Then in 1867 Prince deduced a period of 24 h 5 min 30 s by observing the movement of similar markings. So a period of about 24 hours seemed assured.

As Mercury can never be more than 28° away from the sun in our sky, it is very close to the horizon when it is normally observed, either just before dawn or just after sunset. Because of this, Mercury is seen through a great depth of the earth's atmosphere, which is usually very turbulent relatively close to the ground. This makes it very difficult to see any surface features on a disc that is never more than about 13″ in diameter. In 1882, however, Schiaparelli of the Brera Observatory in Milan decided on the bold new strategy of observing the planet in broad daylight. This meant that he could observe Mercury when it was high in the sky, far away from the atmospheric turbulence near the horizon. It is true that the sun is very bright, and the lack of a dark atmosphere at the observatory significantly reduces the contrast of the image, but Schiaparelli found that the much reduced turbulence more than made up for the brightness of the sun and sky. It also meant that he could now observe Mercury for hours on end, and not be limited to a short observing window every 24 hours either just before dawn or just after sunset.

With his new observing strategy, Schiaparelli soon noticed that Mercury's appearance was not changing significantly over the course of a day, and so it was definitely not rotating in anything like the 24 hours previously thought. Over the period from 1882 to 1889 Schiaparelli produced about 150 drawings of Mercury that allowed him to produce its first map. He described the features that he had detected and sketched as being of a very delicate nature, being light brown in colour against a rosy background. Because of the varying contrast of the markings with time, and with distance from the centre of the disc, Schiaparelli concluded that Mercury must have a cloudy atmosphere. From the movement of the markings he also concluded that Mercury's rotation axis is almost perpendicular to the plane of its orbit, and that its rotation period is the same as its sidereal orbital period of 87.969 days. So Mercury's axial rotation period is locked to its orbital period like the earth's moon. But, because of Mercury's highly eccentric orbit, the sun illuminates about 63% of its surface over the course of a Mercurian year. Interestingly, Kirkwood had predicted this synchronous rotation over twenty years earlier, because of the tidal effects on its crust by the relatively close sun, which would have acted as a brake on Mercury's rotation.

The American astronomer Percival Lowell (1855–1916), observing from Arizona and Mexico in 1896 and 1897, agreed that Mercury's rotation was synchronous, but he thought that the planet's surface markings were much clearer than Schiaparelli reported. As a result Lowell was not sure if Mercury has an atmosphere but, if it does exist, he was sure that it must be very thin.

So by the end of the nineteenth century the idea of a synchronous rotation period was thus gaining acceptance, but there was still no agreement on whether Mercury has an atmosphere or not, with or without clouds.

## 7.7 Venus

Jacques Cassini (1677–1756) had deduced a rotation period for Venus of 23 h 20 min in 1740 by analysing his father's observations of spots on the planet (see Section 4.3), and Mikhail Lomonsov had interpreted the ring seen around Venus during the transit of 1761 as being due to Venus' atmosphere (see Section 5.4). In addition, in 1646 Francesco Fontana had attributed the irregularity of the Venus terminator to mountains. These conclusions about Venus have an uncanny resemblance to the approximately 24 hour rotation period deduced for Mercury in the early nineteenth century, the observation of a ring around Mercury during its transits of 1707, 1736, etc., which had been interpreted as being due to its atmosphere, and Schröter's conclusion that Mercury had high mountains because of the irregularity of its terminator. In some cases the conclusions about Venus had predated those of Mercury, but in the case of the ring it was the other way around. Were these planets really so similar, even though Mercury is so much closer to the sun and so much smaller than Venus? The nineteenth century observations of Mercury have already been described, so those of Venus will now be discussed.

Venus is far easier to observe than Mercury as it is larger and comes nearer to the earth at closest approach, at that time being five times the maximum angular diameter of Mercury. In addition Venus can be observed against a dark sky well above the earth's horizon either before dawn or after dusk.

William Herschel observed Venus on a number of occasions in 1780 and noticed markings on the planet which changed over time. Unfortunately, he found that these changes were too random to allow an estimate to be made of its rotation period. Nevertheless, from these observations he concluded that

Venus 'has an atmosphere..., from the changes I took notice of, which surely cannot be on the solid body of the planet'.[30] As a result Herschel concluded that Venus has a thick atmosphere, which would shield the inhabitants from the excessive heat of the sun.[31]

In 1792 Schröter observed that the brightness of the planet decreases towards its terminator, and that the horns of the crescent Venus are often seen extended beyond the semicircle. These effects he felt sure were due to Venus having an atmosphere. Its existence was confirmed in 1842 when Guthrie observed a thin luminous ring surrounding Venus when it was almost in front of the sun. This ring was again seen by Prince in 1861, Lyman in 1866 and 1874, and later by many other observers. Interestingly, on 26th and 27th September 1878 Venus and Mercury were observed so close together by James Nasmythe that both could be seen at the same time in his telescope. He noticed that, in spite of the fact that the intensity of the sunlight on Mercury is far greater than on Venus, Venus appeared very much the brighter. This was a vivid demonstration of their radically different albedos which, according to Zöllner, were about 0.70 for Venus and 0.13 for Mercury. It was also another strong indication that Venus is almost covered in clouds.

Percival Lowell astonished the astronomical community when he reported to the Boston Scientific Society, in October 1896, that he had seen clear linear markings on Venus which appeared to be permanent. As a result he concluded that Venus' atmosphere was transparent. Lowell and his colleagues at the Flagstaff Observatory in Arizona were the only professional astronomers to record such clearly defined markings, however. Barnard failed consistently to find them using the 36 inch (91 cm) Lick refractor, and Antoniadi (1870–1944) had no more success with the 33 inch (84 cm) Meudon refractor. So the Flagstaff observations were generally considered to be spurious, and the concept of a cloud-covered planet retained.

It was clear that Venus definitely has an atmosphere, but what is it composed of? Until the advent of spectroscopy in the second half of the nineteenth century, this was an impossible question to answer, of course. Naturally, the two key constituents that the early spectroscopists and astronomers hoped to find were oxygen and water vapour, which would indicate that there could be life on Venus.

The first evidence of water vapour in Venus' atmosphere was found spectroscopically during the transits of 1874 and 1882 by Tacchini and Riccò in

---

[30] W. Herschel, *Philosophical Transactions*, 1793, pp. 201–209.

[31] William Herschel took it for granted that Venus would be inhabited, as it is about the same size as the earth and has a thick atmosphere.

Italy and by Young in America. Later, Huggins, Vogel, Janssen and Maunder independently concluded, by comparing the spectrum of Venus with that of the moon,[32] that there is oxygen and water vapour in Venus' atmosphere. In 1894, however, W.W. Campbell (1862–1938) found, using the powerful telescopes at the Lick Observatory, that he could not detect any difference between the spectrum of Venus and the moon when they were at the same altitude.

It was appreciated that, as Venus appeared to be almost completely covered in cloud, the movement of any dark or light patches may not provide a true estimate of the planet's axial rotation period, although any inaccuracy due to this was expected to be small.[33] In 1788 Schröter deduced a rotation period of 23 h 28 min by observing the movement of a 'filmy streak'. Then in December 1789 Schröter observed that the planet's southern horn appeared to be too short, with a bright point just inside the dark region (just like Schröter was to see for Mercury in 1800), which he attributed in 1792 to a mountain an incredible 140,000 ft (40,000 m) high. Timing the reappearance of this bright spot yielded a rotation rate of 23 h 21 min.

Surprisingly, perhaps, Schröter's idea of a mountainous south polar region lasted well into the nineteenth century, even though it had been dismissed out of hand by William Herschel in 1793. For example, in 1878 the French astronomer Leopold Trouvelot commented that 'This surface [of Venus]... resembles that of a mountainous area with numerous peaks, or our polar regions with numerous bright pieces of ice reflecting the sunlight'. Then in 1888 Camille Flammarion (1842–1925) endorsed Fontana's view that the irregular terminator was due to mountains.

Over the next few decades after Schröter, Fritsch (in 1801), Di Vico (in 1841) and Denning (in 1881) all came up with an estimated rotation period for Venus of about 24 hours, although one observer, Hussey, produced a figure of 24 *days*. Then in 1890 Schiaparelli followed up his announcement of the synchronous rotation period of Mercury with a similar conclusion for Venus, giving a rotation period for Venus of 224.7 days. Shortly afterwards Henri Perrotin came to the same conclusion, based on a series of observation between May and October 1890, followed by confirmatory results from a number of astronomers including Percival Lowell, but others stood by the

---

[32] It was assumed that, if the moon did have an atmosphere, it would be so thin as to have no measurable effect on the spectrum of the sunlight reflected or scattered off its surface.

[33] Measuring cloud movements on the earth, instead of measuring the movement of its surface, would only give a maximum error ~5% in its rotation period.

24 hour figure. So at the end of the nineteenth century Venus' rotation period was still an open issue, with the two main alternatives being around 24 hours or the synchronous rate of 224.7 days. The idea of synchronous rotation had its theoretical attractions, of course, as it could easily be explained, as for Mercury, by the long-term effect of tidal friction on the planetary surface.

The presence of life on Venus was also an open issue at the end of the nineteenth century. William Herschel had been convinced, at the end of the eighteenth century, that life exists on Venus, but he also thought that it exists on the sun (beneath the hot photosphere) and on the moon. Although some of Herschel's ideas seemed bizarre, nevertheless the fact that a century later astronomers knew that Venus:

- orbits the sun just inside the earth's orbit
- is about the same size as the earth
- has a substantial atmosphere

and may well have:

- oxygen and water vapour as major atmospheric constituents
- a rotation period the same as the earth's

seemed too much a coincidence for many of them. It looked as though Venus was almost a twin of the earth, so many astronomers thought that it may well be inhabited. In particular, in 1880 Flammarion concluded that Venus 'should, then, be inhabited by vegetable, animal and human races but little different from those which people our planet'.[34] Or, as Sir Robert Ball commented in 1897, 'If water be present on the surface of Venus and if oxygen be a constituent of its atmosphere, we might expect to find in that planet a luxuriant tropical life, of a kind analogous in some respects to life on earth'.[35]

## 7.8 The moon

### Surface and atmosphere

Johann Schröter, one of the key lunar observers of the eighteenth century, made a special study of lunar rilles, recording eleven between 1787 and

---

[34] Camille Flammarion, *Popular Astronomy*, trans. J. Ellard Gore, Appleton, New York, 1907, p. 347.

[35] Sir Robert Ball, *The Story of the Heavens*, new and revised edn. Cassell, London, 1897, p. 143.

1801.[36] Many more rilles were discovered by other observers in the nineteenth century. They had linear, curved or branching tracks, and were thought by most astronomers to be cracks in the surface of the moon caused by surface cooling.

Schröter also recorded what he thought were a number of small surface changes on the moon, particularly in the floor of the craters Posidonius and Cleomedes, and in February 1792 he detected what he thought were signs of lunar twilight. This confirmed him in his view that the moon has an atmosphere, for which he estimated a density of about 3% that of the earth's atmosphere. Schröter also believed that the moon supported vegetation in places, and that it was probably inhabited.

In 1783 and 1787 William Herschel reported that he had observed volcanic eruptions on the moon, but his observations were later attributed to rapid changes in the side lighting of the lunar peaks. He also believed, like Hevelius before him, that 'lunarians' existed but, on Nevil Maskelyne's advice, he watered down his published views on the subject. Then in 1816 Franz Gruithuisen of Munich reported that he had seen clouds on the moon, and six years later he announced that he had detected a city on the southern edge of the Sinus Aestuum. He thought that the rilles were dry riverbeds or roads, and he agreed with Schröter that there was vegetation on the moon in a reasonably dense atmosphere.

Bessel realised that probably the best method of estimating the density of the moon's atmosphere, if it has one, was to try to observe the refraction of starlight just before or just after a lunar occultation. On this basis he estimated in 1834 that the density of the moon's atmosphere could be no more than about 0.2% that of the earth. This was confirmed by Simon Newcomb later in the century, although John Herschel came up with an even smaller figure. By the end of the nineteenth century maximum density figures of about 0.02% of the density of the earth's atmosphere were in favour. In short, it was thought that if the moon has an atmosphere it is very tenuous.

In 1837 Wilhelm Beer (1797–1850) and Johann Mädler (1794–1874) produced a highly influential book entitled *Der Mond*, which almost killed off interest in the moon as a dynamic body. In their view, virtually all of the reported surface changes on the moon could be explained by tricks of the light. As the shadows cast by lunar mountains are jet black and very sharp,

---

[36] Huygens was the first to record a rille in about 1685, but his lunar observations were generally unknown until they were published in 1925.

they concluded that if the moon has an atmosphere it must be very thin. As a consequence the lunar surface must be dry, with no vegetation, and it is geologically dead.

This concept of a geologically dead moon was generally accepted following Beer and Mädler's work. But in October 1866 J.F. Julius Schmidt (1825–1884), the director of the Athens Observatory, shook lunar astronomers when he announced that the small crater Linné in the Mare Serenitatis had disappeared and had been replaced by a white spot. In 1824 Wilhelm Lohrmann (1786–1840) of Dresden had drawn Linné as an 8 km diameter crater in his acclaimed lunar atlas, and in 1837 Beer and Mädler had described Linné in *Der Mond* as a 10 km crater which had an indistinct outline at full moon. In 1867, just after his discovery of its changed appearance, Schmidt mentioned that he had previously observed Linné five times over the period from 1841 to 1843, each time recording it as a crater. So Schmidt concluded that it had changed its appearance from a crater to a white spot somewhere between 1843 and 1866.

On 11th February 1867 Secchi observed a very small crater at the centre of Schmidt's white spot. Then Mädler, who had first observed Linné in 1831, re-examined it on 10th May 1867 to see if it had changed its appearance. He found that it appeared to be exactly the same as in 1831. Nevertheless, Schmidt was convinced that the original crater had been largely obscured by a new eruption of lava, but others were not sure. On the face of it, an 8 to 10 km crater that had been observed between 1824 and 1843 had by 1866 become a white spot of about the same size with a smaller crater at its centre, although Mädler claimed that its appearance had not changed. So what had happened? The key to the mystery appears to be in Schmidt's book of 1878 in which he lists nineteen of his drawings made between 1841 and 1843, on two of which Linné is shown as a bright spot!

The mystery of Linné was finally solved when it became clear that its appearance is strongly dependent on illumination conditions. When the sun is overhead only a white spot can be seen, with a diffuse edge, but a few days earlier or later, when there is some side lighting, the small 2.5 km crater detected by Secchi is seen in the middle of the white spot. When the sun is almost on the horizon, as seen from Linné, the white spot disappears and only Secchi's small crater can be seen. This small crater was too small to have been observed by Lohrmann in 1824, Mädler in the 1830s, or Schmidt in 1841–1843, so the explanation of the various historical observations appears to be as follows.

It appears that Lohrmann and Mädler had both observed Linné when the lighting was almost directly overhead, when it would appear as a white spot. Instead of drawing it as a white spot, however, they had drawn it as a crater. In fact, we know that Mädler drew two other white spots on the moon as craters, so this seems a plausible explanation. Mädler also mentioned in his book that Linné had an indistinct outline at full moon, which we know is the case with the white spot.[37] Furthermore, over the period 1841–1843 Schmidt sometimes recorded Linné as a crater and sometimes as a white spot. So in reality Linné had not changed its appearance in the nineteenth century, but it took some time for this to become clear, and even at the end of the century there were still some astronomers that believed that it had.

From time to time in the nineteenth century, claims were made that other changes had been detected on the moon, probably the best known being Hermann Klein's 1877 observation of an apparently new crater near to the crater Hyginus, but all these claims were hotly disputed. Even if one or two changes had occurred, however, they were clearly only minimal when considering the moon as a whole. Some may have been caused by meteorite impacts, and there may have been occasional minor outgasing episodes, but basically the moon is geologically dead.

In the seventeenth century Robert Hooke had shown that lunar craters could have been produced either by volcanic activity, which he favoured, or by the impact of small bodies (see Section 4.5), although he could not think of a source of such objects. With the discovery of the first asteroids in the early nineteenth century, however, the bombardment idea was resurrected. The Bieberstein brothers, in particular, suggested that lunar craters may have been formed by meteorite impact. Gruithuisen had a similar theory in 1824, as did Proctor in 1873, but what puzzled the meteorite protagonists of the nineteenth century was the fact that lunar craters are almost all circular, whereas most meteorite impacts would have been at an angle to the vertical. So the craters should be elliptical if they have been caused by such impacts. As a result most astronomers of that period thought that the lunar craters were caused by volcanic eruptions, in the following way:

- In the early stages of a volcanic eruption, when the velocity of the ejecta was high, the ejecta would spray out in an umbrella-shaped plume, falling

---

[37] The sun is almost directly overhead at Linné at full moon, as Linné is only just off the centre of the moon as seen from earth.

back to the surface of the moon in a ring centred on the volcanic vent. This created the crater walls.

- As the eruption lost energy, the spray would cease, and the lava would flow slowly onto the surface, cooling as it went and forming the central cone.
- Finally, the lava may continue to flow out of the vent and flood the crater floor to such a depth that the lava covered the central cone, so making the cone invisible.

There were a number of problems with such a theory, however. Probably the most puzzling being that pointed out by the American geologist Grove K. Gilbert in 1892, that lunar craters have floors generally below the height of their surroundings, whereas on earth the floors of volcanic craters are generally higher, and in some cases much higher, than the surrounding area.

In summary, at the end of the nineteenth century most astronomers thought that the moon is geologically dead with, at most, only a few trivial surface changes now taking place. There is virtually no atmosphere and, as a result, all surface water that may have existed in the distant past would have evaporated and be lost to space. With virtually no atmosphere and no water there is no life. It was unclear as to whether the craters which littered the moon's surface were a result of volcanic activity or meteorite impacts, but the former was generally thought to be the case.

## Secular acceleration re-examined

Edmond Halley had discovered that the moon's position in the sky was in advance of where it should be based on ancient eclipse records (see Section 5.5). This so-called secular acceleration of the moon could either be because the moon was accelerating in its orbit, and/or because the earth's rotation rate was slowing down. In 1787 Laplace had shown that the effect could be completely explained by the moon's orbit gradually being reduced in size, caused by a reduction in the eccentricity of the earth's orbit.

In 1754, some years before Laplace had finalised his theory, Kant had published his theory of tidal friction. Kant proposed that tidal friction, caused by the gravitational attraction of the earth on the still molten early moon, had caused the moon's axial rotation to be reduced, so that one side of the moon is now permanently facing the earth. Interestingly, Kant also explained that the moon would have a similar effect on the earth, causing the earth's rotation rate to slow down.[38] Apparently Laplace was unaware of Kant's conclusions when he undertook his work on the moon's secular acceleration. But Laplace's ignorance of Kant's theory did not seem to matter, as his calculated magnitude of the effect caused by the reduced eccentricity of the earth's orbit was $10''.2/\text{century}^2$, which was, within error, the same as the observed magnitude.

In the early nineteenth century various astronomers tried to refine Laplace's theory, also in ignorance of Kant's tidal theory. In particular Peter Hansen calculated a value of $11''.5/\text{century}^2$, compared with an observed value of $12''.2/\text{century}^2$. But in 1853 John Couch Adams showed that some of the second order terms that Laplace had omitted from his calculations, thinking that they were unimportant, were, in fact, significant. Including them reduced the effect of the increase of the moon's orbital rotation, due to the reduction in the eccentricity of the earth's orbit, to just $6''.1/\text{century}^2$, or about half of that observed.

At first Adams' work was heavily criticised, because theory and observations had previously agreed within observational error. However, in 1866 Charles Delaunay (1816–1872), of the Paris Observatory, suggested that the other half of the effect was probably due to tidal friction. The effect of tidal friction would be to slow down the axial rotation of the earth, and increase the angular momentum of the moon, expanding its orbit and slowing down its orbital motion. In terms of the longer days, the month is apparently shortened, however, and the moon's motion apparently accelerated. Unfortunately, it was impossible at that time to produce a reasonably accurate estimate of the effect.

Near the end of the eighteenth century, Laplace had found evidence of an oscillation in the position of the moon in the sky with a period of some hundreds of years, that was superimposed on all other effects. In 1846 Hansen suggested that this was caused by perturbations of Venus on the moon, producing a maximum deviation of $21''$ in the moon's position over a period of 240 years. Delaunay also calculated the magnitude of this effect and disputed Hansen's value, which Hansen was then forced to admit was empirical. To make matters worse, the moon then deviated from Hansen's empirical tables over the next few decades, and by 1880 the moon was lagging $10''$ behind its predicted position.

Simon Newcomb decided to get to the bottom of the problem of the moon's apparent motion, by analysing all ancient eclipse observations, and

---

[38] The earth did not yet have one side permanently facing the moon, as it has much more energy to dissipate, as it is much larger than the moon.

all star occultations by the moon since the first telescopic observations. This led him in 1878 to conclude that the observations could be explained by adding an empirical term of 16″, with a period of 273 years, which was similar in value to Hansen's empirical figure. The cause, however, was still a mystery at the end of the nineteenth century. Newcomb speculated that it could be either because the moon's motion is influenced by something other than the gravitational effects of the sun, earth and planets, or because the earth's day is varying in a periodic manner, in addition to it gradually lengthening due to the tides.

## The origin of the moon

Laplace, in his theory of the origin of the solar system (see Section 6.3), had proposed that all planetary satellites had been produced from the original nebula, as it condensed to form rings of material around the protoplanets. In the second half of the nineteenth century, however, George Darwin worked out a completely different theory for the origin of the earth's moon, although it was clear that his theory was not applicable to the formation of any of the other planetary satellites.[39]

George Darwin, the second son of the famous biologist Charles Darwin, suggested in a paper given to the Royal Society in December 1879[40] that the proto-earth had gradually contracted as it cooled when it was still molten. Its rotation rate had increased as it contracted (according to the principle of conservation of angular momentum), and when its rate had increased to about 3 hours per revolution, it had broken into two unequal parts, one being the earth and the smaller one the moon. Darwin realised that the rotation of the proto-earth would not have been sufficient on its own to cause rupture, and suggested that the breakup had been facilitated by a resonance[41] coupling between the tidal forces acting on the proto-earth from the sun, and the natural oscillation frequency of the molten body.

After breakup of the proto-earth, both the earth's rotation on its axis, and the moon's rotation around the earth, had slowed down because of tidal forces, as both bodies were still molten. This caused the moon's orbit

---

[39] This is because none of the other planetary satellites are as large as the moon, relative to the size of the planet about which they revolve.

[40] Darwin worked out the complete details of his theory over the following two years.

[41] Resonance occurs when the frequency of application of periodic forces acting on a body exactly matches the natural oscillation frequency of that body.

to gradually become larger. Tidal forces of the earth on the less-massive moon locked the axial rotation period of the moon to that of its orbital rotation period around the earth, as suggested by Kant, thus ensuring that the moon always had the same side facing the earth. The gravitational attraction of the earth also caused the moon to have a small earth-facing bulge, which remained facing the earth, even after the moon had almost solidified.

Originally when the moon had just separated from the earth, the time taken for the moon to orbit the earth (i.e. the month) was the same as that taken for the earth to rotate on its axis (i.e. the day) at about three of our current hours. Both the month and day have been getting longer since then, but the ratio of the month to the day has increased from the initial value of 1 to about 27 today. This ratio was thought to have gone through a maximum value of about 29 some time ago, with it being 27 now and reducing. If the earth–moon system could be isolated from all other outside influences, particularly that of the sun, this ratio would probably continue reducing until it was 1 again, with both periods being equal to about 1,400 hours (about 58 of our current days). In this case the earth, as well as the moon, would have its same side permanently facing the other body.

Darwin had shown theoretically that the proto-earth could have been unstable at a speed of about 3 hours per revolution, but it was left to A.M. Lyapunov to show that, in such a case, the body would have naturally broken up into two unequal parts.[42]

In the late nineteenth century it was recognised that there was a possible problem with Darwin's theory, as there would have been a tendency for the earth to break up the moon by tidal forces when it was still very close. This break-up would not happen immediately after separation, however, and the moon could possibly have moved away from the danger area before it had had time to break up. Whether this was possible or not was unclear.

If the earth and moon had once been part of the same body, the internal temperature of the moon today would be much lower than that of the earth, as the moon is much smaller and would, therefore, have cooled down faster. No active volcanoes had been seen on the moon in the nineteenth century, but it was not known whether the centre of the moon was still molten or whether it had solidified. Either way it was clear that the surface of the moon was very stable.

---

[42] The theory of the rotation of a fluid sphere was progressively developed by Maclaurin (1740), Jacobi (1834), and Poincaré (1885), but it was A.M. Lyapunov who showed that it would break into two unequal masses.

## 7.9 **The earth**

Polar motion

In the second century BC Hipparchus had discovered the precession of the equinoxes (see Section 1.4). This is due to the earth's spin axis moving in a conical motion in space at a rate of 50″.3/year, giving a period of about 25,800 years. The physical cause of this precession was a mystery until Newton showed that it is due to the gravitational attraction of the moon and sun acting on the earth's equatorial bulge (see Section 4.7). Then Bradley found in the eighteenth century that the earth's axis is nutating by about ±9″, as it precesses, with a period of 18.6 years (see Section 5.3). So the earth's spin axis nutates, or wobbles slightly as it precesses in space.

Laplace had estimated the oblateness or polar flattening of the earth to be about $\frac{1}{305}$ in the early nineteenth century from its effect on the orbit of the moon (see Section 5.5). But as the nineteenth century progressed it became clear that the shape of the earth is not an ellipsoid, but is an irregular shape caused by the non-symmetrical distribution of mass in the earth's interior.

In 1765 the Swiss mathematician Leonard Euler (1707–1783) had theoretically examined the case of a body like the earth in which the spin axis does not exactly coincide with its shortest diameter. In that case he showed that the spin axis, or poles, should move in a circle relative to the surface of the rigid body in a period of about 305 days or 10 months. It was natural to ask whether the earth's poles move on the surface in that way. If they do, the geographical coordinates of every place on the earth will vary, relative to the earth's spin axis, over this period of about 10 months.

It should be possible to detect this so-called polar motion by accurately observing the positions of a few fixed stars over a year or so, and correcting the observations for the effects of precession and nutation mentioned above. In addition a correction would have to be applied for the effect of the aberration of light of about ±20″, with a period of one year (see Section 5.3), and of stellar parallax, also with a period of one year, if the selected stars are relatively close to the earth. Several attempts were made to observe this polar motion, but all failed until it was discovered by accident in 1888. In fact, the effect was actually in data produced earlier. For example it was in Washington observations from 1862 to 1867, but astronomers were misled into looking for Euler's 10 month period, which turned out to be incorrect, as will now be explained.

In 1862 Hubbard had began a series of observations of the star α Lyrae at the Washington Observatory to determine the constants of aberration and nutation and the parallax of the star. Hubbard died in August 1863, but the observations were continued by Simon Newcomb, William Harkness (1837–1903) and Asaph Hall until the observations were terminated in April 1867. These observations had not taken place over a long enough period to give an accurate estimate of the constant of nutation, because the period is 18.6 years. But they did yield an accurate value for the constant of aberration of $20''.4542 \pm 0''.0144$ (modern value $20''.4955$), and of the parallax of $\alpha$ Lyrae of $0''.134 \pm 0''.006$ (modern value $0''.123$). In addition, Hall deduced a value of $8''.810 \pm 0''.006$ (modern value $8''.794$) for the solar parallax from the value of the constant of aberration. Unfortunately, however, in analysing these results Hall found that the residual errors on each measurement, or even of each group of measurements, of $\alpha$ Lyrae's position was much larger than he had expected. Hall wondered if these errors were due to Euler's predicted polar motion, but in his paper[43] of 6th April 1888 Hall mentions that 'Neither will the theoretical ten-month period [of Euler] in the latitude furnish a satisfactory explanation'. But Hall was partly right and partly wrong, as was soon to become apparent.

In the early 1880s Karl Friedrich Küstner (1856–1936) had decided to make a new measurement of the constant of aberration by observing the position of seven star pairs in the spring and autumn of 1884, and the spring of 1885, from the Berlin Observatory. To his surprise, he found that the mean latitude of the stars had decreased by about $0''.2$ between the spring observations in 1884 and those in 1885. This could not be due to the aberration of light or the star's parallax, as these are both annual effects, so what was the cause? In 1888, after a thorough investigation, Küstner concluded that the latitude of the Berlin Observatory had decreased by $0''.204 \pm 0''.025$ in one year. Two years later he announced that the overall range of his observations over the period spring, autumn, spring was about $0''.5$, but he did not give a period.

In the meantime, the International Commission for Geodesy (ICG) had decided to organise concurrent observations of selected stars at radically different sites to better quantify the effect discovered by Küstner. The first measurements were made from Berlin, Potsdam and Prague in January 1889, followed by observations from Honolulu in 1891, which is about 180° away from the European sites in longitude. If Küstner had discovered a true movement of the poles, then the movement in latitude at Honolulu should be in the opposite direction to that deduced from the European measurements. This is exactly what was observed, and a period was deduced from the European measurements of about 12 or 13 months.

---

[43] A. Hall, *Astronomical Journal*, **8**, 1888, pp. 1–5 & 9–13.

Meanwhile the American astronomer Seth C. Chandler (1846–1913) had been measuring the latitude of the Harvard College Observatory over the period from May 1884 to June 1885, and had found slight changes in his monthly means over time which puzzled him. When Küstner's and the initial ICG results had been published, however, Chandler decided to look for the effect of polar motion himself in this previous Harvard data. As he admitted in his first paper on the subject on 7th November 1891,[44] he had not taken the apparent variations in the monthly means of latitude further, because 'There was no known or imaginable instrumental or personal cause for this phenomenon, yet the only alternative seemed to be an inference that the latitude had actually changed [over the period 1884–1885]. This seemed at the time too bold an inference to place upon record, and I therefore left the results to speak for themselves.' Now, prodded by Küstner's and the ICG's results, Chandler in this first paper deduced that the Harvard residuals seemed to change over a period of about 444 days,[45] with a total range of about 0″.7.

On 23rd November 1891, in his second paper,[46] Chandler concluded that the earth's pole moves in a circle of radius of about 0″.3, which is equivalent to about 30 ft (9 m) on the surface of the earth, in a period of about 427 days. In addition, he compared Küstner's results of 1884–1885 for Berlin with those from the Harvard Observatory (at Cambridge, USA), showing that the magnitude and period of the effect was similar at both. He also concluded that the Harvard latitudes were the smallest in September or October 1884, whereas those at Berlin, which is about 90° east of Harvard, were probably smallest in January or February 1885,[47] i.e. about 4 months later. As a result, he concluded, the polar motion must be from west to east.

In this second paper Chandler also analysed Gyldén's, Nyrén's and other observations at the Pulkovo Observatory of St Petersburg, Russia, over the period from 1863 to 1875. He showed that these were also consistent with the 427 day period. Finally he was able to show that the troublesome residuals in the 1862–1867 Washington measurements, outlined above, were also consistent with the 427 day (14.0 month) period. So the reason that Asaph Hall had not been able to explain these Washington residuals now became clear. They were due to polar motion, but Hall had assumed that its period

---

[44] S.C. Chandler, *Astronomical Journal*, **11**, 1891, pp. 59–61.

[45] In this paper Chandler variously quotes a half period of about 7 months, or 222 days ($7\frac{1}{3}$ months), or from 1st September 1884 to 1st May 1885, which is 8 months.

[46] S.C. Chandler, *Astronomical Journal*, **11**, 1891, pp. 65–70.

[47] This deduction was based on interpolating the measurements at Berlin, as no measurements had been made there in these two months.

would be 10 months, as predicted by Euler, whereas the true period was about 14 months. The obvious question now arose as to why the period found by Chandler was substantially longer than that predicted by Euler.

On 23rd December 1891, just one month after Chandler's second paper, Simon Newcomb came up with the answer. He explained that the reason why Chandler's period was longer than that predicted by Euler, was because Euler had produced his theory for a rigid body, whereas the earth is not rigid. The oceans move freely on the earth's surface, and the body of the earth is elastic, both of these factors causing an increase in the pole's rotation period. But Newcomb did not try to quantify the magnitude of this increase, because of the complexity of the analysis.

Chandler now went further back through observatory records world-wide, and in November 1892 he announced that the observed movement of the pole was due to two superimposed effects, one with a period of about 14 months and one with a 12 month period. The 14 month period, now called the Chandler component, is that predicted by Euler, but modified for a non-rigid earth. The 12 month period, on the other hand, is caused by meteorological effects such as the movement of water and air masses between the northern and southern hemispheres on an annual basis. Seven years later the International Latitude Service was established to track this movement of the poles.

## Magnetic variations

It is well known today, by any reader of maps, that the direction of magnetic north on the earth's surface changes gradually over the years. In fact, in 1666 a compass needle in Paris would have pointed true north,[48] but in 1700 it would have pointed 8° west of true north, in 1750 the angle was 17°, and in 1814 it had reached its maximum western value of 22½°, before the angle started to reduce. But superimposed on this gradual change in direction over the years, there is a diurnal variation which was discovered by Graham in 1722.

The diurnal variation is somewhat surprising in form starting, as it does, with a maximum eastern declination at about 8 am, and reaching a maximum western declination at about 1.30 pm. The direction of the compass needle then returns eastwards, with a temporary reversal in direction between 8 pm and 11 pm, before reaching its maximum eastern declination

---

[48] In London the compass pointed due north about ten years earlier.

again at about 8 am the following morning. The amplitude of the swing is about 10′ at moderate latitudes, but only 2′ or less in the tropics. It is less in winter than in summer.

Although this diurnal movement in declination is generally quite regular, there are occasions when, over large parts of the earth, the needle moves quite randomly during what are called magnetic storms.[49] It was to try to understand these storms in particular, and the variations of the earth's magnetic field in general, that Alexander von Humbolt recommended to the Scientific Congress of Berlin in 1828, that an international network of magnetic stations be established. In the event, Göttingen became the centre of this network which rapidly spread throughout the globe.

In September 1851 John Lamont was analysing the magnetic declinations at the Göttingen and Munich Observatories over the period 1835–1850, when he noticed a remarkable effect. The magnitude of the diurnal change showed a clear periodicity which he estimated to be about 10.3 years. Meanwhile, in 1843 Heinrich Schwabe of Dessau, Germany, had discovered that the number of sunspots on the sun seemed to vary with a period of about 10 years. His data was generally ignored, however, until it was republished in 1851 by von Humboldt in the fourth volume of his book *Cosmos*. Then in March 1852, Sir Edward Sabine, who was unaware of Lamont's discovery of a 10.3 year period in the magnitude of the diurnal magnetic effect, announced that there was a strong correlation between the frequency and severity of magnetic storms and Schwabe's sunspot cycle. This was followed by similar announcements by Rudolf Wolf in Switzerland and Alfred Gautier in France who had independently come to the same conclusion.

Rudolf Wolf then showed, by looking through historical records of sunspots going back over 200 years, that the average period between maxima was about 11.1 years, but with a range varying from 7 to 17 years. The reason for this sunspot cycle was unknown, and no-one was able to detect any associated variation in the solar heat output. However, there was much speculation about a possible link between sunspot numbers and the earth's weather, although there was little evidence of an 11 year weather cycle. Nevertheless, although no such weather cycle could be found, E. Walter Maunder at Greenwich did find in 1890 that sunspots had been virtually non-existent

---

[49] For example, the intense magnetic storm of 25th September 1841 was observed simultaneously in places as far apart as Toronto, the Cape of Good Hope, and Tasmania. Although the observations in Tasmania were not complete, as part of the storm was on a Sunday, the day of rest!

between 1645 and 1715, coinciding with a period of relatively cold winters in the earth's northern hemisphere.

In 1741 Hiorter had observed at Uppsala, near Stockholm, that the compass needle was unsteady at times of strong aurorae. Over subsequent years the linkage between individual magnetic storms and aurorae was more widely reported. For example, James Glaisher observed at the Royal Observatory, Greenwich, that[50] 'The aurora borealis of October 24, 1847, which was one of the most brilliant ever known in this country, was preceded by great magnetic disturbance. On the 22nd October the maximum of the west declination was 23° 10'; on 23rd the position of the magnet was continually changing, and the extreme west declinations were between 22° 44' and 23° 37'; on the night between the 23rd and 24th October, the changes of position were very large and very frequent, the magnet at times moving across the field so rapidly that a difficulty was experienced in following it. During the day of the 24th of October, there was a constant change of position, but after midnight, when the aurora began perceptibly to decline in brightness, the disturbance entirely ceased.'

So there was a clear link between individual auroral displays and magnetic storms, but over the years it was extended to include links between individual auroral displays, magnetic storms and sunspots. For example[51] 'On September 1, 1859, two astronomers, Carrington and Hodgson, were observing the sun, independently of each other, the first on a screen which received the image, the second directly through a telescope, when, in a moment, a dazzling flash blazed out in the midst of a group of spots. This light sparkled for five minutes above the spots without modifying their form, as if it were completely independent, and yet it must have been the effect of a terrible conflagration occurring in the solar atmosphere. Each observer ascertained the fact separately, and was for an instant dazzled. Now, here is a surprising coincidence; at the very moment when the sun appeared inflamed in this region the magnetic instruments of the Kew Observatory, near London, where they were observing,[52] manifested a strange agitation; the magnetic needle jumped for more than an hour as if infatuated. Moreover, a part of the world was on that day and the following one enveloped in

---

[50] From *Cosmos*, Vol. 1, by Alexander von Humboldt, translated by E.C. Otté, 1849, p. 188.

[51] From *Popular Astronomy*, by Camille Flammarion, translated by J. Ellard Gore, Chatto and Windus, 1907, pp. 290–291. Incidentally, this is the first known observation of a white light solar flare.

[52] This is not correct. Hodgson was, in fact, observing some miles away at Highgate.

the fires of an aurora borealis, in Europe as well as in America. It was seen almost everywhere: at Rome, at Calcutta, in Cuba, in Australia, and in South America. Violent magnetic perturbations were manifested, and at several points the telegraph-lines ceased to act.' There are many more examples of spectacular aurorae being linked to specific sunspot groups and to magnetic storms on earth, but the correlation was not perfect. Nevertheless it seemed to be established that solar activity, generally in the form of sunspots, was the origin of aurorae and magnetic storms on earth.

## 7.10 Mars

The dark, triangular-shaped region on Mars known as Syrtis Major and the Martian polar caps had been discovered in the seventeenth century (see Section 4.3), but telescopes of that period revealed no other obvious surface details. Mars has a relatively small disc in telescopes, and the poor quality of the lenses of the time, together with atmospheric turbulence, discouraged all but the most determined of observers.

One of the key observers of the period was Giacomo Filippo Maraldi, who was a nephew of G.D. Cassini. He became an assistant at the Paris Observatory, and observed Mars at every opposition between 1692 and 1719.[53] Maraldi observed Syrtis Major and other dark features which he took to be clouds as they were not always visible. More importantly, however, he made the first thorough investigation of the white Martian polar caps. He observed that the south polar cap disappeared entirely in the autumn of 1719, indicating its non-permanent nature, although it later returned. When the south polar cap was very small he also noticed that it had a slight oscillation at Mars' rotation period, indicating that it was not exactly centred on the south geographical pole.[54]

Very little in the way of useful observations were undertaken of Mars during the remainder of the eighteenth century until William Herschel started

---

[53] Because the orbit of Mars is just outside that of the earth and because it is relatively eccentric, the maximum size of Mars at opposition can be about 26″, but at conjunction it can be as low as just 4″. So the period around opposition was by far the best time to observe the planet. The best periods, of course, were around those oppositions that took place when Mars was near perihelion.

[54] It is easier to see the southern polar cap from the earth, because the southern hemisphere is tilted towards the sun and earth around perihelion. At aphelion, when the northern hemisphere is tilted towards the earth, the earth–Mars distance is almost twice as much.

a series of observations of the planet in 1777. But it was during the very close opposition of the autumn of 1783 that Herschel made his first substantial observations. At this time the south polar cap was rather small and he noticed the slight oscillation previous observed by Maraldi. Herschel had the advantage of better telescopes than Maraldi, however, and was able to deduce that the centre of the south polar cap was about 9° from the south pole. He also found that the inclination of the planet's spin axis to its orbital plane was 28° 42′ (the modern value is 25° 11′).

Southern summers on Mars occur near perihelion, so southern summers are warmer than northern summers, whereas southern winters are colder than those in the north. The same is true for the earth, where perihelion occurs in early January. This effect on Mars is much larger than on the earth, however, owing to the much higher eccentricity of Mars' orbit (0.093 versus 0.017). Nevertheless, Herschel reasoned that the similarities between Mars and the earth are very strong, as their orbits are adjacent to each other, so having a *roughly* similar level of heating from the sun, whilst their axial inclinations are almost identical. So both would have similar seasonal effects, although the Martian year is about twice as long as the earth's, and the higher eccentricity of Mars' orbit ensures that there is a larger seasonal difference between the temperatures in the two hemispheres, as explained above. As a result, he concluded that[55] 'If, then, we find that the globe we inhabit has its polar regions frozen and covered with mountains of ice and snow, that only partly melt when alternately exposed to the sun, I may well be permitted to surmise that the same causes may probably have the same effect on the globe of Mars; that the bright polar spots are owing to the vivid reflection of light from frozen regions; and that the reduction of those spots is to be ascribed to their being exposed to the sun.'

Cassini thought that he had detected the presence of a very large, dense atmosphere on Mars during his observations of 1672, when he noted that a bright star had disappeared 6′ before its occultation by the planet. However, Herschel repeated the observation with two other stars and found no such effect. He suggested, therefore, that Cassini had been unable to see the star, when it was very close to Mars, because of the glare of the planet. Nevertheless, Herschel did notice that the surface markings on Mars appeared to be transient. So he concluded that there was a 'considerable but moderate atmosphere', which was able to support 'clouds and vapours', although it was not as large and dense as the one deduced by Cassini. As a result, he

---

[55] W. Herschel, *Philosophical Transactions*, **74**, 1784, pp. 233–273.

suggested that the inhabitants would 'probably enjoy a situation in many respects similar to ours'.[55]

The idea of an inhabited Mars was by no means new. For example, earlier in the eighteenth century Kant had suggested that as the earth and Mars 'are placed in the middle of our planetary system, so that we may suppose, without improbability, that their inhabitants possess an average condition, in their constitutions as well as in their morals, between the two extremes'. On the other hand, 'the inhabitants of Mercury and Venus are too material to be reasonable, and are probably not even responsible for their actions'.[56]

Johann Schröter first observed Mars in 1785, but he was disappointed to find that he could see little in the way of surface features. In fact, at the next opposition Schröter concluded that he was observing clouds in the atmosphere, rather than the planetary surface, because the features were continually changing their appearance over the course of only a few hours. This conclusion influenced other observers of the early nineteenth century until it was rejected by Georg Kunowsky of Berlin, who was convinced that he was seeing the surface during his observations at the opposition of 1821–1822.

Ten years later, Kunowsky's interpretation was heavily endorsed by Beer and Mädler who observed Mars during its favourable opposition of 1830. They noticed that the dark features were in fixed locations, but that they were frequently covered in mists, thus changing their contrast and outlines. Syrtis Major was clearly seen, as was a small round dark patch that had previously been recorded by Herschel and Schröter. This is now called the Meridian Bay (Sinus Meridiani) as it is used to define the zero meridian of Mars. In fact, using this marking, Beer and Mädler were able in 1837 to deduce Mars' rotation period to within just over one second of its true value.

In 1830 Beer and Mädler observed that the south polar cap was at its smallest extent of about 6° about a month after the summer solstice in the southern hemisphere, adding to William Herschel's suspicion that it is made of ice and snow. Beer and Mädler also noticed during the unfavourable opposition of 1837, when the northern hemisphere was titled towards the sun and earth, that the north polar cap was about 12° at its minimum extent. But whilst the north polar cap was larger than the south polar cap during their respective summers, it appeared to be the other way round during their winters, as predicted by the French astronomer Honoré Flaugergues in 1811. This was attributed to the fact that southern summers are warmer than

[56] Kant's views as described in *Popular Astronomy*, by Flammarion, translated by J. Ellard Gore, Chatto and Windus, 1907, pp. 390–391.

northern summers, whilst southern winters are colder than those in the north.

So much for the polar caps, but what about the remainder of the planet. Why was the planet reddish in colour and what did the dark patches consist of?

In the eighteenth century J.H. Lambert had hypothesised that the generally reddish colour of Mars was due to the reddish colour of its vegetation. But by the end of the eighteenth century most astronomers had concluded that the colour was caused by an atmospheric effect on Mars, similar to that which causes red sunsets on earth. In 1830, however, John Herschel suggested that the reddish colour was the true colour of the surface material. Then later in the century Flammarion reverted to Lambert's hypothesis that the colour was not that of Mars' surface but was due to vegetation. As Flammarion explained,[57] it is unreasonable to 'suppose that it [Mars] has nothing on its surface – no species of vegetation, not the least carpet of moss, neither forests, meadows, nor fields; for, whatever may be the vegetation which clothes this surface, it is this we see, and not the soil.' It was inconceivable, Flammarion maintained, that 'the meteorological circulation which produces on this planet, as on ours, seasons, fogs, snows, rains, heat and humidity . . . act for thousands of centuries on the surface of this world without giving birth to the smallest blade of grass?' No, he concluded, 'either it has a great number of red flowers or fruits of the same colour, or in reality the vegetation itself may be not green, but yellow.'

As far as the dark patches are concerned, John Herschel had noted that they sometimes have a green tint, suggesting to him that they were seas. Some thirty years later, in 1858, Angelo Secchi, the director of the Collegio Romano in Rome, saw the dark areas as generally bluish in colour, sometimes with a touch of green. Secchi interpreted the white polar regions as being due to either snow on the ground or clouds over the poles. Either way, this implied to him that there was water on Mars, leading him to conclude, like John Herschel, that the dark areas were seas that were bordering reddish-coloured continents. In 1860, however, Emmanuel Liais, the French director of the Rio de Janeiro Observatory, suggested that the dark areas were areas of vegetation, whilst the reddish areas were deserts.

Although most astronomers of the 1860s believed like Herschel and Secchi that the dark areas were seas, there appeared to be no obvious way of proving it. Then, John Phillips of Oxford came up with a solution, suggesting

---

[57] *Popular Astronomy*, by Flammarion, translated by J. Ellard Gore, Chatto and Windus, 1907, pp. 385–386.

that, if the dark areas were seas, we should be able to see the spectacular reflection of the sun from their surface. Unfortunately, although his logic was impeccable,[58] no such reflection could be seen.

## Schiaparelli discovers 'canali'

The most favourable opposition since 1845, which was due on 5th September 1877, inevitably attracted a number of observers of Mars, including Giovanni Schiaparelli, the director of Milan's Brera Observatory. The observatory had recently taken delivery of a superb new 8.6 inch (22 cm) refractor, which Schiaparelli was eager to try out on planetary observations. Unfortunately, he found his first observations of Mars somewhat disappointing but, by the time of the opposition, he was able to produce observations that were as good as anyone's. Thus encouraged, he decided a week after the opposition to start an intensive series of observations with the aim of producing a new, more detailed map of the planet. In the process, he measured the geographical coordinates of no less than sixty-two clearly recognisable points on the surface of Mars with a micrometer attached to his telescope.

Schiaparelli's map following this 1877 opposition is always compared with the best alternative, which was produced by the English amateur astronomer Nathaniel Green based on his observations made from Madeira (see Figure 7.12). In fact their maps were quite different in both content and style. Schiaparelli drew hard outlines to all the features, whereas Green, who was an artist, relied more on shading. Or as T.W. Webb put it in *Nature*[59] 'the one [Green] has produced a picture, the other [Schiaparelli] a plan'. Although some of the main features on the two maps were similar, most of the detail was completely different. In particular Schiaparelli drew a network of fine linear features, which he called 'canali', whereas Green showed none of them.

Schiaparelli's map caused a great stir, aided by his use of the Italian word 'canali' to describe the linear features. Unfortunately, English astronomers instantly translated the word to mean 'canals', which implies that they were produced by intelligent beings, whereas Schiaparelli had used the 'canali' to mean 'channels'.[60]

---

[58] Such specular reflections can now be seen in images of the earth's oceans taken from geosynchronous meteorological satellites.

[59] *Nature*, **34**, 1886, p. 213.

[60] Schiaparelli was not the first person to use the word 'canali' to describe a feature on Mars, as in 1858 Secchi christened what we now call 'Syrtis Major' as 'Atlantic Canali'.

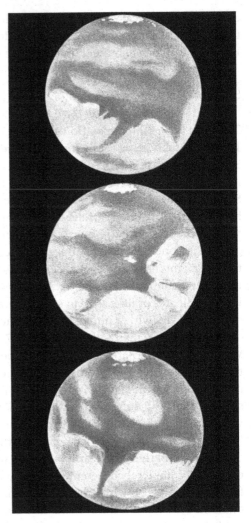

**Figure 7.12** Three drawings of Mars made at Madeira by Nathaniel Green around the opposition of September 1877. There was 2$\frac{1}{2}$ hours between the first two drawings and nine days between the second and third. The south polar cap is clearly seen at the top of each image, with Syrtis Major the triangular (tornado) shaped, dark area near the bottom. (From *The Sun: Its Planets and their Satellites*, by Edmund Ledger, 1882, Plate VII.)

It was assumed that the canali and any other features would be best seen when Mars was at or near opposition on 5th September, as it was then nearest to the earth. But this turned out not to be so. This was because, as Schiaparelli noted, many of the most obvious features were often covered by clouds between September and December 1877. After then, however, the clouds largely disappeared, allowing the features to be clearly seen. But not every astronomer accepted that the features' visibility was being affected

---

In this case, however, the feature was already well known, and from its shape there was no possibility that people could think that Secchi intended to imply an artificial origin.

by clouds, and to them the varying visibility, particularly of the canali, was something of a mystery. In fact, to understand the cause we need to consider the relative positions of the sun, earth and Mars, and the orientation of Mars' axis during this period.

In 1877 the perihelion occurred on 26th August, the opposition on 5th September, and the summer solstice in the southern hemisphere on 2nd October. So in the period from September to December 1877 observers were predominantly viewing the southern hemisphere of Mars, as it was tilted towards the sun and the earth around its summer solstice. This period was also just after perihelion, so the solar heating was close to its maximum. Evidently the sun's heat was causing clouds to form in the southern hemisphere, which gradually dispersed as the planet retreated from perihelion. The problem, which was unsolved at that time, was understanding why the clouds only seemed to form around the period of maximum solar heating, as on earth that is not the case.[61]

Schiaparelli believed that the large, dark areas of Mars were shallow seas, so he was not surprised when he noticed, during the 1879 opposition, that some of their outlines were somewhat different than during the 1877 opposition. In 1879 he also noticed what turned out to be a significant new feature, a very small white patch that he named Nix Olympica (the Snows of Olympus). Amazingly, Schiaparelli also observed that one of the canali now appeared to be double, with two parallel lines in place of the previous one, as a result of a process that he was to call 'gemination'. Whether the canali existed or not, single or double, they were generally very faint and hard to see. So when Schiaparelli produced his new map after the 1879 opposition, he used more shading and less sharp edges, to try to make it more realistic, following the criticism of his previous map.

Amazingly, at the opposition of 26th December 1881 Schiaparelli noticed that even more canali appeared double (see Figure 7.13). Although the planet's disc was much smaller than at the previous two oppositions,[62] he was also able to record more details of the northern hemisphere, as the planet had just passed the northern spring equinox.

In the twentieth century a great deal of scorn has been heaped on Schiaparelli's observations of canali, yet Henri Perrotin and Louis Thollon observed canali using the 15 inch (38 cm) Nice refractor in 1886, and Perrotin

---

[61] We now know that most of the clouds seen on Mars are dust clouds, and not the clouds of water vapour envisaged by nineteenth century astronomers. This was the key to solving the problem.

[62] The apparent diameter of Mars was only about 16″ at the opposition of 1881, compared with 20″ and 25″ at the oppositions of 1879 and 1877, respectively.

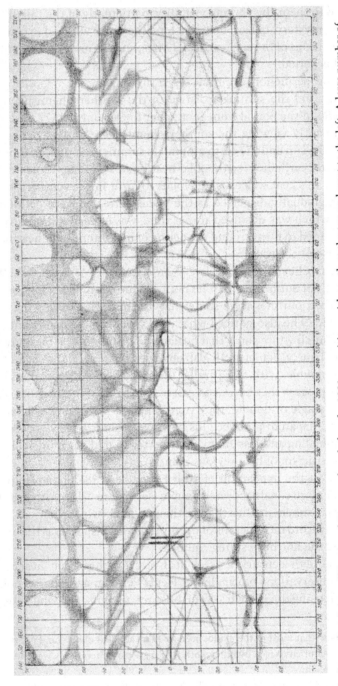

**Figure 7.13** Schiaparelli's map of Mars produced after the 1881 opposition, with south at the top and west to the left. A large number of double canali are seen, mainly in the equatorial regions. Syrtis Major is just to the left of center. (From *The Story of the Heavens*, by Robert Ball, 1897, Plate XVIII.)

observed more canali in 1888 using the new 30 inch (76 cm) refractor of the Nice Observatory. Likewise, Edward Holden, James Keeler and John Schaeberle all observed canali in 1888 using the new 36 inch (91 cm) Lick refractor on Mount Hamilton, and in 1890 they saw even more. Holden and Keeler saw the canali as 'dark, broad, somewhat diffused bands',[63] whereas Schaeberle reported them to be 'narrow lines' with, on 12th April 1890, 'two of the canals doubled'. On the other hand, the respected American astronomers Charles Augustus Young (1834–1908) and Asaph Hall were unable to see any canali at all.

Anyone who has observed a planet through a telescope will confirm how difficult it is to be certain of its surface details, as turbulence in the earth's atmosphere can change the image over fractions of a second. In addition, the brain has to be taught how to interpret images that the eye sees, as any observational astronomer or military observer will confirm, (often referred to as 'training the eye'). Finally, long periods of time at the eyepiece, often in low temperatures, produce fatigue and stress that also affect what one sees. It is not really surprising, therefore, for people to claim to have seen details in an image which are not really there, particularly when, in the case of Mars, some of the observers had a copy of Schiaparelli's map showing them what they should see. Were Schiaparelli's observations real? Only time and better telescopes would tell.

## Lowell's and Pickering's observations

Schiaparelli was prevented by failing eyesight from continuing his observations,[64] but the American millionaire Percival Lowell was determined to continue Schiaparelli's work. Lowell returned to the United States in late 1893, from a period of working in the Far East, and decided to set up an observatory to study Mars during the opposition of October 1894. Lowell, aided by William H. Pickering[65] (1858–1938) and Andrew Douglass, set up a small observatory at Flagstaff, Arizona in May 1894, where they installed

---

[63] E.S. Holden, J.M. Schaeberle, and J.E. Keeler, *Publications of the Astronomical Society of the Pacific*, **2**, 1890, pp. 299–300.

[64] Schiaparelli did not publish any observations that he made after 1890, although he continued observing with more and more difficulty until he was finally forced to stop in 1898.

[65] William H. Pickering, the brother of Edward C. Pickering, the director of the Harvard College Observatory, was given one year's leave of absence from Harvard to assist Lowell.

18 inch (45 cm) and 12 inch (30 cm) refractors, borrowed from the Harvard College Observatory.

William Pickering had observed the 1892 opposition from the Harvard College Observatory's Arequipa Station high in the Peruvian Andes. During this opposition he had noticed that a number of the canali crossed the dark, bluish-green areas of Mars, showing that the latter could not be seas. As a result he concluded that they were probably areas of vegetation, as suggested by Emmanuel Liais some thirty-two years earlier. However, it was just possible that canali may not be transporting water, so he needed another way of checking on whether the dark areas were seas. If they were, then the light reflected off them should be polarised. Pickering examined this light in the pre-opposition period in 1894 and found that there was no evidence of polarisation, confirming his previous conclusion that these dark areas were not bodies of water. In the meantime, Douglass had confirmed that the canali did cross the dark areas. So at this stage it looked as though the dark, bluish-green areas were areas of vegetation.

Lowell, after making some initial observations with Pickering and Douglass in May and June 1894, left Flagstaff Observatory in their charge, returning about two months before opposition. On Lowell's return he noticed that the white, southern polar cap was clearly smaller than before, and it was now surrounded by a dark band, which he called the antarctic ocean. Eventually the 'antarctic ocean' disappeared, but then the canals between this 'ocean' and the equator became more obvious. The southernmost canals became visible first and then 'the other [canals] follow in their order north'.[66] This led Lowell to conclude that the water in the 'antarctic ocean', which had been formed from the melting south polar cap, was gradually being drained northwards. He was unsure, however, whether the colour of the canals was solely due to water or to vegetation in marshy canal beds, or a mixture of both, although he veered to the latter.

In 1892 Pickering had caused a great deal of controversy when he reported seeing a number of lakes on Mars at the junction of some of the canals.[67] Lowell now confirmed the existence of these 'lakes', and reported that their seasonal darkening was consistent with his theory of water flowing along the canals towards the equator. Lowell's imagination was now in full spate, and he suggested that Pickering's lakes were oases in the red desert of Mars. Moreover, he concluded that 'the canals are constructed for

---

[66] Lowell, *Popular Astronomy*, **2**, 1894, pp. 255–261.

[67] See, for example, E.S. Holden, *Publications of the Astronomical Society of the Pacific*, **6**, 1894, pp. 160–169.

the express purpose of fertilizing the oases ... And just such an inference of design is in keeping with the curiously systematic arrangement of the canals themselves.'[68] The reason for their existence was now blindingly obvious to him. They had been built by intelligent beings to transport precious water from the melting polar caps to irrigate the dark areas, which were areas of vegetation. Such a startling conclusion inevitably caused more controversy.

It is inappropriate in a book of this size and subject matter, to go into detail of the great canal controversy around the end of the nineteenth century. Suffice it to say that some respected astronomers were convinced that the canals were there, whether or not they had been made by intelligent beings, whereas other, equally respected astronomers were convinced that they were figments of the imagination caused by wishful thinking in some cases.

## Water vapour, carbon dioxide and the polar caps

In 1867 Jules Janssen (1824–1907) and William Huggins had independently detected water vapour in Mars' atmosphere, using the moon as a comparator. Janssen had gone to the considerable trouble of taking his spectroscope to the top of the 10,000 ft (3,000 m) Mount Etna in Sicily, to get above as much of the earth's atmosphere as possible. This was to try to minimise possible confusion caused by the telluric spectral lines. Then in 1873 Vogel found evidence for both water vapour and oxygen. As in the case of Venus, however, W.W. Campbell, using the 36 inch (91 cm) refractor of the Lick Observatory with an improved spectroscope, could find no evidence in 1894 for water vapour on Mars. In fact, as far as he could determine, the spectrum of light from Mars was no different to that from the moon. This lack of evidence for water vapour was a complete surprise, as there was clear evidence of clouds in Mars' atmosphere, and the polar caps clearly melted considerably during their summer. There appeared to be abundant evidence for water on Mars, so where had the water vapour gone?

The polar caps also provided another puzzle. Calculations showed that the average temperature of Mars, taking into account its distance from the sun and its observed albedo, should be about −34 °C. If that was the case, the polar caps should not melt nearly as much as observed during their spring and summer seasons. In fact, in 1894 the southern polar cap seemed to have virtually disappeared. Clearly the polar caps must be very thin, but

---

[68] Lowell, *Popular Astronomy*, **2**, 1895, pp. 343–348.

astronomers realised that there must be more to it than that. The problem was, what?

In 1898, A.C. Ranyard and G. Johnstone Stoney suggested a possible solution. They proposed that the polar caps may consist of frozen carbon dioxide, rather than frozen water, which could explain their rapid melting at low temperatures, as carbon dioxide has a lower freezing point than water. Unfortunately, there appeared to be a blue melt band (Lowell's 'antarctic ocean') at the edge of the melting ice caps in spring, and, at the low atmospheric pressure on Mars, carbon dioxide would not melt into a liquid. Instead it should sublimate directly into the gaseous state. So water ice still seemed to most astronomers to be a more likely candidate for the constituent of the polar caps, although the temperature problem remained.

The nineteenth century was very much a time when evolutionary theories began to hold sway. Not only did Charles Darwin propose his evolutionary theory for the Origin of the Species in 1859, but theories were developed by Laplace and others for the evolution of the solar system, and measurements indicated that the solar system was still changing. It was also hypothesised by some astronomers that the suitability of various planets for life went through optima at different times.[69] Thus, although the earth has good conditions for life now, Mars was thought to have probably passed its optimum, whereas the optima for Venus and Mercury were still to come when they had cooled down sufficiently.

## Phobos and Deimos

Asaph Hall, of the US Naval Observatory in Washington, decided to use the 1877 opposition of Mars to try to find any satellites of Mars. At that time, the US Naval Observatory had the largest refracting telescope in the world, namely the 26 inch made by Alvan Clark. In the first few nights, Hall searched the regions some distance away from Mars, but on subsequent nights he moved closer and closer to the planet. Then in the early hours of 12th August, when he had almost despaired of finding anything, he observed the outer of the two satellites, now called Deimos. He did not realise at that time what he had observed, however, simply recording the object as a star. But on 16th he saw it again, and realised that it was probably what he was

---

[69] See, for example, G.M. Searle, *Publications of the Astronomical Society of the Pacific*, **2**, 1890, pp. 165–177.

looking for. Then on 17th, whilst waiting for this satellite to reappear, he discovered Mars' second satellite, now called Phobos.[70]

The orbits of the two satellites were found to be very close to Mars, with that of Phobos skimming just 5,900 km above the surface of the planet.[71] From their orbits Hall was able, for the first time, to calculate an accurate mass for Mars, which he reported to be 0.1076 times the mass of the earth (modern value 0.1074).[72]

It was impossible to see the discs of Phobos and Deimos as they were too small. So the only way of guessing their size was to measure their brightness and assume a reflectivity similar to that of some other body in the solar system. Using our moon as a comparator yielded a diameter of about 15 km for Deimos and about 25 km for Phobos, which are very close to their true values. It was thought possible, because of their small size, that these satellites were captured asteroids, and one asteroid (Aethra, 132) was known at the time to have an orbit that crossed that of Mars (see Section 7.2).

There is an interesting postscript to this discovery of the two satellites of Mars, as Jonathan Swift (1667–1745) had related such a discovery over a hundred years earlier in his novel *Gulliver's Travels*, published in 1726. In that book Swift describes how the scientists of Laputa had discovered 'two lesser stars, or satellites, which revolve around Mars, whereof the innermost is distant from the centre of the primary planet exactly three of his diameters, and the outermost five; the former revolves in the space of ten hours, and the latter in twenty-one and a half.'

Swift made this remarkably accurate guess 151 years before Phobos and Deimos were discovered with periods of 7.6 and 30.3 hours. The shortest satellite period known in 1726 was 42 hours for Io. So why did Swift attribute two such very short periods of 10 and $21\frac{1}{2}$ hours for the two satellites, both of which were faster than the rotation rate of Mars?

Swift was not the only person at that time to think that Mars probably had two satellites. In fact, Kepler had come to the same conclusion in

---

[70] The names Deimos and Phobos were suggested by Henry Madan of Eton, England, as they were the sons of Ares, the Greek equivalent of Mars, in the Iliad. Deimos means 'rout' or 'flight' and Phobos 'panic' or 'fear'.

[71] The mean distance of Phobos from the centre of Mars is about 9,270 km, and that of Deimos is about 23,400 km.

[72] An interesting feature of the orbit of Phobos is that it was the only moon then known to orbit its planet faster than the planet rotates about its own axis. Thus to an observer on Mars, he would see one of the moons moving across the sky in one direction, whilst the other moon would be seen to move in the opposite direction.

1610, on the basis that the earth has one satellite and Jupiter four. So Kepler suggested that Mars would probably have two satellites and Saturn six or eight, to give a mathematical series of 1, 2, 4, 6 or 8. In the late seventeenth and early eighteenth centuries, others came to a similar conclusion about Mars having two satellites. At that time, the earth was known to have one satellite, Jupiter four and Saturn five, and it was expected that another planet would be discovered in the large gap between Mars and Jupiter. If that was the case, the numbers of satellites could be one for the earth, two for Mars, three for the unknown planet, four for Jupiter and five for Saturn. So it was considered natural at the time for Mars to have two satellites, but why did Swift propose two with such short orbital periods?

Owen Gingerich points out in his book *The Great Copernicus Chase*[73] that the idea that the satellites may be 3 and 5 planetary diameters from the centre of Mars, may well have been based on the distances from their planets of Jupiter's and Saturn's nearest satellites. Io and Europa were known to be about 3.0 and 4.8 planetary diameters from the centre of Jupiter, and Tethys and Dione about 2.5 and 3.2 planetary diameters from the centre of Saturn. So the distances for Mars' satellites may have been easily guessed, but explaining how Swift came to the very short orbital periods is a much more difficult matter. After all, if Mars has the same density as that of the earth, then a satellite at three planetary diameters would have a period of 24 hours, not 10. According to Gingerich, N.T. Roseveare of Chelsea College, London, suggested to him that Swift may have assumed a density for Mars of 22 times that of Jupiter, as Jupiter's diameter is about 22 times that of Mars. Then the period of Mars' innermost satellite would be about 10 hours. Somewhat far-fetched, possibly, but how else did Swift work out the periods, as he was well aware of the basic physical formulae required? A most interesting conundrum.

---

[73] Cambridge University Press, 1992, pp. 177–183.

# Chapter 8 | THE OUTER SOLAR SYSTEM IN THE NINETEENTH CENTURY

## 8.1 **Jupiter**

Jupiter had been somewhat neglected in the eighteenth century, in spite of its beautiful telescopic appearance with its clearly banded structure of dark belts and bright zones parallel to the equator, and various light and dark spots. The largest and most prominent spot observed prior to the nineteenth century was that discovered by Cassini, which had last been observed in 1713. It was not until the 1870s, however, that a similar prominent spot appeared (see Figure 8.1) in a similar position on the planet, sparking off a new enthusiasm for observing Jupiter.

The first evidence of the possible return of Cassini's spot was the recording of a notch or hollow in the South Equatorial Belt, at about the same latitude as Cassini's spot, by Heinrich Schwabe on 5th September 1831. This notch, now called the Red Spot Hollow, was later observed by Laurence Parsons (1840–1908), the fourth Earl of Rosse, using the 72 inch (1.8 m) reflector, known as the Leviathan of Parsonstown, in 1872 and 1873. There is some evidence that George Hirst (1846–1915) of Sydney, Australia, observed what became known as the Great Red Spot (GRS) at the same latitude in 1876, but his records are somewhat ambiguous.[1] There is no doubt, however, that the GRS was seen by many observers in 1878, having a rose-red colour and stretching for about one-fifth of the planet's diameter.

The earliest observation of the appearance of the GRS seems to have been made on 6th July 1878 by Carl Pritchett (1837–1888), the director of the Morrison Observatory in the United States, followed on 6th August by Louis Niesten (1844–1920) of the Royal Observatory in Brussels. By the following

---

[1] See Thomas Hockey, *Galileo's Planet*, Institute of Physics, 1999, pp. 139–140.

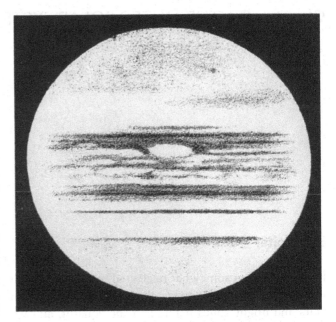

**Figure 8.1** Drawing of Jupiter made in 1878 by G.D. Hirst of Sydney, Australia, showing its longitudinal bands and spots. The newly discovered Great Red Spot is the white oval just above the centre. (From *The Sun: Its Planets and their Satellites*, by Edmund Ledger, 1882, Plate VIII.)

year the GRS had become brick red in colour, standing out quite dramatically against the bright, white South Tropical Zone and some white equatorial spots. George Hough (1836–1909), of Chicago's Dearborn Observatory, observed in 1881 that the GRS rotated around Jupiter in 9 h 55 min 35 s, and that the white spots had periods of rotation from 9 h 50 min 0.6 s to 9 h 50 min 9.8 s, giving a differential velocity between the GRS and these white spots of an astonishing 400 km/h (250 mph). The GRS, which dominated the views of the planet from 1879 to 1882, was measured to be about 40,000 km in length (over three times the diameter of the earth) and about 13,000 km in width. Since 1882 it has varied significantly in visibility and colour, and somewhat in size.

In 1778 Georges-Louis Leclerc, Compte de Buffon, suggested that the continuous, rapid changes in the appearance of Jupiter indicated that it had not cooled down completely since its formation. In 1785 Kant thought the same, but the idea did not catch on. Then in 1853 James Nasmyth (1808–1890) resurrected the idea, seeing the turbulence of the atmosphere as being indicative of a hot surface, whilst numerous, variable light and dark spots were caused, in his view, by volcanoes throwing ash into the atmosphere.

George P. Bond (1825–1865) was an early exponent of the use of photography in astronomy, finding that it could be used to measure stellar intensities. He also applied the technique to the planets, but was surprised to find in 1860 that Jupiter appeared to have an albedo of greater than unity, implying that it emitted more energy than it received from the sun. He was unhappy with this result, but suggested that if later, more accurate measurements showed it to be correct, it could possibly be explained by extensive aurorae on Jupiter. Five years later, however, Zöllner measured an albedo of 0.62, which was a much more reasonable value than Bond's. Even so the white zones on Jupiter must be extremely bright to balance the darker belts and produce such a high overall figure. Zöllner then went on to agree with Buffon's earlier theory, that the rapid changes in Jupiter's cloud bands indicated that it has a high internal temperature.

In the first half of the nineteenth century, some astronomers had observed that the shadows of the Galilean satellites were not completely black, but slightly reddish, supporting the notion that Jupiter was partly self-luminous. Others disagreed, however, seeing the shadows, particularly when near the middle of Jupiter's disc, as being jet black.

Although Jupiter was clearly nothing like as hot as the sun, evidence was beginning to accumulate in the nineteenth century that Jupiter had some similarities to the sun and possibly to some red stars. For example, Cassini, Schröter, and others had observed that Jupiter had a more rapid angular rotation rate at the equator than at middle latitudes, and in 1859 Richard Carrington had shown, by timing the movement of sunspots, that the sun behaves in exactly the same way. Cassini and Schwabe had even suggested that the spots on Jupiter were similar to sunspots in their rate of formation and decay, and Schwabe also thought that they were similar in appearance with detectable penumbrae. Isaac Newton had shown that the density of Jupiter was similar to that of the sun (see Section 4.7), and in 1860 Bond showed that Jupiter exhibited limb darkening, also like the sun.

In the 1860s and 1870s, Secchi, Huggins and Vogel examined Jupiter's spectrum to try to identify the gases in its atmosphere, and hopefully to discover whether it was self-luminous. Unfortunately, the results were difficult to interpret as, in addition to the usual Fraunhofer and telluric absorption lines, there seemed to be a number of absorption lines in the red and yellow part of the spectrum, none of which they could identify, although one had been previously found in the spectra of some red stars. All this evidence, that Jupiter looked in many ways like a miniature sun, persuaded some astronomers that it was partly self-luminous. Others disagreed, however, pointing out that each of the above-mentioned effects could be explained in other ways.

Newton had calculated that the density of Jupiter was about 23.6% that of the earth, and in 1798 Henry Cavendish (1731–1810) had measured the density of the earth to be about 5.5 g/cm$^3$. This implied a density for Jupiter of about 1.3 g/cm$^3$. As a result, the rocky core of Jupiter, if any existed, must necessarily be very small. But what was the rest of Jupiter like? Were we observing the top of a thick atmosphere, possibly over an icy or liquid surface, or were we observing the surface directly? What was the Great Red Spot and the other light and dark spots that appeared from time to time, and what were the bright zones and dark belts? The appearance of the GRS in 1878, and of an exceptionally bright, white, equatorial spot in the following year encouraged a number of astronomers to try to understand the structure of the planet.

Both William and John Herschel had thought that the bright zones were clouds and that the dark belts were areas free of clouds where we could see Jupiter's surface.[2] Some astronomers also thought that the white spots were clouds and the dark spots were holes in the clouds, but the Great Red Spot was a complete mystery. If the bright zones were clouds and the dark belts showed us the surface of Jupiter then, as Proctor pointed out,[3] the bright zones, when seen at the edge of the planet, should be slightly higher than the dark belts. But such an effect had not been observed, so Proctor concluded that we were not seeing the surface of Jupiter in the dark belts, but only some slightly lower clouds.

The relatively long life of the Great Red Spot suggested to astronomers in the early 1880s that it may well be connected with a permanent feature on the surface of Jupiter, but, unfortunately, the clearly defined white, equatorial spots were found to be going around Jupiter about 400 km/h faster than the GRS. Because of this, both the white spots and the GRS could clearly not be connected with permanent surface features, but how could atmospheric phenomena, if that is what the white spots were, move so rapidly whilst retaining their shape for so long?

This dilemma is probably best summarised by Ledger in his Gresham Lectures in 1881 and 1882.[4] 'We must confess' Ledger writes, 'that we were at one time inclined to think that the permanency of the Great Red Spot seemed to indicate, that it might be something which, while coagulating or solidifying, in some way caused a gap or break in the cloudy regions above it, ... and thus increased its own visibility; in fact, that we might be watching

---

[2] Prior to William Herschel it had been generally believed that the dark belts were clouds, but opinion gradually changed in the nineteenth century.

[3] *Other Worlds Than Ours*, by Richard A. Proctor, Longmans, third edn., 1872, p. 131.

[4] *The Sun: Its Planets and Their Satellites*, by Edmund Ledger, Edward Stanford, 1882, pp. 313–314.

in it the gradual formation of a huge continent upon Jupiter. But the recent observations of the [exceptionally bright] equatorial white spot and of some other similar spots tell us, that, if we see in the Red Spot a part of the body of Jupiter itself, we must find some explanation of the more rapid rotation of these other spots. It might not be so difficult to explain a slower rate of rotation in their case, if they are simply atmospheric phenomena, but it seems hard to imagine by what means they can be carried steadily round, with a speed so much more rapid than that with which the globe below them must be rotating, if the Red Spot is actually a part of it and by its movement indicates the true rotation-period.' So is the GRS, the white spots, or neither connected with the surface of Jupiter? If the latter, how can the long-lasting nature of the GRS and white spots be explained, together with their large differential velocity?

Ledger then went on to speculate that, if the exceptionally bright equatorial white spot, rather than the GRS, is connected with a surface feature it 'may be the summit of some snow-clad mountain, which from time to time is more or less clearly seen, or is occasionally altogether hidden by masses of overhanging clouds. Or possibly some volcano there placed may periodically eject vast masses of vapour which may hang over it, and alternately be of such a character as to shine with special brightness by reflected sunlight, or to obscure and darken our view.'

On 24th December 1881, William Denning (1848–1931) of Bristol started an extensive set of measurements of the GRS, of the exceptionally bright, white, equatorial spot, and of the other white and dark spots visible at the time. By 1883 he had concluded that the average period for the GRS was 9 h 55 min 34 s, almost exactly the same as Hough, and that of the equatorial white spot was about 9 h 50 min 5 s. Importantly, however, both spots showed apparently random variations about these figures.[5] In October 1882 Denning also observed an equatorial dark spot, finding that it had the same rotation rate around Jupiter as the equatorial white spot. So whatever the light and dark equatorial spots were, they appeared to participate in the same equatorial motion. Finally, tracking some dark spots in northern latitudes, that he had first observed in October 1880, produced a rotation rate of 9 h 48 min, which was even faster than that of the equatorial spots.

It was clear from Denning's observations that as the period of rotation of the GRS, and of the exceptionally bright, white equatorial spot, were not constant, these spots could not be connected with surface features. In that

---

[5] Observations later in the century by numerous observers showed that the period of the GRS also appeared to vary randomly over longer periods.

case, it was thought that the white and dark spots are probably clouds, but the idea that the GRS is also a cloud system seemed improbable because of its longevity. Then at the turn of the century Hough appeared to solve this problem by hypothesising that the GRS and the small white spots are solid objects floating in the liquid surface of the planet. Incredible though it may seem, this theory was further developed in the twentieth century before being abandoned.

Galileo had noticed variations in the intensity of Jupiter's four satellites as they orbited the planet, then in 1797 William Herschel observed that these intensity variations were consistent from one orbit to the next. As a result he concluded that the axial rotation periods of each of the four satellites were synchronous with their orbital periods. So they always turned the same face towards Jupiter as they orbited the planet, like the moon does with the earth. This seemed perfectly natural, owing to their relative proximity to such a large planet. For example, Callisto, the outer Galilean satellite, is only 13 planetary diameters from the centre of Jupiter, compared with 30 planetary diameters for the distance of the moon from the earth.

Subsequent observations by numerous astronomers in the nineteenth century could find no evidence for synchronous rotation of the Galilean satellites, however. But in the 1870s Engelmann of Leipzig and C.E. Burton independently observed intensity variations on Callisto which correlated with its orbit position, indicating that Herschel had probably been correct at least for Callisto. Then in 1892 William Pickering and Andrew Douglass confirmed this synchronous rotation for Callisto, and rediscovered it for Ganymede, following observations from the Harvard College Observatory at Arequipa, Peru. Nevertheless, at the end of the century it was still considered doubtful that Io and Europa, the two inner satellites, had synchronous rotation. This seemed strange, however, as if any of the four satellites were to have synchronous rotation, it would be expected to be those satellites closest to Jupiter, i.e. Io and Europa, not those furthest away.

Jupiter shows noticeable edge darkening, so in the early stages of a transit the Galilean satellites were seen as bright discs against the relatively dark edge. As they continued with their transit, however, they were gradually lost in the background of the progressively brighter planet. They then appeared bright again as they approached the darker limb of the planet near the end of the transit. Sometimes, however, Ganymede and Callisto appeared dark against the bright central part of the planet in what were described as dark transits, the first of which was observed by Cassini in 1665. Sometimes this effect was attributed to the satellites traversing a particularly bright part of Jupiter, but sometimes it was thought to be due to darkening of the satellites themselves. Io tended to appear dusky when transiting Jupiter, but Europa always appeared white, which was attributed to white clouds in its atmosphere. In fact, all four of the Galilean satellites were assumed in the nineteenth century to have atmospheres, and this was apparently confirmed by Vogel who detected lines in their spectra similar to those observed for Jupiter. This indicated that the atmospheres of the Galilean satellites and Jupiter had similar chemical constituents.

William Cranch Bond (1789–1859) and his son George observed in 1848 that markings on Ganymede were sometimes variable during its transit of Jupiter. So if Ganymede's rotation period was synchronous, this indicated that real changes were taking place on this satellite in a short period of time. Similar observations were later reported by Barnard, Schaeberle and Campbell. As the surface of Ganymede was assumed to be frozen, such rapid changes were presumed to occur in its atmosphere.[6] Secchi also noted that Ganymede appeared to be spheroidal in shape, not spherical. Then in August 1891 Schaeberle and Campbell observed equatorial bands on Ganymede, confirming the similarity between its atmosphere and that of Jupiter.

Meanwhile, on 13th August 1890 Barnard observed a dark transit of Callisto during which it appeared to change colour slightly, at one stage having a slightly brownish tinge, and also change shape from being perfectly circular to being slightly elongated. On 8th September Barnard noticed that Io was elongated and bisected by an equatorial band. In the following year Schaeberle and Campbell observed that Io was elongated in a direction nearly parallel to Jupiter's equator, with the ratio of its major to minor axes being about 5:4. In fact, by the end of the nineteenth century it seemed reasonably well established that all four satellites were spheroidal in shape with significant, and possibly banded atmospheres.

The four Galilean satellites had been the only known satellites of Jupiter for almost three hundred years, when at about midnight of 9th September 1892, Barnard noticed a very faint object very close to the planet. It moved rapidly with respect to Jupiter and the nearby stars, and quickly gave away its identity as Jupiter's fifth satellite, with a period of a little less than 12 hours. At 181,000 km from the centre of the planet, and 110,000 km above its surface, it is very much closer to Jupiter than the moon is to the earth. However, its then estimated size, of about 100–200 km diameter, was more reminiscent of a reasonable sized asteroid than any of its four large Galilean brethren.

---

[6] See J.E. Keeler, *Publications of the Astronomical Society of the Pacific*, **2**, pp. 294–296.

The discovery of this satellite, now called Amalthea, marked the end of an era being, as it was, the last satellite of any planet to be discovered visually.

## 8.2 Saturn

### Rings

Laplace's theory of the nature of Saturn's rings, as consisting of a number of very thin, solid rings (see Section 6.2), generally held sway in the first half of the nineteenth century, and from time to time astronomers apparently discovered subdivisions in the two known rings. But William Herschel was very sceptical about these discoveries, as he could not see the subdivisions himself with his excellent telescopes, and he doubted whether they existed. Nevertheless, as the nineteenth century progressed more and more astronomers announced the discovery of such subdivisions.

For example, Lambert Quetelet, director of the Brussels Observatory, saw the A ring divided into two in December 1823, and Henry Kater (1777–1835), the vice-president of the Royal Society, saw the A ring divided by three dark lines around both ansae on 17th December 1825. On the same evening Kater asked two friends to examine the A ring through his telescope; one friend saw six subdivisions, and the other only one at about the middle of the ring. Kater also thought that he could detect some markings on the A ring on 16th and 17th January 1826, but he was not sure. On subsequent occasions, however, he could see no markings at all, even on the nights of best seeing. In 1826 both John Herschel and Wilhelm Struve (1793–1864) tried to find the subdivisions of the A ring reported by Kater, but without success.

However, on 25th April 1837 Johann Encke, the director of the Berlin Observatory, observed that the A ring was divided into two nearly equal parts, and he also found several subdivisions near the inner edge of the B ring. Then on 28th May he observed the division in the A ring again, but this time he measured its position with a micrometer, finding it to be about one-third of the way from the inner to the outer edge. Further observations of this division were reported by Di Vico and Decuppis in 1838 and Schwabe in 1841. Then on 7th September 1843 Dawes and Lassell both saw the so-called Encke division *outside* the middle of the A ring, with Dawes estimating its width as about one third of that of the Cassini division, and Lassell making it slightly narrower.

In summing up the situation on ring subdivisions in the middle of the nineteenth century G.P. Bond wrote[7] 'We have, then, the best assurance in the number and reputation of those who have described the phenomena in question [i.e. ring subdivisions], that to set aside these appearances by referring them to some optical deception on the part of the observer, or to some defect in his instrument, is an explanation altogether insufficient and unsatisfactory. On the other hand, we know that some of the best telescopes in the world, in the hands of Struve, Bessel, Sir John Herschel, and others, have given no indication of more than one division [i.e. the Cassini division], when the planet has appeared under the most perfect definition.' Bond then concluded that 'the difference [in the observations] is not probably owing to any extraordinary tranquillity or purity of the atmosphere, nor to any peculiarly favorable condition of the eye or instrument, but rather to some real alterations in the disposition of the material of the rings.'

A new part of Saturn's ring system was observed by W.C. and G.P. Bond on the night of 11th November 1850 using Harvard College Observatory's 15 inch (38 cm) refractor. They first described it in their observing notebook as having noticed[8] 'to-night, with full certainty, the filling up of light inside the inner edge of the inner ring of Saturn; also, what is very singular, where the ring crosses the ball... *below* the edge, there is a dark band, no doubt the shadow of the ring. But there is *also a dark line*... above the ring... where the ring crosses the ball.' But it was not until 15th November that one of their observatory assistants, Charles W. Tuttle, suggested that what they were really observing was a dark inner ring. On the same day, the Bonds observed that the inner edge of the new ring was sharply defined in the ansae, whilst its outer edge was not as clear. W.C. Bond thought that he had detected a gap between the inner edge of the B ring and the outer edge of the new ring, but he was not sure. Interestingly, in spite of the fact that the Bonds reported that the inner edge of this new ring was clearly defined, their estimate of the inner diameter of this ring of 26.3" was much higher than the modern value[9] of 23.8", giving a width of 1.5", compared with the current figure of 2.8".[10]

---

[7] G.P. Bond, *Astronomical Journal*, **2**, 1851, pp. 5–8.

[8] W.C. Bond, *Astronomical Journal*, **2**, 1851, p. 5.

[9] All the angular values given in the remainder of this section have been normalised to Saturn's distance at the time of the discovery of the new ring.

[10] Their estimate of the breadth of the A ring of 2.3" is exactly the same as observed today, however.

Unlike the A and B rings, this C or crape[11] ring, as it was called, was very dark and relatively difficult to see. It was clear to the Bonds, however, that it was not as dark where it crossed in front of Saturn, as the shadow of the A and B rings on the planet. This suggested that the C ring was partly transparent.

News of the discovery of the C ring reached England on 3rd December, and was reported in *The Times* of the following day. By that time the English clergyman and amateur astronomer W.R. Dawes (1799–1868), had already discovered it independently. In fact Dawes had observed the ring from his observatory at Wateringbury, near Maidstone, on 25th November. Then on 3rd December another English amateur, William Lassell, confirmed the C ring's existence, when he also saw it through Dawes' 6.4 inch (16 cm) refractor. Dawes estimated its breadth as being 'rather more than two-thirds of the breadth of the outer ring', giving a breadth of about 1.5″ or so, which was similar to the Bonds' figure. Lassell, on the other hand, thought that the C ring extended about half-way between the inner edge of the B ring and the planet (as in Trouvelot's later drawing of Figure 8.2). This gave a breadth of about 2.6″, which is about the same as the modern value of 2.8″.

Amazingly, the Bonds were not the first to have seen the C ring, as J.G. Galle had observed it in the ansae in 1838, using the 9 inch (23 cm) refractor of the Berlin Observatory. However, apparently Galle had not seen the C ring as it crossed in front of Saturn's disc, so he had not realised that he had detected a new ring. Unfortunately, although Galle's observations had been published in the *Transactions of the Berlin Academy of Sciences* in 1838, they had attracted no attention until after the discovery of the C ring by the Bonds and Dawes twelve years later.[12]

Although the Bonds had suspected that the C ring may be partly transparent, they had not seen the edge of the planet through the material of the ring. That was first observed independently by W.S. Jacob at Madras and by Lassell at Malta in the autumn of 1852, whereas G.P. Bond and C.W. Tuttle did not see it until November 1853. In October 1852 Lassell also saw the globe of Saturn through the Cassini division, thus confirming that it was a true gap between the two bright rings.

The discovery of the C ring was something of a surprise, particularly as it could easily be seen in relatively modest telescopes in the year of its

---

[11] So-called as, at that time, a crape was a black veil, or a black band worn on the sleeve or hat at times of mourning. The comparison of the new ring with a crape veil was first made by Lassell in a letter to the Astronomer Royal in December 1850.

[12] It is also possible that William Herschel saw the C ring in 1793, although this is not certain (see Section 6.2).

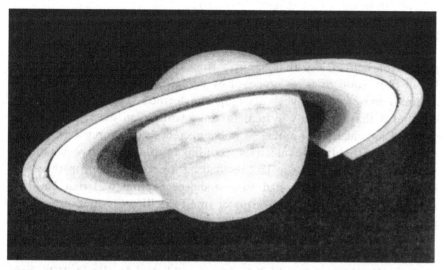

**Figure 8.2** Etienne Trouvelot's drawing of Saturn, made at the Harvard College Observatory in 1874, clearly shows the dark C ring as extending about half-way from the inner edge of the B ring to the globe of the planet. (From *The Sun: Its Planets and their Satellites*, by Edmund Ledger, 1882, Plate IX.)

discovery, whereas it had previously not been clearly detected by the very best observers with much larger and more performant telescopes. This suggested that it was changing its appearance with time. A similar situation seemed to apply to the subdivisions of the A and B rings, which were sometimes seen with relatively modest instruments, but had not been seen by some of the best observers with state of the art instruments. So, as mentioned above, Bond had concluded in 1851 that the A and B rings were also changing their appearance with time.

A number of astronomers now decided to revisit the theory of Saturn's rings, with these new observations to hand, and with improved mathematical tools that had not been available over fifty years earlier when Laplace had developed his theory of the rings.

The first results were quick in coming. For example in 1851 Bond concluded, on the basis of a mathematical analysis of the observed rings, that they cannot be solid, but must be liquid in which there are divisions continually being created and destroyed due to local disturbances. Also, at about the same time, Benjamin Peirce of Harvard proved purely mathematically, with no recourse to observational details, that the rings must consist of,[13]

---

[13] B. Peirce, *Astronomical Journal*, **2**, 1851, pp. 17–19.

'streams of a fluid somewhat denser than water, flowing around the planet.'

Then in 1857 the Scottish physicist James Clerk Maxwell (1831–1879) proved mathematically that Saturn's rings could not consist of a number of thin, solid rings, as proposed by Laplace. Nor could they be fluid, as fluid rings would have broken up and formed satellites. Instead, he concluded that 'the only system of rings which can exist is one composed of an indefinite number of unconnected particles, revolving round the planet with different velocities according to their respective distances.' So J.D. Cassini and his son Jacques had been correct when, in the early eighteenth century, they had suggested (see Section 6.2) that Saturn's rings probably consisted of swarms of small satellites in orbit around the planet.

Interestingly, a few years earlier, Edouard Roche (1820–1883) of Montpellier had analysed the stability of a fluid satellite orbiting close to a planet, but his paper, which had been published locally in 1848, was generally unknown for many years. In it he considered the case of a fluid satellite, of the same density as Saturn, in orbit around the planet.[14] He then showed that the satellite would have been disrupted by tidal forces created by Saturn, if it were closer than 2.44 Saturn radii to the centre of the planet.[15] Roche hypothesised, therefore, that a fluid satellite of Saturn had approached too close to the planet many years ago, that the satellite had been seriously distorted by Saturn's gravitational force, that it had taken the form of an elongated ellipsoid just before breaking up, and that it had finally broken into innumerable small particles which now form the rings of Saturn. This was consistent with the rings of Saturn which were known at that time, as they extended outwards to about 2.27 Saturn radii from the planetary centre, which was less than Roche's 2.44 limit.

Daniel Kirkwood had pointed out in 1857 and 1866 that there were no asteroids with periods that were a simple fraction of Jupiter's period (see Section 7.2), and he had attributed this to a resonance between Jupiter and the asteroids which forced then out of such orbits. Then in 1867 he turned

---

[14] Roche simplified his analysis in choosing to analyse a fluid satellite. Nevertheless, if the encounter of the satellite with Saturn had occurred early in the history of the solar system, the satellite would have been completely molten anyway.

[15] The situation for a solid or partially solid satellite is more complex, however, with their ease of fragmentation depending on size, amongst other things. The smaller solid bodies being able to withstand fragmentation the better.

Table 8.1 *Data on those of Saturn's satellites known before 1900,*
*compared with that for particles in the Cassini division*

| Satellite[a] or division | Year discovered | Discovered by | Sidereal period (in days)[b] | Period relative to that of particles in the Cassini division[c] |
|---|---|---|---|---|
| Cassini div. | 1675 | J.D. Cassini | 0.490 | 1 |
| Mimas | 1789 | W. Herschel | 0.942 | 1.92 or approx. 2 |
| Enceladus | 1789 | W. Herschel | 1.370 | 2.80 or approx. 3 |
| Tethys | 1684 | J.D. Cassini | 1.888 | 3.85 or approx. 4 |
| Dione | 1684 | J.D. Cassini | 2.737 | 5.59 or approx. 6 |
| Rhea | 1672 | J.D. Cassini | 4.518 | |
| Titan | 1655 | C. Huygens | 15.945 | |
| Hyperion | 1848 | G.P. Bond | 21.277 | |
| Iapetus | 1671 | J.D. Cassini | 79.331 | |
| Phoebe | 1898 | W. Pickering | 550.48 | |

[a] All the satellites have orbital inclinations of $<2.0°$, except for Iapetus ($15°$) and Phoebe ($150°$).

[b] The ratio of the periods of Tethys to Mimas and of Dione to Enceladus are 2.00 to 1, to the second place of decimals.

[c] The reciprocal ratios are 0.52 which is approx. 0.5 or $\frac{1}{2}$ for Mimas, 0.36 (0.33 or $\frac{1}{3}$) for Enceladus, 0.26 (0.25 or $\frac{1}{4}$) for Tethys, and 0.18 (0.17 or $\frac{1}{6}$) for Dione.

his attention to Saturn, and the effect of the inner satellites on particles in Saturn's rings. Kirkwood pointed out that any particles in the Cassini division would have periods of about one-half that of Mimas (see Table 8.1), one-third that of Enceladus, one-quarter that of Tethys, and one-sixth that of Dione. As a result of these resonances he deduced that the Cassini division would be cleared of particles.

Only Iapetus, of Saturn's satellites known in the 1880s, has an orbit that is appreciably out of the ring plane. So only Iapetus could provide the opportunity to estimate the transparency of the C ring by observing its eclipse by the ring. The German-born astronomer Arthur Marth (1828–1897), realising this, examined the relative positions of the ring and satellite, as seen from the earth, and predicted that an eclipse of Iapetus by the C ring would occur on the night of 1st–2nd November 1889.

Surprisingly, the American Edward Emerson Barnard seems to have been the only astronomer to have observed this eclipse. Using the 12 inch (31 cm) refractor on Mt Hamilton, he first observed Iapetus emerging from Saturn's shadow at about 2.38 am local time. It was shining at its normal intensity, as far as he could determine visually,[16] until at 3.47 am it entered the shadow of the C ring. Its intensity gradually decreased as it entered further and further into the shadow of this ring, until Iapetus suddenly became invisible when it entered the shadow of the B ring. Shortly afterwards the event was overtaken by daylight, putting an end to further observations. So Barnard had proved what had long been suspected, that the C ring was partially transparent. In addition he found no evidence for another ring between the C ring and the planet, and no evidence for a gap between the C and B rings, which had sometimes been reported.

A few years later, James E. Keeler (1857–1900) attempted, using the 13 inch (33 cm) equatorial telescope of the Allegheny Observatory in the United States, to measure the relative velocities of the different parts of Saturn and its rings. To do this he photographed their spectrum with the slit of the spectrograph aligned along Saturn's equator. Those parts of the system moving towards the earth would show a blue Doppler shift of the spectral lines, and those moving away would show a red shift, the amount of the shift being directly related to the velocity along the line-of-sight. His first attempt in 1893 was a failure, due to both atmospheric and instrumental problems, but his next attempt two years later was a success. Keeler obtained the first successful photograph of the spectrum of the planet and its ring system on 9th April using a two hour exposure, and obtained another such photograph the following night.

A diagrammatic representation of his results is shown in Figure 8.3, which clearly shows that at 'A' the edge of the globe is moving towards us, whilst at 'B' it is moving away. More importantly, however, the outside of the main ring[17] at 'a' is moving towards us with a slightly lower velocity than the inside of the main ring at 'b'. If the main ring was rigid, the linear velocity at 'a' would be higher than that at 'b', and so Keeler had finally proved that the main ring is not solid.[18] His photograph was not accurate enough, however, to produce reliable estimates of the velocity at the inner

---

[16] Barnard compared the intensity of Iapetus with that of Tethys (magnitude $10\frac{1}{2}$) and Enceladus (magnitude 12).

[17] Keeler talks about 'the ring' in his first paper on the subject (*Astrophysical Journal*, **1**, 1895, pp. 416–427), by which he means the A and B ring together. I have used the words 'main ring' to signify this in the above discussion.

[18] This proved that the ring is not solid, but it did not prove that the ring system was not made up of a large number of thin, solid rings, as proposed by Laplace. The proof of that still relied on Maxwell's theoretical work of 1857.

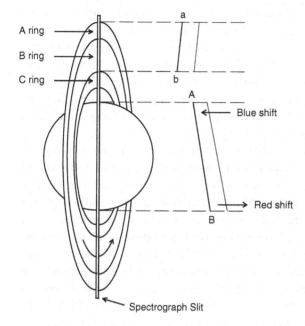

A ring

B ring

C ring

a

b

A

Blue shift

Red shift

B

Spectrograph Slit

**Figure 8.3** A diagrammatic representation of the slope of the spectral lines ab and AB for the rings and Saturn, respectively, as photographed by Keeler. The slope of the lines finally proved that the rings are not solid (see text for explanation). (The C ring was not bright enough to produce a spectrum in Keeler's equipment.)

and outer edges of the main ring. But it did yield a velocity of the limb of the planet of $10.3 \pm 0.4$ km/s, which was consistent with its known equatorial rotation rate of 10 h 14 min, giving a limb velocity of 10.29 km/s. In addition Keeler's estimate of the mean velocity of the main ring of $18.0 \pm 0.3$ km/s was consistent with the velocity of a particle in orbit around Saturn which, according to Kepler's third law, produced a figure of 18.78 km/s.

Later in 1895 W.W. Campbell, of the Lick Observatory, also measured the velocity of rotation of Saturn and its rings using the same method as Keeler. His values were 9.8 km/s for the limb of Saturn, 17.4 km/s for the middle of the main ring, and 3.1 km/s for the excess velocity of the inner edge of the main ring over its outer edge. The theoretical value for the latter parameter was 3.87 km/s. Campbell noted that, although his measurements were reasonably close to the expected values, they were all too low. He attributed this to the problem of keeping Saturn and its rings correctly aligned on the spectroscope slit.

## The satellites

A review of the distances from Saturn of its satellites known in 1800 showed that there was a large gap between Titan and Iapetus, with Titan's mean distance being about 1.2 million km and Iapetus' about 3.6 million km. So it was no great surprise when an eighth satellite, now called Hyperion, was discovered in the gap in 1848. Hyperion was first observed as an estimated 17th magnitude[19] object by G.P. Bond on 16th September, and was observed again by W.C. and G.P. Bond on 18th. As time went by it became obvious that they had discovered a new satellite of Saturn, which by mid October they had determined orbited the planet in about 21 days, in an orbital plane very similar to that of the rings.

Meanwhile in England, William Lassell happened to be looking for Iapetus on 18th September 1848 when he noticed[20] 'two stars exactly in the line of the interior satellites'. He was unsure as to which object was Iapetus, however, so he carefully recorded both of their positions, and looked for the two objects again on the following night. On 19th he was surprised to find that both objects seemed to be accompanying Saturn in its movement, since the previous night, relative to the starry background. The brighter outside object had moved away from the line of the interior satellites (and that of the rings) since 18th, as he expected Iapetus to do because of its orbital inclination of about 15°, whereas the other object was still in line and appeared to be slightly nearer to Saturn. He instantly suspected that the latter was an eighth satellite of Saturn, and confirmed this as he continued to observe its motion during the night of 19th. So, astonishingly, observers on both sides of the Atlantic had independently observed this new satellite, Hyperion, within two days of each other, and both had concluded that it was a satellite of Saturn on the same day, namely 19th September.

Unlike the case of Hyperion, however, W.H. Pickering's discovery of Phoebe, the next satellite of Saturn to be discovered, was no accident, but was the result of a deliberate attempt to uncover new satellites of Saturn using photography. To do this Pickering exposed four photographic plates for two hours each, on 16th, 17th and 18th August 1898, using a 24 inch (61 cm) f/6.7 lens at Harvard's Arequipa Station. The plates were then observed in pairs, one on top of the other, with the position of Saturn exactly superimposed. As a result, each of the background stars, all 100,000 of them, was seen as a pair of points, separated by the movement of Saturn between the two exposures. The satellites of Saturn, however, were each seen as a pair of points, with their separation depending on how far they had moved in their orbit around Saturn over the same period. In the case of the new satellite

---

[19] Bond was way out in his magnitude estimate, as Hyperion's magnitude should have been about 14.2, not 17, as Saturn was very close to opposition.

[20] W. Lassell, *Monthly Notices of the Royal Astronomical Society*, **8**, p. 195.

Phoebe, however, it was registered as a single point, as it had not moved appreciably over this period of time. Phoebe was the first satellite of any planet to be discovered photographically.

Because of Phoebe's great distance of over 13 million km from Saturn, it took some time to determine its orbit, which turned out to be retrograde. That is, it orbits Saturn in the opposite direction to all the other satellites, and to the direction of rotation of the planet. This indicated that it is a captured object, rather than having been originally part of Saturn's system. Its large orbital eccentricity (of 0.163) added credibility to that idea.

## The planet

It had been difficult in the eighteenth century to measure the rotation period of Saturn, owing to the lack of clear markings on its visible surface. Nevertheless, William Herschel had managed to estimate a rotation period of 10 h 16 min 0.4 s (see Section 6.2), which is only about two minutes longer than the true equatorial period. This rapid rotation rate, coupled with Saturn's very low density of about 0.7 g/cm$^3$, results in a relatively large polar flattening of about 10% that had also been measured by William Herschel.

No significant progress was made in trying to understand the rotational behaviour of Saturn in the nineteenth century, until Asaph Hall noticed a well-defined, white equatorial spot on 7th December 1876. Subsequent observations showed that it had a rotation period around Saturn of 10 h 14 min 23.8 s, with an estimated error of about 2.5 s. However, Hall was cautious enough to point out that this was the rotation period of the spot, and not necessarily that of the planet as a whole. It was thought most likely, in view of the very low density of Saturn, that we are observing Saturn's atmosphere, and this was the rotation period of a spot in that atmosphere.

Naturally, measuring the period of just one spot could not tell us if the atmosphere is static with respect to the surface of Saturn, assuming that such a surface exists, or whether the atmosphere moves as a whole with respect to that surface. Equally, it was unclear if parts of the atmosphere move at different velocities, as on the sun, where the photosphere moves more slowly at higher latitudes than at the equator, or on Jupiter, where even at the same latitude the atmospheric spots move at slightly different velocities.

No more spots appear to have been noticed on Saturn until 1891 when a number of bright equatorial spots were observed independently by Arthur Marth and W.F. Denning, who deduced rotation periods of 10 h 14 min 21.8 s and 10 h 14 min 26.6 s, respectively. More equatorial spots were seen in the following year, then in 1893 a number of small dark spots were seen in moderate, northern latitudes. An English amateur astronomer, Arthur Stanley Williams (1861–1938), studied these northern spots over many rotation periods, finding that the eleven most long-lasting had periods ranging from 10 h 14 min 28 s to 10 h 15 min 1 s, depending on the longitude of the spots, with those on opposite sides of the planet clearly moving at different velocities. Williams also noted that the rotation period of the equatorial spots appeared to have reduced by over a minute since 1891, so both the equatorial and the northern hemisphere spots appeared to be phenomena within the atmosphere, that showed a composite of both the general and local movement of the atmosphere across our line of sight. What the rotation rate of Saturn's solid or liquid core was, if such existed, was still anyone's guess, however.

Secchi and Huggins had observed the spectrum of Saturn in the 1860s, and had found a red absorption band at 6,180 Ångströms (618 nm) of unknown origin, similar to that that they had observed in Jupiter's spectrum. In 1867, Janssen also found what he thought were water vapour bands in Saturn's spectrum, but these do not appear to have been seen by anyone else. So the true nature of the atmospheric constituents of Saturn, like that of all the other planets, was still unclear at the end of the nineteenth century.

## 8.3 Uranus

### Surface markings

Buffham noticed a number of spots on Uranus over the period 1870–1872 using a 9 inch (23 cm) refractor. As a result of their apparent movement, he tentatively suggested a rotation period of about 12 hours, with the equator inclined at a significant angle to the plane of the satellites' orbits. Schiaparelli also saw faint spots in 1883, using a similar sized refractor, but they were too faint to allow him to determine the direction of the planet's rotation, let alone to estimate its period. Then in the same year Charles A. Young, using a much more powerful 23 inch (58 cm) refractor, found a number of faint bands[21] 'similar to the belts of Jupiter viewed with a very small telescope' making a considerable angle with the line of the satellites.[22] This was puzzling as Uranus' polar flattening, which had been first observed by William Herschel at the end of the eighteenth century (see Section 6.1),

---

[21] Charles A. Young, *General Astronomy*, Ginn & Co., Boston, 1888.

[22] The orbits of Uranus' satellites were virtually edge-on, as seen from earth, at this time.

appeared to imply that its axis of rotation was roughly perpendicular to the line of its satellites.

In 1884, the brothers Paul and Prosper Henry, using the 15 inch (38 cm) Meudon refractor, observed bands on Uranus inclined at about 40° to the plane of the satellites' orbits. But in 1889 Perrotin, using the 30 inch (76 cm) Nice refractor, observed an angle of only 10°. These contradictory results, from careful observers using state-of-the-art instruments, was most disappointing to astronomers of the period. But it was, in the end, to require the use of late twentieth century technology to satisfactorily define the orientation of Uranus' spin axis and the value of its rotation period.

## Planetary spectrum

Angelo Secchi examined the spectrum of Uranus in 1869, and surprisingly found that it was quite unlike that of Jupiter or Saturn. For Uranus there was a broad absorption band in the blue, one in the green, and one in the yellow that was so broad that there was virtually no yellow light in Uranus' spectrum. This led Secchi to suggest that[23] 'If this spectrum [of Uranus] is purely due to reflected sunlight (which might perhaps be questioned), it must undergo a considerable modification in the planet's atmosphere.'

Two years later, William Huggins also examined Uranus' spectrum but, unlike Secchi, he did not find such a dramatic loss of yellow light. Nevertheless, he confirmed Secchi's absorption bands in the blue and green, and added four more to the list. Huggins found that the blue absorption band was very close to the hydrogen β line at 4,861 Å (Ångström), and the green one had a wavelength of about 5,440 Å. Of the remaining four bands, three appeared to be very close to, but not coincident with, those of air.

Secchi and Huggins had been unable to see any Fraunhofer lines in Uranus' spectrum, because the spectrum was not bright enough to allow a narrow enough slit to be used. In 1889, however, Huggins was able to photograph the ultraviolet spectrum of Uranus, with an exposure of two hours. This showed all the main Fraunhofer lines in that waveband, clearly showing that most of the light, at least, was reflected sunlight.

## Satellites

William Herschel had discovered two satellites of Uranus, later called Titania and Oberon, in orbits inclined at about 101° to the orbit of Uranus.

---

[23] A. Secchi, *Comptes Rendus*, **68**, p. 761.

This made them the first satellites in the solar system known to have retrograde orbits (see Section 6.1). Then, some sixty years later, William Lassell thought that he had discovered two more Uranian satellites in a series of observations from September to November 1847. But his observations were too sporadic to yield anything approaching clearly defined orbits. Meanwhile, Otto Struve had been observing the motion of Titania and Oberon from October to December 1847, to obtain a more accurate mass for Uranus. He also found another satellite of Uranus but, again, he did not have enough results to estimate its orbit. Dawes then analysed Lassell's and Struve's observations, and concluded that they had, between them, observed three new satellites of Uranus, which he labelled 'a', 'b' and 'c'. Lassell, he concluded, had seen 'a' with a period of about 2 d 2 h 40 min, whilst Struve had seen 'c' with a period of about 3 d 22 h 10 min. In addition, 'b' had been seen just once, by Lassell on the very clear night of 6th November 1847.

There matters rested until October 1851 when Lassell unambiguously discovered the two satellites of Uranus that are now called Ariel and Umbriel.[24] He first observed them, together with Titania and Oberon, on 24th October with his 24 inch (61 cm) reflector. He again saw them on 28th and 30th October and on 2nd November, declaring that[25] 'The observations are all perfectly well satisfied with a period of revolution of almost exactly four days for the outermost [Umbriel], ... and 2.5 days for the closest [Ariel]... They are therefore both considerably within the nearest [Titania] of the two bright satellites...'

Lassell continued observing Uranus and its retinue until 22nd December 1851, calculating improved periods for Ariel and Umbriel. These indicated that he had seen both satellites previously on 6th November 1847, and, using those observations and those of 2nd November 1851, he calculated periods for Ariel and Umbriel of 2.51170 days or 2 d 12 h 15 min 51 s, and 4.14448 days or 4 d 3 h 28 min 3 s. By 1853 he had modified these to 2.52038 days and 4.14454 days, which are correct to at least the fourth place of decimals. So he had definitely seen both of the inner satellites on 6th November 1847. In fact, it turns out that the first observation of Ariel had been made by Lassell on 14th September 1847, and the first observation of Umbriel had been by Struve on 8th October of the same year. So Dawes' identification of three new satellites, using Lassell's and Struve's early observations, had been incorrect.

---

[24] The names Ariel and Umbriel, which were suggested by Sir John Herschel, are two mythological characters in Pope's *The Rape of the Lock*.

[25] W. Lassell, *Monthly Notices of the Royal Astronomical Society*, **11**, 1851, p. 248.

## 8.4 Neptune

The problems of trying to see markings on Neptune,[26] to deduce its rotation period, are even worse than for Uranus, as Neptune is about 50% further away from earth. In 1883, however, Maxwell Hall, observing from the Kempshot Observatory on Jamaica, noticed a regular variation in Neptune's brightness from about magnitude 7.6 to 8.3, with a period of about 7 h 55 min. These variations only lasted for just over two weeks, however, before recommencing about a year later with a similar period. Other observers were unable to find any such variations, although in 1899 T.J.J. See said that he was able to see equatorial belts on Neptune with the Washington 26 inch (66 cm) refractor.

Neptune's satellite Triton, which had been discovered by Lassell in 1846 (see Section 6.5), was observed in 1847 and 1848 by W.C. Bond at the Harvard College Observatory in the United States, and by Struve at the Pulkovo Observatory in Russia. As a result it was concluded that Triton has an orbit inclined at about 30° to the ecliptic. However, at that time, because of its large orbital inclination, Triton was not eclipsed by Neptune as seen from earth, and so it was impossible to know whether Triton moves in a prograde or retrograde orbit. A few years later it became clear that its motion is retrograde.

Triton was seen to be very bright, considering how far away it is from the sun. As a result, it was thought that it may be the largest satellite in the solar system. It is also quite close to Neptune in its retrograde orbit, and so it was thought probable that Neptune's axial spin would also be retrograde.

Both Neptune and Uranus are moving in their orbits around the sun in the same direction as all the other planets, and in roughly the same orbital plane. So it appeared as though the solar system had formed as a consistent whole. In that case it seemed unlikely that Neptune and Uranus would have been formed spinning in a different sense to all the other planets whose spin direction was known at that time. It appeared as though something had collided with Uranus to cause it to spin on its side, and it was also suggested that a collision had caused Neptune to have a retrograde axial spin direction.

The discovery of Triton allowed the first reasonably accurate estimate to be made of the mass of Neptune. It turned out to be about 17.1 times the mass of the earth. Various observers also measured its diameter, normalised

---

[26] Neptune's discovery is covered in Section 6.5.

to its mean distance from the sun, varying from 2.99″ by Challis in 1847 down to 2.20″ by Struve in 1893. The accuracy was apparently improved by Barnard, who used the 36 inch (91 cm) Lick refractor over the period 1894–1895, who measured its diameter to be 2.43″. This corresponded to a diameter of 52,900 km, and implied a density of about 1.32 g/cm³, or similar to that of Jupiter and Uranus.

# Chapter 9 | QUIET INTERLUDE – THE TWENTIETH CENTURY PRIOR TO THE SPACE AGE

## 9.1 **Pluto**

Irregularities in the orbit of Uranus had led to the discovery of Neptune in September 1846 (see Section 6.5) but, within a week of its discovery, Urbain Le Verrier was speculating on the possible existence of planets even further from the sun. Le Verrier first made such a suggestion to Galle on 1st October 1846, but even then he realised that finding such so-called 'trans-Neptunian' planets would be considerably more difficult than finding Neptune itself. For a start, it was clear that it would take many decades for any discrepancy to appear between the observed and calculated orbit of Neptune, as its orbital period was 165 years, or about double that of Uranus. In addition, it was likely that any such trans-Neptunian planets would appear to be even dimmer than Neptune, and maybe considerably so, owing to their greater distance from the sun. However, if a trans-Neptunian planet was large enough and close enough to the sun, it may have a measurable effect on the orbit of Uranus, as well as that of Neptune.

Some thirty years after the discovery of Neptune, David P. Todd, an assistant at the US Naval Observatory in Washington, analysed Uranus' orbit and concluded in 1877 that it was being affected by a trans-Neptunian planet. Todd calculated that the new planet was orbiting the sun at a distance of about 52 AU and should, at that time, be in the constellation of Virgo. Unfortunately, his search for it with the US Naval Observatory's 26 inch (66 cm) refractor was unsuccessful.

Two years later Camille Flammarion suggested that an analysis of the orbits of the periodic comets may indicate the presence of an unknown planet. Shortly afterwards the Scottish astronomer George Forbes made such an analysis and concluded that there were two possible trans-Neptunian planets. Likewise Hans-Emil Lau of Copenhagen concluded that there were two such possible planets by analysing the residuals of Uranus' orbit. Observational searches yielded no such planets, however.

It was gradually becoming clear that the search for a trans-Neptunian planet would benefit from the aid of some new technology which could quickly detect if any one of the many thousands of stars visible through a telescope was not really a star but a planet. The obvious technology was that of photography. After all, in 1891 Max Wolf had discovered an asteroid using photography, and that discovery had been followed by a veritable avalanche of further asteroid discoveries. Similarly in 1898 W.H. Pickering had been the first to use photography to discover a planetary satellite when he found Phoebe, a satellite of Saturn. It was natural, therefore, that photography would be used in the late nineteenth and early twentieth centuries to search for a possible trans-Neptunian planet.

By the early years of the twentieth century, Percival Lowell was tiring of the controversy about Martian canals (see Section 7.10), and was looking for a new area of research. It so happened that he had been interested in the idea of finding a new planet ever since Benjamin Peirce, his tutor at Harvard, had claimed that the discovery of Neptune was a fortunate accident. So in 1905 Lowell recruited an assistant, William Carrigan, to analyse the residual motions of the outer planets, to try to predict the position of any trans-Neptunian planets. Without waiting for the results of Carrigan's analysis, however, Lowell and his assistants at his observatory at Flagstaff, Arizona, started a photographic search of the region around the ecliptic, which is probably where such a planet would be located, using a 5 inch (12.5 cm) Brashear refractor. This telescope had a relatively wide field of view of about 5°, which was almost ideal for such a planetary search, although exposures of about three hours were required to get down to a magnitude of 16.[1] In all, 440 plates were exposed by September 1907 in this first search.

Although much useful work was done in this first search, neither the analytical efforts of Carrigan, nor those of the photographic team, had yielded any concrete results. To make matters worse, in November 1908 Lowell attended a presentation by William Pickering of the Harvard College Observatory, in which he described his search for the unknown planet. Lowell was horrified to discover that Pickering appeared to be further down the road

---

[1] Lowell expected the magnitude of the trans-Neptunian planet to be about 13, so such an exposure should show it up very clearly, even if it were to be a little dimmer than expected.

than he was, so he dismissed Carrigan, on the grounds that he was being too meticulous and slow, and decided to lead the calculations himself.

Lowell started his analysis in 1910, assisted by Elizabeth Williams and, in the following year, he requested Carl Lampland (1873–1951), his assistant observatory director, to begin a new photographic survey using the observatory's 42 inch (107 cm) reflector, which had been installed the previous year. This reduced the exposure time from three hours to seven minutes to reach magnitude 16, but it was at the expense of a smaller field of view of about 1°. On Lampland's advice, Lowell bought a Zeiss blink comparator to facilitate comparing plates, looking for the movement that would give the planet away.

The small field of view of the 42 inch soon proved to be impracticably small, however, and in April 1914 the search was switched to a Brashear 9 inch (23 cm) photographic refractor that Lowell had borrowed from Swarthmore College's Sproul Observatory. Lowell published his analytical conclusions in 1915, in which he predicted that Planet X, as he called it, would have a mass of 6.6 times that of the earth, would have an orbital radius of about 43 AU, and would probably be found in Gemini, near its border with Taurus. In the event, Lowell and his staff never found Planet X before the search was put on hold following Lowell's death on 12th November 1916.

The Harvard astronomer William Pickering also tried to find Planet O, as he called it, publishing various predicted positions between 1908 and 1928, but he could find no such planet on his photographs. In addition, in 1911 he predicted that there were three more planets beyond Planet O. One of them, Planet Q, was in a highly elliptical polar orbit, with an orbital period of about 26,000 years, and with a mass of about 63 times that of Jupiter. Pickering's predictions were never taken seriously by the astronomical community, however, as his analysis was not very thorough and he kept changing his mind radically on his results.

After Lowell's death in 1916, the operation of his observatory had been put on a minimum cost footing, whilst the ownership of his estate could be sorted out, following an attempt by his wife Constance to have his will overturned. It took until 1927 for the litigation to be completed. Then Lowell's brother, Abbott Lawrence Lowell, who was president of Harvard, agreed to provide funds for a 13 inch (33 cm) photographic refractor, to enable the search for Planet X to be recommenced. This new telescope had a field of view in excess of 12°, and it could record stars as dim as magnitude 17 after an exposure of one hour. Unfortunately, there was no one available at the observatory to undertake the search.

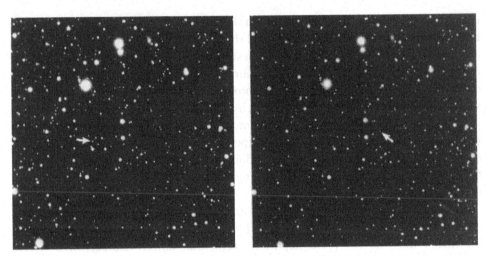

**Figure 9.1** Pluto is marked by an arrow in these discovery plates of 23rd January 1930 (left) and 29th January 1930 (right). It has clearly moved over the time between the two images, indicating its planetary nature. (Lowell Observatory Photograph.)

It so happened that about this time a keen, young amateur astronomer called Clyde Tombaugh (1906–1997), who made his own telescopes, had contacted the observatory to ask for an opinion about some drawings that he had made of Mars, Jupiter and Saturn. However, Vesto Slipher (1875–1969), the director of the observatory, was so impressed with the precision of Tombaugh's drawings and by his obvious determination, that he offered him a job as an assistant observer on a trial basis. He was to undertake the photographic search for Planet X.

After some preliminary work, Tombaugh started the search on 6th April 1929 by photographing the region of Gemini. But he had a number of problems at the beginning, with both the drive of the new 13 inch refractor, and the plates which cracked in the intense cold. He solved these problems, and by January 1930 he had photographed the whole of the zodiac and was back at Gemini again. So far the planet had not been found but that was soon to change. Tombaugh took three plates of the Delta Geminorum region, on 21st, 23rd and 29th January, and began blinking the plates on 15th February. Three days later, at about 4 pm, he saw a 15th magnitude object move when he blinked the latter two plates (see Figure 9.1). He checked the plate of 21st and found that the object moved there as well. He had found Planet X!

The night of 15th February was cloudy, but on the next night Tombaugh took another photograph of the planet, which had moved as expected. Then on 20th February Tombaugh, Slipher and Lampland observed the object visually using the observatory's 24 inch (61 cm) refractor, and were surprised to find that it showed no disc, so the planet was either very small or very far away. For some days Tombaugh worried that he may have found the satellite of a planet, and not the planet itself, but plates taken with the 42 inch reflector quickly dispelled that possibility.

Vesto Slipher insisted on more evidence before making a public announcement; the delay giving the observatory a head-start in calculating the orbit of the new planet. Just over three weeks later, he was ready to announce the discovery of the planet, sending a telegram on 12th March to Harlow Shapley at the Harvard College Observatory, the official clearing house for such discoveries. The discovery was made public on the following day, being the 75th anniversary of Lowell's birth, and the 149th anniversary of William Herschel's discovery of Uranus.

Slipher's next priority was to establish the orbit of Planet X, and here the Lowell Observatory was very unpopular, as it released only one position of the planet, and not the series of positions that had been measured over almost two months. In spite of this, the first orbital estimate was published on 7th April by Armin Leuschner, Ernest Bower and Fred Whipple of the University of California, who had deduced a distance of 41 AU and an orbital inclination of 17°, both very close to the truth, although their orbit was highly eccentric. The Lowell Observatory produced their first estimate on 12th April, but it was wildly out, being almost parabolic with an orbital period of 3,000 years. It was not until the British astronomer Andrew Crommelin, of the Royal Observatory of Belgium, announced on 9th May that the observatory had photographed Planet X on 27th January 1927, that a reasonably accurate orbit could be calculated.

Although Constance Lowell, Percival's wife, had severely disrupted the work of the observatory with her law suit after her husband's death, she still wanted to share the glory of Planet X's discovery. As a result, she suggested various names for the new planet, including Zeus, Percival, Lowell, and Constance, but each of these suggestions were rejected. Slipher preferred Minerva, but that was already the name of an asteroid. The observatory considered many other names, some of their own choosing, and some sent in by the enthusiastic public. Amongst them was Pluto, which the observatory was considering, when it was also suggested by Venetia Burney, an eleven year-old schoolgirl from England, who thought that the name of the god of the underworld was suitable for a planet so far from the sun.[2] The observatory

---

[2] Although it had been forgotten at the time, the French astronomer P. Reynaud had been the first to suggest as long ago as 1919 that Lowell's Planet X should be called Pluto.

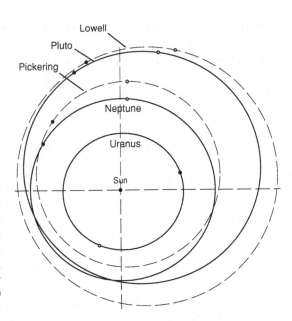

**Figure 9.2** The true orbit of Pluto is compared in this diagram with that predicted by Lowell in 1914 (published in 1915), and by Pickering for Planet O in 1928. The open circles show the positions of the planets (real and predicted) in 1900, and the closed circles show them in 1930. It is clear that Lowell's orbit was much closer to that of Pluto, than was Pickering's, particularly for the period around 1930 when Pluto was discovered.

agreed, and Slipher officially proposed on 1st May 1930 that the planet should be called Pluto.

At the end of June 1930 it was found that Pluto had been recorded on two plates taken on 19th March and 7th April 1915 by Lowell's assistant Thomas Gill, but it had been missed because the plates contained too many stars for the fainter-than-expected planet to be noticed. It had also been recorded on two plates taken by Milton Humason on Mount Wilson in 1919, in support of William Pickering's search, but again it had been missed as on one plate it had fallen on a plate defect, and on the other it was too close to a bright star. Tombaugh had also unknowingly photographed Pluto before. It was found on plates taken on 11th April and 30th April 1929, just after he started his search, but the first plate had cracked, and Pluto was too faint to be easily seen on the second.

Was Pluto the Planet X predicted by Percival Lowell? Yes and no. Yes, because his orbital prediction was quite accurate (see Figure 9.2), and no, because Pluto was much smaller than he had calculated, and was far too small to have perturbed Uranus by anything like the extent that he had assumed.

Pluto did not show a disc in even the largest telescopes in 1930, and it did not appear to have a satellite, making it impossible to produce an accurate estimate of its mass. Even if Pluto was made completely of iron, however, its mass could not be much more than that of the earth, and it was considered more likely that it was somewhat lighter. Amazingly, over the years since its discovery, Pluto's estimated mass has been gradually reduced, until today it is only 0.002 times that of the earth, or 20% that of the moon.

As Pluto is so small, its discovery was a fortunate accident, and if something really is disturbing Uranus to the extent deduced by Lowell, it must be a larger planet than Pluto. In the meantime, Clyde Tombaugh had been asked by Slipher to continue searching for more trans-Neptunian planets. Over the next thirteen years he covered most of the sky visible from the Lowell Observatory, and, as a result, blinked over 40 million objects down to magnitude of about 16.5, without finding another planet. He found many other objects, however, including over 700 asteroids.

Pluto's orbit is highly eccentric, with an aphelion of 30 AU and a perihelion of 49 AU, so that, at its closest to the sun, its orbit lies just within that of Neptune. In addition, Pluto's orbit is inclined to the ecliptic at an inclination of 17°, which is by far the largest inclination of any planet. In fact, Pluto's large orbital eccentricity and inclination are very reminiscent of the orbits of some of the outer satellites of Jupiter and Saturn. A similarity that Lowell had predicted in 1915.

Many years later, in 1955 Merle Walker and Robert Hardie of Vanderbilt University used a photomultiplier attached to the 42 inch (107 cm) reflector of the Lowell Observatory in an attempt to measure any light fluctuations of Pluto. Not only would this enable its rotation period to be measured, assuming that its atmosphere or surface were not uniform in reflectivity, but it may also give a first indication about the physical condition of the planet. Much to their surprise, they found that Pluto fluctuated in intensity by more than 10%, which is more than any other planet except Mars, over a period of 6 d 9 h 17 min. This period was absolutely regular, indicating that it is the axial rotation period of the planet. The length of the period was something of a surprise, however, as it is much more than that of any other planet outside the earth's orbit. Walker and Hardie also found that the shape of the light curve was remarkably consistent from one spin cycle to the next, indicating that either Pluto's atmosphere, if it has one, has a remarkably stable cloud structure, or, more likely, that they were measuring a surface that has a great variation of reflectivity from place to place.[3]

---

[3] Although Pluto has not yet been visited by a spacecraft, post-1960 work on Pluto is discussed later (see Section 11.5).

## 9.2 Mercury

### General

The late nineteenth and early twentieth centuries saw a radical change in emphasis in astronomy, as both spectroscopy and photography were used, for the first time, to examine and record astronomical objects. Spectroscopy enabled the compositions and relative velocities of stars to be determined, and long exposure photographs enabled astronomers to detect faint stars, and to objectively record the structure of galaxies for the first time. Unfortunately, neither technology proved of major value to planetary astronomy in this period,[4] with the result that at the start of the space age in 1957 most professional astronomers had given up planetary research, as other fields of astronomy were proving far more rewarding. Taking instruments above the earth's atmosphere on spacecraft, and sending some of them to the planets, was to change the balance between planetary and non-planetary research yet again later in the twentieth century. But, in the early part of the century, which is the subject of this chapter, planetary astronomy had almost stagnated, except in the areas of theory and in some limited spectroscopic observations of the outer planets.

### Axial rotation period

The idea of a synchronous rotation period for Mercury had been gradually gaining acceptance at the end of the nineteenth century (see Section 7.6), although some astronomers still favoured an approximately 24 hour period. In 1929 Eugène Antoniadi, a Greek astronomer who spent most of his working life in France, confirmed this synchronous rotation period after five years of intensive observations with the 33 inch (85 cm) Meudon refractor, during which he also saw evidence of occasional dust clouds. Later, Bernard Lyot (1897–1952) and Audouin Dollfus (b. 1924), working at the Pic du Midi Observatory in the French Pyrenees, independently confirmed the synchronous rotation period, the latter quoting an agreement to within one part in a thousand.

A synchronous rotation period was generally accepted by astronomers and was usually stated as a fact until, in 1962, W.E. Howard and his colleagues at Michigan found that the night side seemed to be warmer than it should be

---

[4] Except in discovering very dim objects like Pluto and some planetary moons.

if it were permanently in shadow. Then, in 1965, R. Dyce and G. Pettengill measured the Doppler shift of radar signals sent from earth using the Arecibo radio telescope in Puerto Rico. They found that Mercury's rotation period was not 88 days, but 58.65 days, or exactly $\frac{2}{3}$ of the planet's rotation period around the sun, so Mercury rotates exactly $1\frac{1}{2}$ times on its axis per Mercurian year. This means that, at perihelion, when Mercury is at the closest point in its orbit to the sun, there are only two positions on Mercury's equator that can be directly facing the sun. These sub-solar points, which are on opposite sides of the planet, are called the hot poles of Mercury, alternating as sub-solar points on successive perihelia.

Why did Antoniadi, Lyot, and many other respected astronomers, erroneously believe in Mercury's synchronous rotation period? It was because the period between successive appearances of Mercury in the same phase as seen from earth, that is its synodic period, of 115.88 days, is almost exactly twice its axial rotation period of 58.65 days. As a result, we see almost exactly the same part of Mercury around successive greatest elongations, when it is best placed for observation. This gave the erroneous impression that the same face is always facing the sun.

Peter Goldreich explained the linkage between Mercury's day and its year by a process of spin–orbit coupling. Mercury's orbit is highly eccentric, with a perihelion of 46 million km and an aphelion of 70 million km, and so the gravitational attraction of the sun at perihelion is over twice that at aphelion (gravitational attraction being proportional to the square of the distance). Goldreich suggested that Mercury originally had a fast rotation and, as a result, an equatorial bulge had been formed. As the sun's gravitational pull is much higher at Mercury's perihelion, compared with aphelion, it slowed the planet's axial rotation, so that the long axis of the bulge pointed at the sun at every perihelion. The situation stabilised with alternate ends of the long axis pointed at the sun at successive perihelia.

## The surface

Gustav Müller and Paul Kempf's photometric measurements of Mercury indicated in 1893 that, if it has an atmosphere, it is largely transparent, and that Mercury's surface is somewhat darker and rougher than that of the moon. Then, in 1924, Lyot deduced, by measuring the variation in the polarisation of light scattered from its surface as a function of illumination and viewing angles, that Mercury is covered in volcanic ash. At about the same time, Seth B. Nicholson (1891–1963) and Edison Pettit measured the temperature of Mercury using an evacuated thermocouple[5] attached to the Mount Wilson 100 inch (2.5 m) reflector. They found a value of about 690 K (417 °C) at full phase (i.e. with vertical illumination) and about 600 K at half phase.

## 9.3 Venus

### Axial rotation period

It was generally accepted at the end of the nineteenth century that, as Venus seemed to be virtually covered in clouds, any rotation period deduced from visual observations would probably be that of its cloud layer, rather than of its surface. The difference between the planetary rotation rate and that of its clouds was expected to be small, however, as it was expected that the clouds, like those on earth, would move relatively slowly across its surface.

At the end of the nineteenth century there were two competing schools of thought as far as Venus' rotation period is concerned (see Section 7.7). One considered it to be about 24 hours, and the other that it was the same as its orbital period of 224.7 days. Although the latter figure was probably the favourite at the end of the century, it was by no means universally accepted.

In 1900 the Russian astronomer Aristarkh Belopol'skii used spectroscopy at the Pulkovo Observatory to measure Venus' axial rotation period, by observing the Doppler shift of light from its limb.[6] Using this technique, he deduced a tentative figure of 24 h 42 min, which eleven years later he amended to 35 hours. In the meantime Vesto Slipher, of the University of Indiana, using a similar system concluded, from the imperceptible slope of the spectral lines that he observed, that the rotation period must be at least three weeks and was most likely synchronous.

In 1921 W.H. Pickering proposed an entirely new period of 2 d 20 h based on the movement of faint markings on the planet. Three years later H. McEwen, the director of the Mercury and Venus section of the British Astronomical Association, supported Pickering's period, but in the following year W.H. Steavenson deduced a period of 8 days, although he did point

---

[5] A thermocouple is a device in which a voltage is produced by the junction of two dissimilar metals, with the magnitude of the voltage being dependent on the temperature of the junction.

[6] This was similar to the method used by Keeler a few years earlier to estimate the rotation period of Saturn and its rings (see Section 8.2).

out that a sub-multiple of that figure would also be compatible with his observations. Then in 1924 W. Coblentz (1873–1962) and Carl Lampland measured the temperature of Venus' light and dark hemispheres as about 320 and 260 K, respectively.[7] This temperature of the dark hemisphere indicated that it cannot be in permanent darkness, and so the rotation period cannot be synchronous. So the simple split, that had existed at the start of the twentieth century, between astronomers who believed in either a 24 hour period or a synchronous period, had been abandoned twenty years later. Instead, some astronomers were beginning to favour an intermediate figure, although the value of that figure was by no means clear.

The situation was no less confusing by 1957 when Charles Boyer, a French amateur astronomer living in the Congo, took a series of photographs in ultraviolet light which was known to emphasise cloud details.[8] Boyer used only a 10 inch (25 cm) reflector, but he found a distinctive V-shaped marking (with the V on its side), which reappeared every 4 days. When this was announced, Henri Camichel found the same period on photographs taken in 1953, and so the rotation period of Venus' clouds, at least, appeared to be 4 days retrograde (i.e. from east to west).[9]

Although Venus is covered in clouds, radar signals have the advantage over visible light in that they can penetrate cloud. So, if radar contact could be made with Venus, not only could its distance, and hence the value of the astronomical unit, be determined, but an analysis of the frequency distribution of the return signal would also enable the speed of rotation of its surface beneath the cloud to be determined. Radar signals had first been bounced off the moon by the US Army Signals Corps in 1946, but Venus, even at inferior conjunction, was a hundred times further from the earth than the moon, and the strength of radar return signals diminishes as the fourth power of

---

[7] Pettit and Nicholson had also measured the night and day side temperatures of Venus using the 100 inch (2.5 m) Mount Wilson reflector between 1923 and 1928. In their initial analysis, published in 1924, they concluded, like Coblentz and Lampland, that the night side was about 70 K cooler than the day side. But in a reanalysis of their whole series of measurements, published in 1955, they deduced a figure of 240 K for the night side and 235 K for the day side. Because of the scatter of their results, however, Pettit and Nicholson concluded that there was no discernible difference between the temperature of the night and day sides of the planet.

[8] The benefit of using ultraviolet photographs to image the clouds of Venus was discovered by W.H. Wright in 1924 whilst photographing the planet through different colour filters.

[9] Although Boyer incorrectly attributed this 4 day period to that of the planet as a whole.

the target's distance. So it took some time for the technology to advance sufficiently for a radar contact to be made with Venus.

The first successful radar returns from Venus were announced in March 1959 by Bob Price and Paul Green of MIT's Lincoln Laboratory, following an analysis of their previous year's radar experiments. In these they had used the Lincoln Laboratory's 25 m diameter antenna at their Millstone Hill Radar Observatory. Unfortunately, the return signal from Venus was just on the limit of detectability, but two of the four observing runs apparently produced detectable returns. However, an attempt in September 1959 by Bob Price and his colleagues at the Lincoln Laboratory to repeat the experiment with an improved radar system failed. This led some astronomers to doubt whether the 1958 signals really had been radar returns from Venus.

It was not until 10th March 1961, in fact, that the first unambiguous radar return signals from Venus were detected by Richard Goldstein, using the Jet Propulsion Laboratory's (JPL's) Goldstone tracking station in California. These results, which produced a different value of the astronomical unit from that of the Price and Green work, caused Price to re-examine their 1958 tapes. In doing so he found that the reported return signals of 1958 and 1961 had an inexplicable timing difference of 2.2 milliseconds. As a result MIT retracted their 1958 results and acknowledged priority to JPL for being the first, in 1961, to achieve a radar contact with Venus.[10] But at this stage the rotation of Venus had still not been detected by radar.

Then in an internal report of May 1962, Roland Carpenter of JPL announced his detection of Venus' rotation, after analysing the results of JPL's radar campaign of the previous year. Surprisingly, Carpenter found the rotation to be retrograde, but he was unsure of the actual rotation period. A further JPL radar campaign was undertaken in 1962 from which Carpenter deduced a period of $250 \pm 40$ days retrograde.

In the meantime, William B. Smith of MIT had also concluded that Venus' rotation was retrograde, based on the results of the Lincoln Laboratory radar campaign of April–June 1961. Unfortunately, however, his paper making this announcement, which was published in *The Astronomical Journal* of February 1963, was watered down at the request of his supervisor, Paul Green, to suggest that the motion was possibly retrograde. Green was

---

[10] Later analysis showed, in fact, that MIT had, on 14th September 1959, been the first to detect radar returns from Venus, although the signals had been initially lost in the noise. Further analysis showed that MIT had also detected Venus radar returns on 6th March 1961, four days before JPL's successful contact.

only too well aware of the embarrassment caused at MIT by the retraction that had had to be made following the erroneous Venus radar claims of 1959, and he did not want to have to make another retraction so soon. The JPL people had no such problems, however, so Carpenter and Goldstein's paper, which was published in *Science* the month after Smith's paper, was quite explicit in announcing Venus' retrograde rotation. Because of this the JPL team are usually given credit for discovering the retrograde spin direction of Venus.

Carpenter's period of about 250 days retrograde was confirmed in 1965, using the new Arecibo radio telescope, when a more accurate period of 243 days retrograde was measured. Then in 1970 the International Astronomical Union (IAU) accepted a rotation period of 243.0 days, which implied that the length of the Venusian day is 116.69 terrestrial days.

## The atmosphere

It was not clear in the late nineteenth century as to whether there was oxygen and water vapour in Venus' atmosphere or not (see Section 7.7). In 1908, Vesto Slipher took spectra of Venus and looked for the Doppler shift in the oxygen and water vapour spectral lines that should have been caused, if they were produced on Venus, by the difference in orbital velocities of the earth and Venus. This would vary, being largest when Venus was at greatest elongation. He found no such effect, however, and concluded that the oxygen and water vapour lines that he had observed were due to the earth's atmosphere and did not originate in Venus' atmosphere. So there was no oxygen or water vapour in the upper part, at least, of Venus' atmosphere.

In 1932 Walter S. Adams (1876–1956) and Theodore Dunham, Jr. (1897–1984) used a diffraction spectrograph, operating in the infrared, on the 100 inch (2.5 m) Mount Wilson reflector. They could find no evidence for the oxygen and water vapour on Venus, but carbon dioxide was clearly present. In 1939 Rupert Wildt (1905–1976) of Princeton University Observatory calculated that the greenhouse heating caused by this carbon dioxide could produce a surface temperature as high as 408 K, which is above the boiling point of water on earth (i.e. at an atmospheric pressure of 1 bar).

Wildt had also suggested in 1937 that, if the atmosphere of Venus has very small amounts of oxygen, there would be no ozone layer to shield the lower parts of the atmosphere from the sun's ultraviolet light. Under these conditions, ultraviolet light from the sun would cause carbon dioxide and water vapour (if it existed) to recombine to form oxygen, that would be absorbed by the rocks, and formaldehyde ($CH_2O$). Wildt thought, therefore, that the clouds on Venus could consist of formaldehyde and water droplets, although he could find no observational evidence for the presence of this formaldehyde.

In 1954 Donald Menzel (1901–1976) and Fred Whipple of the Harvard College Observatory resurrected the popular nineteenth century view that there were oceans of water on Venus, with clouds of water vapour. Menzel and Whipple based their idea that the clouds consisted of water vapour on the polarisation measurements of Venus' light by Bernard Lyot in 1929, which water vapour fitted well.[11] In fact they envisaged Venus as being in a state similar to that of the earth millions of years ago, but Venus would be hotter as it is closer to the sun. But how hot was its surface? If it was hotter than the boiling point of water, the surface must be completely dry and desert-like, but if it was cooler Menzel and Whipple may be correct.

Substantial new evidence for a very high surface temperature was produced in 1956 by Mayer, McCullough, and Sloanaker, using the US Naval Research Laboratory's 15 m diameter radio telescope. An analysis of the 3.15 cm radio emissions from Venus indicated a surface temperature of about 600 K (or about 300 °C). Although a high temperature of 350 to 400 K had been expected, this new value seemed unreasonably high. It appeared to demolish Menzel and Whipple's idea about the surface conditions, but many astronomers doubted the validity of these temperature measurements, arguing that some of the 3.15 cm radio emission could be caused by non-thermal processes. As a result there was in 1956 still a very big question about the surface conditions of Venus. This question was compounded a little later when Gerard de Vaucouleurs of the Harvard College Observatory deduced a surface atmospheric pressure of the order of five bars (i.e. five times that on earth). Then shortly afterwards Carl Sagan, who at that time was a doctoral student at the University of Chicago, deduced a surface pressure of an incredible 100 bars.[12] So as the space age got under way it was still not known whether Venus was hot and humid, with oceans of water and clouds of water vapour, as envisaged by Menzel and Whipple, or whether it was so hot that the surface was one great desert, with no water to be found anywhere.

---

[11] At first sight, this suggestion seemed to be inconsistent with Venus' spectral results that showed no obvious water vapour lines. This lack of water vapour lines was put down to the fact that Venus' spectrum was largely that of its atmosphere above the clouds, which was too cold for water vapour.

[12] As a comparison, this is the pressure about 1,000 m below the surface of the earth's oceans.

And what was the surface pressure? Was it really one hundred times that on earth or was it a much more believable 5 bars? In 1960 both concepts seemed equally valid.

## The surface[13]

It was impossible to estimate the physical diameter of Venus from earth using observations in visible wavelengths because of its permanent cloud cover, and it was even difficult to measure the diameter at cloud-top heights because there is no clear boundary between the cloud tops and space. In 1962 Martynov estimated an optical diameter (i.e. at cloud-top height) of 12,200 ± 60 km, and two years later de Vaucouleurs produced an estimate of 12,240 ± 14 km. Then in 1965 Kuzmin and Clark produced the first estimate of Venus' true physical diameter of 12,114 ± 110 km using a radio interferometer, which was refined a few years later to 12,104 ± 4 km using radar. So the cloud tops appeared to be about 70 km above the surface of the planet, compared with a maximum of about only 10 km for clouds on earth.

Radar observations of the surface of Venus begun in 1962 showed three regions, which Richard Goldstein labelled α, β and δ, of greater than average reflectivity. Then in 1967 Ray Jurgens of Cornell University found another high reflectivity feature, using the Arecibo radar, which he named Maxwell, after the well-known physicist. This feature was later found by spacecraft (see Section 10.5) to be a key geological formation on the planet.

In 1972 radio interferometric measurements by Goldstein showed clearly defined craters on Venus for the first time. This was something of a surprise as it had been thought that Venus' thick atmosphere would have broken up incoming objects, or at least slowed them down sufficiently, to avoid them leaving large craters. Further observations from earth eventually resolved a number of shallow, circular craters with diameters ranging from 35 to 150 km.

Meanwhile, measurements of Venus' surface relief had begun at the Haystack and Arecibo radio observatories in 1967. These showed that there

---

[13] Mariner 2 returned the first useful spacecraft data on Venus in 1962 (see Section 10.5), providing information on its temperature, cloud top heights, etc., but not on the topography of its surface, as it had no on-board radar system. Such a system was first carried to Venus by the PVO spacecraft in 1978. As a result I have extended the coverage in this section on the surface structure of Venus to the mid 1970s, compared with the end of the 1950s for the other Venus topics above, to outline the situation just before the PVO spacecraft arrived.

was a large area, covering about 6,000 km in longitude and 500 km in latitude, that was about 3 km above the average datum level of the surface. This turned out to be the first observation of the region now known as Aphrodite Terra.

## 9.4 Origin of the solar system

### Age of the solar system

Physicists had been mystified in the nineteenth century as to how the sun produced its heat. The generally accepted theory in the second half of the nineteenth century, that had been first proposed in 1853 by the German physicist Hermann von Helmholtz (1821–1894), was that the heat was produced by gravitational contraction. The problem was that Helmholtz had calculated that this mechanism would only have enabled the sun to produce heat at its current rate for about 22 million years. Unfortunately, this proved to be considerably less than the estimated age of the earth, which, according to geological and other evidence, varied from about 100 million years, according to the distinguished English physicist Lord Kelvin (1824–1907),[14] to about 400 million years according to George Darwin.

This inconsistency between the estimated age of the sun, and that of the earth, was made worse in the early twentieth century following the early work on radioactivity by Marie and Pierre Curie. In 1905, the American Bertram Boltwood (1870–1927) suggested that, as lead is a radioactive decay product of uranium, measuring the proportions of uranium and lead in rocks should enable the ages of rocks to be determined. He proved this principle by measuring the ratio of lead to uranium in samples of known relative age. Then, knowing the half-life (i.e. the rate of decay) of uranium, he was able to determine that his oldest rock specimen was about 2.2 billion years old.[15] This indicated that the solar system as a whole was probably a few billion years old, far more than the estimated age of the sun.

The first clue of an alternative energy source for the sun, to that generated by gravitational contraction, was provided in 1905 by the brilliant young German physicist, Albert Einstein (1879–1955), when he published

---

[14] Lord Kelvin was born William Thomson and was knighted in 1866. He worked out the age of the earth from the time required for its heat flow to reduce to its present level, assuming that it was originally completely molten.

[15] I have used the American definition of a billion as 1,000 million in this book.

his special theory of relativity. In this he proposed the equivalence of mass and energy, i.e. that mass can be transformed into energy according to the relationship $E = mc^2$, where $E$ is energy, $m$ is mass, and $c$ is the velocity of light. It was immediately realised that, if the whole mass of the sun could eventually be transformed into energy in this way, it would enable it to continue shining at its current rate for a few thousand billion years. This clearly eliminated the inconsistency between the estimated age of the sun and that of the earth. The exact mechanism for this complete transformation of mass into energy was unclear, however.

Twenty years later, Arthur Eddington (1882–1944), working at Cambridge University, outlined an alternative to the theory of the complete annihilation of matter when he suggested that four hydrogen nuclei (i.e. protons) and two electrons could combine to form a helium nucleus. He calculated that the energy released in such a process, in which the sun would lose about 1% of its mass, would be enough to keep the sun shining for a few ten billion years. Then in 1938 Hans Bethe (b. 1906) in America and Carl von Weizsäcker (b. 1912) in Germany both independently proposed a fusion theory in which hydrogen nuclei are transformed into helium nuclei with carbon as a catalyst. On this basis, Bethe estimated that the sun could produce energy at the present rate for well over 10 billion years.[16]

So the inconsistency between the age of the sun and that of the earth had clearly been resolved by 1940. Unfortunately, in 1929 an inconsistency had arisen between the estimated age of the earth and the age of the universe, following the work by Edwin Hubble (1889–1953) and Milton Humason (1891–1972) on the distances and recession velocities of galaxies.

Hubble and Humason had, in fact, shown that the galaxies were moving away from each other at a rate of about 500 km/s per million parsecs.[17] This parameter is now called the Hubble constant $H_0$. So those galaxies that are 1 million parsecs away from us have a recession velocity of 500 km/s, those that are 2 million parsecs away have a recession velocity of 1,000 km/s, and so on. Now, if the galaxies had been moving away from each other forever at this rate, they would all have been in the same place about 2 billion years ago, when the expansion of the universe started in what we now call the Big Bang. Unfortunately, this age of the universe of about 2 billion years,

---

[16] Further work has shown that a process based on proton–proton collisions, as proposed by Charles Critchfield, is dominant in the Sun, rather than the carbon cycle proposed by Bethe and von Weizsäcker. Although this produced a lower estimated age for the sun than 10 billion years, it never again fell below the estimated age of the earth.

[17] A parsec is a unit of distance, equivalent to 3.26 light years.

was uncomfortably close to Boltwood's minimum age for the earth of about 2.2 billion years. By the late 1940s, in fact, the situation had deteriorated even further as the minimum age of the earth, deduced from radiometric dating, was nearing 3.0 billion years, whereas the estimated age of the universe was still about 2.0 billion years. So attempts were made in the late 1940s and early 1950s to improve the age estimates for both the earth and the universe.

Progress came thick and fast in the early 1950s. First of all, Walter Baade (1893–1960) announced to the 1952 meeting of the IAU in Rome, a new value of the Hubble constant, which was half of that deduced by Hubble and Humason.[18] As a result, the estimated age of the universe was increased by a factor of two to about 4.0 billion years, compared with an age of the earth which had crept up by that time to about 3.5 billion years. So at last the ages of the earth and of the universe appeared to be compatible with each other. But that compatibility did not last very long, as in the next year Patterson's team at Caltech announced a more refined estimate for the age of the earth of about 4.5 billion years. This was based on lead isotope ratio measurements of the earth's oceans, crust and meteorites. By 1958, however, the estimated age of the universe had increased once more to about 10 billion years. This permanently resolved the inconsistency with the age of the earth, which, even today, is still reckoned to be about 4.5 billion years.

## Collisions and close encounters

Towards the end of the nineteenth century doubts began to be expressed about Laplace's nebula contraction theory of the origin of the solar system (see Section 6.3). The angular momentum of the primeval nebula would have been very similar to that of the current solar system, because of the principle of the conservation of angular momentum. But Babinet had shown in 1861 that, on this basis, the primeval nebula could not have had enough angular momentum to cause it to spin off material in the way that Laplace had suggested.

As an alternative, the English astronomer Richard Proctor (1837–1888) suggested in 1870 that the solar system had been formed when another star

---

[18] Baade realised that the Cepheid variable stars, which Hubble and Humason had used to determine their value of the Hubble constant, are of two types. Unfortunately, Hubble and Humason had thought the Cepheids were all of the same type, and this had led them to produce incorrect estimates of their absolute intensities, and hence their distances and those of their host galaxies.

had had a glancing collision with the sun. However, if the planets had been formed in this way, how had the planetary satellites been formed? As the planetary satellite systems appeared to mimic the solar system, it was expected that they had been formed in the same way, and yet the probability that the Earth, Mars, Jupiter, Saturn, Uranus and Neptune had all received glancing blows was infinitesimally small. Nevertheless, Uranus does spin on its side and Neptune seemed to have a retrograde axial rotation (see Section 8.4), so maybe these two planets, at least, had been subjected to some sort of collision.

In 1898 the Cambridge mathematician W.F. Sedgwick suggested that the star didn't need to hit the sun, as if the star had passed close by the sun, it could have drawn out a large amount of material by tidal attraction.[19] The English astronomer Sir James Jeans (1877–1946) put forward a similar idea in 1901, and worked out the mathematics in 1916. He showed how a long tongue of gas that had been pulled out of the sun would break up into individual gas clouds. The small clouds would dissipate, because of their low gravitational attraction, but the larger ones would condense into a series of gaseous spheres, similar in mass to that of the planets.

Jeans further suggested that, whilst the gaseous planets were cooling, their orbits around the sun would still be highly eccentric, and so the sun would, by tidal attraction, cause material to be pulled out of them, which would condense to form the planetary satellites. The smaller of the planetary spheres would cool to a liquid state more rapidly than the larger ones. The more liquid a sphere, the more difficult it is to break it up, and so the smaller planets resisted the sun more successfully than the larger ones, hence Mercury and Venus have no satellites, and Jupiter and Saturn have the most. Mars does not fit in with this theory, however, as it is appreciably smaller than Venus, and so should have no satellites also. As a result, Jeans was obliged to propose that Mars had initially been much larger than it is now, but had lost a considerable amount of mass early in its lifetime.

The current planets are progressively larger the further they are away from the sun, until Jupiter is reached, and then they become smaller with increasing distance.[20] Jeans thought that this was too much of a coincidence

---

[19] This is similar to Buffon's theory of the late eighteenth century, in which he proposed that matter had been dragged out of the sun by a passing comet (see Section 6.3). We now know that a comet is much too light to do such a thing, of course, but another star could have done so.

[20] Again Mars is an exception, unless, as Jeans suggested, it was originally much larger than it is now.

and explained the size distribution by proposing that the tongue of gas pulled out of the sun was thickest in its middle than at either end. He also proposed that the encounter with the star took place in a plane a few degrees from the equatorial plane of the sun, which also explains why the planetary orbits are not quite in that equatorial plane. The newly formed planets would have to plough their way through all sorts of gas, dust and condensates left over from the original event, and this would have caused the planetary orbits to become more circular, explaining why most planetary orbits today have a very low eccentricity. The density of this debris would be least both nearest and furthest away from the sun, which explains why the planets with the most eccentric orbits are Mercury and Pluto.[21]

Jeans calculated that such close encounters with the sun and stars would be extremely rare, so that only about one star in 100,000 would today have a planetary system.[22] This was quite a change from the nineteenth century view, where planets were thought to be present around the majority of stars.

Thomas Chamberlin and Forest Moulton of the University of Chicago proposed an alternative star encounter theory in 1905, based on the idea that the sun originally had much larger prominences than now. They then suggested that these prominences had been enormously amplified by a passing star, which caused the sun to eject a great number of small clouds of gas. These cooled to form small solid bodies that they called 'planetesimals', which in turn cooled and coalesced to form the present planets. Unfortunately, Jeans showed that the small clouds of gas would have dissipated before they had had time to cool to form the planetesimals. It may appear that Jeans' theory had the same problem in explaining the formation of the planetary satellites, but the gas at that stage would have been much cooler than the gas ejected from the sun in Chamberlin and Moulton's theory, and so it would have been much less likely to dissipate.

The rapid explosion of similar theories in the early part of the twentieth century gradually fizzled out, as every new theory was found to have serious drawbacks. For example, Lyman Spitzer (1914–1997) of Princeton University analysed Jeans' close encounter theory in 1939, and showed that the filament

---

[21] Pluto's existence was unknown when the theory was first proposed, but Pluto's size, as understood in the decade after its discovery, fitted the theory well.

[22] In fact, Jeans overestimated the average age of the stars by about a factor of 1,000. With the correct age inserted into Jeans' calculations, they show that only about one star in 100 million would today have a planetary system, which is of the order of only a few stars per galaxy.

of gas drawn out of the sun would not have condensed, but would have formed a permanent gaseous nebula surrounding the sun.

## Condensing nebulae re-examined

As mentioned above, Laplace's theory had been largely dismissed in the latter decades of the nineteenth century because of problems with conservation of angular momentum. In the twentieth century, however, it became clear that nebulae, typical of those where stars form, are turbulent, with streams of gas moving at velocities of the order of 10 km/s. The Irish mathematician William McCrea (1904–1999) showed that, if a dense nebula with streams of gas had condensed to form the solar system, it would have had more than enough angular momentum to produce the solar system that we currently see. The key point is that the gas molecules in these nebulae are not moving completely at random, and so the nebula has a net angular momentum, even though it does not appear to be rotating. Given that there was enough angular momentum in such a nebula, how could the nebula condense to form planets?

Carl von Weizsäcker suggested in 1943 that cells of circulating convection currents, or vortices, formed in the solar nebula after the sun had condensed. These vortices then caused small chunks of material to condense into planetesimals in the regions bordered by vortices rotating in opposite senses. The planetesimals then grew to form the planets by accretion. The planetary satellites were formed analogously from small nebulae surrounding each of the planets. Unfortunately, shortly afterwards Subrahmanyan Chandrasekhar (1910–1995) and Gerard Kuiper (1905–1973) showed that the vortices would not have been stable enough to allow planetesimal condensations to take place.

Starting in 1948, Gerard Kuiper developed an alternative theory that had its origins in his earlier work with Otto Struve (1897–1963) on double stars. Kuiper pointed out that double and multiple star systems are not unusual, as previously believed, but are rather common, and suggested that planetary systems would be found around those stars whose initial nebula was not large enough to form a companion star. In Kuiper's theory the planetesimals formed in a region of gravitational instability in the nebula, rather than as a result of turbulence.

Kuiper examined the statistics of the condensing nebula, and concluded that any condensing nebula must originally have had, after the sun had condensed, about 100 times more mass, largely in the form of hydrogen and helium, than that currently residing in the planets, otherwise it would have dispersed rather than condensed. So only about 1% of the nebula condensed to form the planets, the remainder being gradually lost to the solar system.

Kuiper, like most others before him, had considered that the original solar nebula was relatively hot, whereas the distinguished chemist Harold Urey (1893–1981) thought that it was initially cold (see Section 9.6). Urey then used the planetesimal theory in the 1950s to explain the depletion of heavy inert gases on the earth compared with the sun. He reasoned that, if the earth had formed directly from the same nebula as the sun, the earth's gravity would also have been sufficient to retain the heavy inert gases, but if the solar nebula had produced only small bodies, the planetesimals, their gravity would have been insufficient to retain these gases.

In parallel with Urey's work, the Soviet theorist Viktor Safronov investigated how planetesimals could aggregate to produce planets. Safronov showed how planets could form with almost circular orbits, provided they were formed by numerous collisions between small bodies. So, although there was not, in the late 1950s, a generally accepted theory of the origin of the solar system, the concept of planets being built by the collision and merging of planetesimals appeared to be the most promising.

## 9.5 **The moon**

### The surface

The question as to whether there is life on the moon was still unresolved until well into the twentieth century. William Pickering suggested in 1903 that there may be plants, snow and river beds on the moon. Then in 1921 he suggested that there may be a low form of vegetation that completes its life cycle in the 14 days of sunlight, being fed by carbon dioxide released from rock fissures. But over the decades the idea of there being such relatively simple lifeforms was mostly abandoned. Nevertheless, it was still thought possible in the 1960s, when the Americans were planning their manned lunar landings, that there may be some very elementary forms of life, like bacteria, on the moon. As a result the early American lunar astronauts were put into quarantine on their return to earth, just in case they had picked up some forms of life on their spacesuits.

In 1924 Bernard Lyot, of the Meudon Astrophysical Observatory near Paris, started an extensive series of measurements of the polarisation of light scattered by the moon and planets, and concluded in 1929 that the surface of the moon was probably covered by volcanic ash. This conclusion was consistent with the 186 K fall in temperature for the lunar surface during the 1927 total lunar eclipse observed by Pettit and Nicholson at the Mount Wilson Observatory. In the 1950s Thomas Gold of Cornell University and others concluded that the lunar maria are covered in volcanic ash or dust up to a few metres deep, but other estimates put the thickness at no more than a few centimetres. If Gold was correct, however, this would cause major problems with any spacecraft landings.

In the late nineteenth and early twentieth centuries it was still not clear whether the lunar craters had been formed by the impact of solid bodies or by volcanic eruptions. Although the volcanic theory was generally favoured at that time, a number of astronomers still favoured the impact theory. However, one of the problems that the adherents of the impact theory had to solve was why most of the craters are circular, even though the impacts should have occurred at all angles of incidence to the surface. G.K. Gilbert, a renowned geologist, suggested that the impacting bodies were small natural earth satellites, which would have had only small velocities relative to the moon, and so would have fallen onto the moon almost vertically. After the First World War, however, it was realised that the shape of lunar craters resembled shell craters, and that, as craters are formed by the shock wave of the impact or explosion, a non-vertical impact can still produce a circular crater. Later calculations showed that the shock wave produced by a 1 km meteorite impacting the moon at a typical velocity of 30 km/s could, for example, have produced the 100 km diameter crater Copernicus.

Reports of small changes on the moon continued during the twentieth century. In particular, Dinsmore Alter took some photographs that appeared to show a temporary local obscuration in the large walled plain of Alphonsus. On 3rd November 1958, Nikolei Kozyrev, of the Crimean Astrophysical Observatory, photographed the spectrum of a bright red patch that appeared briefly in the same area, and, on 30th October 1963, Greenacre and Barr of the Lowell Observatory saw a disturbance in the Aristarchus area. So, although the surface of the moon looks virtually invariant, it does not appear to be completely dead.

## Origin of the moon

George Darwin's theory of the origin of the moon as being spun off from a rapidly rotating, molten earth was shown, in the early part of the twentieth century, to be untenable. In particular, Forest R. Moulton of the University of Chicago pointed out, in 1909, that the viscosity of the proto-earth would have been high enough to stop it from breaking up, given the known angular momentum of the earth–moon system. If, on the other hand, in some unknown way the rotation rate had been high enough to achieve separation, so much angular momentum would have been transferred from the earth to the moon, that the moon would have escaped completely from the earth's gravitational pull, and would have become another planet of the sun. In other words, either the angular momentum of the earth–moon system was the same as it is today, in which case the separation could not have taken place, or the angular momentum was high enough to allow separation, in which case the moon would have escaped from the earth's gravitational attraction. As a result, by the middle of the twentieth century it was generally assumed that the earth and moon had formed in some way as a pair out of the original solar nebula (see Section 9.6).

However, an alternative theory was put forward in 1955 by Gerstenkorn, who suggested that the moon may have been captured by the earth. Although this was feasible, the moon would have had to approach the earth down a narrowly defined corridor. If the moon had approached too fast, it would not have been captured by the earth, whereas if it had been too slow, it would have collided with the earth and have been fractured. So, although it was not impossible, the probability of a successful capture would be very low.

## 9.6 The earth

### Internal structure

It was generally thought at the end of the nineteenth century that the proto-earth, after condensing from the solar nebula, had then cooled from a gaseous to a fluid state. As it continued cooling, it contracted and speeded up its rotation. Eventually a crust formed on the surface of the still-cooling earth, but because the earth was spinning, the crust set with a slight equatorial bulge, which has resulted today in the earth's equatorial diameter being some 44 km larger than the polar diameter.

It was known in the nineteenth century, by measuring the temperature in deep mines, that the temperature inside the upper part of the earth increased at the rate of about 1 °C for every 30 to 40 m of depth. At this rate,[23] rocks

---

[23] It is now known that the rate of increase of temperature with depth is not constant, but reduces with depth, and that pressure has a significant effect on the melting point of rocks.

would be molten at depths of about 40 km or so, clearly indicating that the earth's interior is molten. This idea of a molten interior was strengthened by the existence of volcanoes, which eject molten lava at temperatures of between 800 °C and 1,500 °C. Although the thickness of the solid crust was unknown at that time, it was clear that the solid skin could not be too thin, otherwise it would be constantly being distorted and fractured by the tidal action of the sun and moon.

The first tentative steps in outlining the interior structure of the earth were taken by Kelvin and Tait towards the end of the nineteenth century. But it was Emil Wiechert (1861–1928), a German seismologist, who produced the first numerically based analysis. In 1897, Wiechert suggested that the earth consists of a dense metallic core, mostly of iron, with a uniform density of 8.2 g/cm$^3$, surrounded by a lighter rocky layer, now called the mantle, with a uniform density of 3.2 g/cm$^3$. Dynamical considerations, based on a mean moment of inertia of the earth $0.3336mr^2$, where $m$ is the mass of the earth and $r$ is its radius,[24] led him to conclude that the radius of the core should be about $0.78r$. Such a considerable change in density at the interface between the core and mantle at $0.78r$ should have been evident seismologically, but no such evidence of this interface was found at that time.

However in 1906, the English geologist and seismologist Richard Oldham (1858–1936) showed, by studying the seismic records of earthquakes, that the earth definitely has a core. Then in 1914, the German-born geophysicist Beno Gutenberg (1889–1960) showed that the interface between the core and mantle, now called the Wiechert–Gutenberg discontinuity, is at about $0.545r$ from the centre of the earth, and not the $0.78r$ of Wiechert. So the mantle extends down to about $0.455r$, or about 2,900 km below the earth's surface where it interfaces with the core.

Evidently with a smaller core than predicted by Wiechert, the density of the mantle must be higher than he expected, as the overall density of the earth was well known. That is indeed so, as Wiechert had not taken into account the effect of pressure inside the earth on the density of materials. Because of this pressure, the average density of the mantle material was expected to be about 4 g/cm$^3$, rather than the 3.2 g/cm$^3$ previously assumed. In addition, the density of iron in the core was expected to increase from

---

[24] If the earth were a sphere of uniform density, its moment of inertia would have been exactly $0.4mr^2$. The lower figure of $0.3336mr^2$, which was based on observational evidence, shows that the density of the earth is higher towards the centre than the outside.

about 8 to about 12 g/cm$^3$. These higher density figures were consistent with a core radius of 0.545$r$.

In 1909, shortly after the discovery of the core, the Croatian geophysicist Andrija Mohorovičić (1857–1936) discovered the boundary between the earth's crust and its mantle by studying the records of the Croatian earthquake of October 1909. It was later found that this boundary, which is now called the Mohorovičić discontinuity, varies in depth from about 5 km under the deep oceans to about 70 km under some of the mountain ranges, the average depth being about 35 km.

The English geophysicist Harold Jeffreys tried, in 1934, to link the structure of the earth to that of the other terrestrial planets and the moon. As a baseline, he suggested that the earth consists of a core of liquid metals, mostly iron, surrounded by a shell of silicates, of uncompressed density 3.3 g/cm$^3$, about 2,900 km thick. He thought that the silicates were probably mostly olivine, $(Mg,Fe)_2SiO_4$. Jeffreys further proposed that the other terrestrial planets and the moon have a similar structure, but with the dense metallic core being smaller as the overall densities decrease from Earth (5.5 g/cm$^3$) to Venus (4.9 g/cm$^3$) to Mars (4.0 g/cm$^3$) to Mercury (3.8 g/cm$^3$)[25] and to the Moon (3.3 g/cm$^3$). In the case of the Moon he thought that the core is probably non-existent.

Jeffreys, in making this proposal of radical different core sizes for the planets, realised that it created problems for theories of the formation of the solar system. In particular, it was difficult to explain how the planets smaller than the earth, with smaller cores, had apparently retained a higher percentage of the lighter constituents in their mantles.

A detailed analysis of seismic record in 1936 led the Danish seismologist Inge Lehmann to conclude that there was a seismic discontinuity inside the core. This indicated that the core was separated into an inner and outer core at a radius of about 1,200 km. A little later it was hypothesised that, although the outer core was fluid, the inner core was solid, because of the very high pressures prevailing there.

In 1936 Bullen tried to update Wiechert's analytical model of the earth by taking compressibility into account. He adopted a value of 3.3 g/cm$^3$ for the density of the earth at a depth of 35 km, and of 5.0 g/cm$^3$ just outside the core, at a depth of 2,900 km, with a core density ranging from 9.9 to 12.3 g/cm$^3$. Now there was, at that time, seismological evidence that there was a density discontinuity in the mantle at a depth of about 300 to 400 km,

---

[25] The density of Mercury is now known to be about 5.4 g/cm$^3$.

in what came to be known as the 20° discontinuity.[26] Bullen assumed a depth of 350 km for this discontinuity, and calculated that the density would have to increase from 3.6 to 4.0 g/cm$^3$ over it. Shortly afterwards, Jeffreys redetermined the depth of the discontinuity as 480 km, which required a density jump of from 3.69 to 4.23 g/cm$^3$. The problem was that there was no suitable material known at the time that would have a density of 4.23 g/cm$^3$ in such an environment.

Bernal had pointed out in 1936 that olivine, the favoured material for the mantle, may change its crystal structure under pressure, which could result in it having the required density at that depth. Unfortunately it was not possible at that time to produce such pressures in the laboratory, and so it was not until many years later that such experimental evidence was forthcoming. It turned out that α-olivine transforms to β-spinel at a pressure corresponding to a depth of about 400 km. This in turn transforms to γ-spinel at the pressure at about 500 km, and this changes its structure yet again at the pressure experienced at about 660 km. In the meantime, seismological evidence had been building up that there was actually a transition zone, covering depths from about 400 to 660 km, where there were various complex density changes which seemed to correlate with this behaviour of olivine under pressure.

Now that the basic structure of the earth appeared to have been established, some scientists started to question the constituents of the core. In the 1930s it had been generally accepted that the core was largely made of iron, with some nickel content, but in 1941 Werner Kuhn and Arnold Rittmann presented a different view. They suggested that the core and most of the mantle are rich in hydrogen, as the earth should, in their view, have almost the same chemical constituents as the early sun. Arnold Eucken, the distinguished German physical chemist, disagreed, however, pointing out in 1944 that the earth must have a largely iron core as, in his view, iron would have condensed first in the high temperature, early solar nebula.

In Kuhn and Rittmann's model, the earth consists of a thin solid crust, surrounding a shallow magnetic zone of molten silicates, followed by a chemically homogenous interior rich in hydrogen. However, the theory ran into difficulties in trying to explain the apparent discontinuity in the properties of the earth at a depth of about 2,900 km. Kronig, de Boer and Korringa

---

[26] The basic structure of the interior of the earth was determined by examining the behaviour of seismic waves emerging from the surface. The 20° discontinuity was so called as there was a sudden change in the behaviour of the waves when the emerging angle reached 20°.

suggested in 1946 that this discontinuity could possibly be explained by a phase change in the properties of hydrogen at high pressure. In particular, they calculated that hydrogen would increase its density from about 0.4 to 0.8 $g/cm^3$ in changing from its normal molecular form to a metallic form at a pressure of about $7 \times 10^5$ atmospheres (bars). Unfortunately, this proposed structure of the earth's interior did not appear to be compatible with its mechanical properties; in particular with the variation of viscosity with depth calculated from earthquake data, and with the calculated pressure inside the earth.

In 1948, the British geophysicist William Ramsey suggested an alternative model for the earth, and the other terrestrial planets, in which the internal pressure caused the silicates of the interior, rather than the hypothesised hydrogen, to become metallic. So these terrestrial planets would each have a compressed silicate core, not an iron one. If Ramsey was correct, it would solve Jeffreys problem about how to explain that the smallest planets, with the smallest cores, have the largest percentage of the lighter constituents in their mantles as, in Ramsey's theory, the core is made of the same basic material as the mantle; it is just in the metallic state. The largest planets with the greatest internal pressures naturally have the largest cores, in this case.

The idea that the core may not be made of iron required a re-evaluation of the evidence on which the previous hypothesis of an iron core was founded. In fact, the evidence at that time was by no means overwhelming. It was based mainly on the fact that iron's density fitted the calculated density of the core, iron is an important constituent of meteorites, and iron is one of the commonest heavy elements in the stars. Seismic evidence also showed that the outer core, at least, must consist of a liquid metal, and iron appeared to fit the bill well at the temperatures and pressures envisaged.

The question of radioactive heating has now to be considered, to take matters forward.

## Urey's cold accretion theory

Early in the twentieth century it became clear that radioactive heating could play an important part in keeping the interior of the earth hot. In fact, in 1906 Lord Rayleigh (1842–1919)[27] showed that the amount of heat produced by the radioactive decay of radium in the earth's crust is sufficient to account

---

[27] Lord Rayleigh was born John William Strutt. He became Lord Rayleigh on the death of his father in 1873.

for the entire flow of heat from the earth. Then in 1949 Harold Urey, a Nobel Prize winner for chemistry, calculated, using radioactive abundances from meteorites, that the interior of the earth was heating up, rather than cooling down. This led Urey to propose a radical new theory for the origin of the earth that came to be called the cold accretion theory.

Urey did not think that the moon had once been part of the earth, but that the earth and moon had formed separately as a pair of objects from the condensing solar nebula. However, it seemed strange that the density of the moon was so much less than that of the earth, if they had formed together as part of a common system. Urey also believed that the only way that the moon could have retained its earth-facing bulge was if it had been formed at a cold temperature. So in Urey's new theory the earth and moon had been cold when they had been formed, not molten as previously thought. As a result, Urey thought that the original solar nebula was cold, whereas Eucken and others had thought that it was hot.

Urey hypothesised that, because the initial solar nebula was cold, lighter elements like silicates would have condensed in the original nebula before denser elements such as iron. At the end of the first stage of planetary growth, when only light elements were available, the proto-earth happened to be much larger than the proto-moon. So, when in the second stage denser elements became available, the earth's rate of growth was much larger than the moon's, because it had the higher gravitational attraction. As a result, although both bodies had formed with a silicate rich core, the heavier earth had picked up substantially more heavier elements like iron as the solar nebula evolved. Over time, both the earth and moon were being heated by radioactive decay, but, in the case of the earth, the heavy iron was gradually finding its way from the surface to the centre, further heating the earth. This movement of massive amounts of iron towards the centre also caused convection currents in the earth's mantle, which resulted in the formation of folded mountain chains.[28] As the moon did not have this massive transfer of iron from the surface to the core, however, there had been no convection, which is why there are no folded mountains on the moon.

Urey's standing as a Nobel Prize winner meant that his ideas on the formation of the earth were taken very seriously. So seriously, in fact, that they were discussed extensively at a conference of earth scientists held at Rancho Santa Fe, California, in January 1950. This conference rejected the

---

[28] Urey borrowed this idea of convection currents in the earth's mantle creating folded mountain chains from other geophysicists.

Kuhn–Rittmann and Ramsey models of the earth's structure,[29] and concluded that the earth has an iron core. Urey's idea that the separation of the heavier and lighter elements over time had helped to start convection within the earth, which was still continuing,[30] was generally accepted, although there was considerable disagreement on the temperature of the earth during its formation. Nevertheless, Urey's claim that the earth had never been completely molten was generally accepted.

At about this time, Whipple told Urey that most astronomers believed that Mars has neither high mountains nor volcanoes, which Urey took as evidence that there had been no convection on Mars. In that case, he argued, its internal temperature must have been appreciably less than that of the earth, so its iron could not have settled towards its centre. As a result, Mars does not have an iron core. In fact, he went even further than this and postulated that Mars, like the moon, was chemically homogenous.

Urey, like others before him, felt that the clues to the formation of the planets and moons of the solar system lay in meteorites. At that time it was thought that meteorites may once have been part of larger solar system bodies which had been broken up, probably by a collision. Alternatively, meteorites may be what was left over after planetesimal formation in the early solar system.

Thus it was that in 1951 Urey started a research programme to measure the natural abundances of elements in meteorites, hoping to confirm that they, at least, had been formed at low temperatures. He also hoped to show that they had subsequently not been subject to high temperatures, except for surface heating during their descent through the earth's atmosphere. In the latter he was successful when he found iron sulphide in meteorites, which indicates that they could not have been heated above about 600 K after the iron sulphide's formation. This appeared to rule out the idea that they had originated in the explosive break up of one or more larger bodies.

---

[29] Ramsey's model was finally rejected shortly after this conference when Eugene Rabe found that Mercury's density was 5.46 g/cm$^3$. This was much higher than previously thought and was higher than that of Venus and Mars, which are larger planets. In addition, Bridgman's experiments at ultra high pressures failed to find the phase change in silicates predicted by Ramsey.

[30] As evidence for this, Urey pointed out that changes in the length of the earth's day seemed to indicate that the moment of inertia of the earth is still decreasing with time. This is because, in his view, the heavier elements are still moving towards the earth's centre.

Although Urey was gradually building up evidence for a degree of cold accretion in the early solar system, as the space age started his was just one of a number of theories for the origin of the moon and planets, although it was one of the most thought out. However, even though the Rancho Santa Fe participants did not agree, alternative models based on an initial high temperature solar nebula, as espoused by Eucken, Kuiper[31] and others, were still flourishing. Soon, spacecraft would be able to return geological samples from the moon, and these were eagerly awaited to indicate whether the moon had formed as a hot or cold body. With no significant atmosphere on the moon, there is no weathering. So the rocks should have changed little over the majority of the moon's lifetime, except for the effect of the occasional meteorite impact, heating by the sun, and the impact of any elementary particles from space.

## Dynamics

It had been shown by Adams in the nineteenth century that about half of the observed secular acceleration of the moon could be explained by the moon's orbit being gradually reduced in size, because of a reduction in the eccentricity of the earth's orbit (see Section 7.8). Then in 1866 Delaunay had suggested that tidal friction, which was gradually slowing down the earth's rotation rate, could probably account for the other half of the effect. But in 1909 Thomas Chamberlin of Chicago concluded that tidal friction would not be enough to do this. However, Geoffrey Taylor analysed the ocean geometry in more detail in 1919, and found that shallow inland seas were much more important than the oceans in slowing down the earth. The Irish Sea, for example, had a tidal dissipation of about thirty times the amount for the whole of the open ocean. Then in the following year Jeffreys showed that the energy dissipated in all the shallow seas that he considered, could explain about 80% of the observed effect, with some $\frac{2}{3}$ (of the 80%) being due to the Bering Sea.

Newcomb had shown, in 1878, that the moon's position appeared to deviate by a small amount in a cyclical fashion with a period of 273 years,

---

[31] During the 1950s there was a considerable amount of bad feeling between Kuiper and Urey, because of their competing hot/cold theories for the origin of the earth and moon, the details of which are beyond the scope of this book. Interested readers will find a summary of their dispute in *Solar System Astronomy in America* by Ronald Doel, Cambridge University Press, 1996, pp. 134–150.

but the cause was unknown. Newcomb had thought that it could be due to variations in the earth's rotation rate, but he dropped the idea in 1903. If the effect was due to variations in the earth's rotation rate, it should be evident in the apparent movement of the sun and planets, as well as that of the moon and, in 1914, the Anglo-American astronomer Ernest W. Brown showed that there was such an effect in the apparent motions of the sun and Mercury. This was later confirmed by Glauert at Cambridge, by Harold Spencer Jones at Greenwich, and by Willem de Sitter (1872–1934) at Leiden who noticed a similar effect for the satellites of Jupiter. The earth was showing random irregularities in its axial rotation rate of from +0.0034 s/revolution in 1870, to −0.0045 s/revolution in 1910. Clearly irregular deviations could not be predicted in future, and so the dream of centuries that it would eventually be possible to predict all the observed motions of the moon was clearly not feasible. To make matters even more complicated, in 1949 the Belgian N. Stoyko found a seasonal variation in the earth's rotation period, super-imposed on all other variations, which was attributed in the following year to seasonal changes in the distribution of the atmosphere. Evidently the earth is a most inaccurate clock.


## Ionosphere

It was obvious from the movement of the snow line with the seasons, that the temperature of the air decreases with altitude. But in the seventeenth century Blaise Pascal (1623–1662) showed that atmospheric pressure also decreases with altitude by measuring atmospheric pressure at the top of the Puy de Dôme in France. Little more was known about the physical structure of the earth's atmosphere, however, until in the 1890s Hermite and Besançon used instrumented balloons, called balloon sondes, to probe the upper atmosphere. In particular, they found that at very high altitudes atmospheric temperature reversed its decreasing trend, and started increasing with increasing altitude, in a region that we now call the stratosphere.

As early as 1839, Karl Friedrich Gauss had suggested that the daily variations in the earth's magnetic field (see Section 7.9) could be due to electric currents in an electrically conducting layer, high in the earth's atmosphere. Various theories were put forward to explain the possible existence, and the daily movement of this electrically conducting layer, until in the 1890s Arthur Schuster suggested that the layer could be caused by ionising radiation. But, at this stage, the existence of such an electrically conducting layer was still unproven.

A new field in the investigation into the earth's upper atmosphere was started in 1888 when Heinrich Hertz (1857–1894) demonstrated the propagation of what we now call radio waves in air. Then by the summer of 1896 Guglielmo Marconi (1874–1937) was demonstrating radio communication over a distance of a few hundred metres to the British Post Office. Three years later Marconi had managed to transmit radio waves over 140 km, and in December 1901 he claimed to have achieved transatlantic radio communications between Poldhu in Cornwall to St Johns, Newfoundland. There was some doubt at the time about whether the link had really been made, but two months later he proved, beyond doubt, that radio waves can be successfully transmitted across the Atlantic.

This demonstration of long distance radio communication by Marconi was a big surprise at the time, as radio waves were thought, like light, to travel in straight lines. In the same way that different densities of air can refract light, it was possible that refraction in the earth's atmosphere could cause some bending of radio waves, but not to the extent of allowing communications over such a vast distance as the Atlantic Ocean. In fact, the correct explanation was proposed in 1902 independently by Oliver Heaviside (1850–1925) in the UK, and Arthur Kennelly (1861–1939) in the USA. They suggested that radio waves were being reflected off an electrically conducting layer about 100 km above the earth's surface.

William Eccles realised that the radio waves would be seriously attenuated if they penetrated very far into the conducting layer. As a result he postulated in 1912 that this layer, which he named the Heaviside layer, consists of ions rather than electrons. In 1924, however, Joseph Larmor suggested that the radio waves were refracted back to earth by the conducting layer, which he thought consisted of electrons.

However, at this stage, the existence of an electrically conducting layer was still not generally accepted. But at this time experiments were started on both sides of the Atlantic to investigate the existence, height and properties of this hypothesised layer.

In 1924, Edward Appleton (1892–1965) and Miles Barnett, working in Bournemouth, England, started using a BBC transmitter, when it was available after midnight, to transmit signals of gradually changing frequency. The signals, which were received by their receiver some 100 km away, varied in strength because of interference between the signals received via what is now called the E or Heaviside layer, at an altitude of about 100 km, and those received directly along the ground from the transmitter. At dawn the signals from the E layer reduced considerably as the layer broke up, but they were replaced by signals from a higher layer, which Appleton measured to be at an altitude of about 250 km. This is now called the $F_2$ or Appleton layer.

In parallel, in the United States, Gregory Breit and Merle Tuve also started in 1924 to try to determine the height of the conducting layer using the more direct method of pulse-echo sounding, which is the basis of radar.[32] The pulse delay indicated the height of the conducting layer, and the amplitude of the signals its efficiency.

This book is not the place, unfortunately, to relate the gradual understanding of the structure of the ionosphere achieved by Appleton, Breit, Tuve, Ratcliffe, Hulburt, Chapman, and many others. In essence they found that there are three layers, designated the D, E and F layers. The D layer is centred at about 80 km altitude, the E layer at about 120 km, and the F layer at about 300 km. In 1926 this whole region of electrically charged particles in the earth's upper atmosphere was given the name of 'ionosphere' by Robert Watson-Watt.

The D layer is different from the other two layers in the ionosphere, in so far as it largely disappears at night. During the daytime the weakly reflective D layer absorbs medium radio frequencies and the lower frequencies of the short-wave bands. The strongly reflective E layer maintains its reflectivity until about four or five hours after sunset, but the F layer, which is also strongly reflective, retains its reflectivity for a full twenty-four hours a day. However, during daytime, two F layers can be detected, a lower $F_1$ layer centred at about 220 km, and an upper $F_2$ layer centred at about 380 km. At night they become a single F layer at about 300 km.

Clearly the sun is having a major effect on these layers, apparently creating the D and E layers, and changing the configuration of the F layer. But what exactly is the mechanism for this?

The general view in the 1930s was that ultraviolet solar radiation, with a wavelength below about 175 nm, causes molecular oxygen, $O_2$, to become dissociated above altitudes of about 90 km, with dissociation being complete above about 140 km, thus creating the E and F regions. Some physicists speculated that radiation at the Lyman-$\alpha$ wavelength of hydrogen at 121.6 nm was the source of this ionising radiation, whereas others, in particular Lars Vegard and E.O. Hulburt, speculated in around 1938 that the source was X-rays from the sun, emitted particularly during solar flares.

---

[32] Such an early use of radar may be a surprise, but the first person to patent a radar-like system was Christian Hulsmeyer as long ago as 1904. However, it was Robert Watson-Watt who was largely responsible for the early development of radar, gaining his first patent in 1919.

In fact, in the 1930s Howard Dellinger (1886–1962), of the American Bureau of Standards, had showed that problems with the reception of short-wave radio waves on earth were often, but not always, associated with solar flares seen on the sun a short time before. So he suggested that some unknown phenomenon was producing solar flares and generating invisible radiation that modified the earth's ionosphere, causing problems with radio reception. In 1939 T.H. Johnson, of the Bartol Research Foundation, and Serge Korff, of New York University, observed that the ionisation of the earth's atmosphere was quickly disrupted at the time of solar flares, and suggested that the cause of these atmospheric disturbances was solar X-rays.

## Continental drift

The English philosopher Francis Bacon (1561–1626) had drawn attention in his *Novum Organum*, published in 1620, to the fact that the shore lines on both sides of the Atlantic Ocean seemed to have a similar shape, as if one could fit into the other. In about 1800 Friedrich Alexander von Humboldt, the German naturalist, went one step further, and suggested that the land on either side of the Atlantic Ocean had once been joined together. Then in 1858 Antonio Snider-Pellegrini pointed out that there were identical fossil plants in both North American and European coal deposits, which he took as evidence that the two continents had once been united. Fifty years later the American geologist Frank Taylor suggested that the mountain belts of Asia and Europe were the result of the movement of the continents, but his idea was generally ignored.

Alfred Wegener (1880–1930) was the first to take these ideas further and to develop a reasonably detailed theory of continental drift, which was the forerunner of today's theory of plate tectonics. In 1912, he suggested, based on geological, paleontological and biological evidence, that there had originally been only one continent, which he called Pangaea. Then in the late Palaeozoic era, about 200–250 million years ago, this continent had broken apart, with South America/Southern North America moving away from Africa/Southern Europe, and with Antarctica/Australia moving away from Africa/India.[33] Later, about 100 million years ago, Northern North

---

[33] The / in this sentence indicates that these landmasses were connected together at that time.

America separated from Northern Europe, India from Africa, and Australia from Antarctica.[34]

Although Wegener amassed much supporting evidence for his theory it was not generally accepted, and by the 1930s it had been rejected by most geologists. Thus matters rested for another twenty years, until radical, new data caused Wegener's ideas to be reconsidered (see Section 10.1 later).

## 9.7 Mars

### The surface

The twentieth century dawned with the argument about the possible existence of canals on Mars, which had been started by Schiaparelli and continued by Percival Lowell, still in full swing. At times the disagreements between astronomers became more personal than scientific, which was a pity because the observers were genuinely trying to establish the facts.

In the 1890s Eugène Antoniadi (1870–1944) had, as Flammarion's assistant, recorded many canals on Mars, but in 1902 he gave up his post, and turned his attention to architecture instead. In 1909, however, he returned to astronomy and started a series of observations of Mars using the 33 inch (85 cm) Meudon refractor. Antoniadi rapidly changed his mind on the canals and concluded that what had been seen by Schiaparelli, Lowell, himself, and others as linear features, could be resolved at very high magnifications, during periods of excellent seeing, into a series of small spots. When the seeing was poor, however, he found that the spots often merged to give the appearance of linear features.

The respected American observer E.C. Slipher disagreed with Antoniadi's conclusions, reiterating in 1921, 1931 and again in 1940 that, in his view, there were definitely linear markings on Mars. Robert Trumpler (1886–1956) also observed linear markings with the 36 inch (90 cm) Lick refractor at the favourable opposition of 1924. In addition, Lyot's photographs, taken during the 1941 opposition at the high altitude Pic du Midi Observatory, also appeared to show linear features. On the other hand Dollfus, who observed Mars at the same observatory in 1948, agreed with Antoniadi that, under exceptional viewing conditions, the apparently linear markings appeared to break up into a series of very small spots.

---

[34] Wegener saw this as a continuous process, and so these timescales are only very approximate.

Antoniadi also confirmed, during his extensive observations starting in 1909, that the large dark bluish-green areas on Mars changed in both colour and shape with the seasons, and agreed with Lowell that they were probably areas of vegetation. In addition to these seasonal changes, however, there were non-seasonal changes in some of the dark areas that lasted for a few years. Solis Lacus, for example, changed its size and shape dramatically in 1926, but by 1928 it had resumed its previous appearance.

In 1909 the Russian astronomer Gavril Tikhov compared spectra of the various dark bluish-green regions of Mars with those of localities on earth, and concluded that some Martian areas had vegetation similar to some sub-arctic regions on earth. Vesto Slipher repeated the measurements of these dark areas of Mars at the Lowell Observatory in 1924, looking for evidence of chlorophyll. None could be found, although a number of astronomers then speculated that the dark areas of Mars could be covered in a very basic form of life, like moss or lichens, some of which were known to have no chlorophyll on earth. These basic forms of vegetation were also consistent with organisms that have to survive with relatively small amounts of water, as the polar ice caps, for example, seemed to be very thin. In 1948, and again in 1956, Gerard Kuiper confirmed the lack of chlorophyll in the dark areas of Mars, when he found that chlorophyll's characteristic infrared absorption lines were missing from the spectrum of those areas.

Over the period from 1954 to 1956 the Michigan astronomer Dean McLaughlin developed his idea that Mars has active volcanoes, and that the dark bluish-green areas are ash deposits that have changed their appearance as the ash is blown about in the wind. At first Kuiper completely dismissed McLaughlin's theory, because the lack of water vapour in the Martian atmosphere indicated, in his view, that there are no active volcanoes. Urey disagreed, however, pointing out that the dry silicate sands of Mars would have a considerable capacity to absorb water from the atmosphere. The massive dust storm of 1956 then reminded Kuiper how powerful the wind-driven sand was in changing the appearance of the planet. As a result, he suggested that the dark areas are dust-covered lava fields that change their appearance as their covering dust layer is blown about by seasonal winds.

So much for the dark bluish-green areas, but what about the lighter reddish ones? In 1924 Bernard Lyot measured the polarisation of light scattered by Mars, and concluded that the surface was probably covered with limonite sand (a hydrated iron oxide) which was disturbed by the atmosphere from time to time to produce dust storms. Kuiper disputed Lyot's findings in 1956 on limonite sand, concluding that the yellow regions were covered in felsitic rhyolite, which is an igneous rock.

Temperature

Coblentz and Lampland began an extensive series of infrared observations of the planets in 1922 using the 42 inch (107 cm) reflector of the Lowell Observatory. Their initial temperature estimates were somewhat unsatisfactory, but these were put on a much firmer footing by Donald Menzel using a new analysis technique in 1923. In 1926 Menzel, Coblentz and Lampland concluded that, during the 1924 opposition of Mars, the equatorial, noon-time, surface temperatures for the light coloured areas were a little lower than for the dark areas, with both being generally above 0 °C (273 K). The south polar cap surface temperature varied from about −100 °C about two months before the summer solstice in the southern hemisphere, which occurred on 6th October 1924, to about −15 °C about two weeks after the solstice. These temperatures were clearly too high for carbon dioxide to exist in solid form at the atmospheric pressure expected at the surface of Mars. So Menzel, Coblentz and Lampland concluded that the south polar cap is probably composed of water ice and snow. In addition they found on 12th September that the surface temperature of the unilluminated part of Mars' equator was about 85 °C below that of the subsolar point, indicating that Mars has very large diurnal temperature fluctuations.

Coblentz, Lampland and Menzel repeated their observations and analysis using an improved vacuum thermocouple during Mars 1926 opposition, finding that the temperatures were somewhat higher than those recorded during the 1924 opposition. This was attributed partly to the fact that the 1926 observations were made about two months after the summer solstice in the southern hemisphere, whereas in 1924 most of the measurements had been made before its summer solstice. They reasoned that, as on earth, the maximum temperatures on Mars would occur after the summer solstice at medium and high latitudes, which was why the 1926 temperatures were higher. In particular, they found that the average noon surface temperature around the 1926 opposition ranged from about −30 °C at 55° N latitude, to +35 °C at the subsolar point at 15° S, to 0 °C at the south pole. The south pole being warmer than mid-northerly latitudes because it was in permanent sunlight at that time of year.

The atmosphere

Spectroscopic evidence in the nineteenth century had indicated the possible presence of oxygen and water vapour in the Martian atmosphere. The spectroscopic results were not consistent, however, and this lack of consistency continued well into the twentieth century. For example, in 1908 Vesto Slipher and Frank Very found evidence for both oxygen and water vapour from observations at Flagstaff's Lowell Observatory. Then, in the following year, Campbell carried his observing equipment to the top of Mount Whitney, at about 14,500 ft (4,400 m) altitude, to get above as much of the earth's water vapour as possible. He found no evidence of water vapour or oxygen on Mars. But in 1925 Walter Adams and Edward St John (1857–1935) at Mount Wilson found evidence for both oxygen and water vapour, although Adams and Theodore Dunham could find no evidence for oxygen when they used a new and much more accurate spectroscope, with the 100 inch (2.5 m) Mount Wilson reflector, eight years later. They were also unable to detect any water vapour during observations in 1937, 1939 and 1943.

Eventually in 1947 Gerard Kuiper, at the McDonald Observatory in Texas, found clear evidence for a small amount of carbon dioxide. Then in 1963, a trace amount of water vapour was discovered by Andouin Dollfus, who was observing from the Jungfraujoch in the Swiss Alps, and by Hyron Spinrad, Guido Münch and Lewis Kaplan of JPL, using the 100 inch (2.5 m) reflector on Mount Wilson. The amount of water vapour was tiny, however. For example, if it had all condensed on the surface of Mars it would have produced a layer of water only about 10 microns thick.[35]

The lack of oxygen on Mars was something of a mystery, considering how much there is in the earth's atmosphere. Wildt explained this by hypothesising that oxygen had once existed in the Martian atmosphere, but that it had been changed from the normal form of $O_2$ to $O_3$, or ozone, under the action of ultraviolet light. This ozone had oxidised the Martian surface and had been used up in the process, hence its observed absence today.

The surface atmospheric pressure on Mars was estimated by Percival Lowell in 1908 to be about 87 millibars, based on the planet's measured albedo, and in 1926 Donald Menzel estimated it to be about 65 millibars, by comparing Mars' visual and photographic albedos. Three years later, Lyot produced an estimate of about 25 millibars, by comparing the polarisation of light from Mars with that from the moon. Then in 1948 Dollfus produced a figure of 80 millibars using a Lyot polarimeter. Other methods were used both before and after the Second World War to produce figures ranging from

---

[35] A micron (μm, or micrometre) is one thousandth of a millimetre.

about 50 to 120 millibars, with a best estimate by de Vaucouleurs in 1954 of 85 millibars. So by the start of the space age in 1957, the surface atmospheric pressure of Mars was reckoned to be about 85 millibars, or a little less than 10% of that at the surface of the earth. Then in 1963, two years before the first spacecraft reached Mars, Kaplan, Münch and Spinrad calculated a much lower value of only 25 ± 15 millibars, based on the measured spectrum of Mars in the infrared, with a partial pressure for carbon dioxide of about 4 millibars. No oxygen was detected, giving an upper limit for the amount of oxygen in the Martian atmosphere of an order of magnitude less than the carbon dioxide abundance.

The earth's atmosphere, when dry, consists of about 75.5% nitrogen, 23.1% oxygen, 1.3% argon and 0.05% carbon dioxide, by mass. Amounts of water vapour vary, but the average is about 1%. Oxygen had not been reliably detected on Mars by the middle of the twentieth century, and nitrogen is notoriously difficult to detect spectroscopically, whilst argon is an inert gas. Carbon dioxide and water vapour had only been detected on Mars in very small quantities. So it was expected in the 1950s that the Martian atmosphere would probably consist of about 98% nitrogen, 1% argon, and a total of 1% for carbon dioxide, oxygen and water vapour. However, an alternative model where there is more argon than nitrogen had also been suggested. So it was generally agreed, before the first spacecraft arrived at Mars, that either nitrogen or possible argon would account for well over 50% of the Martian atmosphere.

Schiaparelli had concluded, in 1877, that the poor visibility of surface features in the last few months of that year had been the result of extensive clouds in the Martian atmosphere (see Section 7.10). These so-called yellow clouds[36] were well observed in 1909 by Fournier and Antoniadi, when they appeared to cover almost the whole of the planet. In 1924 Antoniadi again noticed that yellow clouds almost covered the whole of Mars to such an extent that the planet appeared yellow, rather than its usual red colour. Antoniadi also noted that, like Schiaparelli's clouds, the yellow clouds that he observed tended to occur around perihelion, when the solar heating is greatest, and from this he concluded that they probably consist of fine dust thrown into the atmosphere by thermally generated winds. Some thirty years later de Vaucouleurs measured their velocity, which he found to be typically as high as 60 to 90 km/h (35 to 55 mph) at first, gradually reducing after a few days to about 5 to 30 km/h (3 to 20 mph).

---

[36] So called because they appear bright when seen through a yellow filter.

In 1924, W.H. Wright (1871–1959) took photographs of Mars from the Lick Observatory through different colour filters using the 36 inch (90 cm) Lick telescope, and showed that the planet was about 150 km larger in diameter in violet light than it was in red light. As an atmosphere would scatter violet light much more than red light, this difference in diameter was considered to be proof of the existence of an atmosphere at least 75 km deep, which is more extensive than that on the earth. At first sight, this may appear to be inconsistent with the lower atmospheric pressure at the surface of Mars, compared with that of the earth. But Mars has a lower surface gravity, so the atmospheric pressure on Mars would reduce more slowly with altitude than on the earth. As a result, the atmospheric pressures on Mars and the earth were expected to be the same at about 29 km altitude, based on an 87 millibar surface atmospheric pressure for Mars. Unfortunately, however, when Kuiper examined Mars through different coloured filters in 1952, he was unable to detect any differences in diameter.

The reasons for Wright's and Kuiper's different results were unknown at the time, but they seemed to be connected with variations in the so-called violet layer on Mars. This violet layer sometimes obliterated surface detail in the ultraviolet, but sometimes allowed it to be seen more clearly. The effect has subsequently been explained by preferential scattering of atmospheric dust.

## The polar caps

The rapid melting of the polar ice caps puzzled nineteenth century astronomers, as Mars should be colder than the earth as it is further from the sun. Ranyard and Stoney's suggestion that the polar caps may consist of frozen carbon dioxide could explain this, but this idea ran into problems of its own (see Section 7.10). So at the end of the nineteenth century the polar caps were generally thought to consist of a very thin layer of water ice. The temperature estimates of the 1920s, discussed above, produced the same conclusion. Then in 1952 Kuiper confirmed that the caps must consist of frozen water, rather than frozen carbon dioxide, as both water ice and the polar caps appear nearly black in the infrared, whilst carbon dioxide ice appears white. So, prior to the arrival of spacecraft, the caps were thought to consist of very thin water ice, or possibly a thick layer of hoar frost.

## 9.8 Internal structures and atmospheres of the four large outer planets

Internal structures

It was generally thought, in the early years of the twentieth century, that Jupiter and the other three outer planets of Saturn, Uranus and Neptune are largely gaseous because of their low densities. Jupiter, in particular, was also thought to be quite hot because it is the largest planet, and so has probably not had time to cool down completely since its formation. In support of this idea, some astronomers thought that the shadows of the Galilean satellites on the surface of Jupiter were not completely black, indicating that Jupiter is slightly self-luminous (see Section 8.1).

This idea that the four outer planets, and Jupiter in particular, are hot gaseous planets, was challenged by Harold Jeffreys, in 1923, who suggested that they are cold solid bodies, made of very low density materials, whose original heat has virtually completely disappeared. He also pointed out that the atmospheres of Jupiter and Saturn need not be very deep to explain the observed differential rotation of the various spots on their visible discs. In particular he calculated, using the earth's monsoons as an analogy, that an atmospheric thickness of the order of $0.01R_J$ (where $R_J$ is the radius of Jupiter) would be sufficient for Jupiter, and of about $0.05R_S$ (where $R_S$ is the radius of Saturn) would be sufficient for Saturn. Jeffreys also noted that the densities of Io and Europa, the innermost of Jupiter's four Galilean satellites, are about twice that of Titan, Saturn's largest satellite, and that the density of Jupiter is about twice that of Saturn.[37] He then calculated that, if the density of the solid cores of Jupiter and Saturn were the same as these, their large satellites, the depths of the planetary atmospheres would be about 20% of their radii. So they could still be largely solid.

In the same year, 1923, Donald Menzel reanalysed the radiometric measurements made of Jupiter and Saturn by Coblentz and Lampland in 1914 and 1922, and concluded that their cloud-top temperatures are about 160 K (−110 °C). Although this temperature is very cold, it is still higher than the temperatures of 120 K and 90 K for Jupiter and Saturn, respectively, that would be maintained solely by incident solar radiation. So Jeffreys was

---

[37] At that time, the densities of Io, Europa and Titan were thought to be about 2.7, 2.6 and 1.4 g/cm$^3$, respectively, with densities of about 1.36 and 0.63 g/cm$^3$ for Jupiter and Saturn. Modern values are 3.5, 3.0 and 1.9 for the satellites, and 1.33 and 0.69 for the planets.

correct, in so far as these two planets were cold on the outside, on the other hand they did appear to have some source of internal heat.

Jeffreys extended his work on the internal structure of Jupiter and Saturn in 1924 by analysing their moments of inertia. These he calculated to be about $0.265mr^2$ and $0.198mr^2$, respectively, indicating that both planets are much more condensed towards the centre than the earth which has a value of $0.3336mr^2$ (see Section 9.6). Given these moments of inertia, and the known overall densities of the two planets, Jeffreys calculated that the densities of the outer layers of Jupiter and Saturn must be much less than 0.90 and 0.31 g/cm$^3$, respectively, assuming that their densities gradually increase with depth. He also explained that these limiting densities of 0.90 and 0.31 g/cm$^3$ must be valid for some considerable fraction of the planetary volumes.

Jeffreys then pointed out that these density figures, especially the figure for Saturn, together with Menzel's cloud-top temperature of 150 K,[38] are highly restrictive as far as the structure of the two planets are concerned. The only plausible constituents of Saturn's outer volume that have densities appreciably below 0.31 g/cm$^3$ are hydrogen and helium, but they are still gaseous well below 150 K. So Saturn, at least, must have an extensive atmosphere, and it must consist of large quantities of hydrogen and/or helium. Jeffreys also acknowledged that Menzel's temperature of 150 K for Jupiter and Saturn was too high to be just due to solar heating, and so there must be a significant amount of heat being emitted from inside these two planets. He continued to maintain, however, that the original heat at formation must have largely disappeared by now, and concluded that this extra internally generated heat may be due to the decay of radioactive elements located within about 300 km of the surface.[39]

To develop his theory of planetary structures further, Jeffreys assumed that the atmospheres of Jupiter and Saturn are of negligible density, surrounding a layer of ice and solid carbon dioxide of uniform density 1.0 g/cm$^3$, which in turn surrounds a rocky core of uniform density 3.0 g/cm$^3$ (which is similar to the densities of Io and Europa). With these assumptions, the moments of

---

[38] Jeffreys apparently used a pre-publication figure of 150 K from Menzel, rather than the actual figure of 160 K in Menzel's paper.

[39] It was assumed by many astronomers that Jeffreys, in his paper of 1924, was suggesting that such radioactivity was the source of the internally generated heat. He made it clear in 1926 and 1938, however, that he was not making such a suggestion, as the amount of radioactive material required seemed unacceptably high. Menzel's lower temperature estimates of 1926 (see below) made the radioactivity suggestion more plausible, however.

inertia and densities of Jupiter and Saturn imply that the depth of the atmospheres on the two planets is about $0.09R_J$ and $0.23R_S$, respectively. He finally concluded that these atmospheres probably consist mainly of hydrogen, nitrogen, oxygen, helium and maybe methane ($CH_4$), with clouds possibly consisting of solid carbon dioxide.

Interestingly, in the following year, 1925, the young Cecilia Payne of Harvard College Observatory (1900–1971) showed in her PhD thesis that hydrogen and helium are the most abundant elements in stellar atmospheres. Although she withdrew her conclusion when challenged by Henry Norris Russell (1877–1957), he himself finally accepted in 1929, after much agonising, that there are large amounts of hydrogen in the sun and stars. So the idea that Jupiter, Saturn, Uranus and Neptune may also have large amounts of hydrogen in their structures became highly plausible in the late 1920s, as the planets were assumed to have formed from the original solar nebula. The lack of large amounts of hydrogen in the terrestrial planets[40] was thought to be due to the fact that these planets are closer to the sun, and therefore receive more heat per unit surface area than the larger planets, and to their smaller size, implying a lower gravitational force to retain their gaseous atmospheres. Hydrogen, being the lightest gas, is lost first.

Meanwhile, in 1926 Menzel reanalysed the radiometric measurements of Jupiter, Saturn and Uranus made by Coblentz and Lampland in 1924, and concluded that their cloud-top temperatures were about 140 K, 120 K and 100 K, respectively, rather than the 160 K figure that he had deduced for Jupiter and Saturn previously.[41] As a result, he concluded that any internally generated heat, if such existed, should be rather low.

It was difficult in the 1920s to take proper account, in developing the theory of stellar and planetary structures, of the effects of ultra high pressure on the materials of which they are composed. In 1926, Ralph Fowler (1889–1944) applied the theory of degenerate matter, which is a state of matter under ultra high pressure, to white dwarf stars.[42] The theory of degeneracy, as applied to stars and planets, was developed further by Chandrasekhar and Kothari[43] in the 1930s, and in 1938 Rupert Wildt tried to

---

[40] The terrestrial planets are Mercury, Venus, the Earth and Mars.

[41] In fairness to Menzel, he described his 160 K figure as provisional when he published it in 1923.

[42] Atoms are stripped of their electrons when matter is subjected to extreme pressures, as in white dwarf stars.

[43] Although Kothari's concept of degeneracy was somewhat flawed, his work was important in contributing to the debate on the properties of matter at very high pressures.

see if the pressure inside the large outer planets may be sufficient to produce degeneracy.

The model of Jupiter and Saturn used by Wildt consisted of a dense core, similar to that of the terrestrial planets, surrounded by a thick layer of ice, which is in turn surrounded by a layer of highly compressed condensed gases, mainly solid hydrogen. Wildt assumed the densities of these layers to be 6.0, 1.5 and 0.25 $g/cm^3$, respectively, and then calculated dimensions of the layers from the overall densities and moments of inertia of the two planets. This gave radii of $0.43 R_J$ and $0.26 R_S$ for the cores, and $0.82 R_J$ and $0.66 R_S$ for the outside of the ice layers. The resulting pressures at the base of the condensed gas layer were about 1 million bar for Jupiter and 650,000 bar for Saturn, but at the centre of the core they were about 60 million bar for Jupiter and 15 million bar for Saturn, which Wildt concluded were probably just below the pressure required to produce degeneracy.[44] He did speculate, however, that maybe some of the hydrogen that was under very high pressure could be in the metallic hydrogen form proposed theoretically by Wigner and Huntingdon in 1935. Simultaneously, Kothari came up with the same suggestion.

Kronig, de Boer and Korringa suggested in 1946 that hydrogen in the earth's interior could become metallic under high pressure, thus creating a core of metallic hydrogen (see Section 9.6). Unfortunately, that did not appear to be compatible with the observed mechanical properties of the earth's interior. But in 1948 Ramsey used the same compressibility concept in suggesting that the earth's core consists of silicates which have become metallic under pressure.

In 1950, Ramsey investigated the behaviour of hypothetical, non-rotating planets[45] composed completely of hydrogen, which he calculated became metallic at a pressure of $8 \times 10^5$ bar. In so doing, the density of solid hydrogen increases at the transition from the molecular to the metallic phase from 0.35 to 0.77 $g/cm^3$. Ramsey calculated that, on this basis, such a planet could only have a core of metallic hydrogen if the planet's mass exceeds $88 M_E$ (where $M_E$ is the mass of the earth). At Jupiter's mass of $317 M_E$, the hydrogen planet would have a radius of 79,400 km and a mean density of 0.90 $g/cm^3$, compared with Jupiter's actual values of about 69,900 (mean

---

[44] It is now known that even matter at the centre of the sun is not degenerate, as considerably higher pressures and densities are required than exist even there.

[45] The centrifugal force, caused by a planet's rotation, has the effect of decreasing the effect of gravity and, as a result, reducing the internal pressure of a planet. Ramsey consciously ignored this effect in his 1950 paper to simplify his analysis.

radius) and 1.33 g/cm$^3$. Considering the assumptions made, this agreement is surprisingly good, so clearly the concept of a hydrogen planet is a good place to start in trying to deduce a model for Jupiter's internal structure.

Wildt's 1938 model for the internal structure of Jupiter, and the other outer planets, implied that there was about three times as much hydrogen, in percentage terms, in Saturn as in each of the other three outer planets. This seemed unlikely, assuming that they all formed from the original solar nebula. The model also made no allowance for compressibility, which for these large planets must be considerable. So in 1951 Ramsey modified his model of hypothetical hydrogen planets to include helium, and to take account of rotation. As a result, he found that Jupiter and Saturn would be composed of 76% and 62% of hydrogen, by mass (compared, for example, with 74% for the sun), assuming a uniform mixing of the hydrogen and helium. The central pressures were calculated to be 32 and $6 \times 10^6$ bar, for Jupiter and Saturn, respectively, with central densities of 3.66 and 1.91 g/cm$^3$; the metallic cores of the two planets containing 92% and 67% of their total mass.

Unfortunately, Ramsey's new models of Jupiter and Saturn produced moments of inertia that were somewhat too high. So, in the following year, he and Miles analysed the effect of adding heavier elements, which were partly concentrated in the core and partly distributed uniformly throughout the structure. This yielded better moments of inertia, but had little effect on the proportions of hydrogen, which varied from 76% to 84%, by mass, for Jupiter, and from 62% to 69% for Saturn, depending on the assumptions made.

In 1951 Ramsey had also considered the likely structure of Uranus and Neptune, and pointed out that they are about an order of magnitude denser than for hydrogen planets of the same mass.[46] As a result, he concluded that they must have lost most of their original hydrogen and helium, and now be composed mainly of water, methane and ammonia, together with terrestrial materials.

Ramsey also discussed, in his 1951 paper, the problem that Neptune was heavier than Uranus (17.3 versus 14.6$M_E$), but smaller in diameter (44,600 versus 51,000 km), which was difficult to explain if they have the same chemical composition. Ramsey pointed out, however, that this effect may be because one of the constituents is metallic at a pressure somewhere between that at the centre of Neptune and Uranus. He suggested

---

[46] Ramsey found that a hydrogen planet with a mass of that of Uranus or Neptune would have a density of about 0.17 g/cm$^3$, compared with Uranus' and Neptune's actual densities of 1.3 and 2.2 g/cm$^3$, respectively.

that metallic ammonium ($NH_4^+$) may be the cause, and in 1954 Bernal and Massey, of University College, London, showed theoretically that the transition pressure for the phase transition from mixed crystals of ammonia ($NH_3$) and hydrogen to metallic ammonium was almost certainly less than about 250,000 bar. If that was the case, however, it would not solve the problem posed by Ramsey, as ammonium would be metallic for the majority of the structure of both planets.

In 1961 William Porter of Bucknell University, Pennsylvania, used Bernal and Massey's theoretical data on the pressure–density relationship for metallic ammonium to construct a model for Uranus and Neptune. He assumed that both planets consisted mostly of solid ammonium, plus a smaller amount of heavier elements. Initially, the moment of inertia of Uranus turned out to be too high, so Porter added a small amount of solid hydrogen to the outer regions of the planet. As a result, he found that the most realistic model for Uranus consisted of 84% ammonium, 14% heavier elements and 2% hydrogen by mass, and that for Neptune consisted of 74% ammonium and 26% heavier elements.

## Atmospheres

Absorption lines of unknown origin had been detected in the nineteenth century in the spectra of Jupiter, Saturn and Uranus, but it was not until new photographic emulsions became available in the early twentieth century, that were sensitive to the red end of the spectrum, that real progress could be made.[47]

The first attempt of the new century to record and understand the spectra of Uranus and Neptune was carried out by Vesto Slipher at the Lowell Observatory over the period 1902–1904. His photographs of the spectra of the two planets, which covered the range from about 464 to 591 nm (nanometres),[48] showed enhanced solar absorption lines at 486.1 nm and 587.6 nm, which Slipher took to be due to hydrogen and helium, respectively. There were also three absorption bands, of unknown origin, at 510, 543 and 577 nm, all of which were stronger in Neptune's spectrum than in Uranus'. In addition,

---

[47] Photographic plates in the nineteenth century were basically insensitive to red light, and it was not until 1904 that panchromatic plates became available, which had a spectral sensitivity similar to that of the eye. Infrared sensitive plates did not become commercially available until 1919.

[48] 1 nm = $10^{-9}$ m or 10 Ångströms.

Slipher observed an absorption band visually at 618 nm, which also seemed to be stronger in Neptune's spectrum.

In 1905 and 1906, Slipher concentrated on photographing the spectra of Jupiter and Saturn. Comparing these with the earlier spectra of Uranus and Neptune showed strong similarities as well as differences between the spectra of the four outer planets. In particular, he found a strong absorption band in the red at 619 nm in all four spectra, and noted that this band, and that at 543 nm, gets progressively weaker in going from Neptune to Uranus to Saturn to Jupiter. He also noted that the band at 577 nm is stronger for Uranus than for Saturn or Jupiter, whilst one at 646 nm is much stronger for Jupiter than Saturn. In general, however, he observed that the spectrum of Jupiter is similar to that of Saturn, in both the existence and intensities of their absorption bands. In addition, that of Uranus is more like that of Neptune, although there are some bands in Neptune's spectrum that are not in the spectrum of any of the other three planets.

Slipher experimented with various sensitising dyes for his photographic plates in 1906 and 1907, and was able to extend their sensitivity up to about 750 nm in the near infrared. In doing so he was able to record strong absorption bands near 719 and 726 nm in the spectra of Jupiter and Saturn, and show that the spectra of Uranus and Neptune are weak above about 690 and 675 nm respectively. Finally, at a meeting of the Astronomical Society of the Pacific in 1931, he reported numerous absorption bands in the spectrum of Jupiter between 702 and 863 nm, together with a wide and very intense absorption band above about 882 nm.

The interpretation of all these bands proved very difficult. For example, in 1928 McLennan, Ruedy and Burton suggested that one of Jupiter's red absorption bands could be due to liquid water, although this seemed highly unlikely, considering Jupiter's measured temperature of 140 K. Menzel proposed, as an alternative, that the band could possibly be due to ice crystals, whose structure had been changed by high pressure in Jupiter's atmosphere, but this attribution was highly speculative.

Finally, the breakthrough in understanding the spectrum of the four outer planets came in 1931, when Rupert Wildt at Göttingen noticed that two of the absorption bands that he had detected in the spectra of Jupiter and Uranus, centred at about 785 and 899 nm, appeared to correlate with absorption bands at about 792 and 880 nm in the spectrum of ammonia ($NH_3$), as measure by Badger and Mecke two years earlier. Encouraged by this, Wildt reviewed Slipher's results, and in 1932 found that six of Slipher's lines in the absorption band at about 646 nm also appeared to coincide with lines in the spectrum of ammonia, as determined by Badger in 1930.

At that time the only known methane band below about 1 μm (micron)[49] was that at 886 nm, which Dennison and Ingram thought was the third harmonic of the methane band at 3.3 μm. If this was so, Wildt calculated that there would be other harmonics at 726, 620, and 544 nm in the methane spectrum. These seemed to correspond to the planetary absorption bands at 726, 619 and 543 nm. So Wildt attributed these bands, and the Jupiter band above 882 nm, to methane.

As a result, in 1932 Wildt appeared to have found evidence for ammonia and methane in the spectra of the four outer planets. The evidence for methane on Uranus and Neptune was circumstantial, however, as no absorption band had yet been reliably detected at 726 nm, and the Uranus' band centred on 899 nm detected by Wildt could be due to either ammonia or methane. Clearly what was now required was experimental confirmation that the three bands at 726, 619 and 543 nm exist in the spectrum of methane, and that the 726 nm absorption band, and hopefully that at 886 nm, exists in the spectra of Uranus and Neptune.

In 1933, Mecke of Heidelberg responded to Wildt's request for experimental evidence, by detecting absorption bands in the spectrum of methane at 890, 860, 840, 784, 725, and 620 nm, the latter two of which appeared to be two of Wildt's hypothesised lines. In parallel, Theodore Dunham used the 100 inch (2.5 m) Mount Wilson reflector to measure the spectrum of Jupiter and Saturn in 1932–33, and compare these with 'laboratory' spectra produced by some 40 m of ammonia at atmospheric pressure. As a result, Dunham proved convincingly that sixty-nine of Jupiter's absorption lines around 645 and 792 nm are due to ammonia and, although Saturn's lines are undoubtedly weaker, again he confirmed the presence of ammonia. In addition, the existence of methane on both planets was confirmed by the correlation of eighteen lines in its absorption spectrum at around 864 nm.

Mecke's work on the methane spectrum called into question Wildt's attribution to ammonia of the bands at 785 and 899 nm that he had recorded in the spectra of Jupiter and Uranus. Ammonia was known to have bands at 792 and 880 nm, and now Mecke had found methane bands at 784 and 890 nm, so maybe the two bands that Wildt had observed on Jupiter and Uranus were due to methane and not ammonia.

The answer was provided by Adel and Slipher in 1934, who showed theoretically that methane should have absorption bands at 886, 725, 619, 543, 486, and 441 nm. Adel and Slipher were also able to confirm the existence of all but the 441 nm band experimentally, using a 45 m path length of methane

---

[49] 1 μm = $10^3$ nm or $10^{-6}$ m.

at a pressure of 40 atmospheres (bars). So the enhancement of the solar line at 486 nm in Uranus' and Neptune's spectra, as recorded by Slipher in 1902–1904, was not due to hydrogen, as he had supposed, but to methane.[50] By 1934, Slipher had also recorded a line at 441 nm in Neptune's spectrum, but not in that of Uranus. It was now seen to be clearly due to methane, but the line's absence for Uranus indicated that there was more methane in Neptune's atmosphere than in Uranus'.

Furthermore, Adel and Slipher found theoretically and experimentally that methane has absorption bands at 874, 861, 788, 782, 720, 702, 668 and 662 nm. In addition, they showed theoretically, but could not observe experimentally, methane bands at 576 and 509 nm. So, rather than ammonia being the predominant gas spectroscopically on the four outer planets, the predominant gas is methane. It was also clear, from their work, that the methane concentrations reduced in going from Neptune to Uranus to Saturn to Jupiter, which explained the relative intensities on the four planets of the bands observed at 510, 543, 577 and 619 nm.

Interestingly, the band at 646 nm had been observed by Slipher to be stronger for Jupiter than Saturn, but it had not been observable at all for Uranus and Neptune. This band, and the one at 792 nm, which had also not been observable for Uranus and Neptune, were now the only two bands clearly due to ammonia in the spectra of the four planets. The 646 nm band indicated that there was probably more ammonia on Jupiter than Saturn, with no ammonia at all on either Uranus of Neptune. Russell pointed out that this was quite plausible as ammonia would be completely frozen out of the atmospheres of Neptune and Uranus, because of their very low temperatures, and partially frozen out of Saturn's atmosphere, as it has an intermediate temperature between that of Uranus and Jupiter.

Gerhard Herzberg pointed out in 1938 that, although the homonuclear diatomic molecules of hydrogen and nitrogen ($H_2$ and $N_2$) have no dipole moment, they will have a quadrupole moment, which varies during the vibration of the molecule. As a result, they will have a quadrupole rotation–vibration spectrum which, although faint, should be observable in the radiation from the outer planets. Herzberg deduced, from the work of James and Coolridge, that the most likely $H_2$ lines to be detectable were those at 850, 828 and 815 nm. These were not the strongest $H_2$ lines predicted, but it was thought that the strongest lines would be masked by either

---

[50] The hydrogen line would only have appeared in Uranus' spectrum at high temperatures, and it was now known that the upper layer of the Uranian atmosphere was cold.

strong telluric water vapour bands or methane bands from the outer planets. Herzberg also considered the case of similar quadrupole lines in $N_2$, but concluded that they would be much more difficult to detect, mainly because of the presence of nitrogen in the earth's atmosphere in substantial quantities.

None of the predicted molecular hydrogen and nitrogen ($H_2$ and $N_2$) lines were observed in the spectra of the outer planets until 1949, when Kuiper reported detecting a band some 3 to 4 nm wide, centred on 827 nm, in the spectra of Uranus and Neptune, together with a series of lines between 747 and 757 nm. He was unable to attribute any of these lines to known elements, but in 1952 Herzberg suggested that the 827 nm line was due to molecular hydrogen. Herzberg backed up this suggestion by detecting a 4 nm wide line, centred on 826 nm, that had been produced in the laboratory by an 80 m path length of hydrogen, at a pressure of 100 bar and temperature of 78 K. Then in 1957[51] Kiess, Corliss, and Kiess detected all three molecular hydrogen lines at 850, 827 and 815 nm in Jupiter's spectrum. In addition they detected the next line in the series at 805 nm.

## 9.9 Jupiter

It has been convenient, so far, to discuss the four large outer planets as a group, but significant progress was also made at the level of the individual planets, which will now be outlined.

### Radiation belts

During 1955, Bernard Burke and Kenneth Franklin, of the Carnegie Institute in Washington, were experimenting with a Mills Cross radio telescope when they noticed intermittent bursts of energy, at a frequency of 22.2 MHz, in an unexpected part of the sky. Because the radio telescope array was difficult to align, Burke and Franklin had fixed its declination at about 23°, and left it there for about two months around midsummer. They had expected to pick up radio emissions from the sun and the Crab Nebula, but the new, unknown source of intermittent bursts seemed to be about two hours behind the Crab Nebula, and getting closer every day. It soon became evident, from its location and movement, that it was Jupiter.

---

[51] The observations were made in 1957, but the results were not published until 1960.

Immediately after this serendipitous discovery was announced, Alex Shain, of the CSIRO[52] in Sydney, Australia, decided to search through old records of cosmic noise, which he had made over the period 1950–1951, to see if any showed evidence of these radio bursts. The high frequency records showed nothing unusual, but those at 18.3 MHz clearly showed the signals from Jupiter, which had been previously attributed to local interference.[53] Analysis of these records showed that the source had a periodicity of 9 h 55 min 13 s, which was the same as that of Jupiter's South Temperate zone, whilst the longitude of the source seemed to match that of a prominent white spot in that zone.

In 1958, Cornell Mayer and his colleagues at America's Naval Research Laboratory (NRL) deduced an effective temperature of 145 K for Jupiter from its radio emissions at 10 GHz. This indicated that the emissions are largely thermal. But in the following year Russell Sloanaker deduced a temperature of 600 K at 3 GHz, whilst Frank Drake and Hein Hvatum came up with a figure of about 50,000 K at 400 MHz. Clearly the emissions at these lower frequencies are non-thermal.

Drake and Hvatum then suggested that the radio emissions at these lower frequencies are caused by synchrotron radiation generated by relativistic electrons,[54] which are trapped in an intense magnetic field surrounding Jupiter. It appeared as though there was something around Jupiter analogous to the radiation belts around the earth, which had been discovered by Van Allen in 1958 using the early Explorer spacecraft (see Section 10.1). This hypothesis was verified in 1960, when Venkataraman Radhakrishnan and James Roberts, who were working at Owens Valley Radio Observatory in California, found that Jupiter's radio signals are polarised and come from an area about three times the size of the planet.

A few years later, Keith Bigg found that the intensity of the bursts was strongly related to the orbital position of Io, the closest of Jupiter's Galilean satellites, which appeared to disturb the electron emission in Jupiter's radiation belts. At about the same time, James Roberts, who had returned to Australia, and Max Komesaroff observed Jupiter using the Parkes radio telescope. They found that the plane of polarisation of the radio signals oscillates through an angle of about ±10°, indicating that Jupiter's magnetic axis is inclined at about 10° to its axis of rotation.

---

[52] Commonwealth Scientific and Industrial Research Organisation.

[53] It later transpired that Jansky had also, unknowingly, recorded Jupiter's radio emissions in the early 1930s.

[54] Relativistic electrons are electrons travelling at near the speed of light.

## The satellites

The discs of Jupiter's four Galilean satellites each subtend less than 2″.0 at the distance of the earth, even at opposition. So it is hardly surprising that our knowledge of their physical characteristics was sketchy until the first visits by planetary spacecraft in the 1970s. In fact, for many years the density of Callisto was thought to be only about 1.0 g/cm³, instead of its current-known value of 1.86 g/cm³, thus giving a completely false impression of its structure.[55]

At the end of the nineteenth century, it was thought that all four Galilean satellites probably had atmospheres, as Vogel had apparently detected bands, similar to those on Jupiter, in their spectra. In addition Schaeberle and Campbell had reported observing bright and dark bands, similar to those on Jupiter, parallel to Ganymede's equator, and Barnard had noticed changes in the appearance of Callisto and Io that were attributed to atmospheric changes (see Section 8.1).

This consensus on whether each of the four satellites had an atmosphere began to change in the twentieth century, however. It had been clear in the nineteenth century that as these four satellites had relatively low surface gravities, they could only have retained an atmosphere if they had been very cold from soon after their formation. In addition, many of the markings, which were observed from time to time in both centuries, were put down to observations of their real surfaces.

At the end of the nineteenth century, it was thought that Ganymede and Callisto exhibited synchronous rotation. Then in 1914 the German astronomer Paul Guthnick (1879–1947) found that the brightness of all four Galilean satellites varies in a regular way, correlating with their position in orbit, thus showing that the axial rotation of all four is synchronous.

Barnard discovered in about 1900 that the poles of Io, the innermost of the Galilean satellites, appear to have reddish caps. Then in 1926 Joel Stebbins (1878–1966) found that Callisto's brightness varied significantly with phase, indicating that it had a rough surface, like the moon.

Jupiter was known at the end of the nineteenth century to have five satellites, namely the four Galilean satellites plus Amalthea (see Section 8.1). Four more satellites were discovered photographically between 1904 and 1914, and by the time that the first spacecraft arrived at Jupiter in 1973, three additional satellites had been discovered, making a grand total of twelve. Of

---

[55] Cecilia Payne-Gaposchkin in her *Introduction to Astronomy*, Eyre and Spottiswoode, 1956, quotes a density as low as 0.6 g/cm³ for Callisto.

these twelve, the four Galilean satellites and Amalthea orbit relatively close to Jupiter, then come three small satellites orbiting at about 11 million km from the planet, followed by four small satellites orbiting at about twice the distance. The latter group of four all orbit Jupiter in the opposite direction to the others, in a so-called retrograde sense. In fact this, and the closeness of Jupiter to the asteroids, led Forest Moulton to suggest that the outside group were captured asteroids.

A.B. Binder and Dale Cruikshank, of the University of Arizona, found in the 1960s that Io is a few percent brighter than usual for 15 minutes or so after it emerges from Jupiter's shadow. They suggested that this may be because Io has an atmosphere and that, during the cold eclipses, clouds or frost condense which then disperse when Io is illuminated by the sun once more. However, later observations showed that, for some reason, this brightening only occurs for about half of the eclipses.

## 9.10 Saturn

### Differential rotation

A prominent white spot was first observed on Saturn by Edward Barnard, of the Yale Observatory, on 15th June 1903 at latitude of about 36° N. He re-observed it on both 23rd and 24th June, noticing that it appeared to rotate about the centre of Saturn much slower than the normal rate of about 10 h 14 min, but Graff of Hamburg was the first to publish a rotation period. His estimate of 10 h 39 min was initially treated with some scepticism, however, as it was so much slower than the normal period, but subsequently other astronomers deduced a similar value. As the English amateur Stanley Williams (1861–1938) observed at the time, this implied that there is an equatorial current on Saturn, similar to that on Jupiter, that is in the direction of the planet's rotation, and which has a large velocity relative to the atmosphere at higher latitudes. On Jupiter the relative velocity is about 400 km/h (see Section 8.1), whereas on Saturn he deduced an astonishing relative velocity of about 1,400 km/h.

Saturn is almost twice as far away from the sun as Jupiter, and so it receives appreciably less solar heat. In addition, Saturn is smaller than Jupiter, and so it should have cooled down faster after its formation from the solar nebula. It was therefore difficult to explain why the equatorial current on Saturn was faster than that on Jupiter. Stanley Williams speculated that it may be connected with the fact that Saturn is much less dense than Jupiter.

At this time, however, the rotation period of the cores of Jupiter and Saturn were unknown. So, if these were the same as that of the so-called equatorial currents, we would really be observing a slower rotation period at medium latitudes, rather than a faster one in the equatorial region.

A prominent *equatorial* white spot on Saturn had been seen by Asaph Hall in 1876 (see Section 8.2), and another equatorial white spot was discovered by the English amateur astronomer and comedian, Will Hay, in August 1933. Like the 1876 spot, this new one had a rotation period of about 10 h 14 min. Then in 1960 J.H. Botham, a South African amateur, discovered another prominent white spot, this time, like the 1903 spot, in northern latitudes,[56] with a similar period to the 1903 spot of about 10 h 40 min.

So a pattern seemed to be emerging, with these prominent white spots appearing, on average, every 28 years, alternating between equatorial and northern latitudes. In all cases, the spot became gradually elongated before disappearing after a few weeks. Interestingly, the sidereal period of Saturn is about 29 years, and every time the white spot appeared, the north pole of Saturn was tilted at virtually its maximum angle to the sun. This may be just a coincidence, of course, but the correlation appeared to be significant.

## The rings

In the nineteenth century, Saturn was known to have three rings (designated A, B and C), but in 1908 Schaer of Geneva reported observing a very faint ring outside of the A ring. Observers at the Greenwich Observatory searched for this new ring, and found some evidence of its existence, although the evidence was by no means conclusive. Barnard looked for it the following year with the 40 inch (102 cm) Yerkes refractor, but without success, which convinced most people that the previous observations were erroneous. Observations of this very faint ring would not go away, however, as they continued to be reported from time to time during subsequent years. Then in 1966 Walter Feibelman was able to photograph a very faint ring (now called the E ring), at the Allegheny Observatory in Pennsylvania, extending from outside of the A ring outwards to beyond the orbit of Dione, which is about 380,000 km from the centre of Saturn.

Three years later, Pierre Guerin claimed to have discovered another faint ring, this time inside the C ring. A ring, now called the D ring, was later confirmed inside the C ring by the Voyager spacecraft (see Section 11.2), but

---

[56] The 1960 spot was at about 60° N, compared with 36° N for the 1903 one, however.

it is unlikely that Guerin really detected this D ring, as it appears to be too faint to be seen from earth.

The C ring had been shown in the nineteenth century to be transparent. Then in 1917 the A ring was independently found to be transparent by the English amateur astronomers Maurice Ainslie and John Knight, who observed a seventh magnitude star as it passed behind the ring. Ainslie noted that the star showed two increases in brightness during its passage behind the A ring, once at the Encke division and once between that and the outer edge of the ring. Knight, on the other hand, reported that the star's brightness varied continuously during its occultation, indicating that the A ring is not uniform. In 1920, similar variations in transparency were observed by the South African amateur William Reid and three colleagues when the B ring passed in front of a star.

Markings on the rings were seen by a number of observers in the late nineteenth and early twentieth centuries. For example, Etienne Trouvelot (1827–1895), a French astronomer working in America, carried out a series of observations of the rings over the period 1873–6, detecting, from time to time, what he described as dark angular markings on the inner edge of the A ring at both ansae. Antoniadi saw similar markings in 1896, which he considered radially oriented, on the inner edge of the A ring at both ansae. Further observations of such markings were reported by Rudaux (d. 1947) and Maggini (1890–1941) in the early twentieth century. Then in 1955 Guido Ruggieri noticed pale, clear radial streaks at both ansae, from the inner to the outer edge of the A ring. Unfortunately, in view of later developments (see Section 11.2), after further work he concluded that they were an optical illusion.

Unexpectedly strong radar echoes were received from Saturn's rings in 1972 and 1973, when a 400 kW beam was aimed at Saturn.[57] These strong echoes indicated that particles in the rings may be up to a metre in diameter, with the majority falling into the 5 to 30 cm range. Some of the particles were thought to be metallic because of their high reflectivity. Radio and infrared measurements indicated that the rings also contained water ice and silicate particles, with a mean radius of about one centimetre, although later radar data tended to disprove the idea of a silicate component.

## The satellites

The Spanish astronomer J. Cornas Solá noticed in 1908 that Titan, Saturn's largest satellite, showed a pronounced limb darkening effect, and suggested

---

[57] No return signals were detected from the planet itself.

that this meant that it had an atmosphere. No clear, independent evidence was obtained for an atmosphere, however, until the winter of 1943–1944 when Gerard Kuiper photographed the spectra of the ten largest satellites of the solar system (including the moon). He found definite evidence of the 619 and 726 nm methane bands for Titan, and some evidence of the 619 nm band for Triton, Neptune's largest satellite, but he could find no such bands for Jupiter's four Galilean satellites, or for Saturn's satellites Tethys, Dione or Rhea. Interestingly, in the paper giving his results, Kuiper also calculated an index showing, theoretically, the likely stability of planetary or satellite atmospheres against dispersion, over time. This index was based on the escape velocity for such a body, which is dependent on its mass and radius, and on its surface temperature, which is dependent on its distance from the sun. The results gave him, in order of reducing stability, or likelihood of retaining an atmosphere, of Earth, Venus, Triton, Mars, Titan, Ganymede, Callisto, Io, Europa, Mercury, Moon, Rhea, Dione and Tethys, with the likely boundary between atmosphere and no atmosphere between Titan and Ganymede.

Andouin Dollfus, at the Pic du Midi Observatory, reported in 1967 the discovery of the first new satellite of Saturn this century, for which he suggested the name Janus. The new satellite had a magnitude of about 14, and was found to orbit Saturn just outside the rings. Normally the brightness of Saturn's rings would have made it impossible to record such a dim object so close to them. But he had detected it on photographs taken when Saturn's rings were virtually edge on as seen from earth, at the so-called ring-plane crossing. At such a time, the rings are of minimum intensity, enabling dim objects to be detected.

Confirmatory observations of this new satellite were reported almost immediately by Richard Walker, of the US Naval Observatory. Unfortunately, the rings rapidly brightened shortly after these observations, and so further views of this satellite proved impossible. Although the evidence for its existence was reasonably good, there was still some doubt, however, and so the next ring-plane crossing in 1979 was eagerly awaited.

In the meantime, in 1976 Stephen Larson and John Fountain, of the University of Arizona, re-examined Walker's photographs, together with some taken at about the same time at the Catalina Observatory in Arizona. As a result, they came up with the startling conclusion that there was not one new satellite, but two in almost the same orbits, with periods of 16.6 and 18.0 hours. In fact, they had found both objects on one of Walker's plates, one on one side of Saturn and one on the other.

The ring-plane crossing of 1979 was to yield more data on these two satellites, and to produce evidence of even more satellites of Saturn. But that

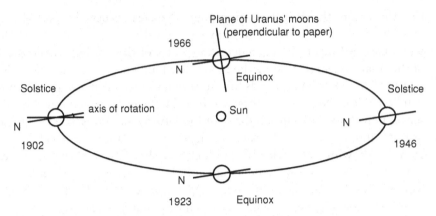

**Figure 9.3** Because Uranus spins almost on its side, there are periods twice per orbit (in 1902 and 1946 in this diagram) when one of its poles points in the general direction of the sun, and hence of the earth. Only around intermediate positions (centred on 1923 and 1966) did the planet's spin produce measurable Doppler shifts as seen from earth.

was after the first spacecraft had visited the planet, and so these discoveries will be described later (see Section 11.2).

## 9.11 Uranus

### Rotation period

Uranus is so far away, and its surface markings so difficult to see, that attempts to use these to determine its spin period were almost bound to fail. However, James Keeler had been successful in determining the rotation period of Saturn's rings, by measuring the Doppler shift of its spectral lines (see Section 8.2). So it was inevitable that such a technique would be used to determine Uranus' rotation period, once the telescopic and spectroscopic equipment was up to the task.

Because Uranus almost spins on its side, there are two times in its orbit when the poles are pointing almost at the earth (see Figure 9.3). Under these circumstances, there will be no significant movement of parts of its surface towards or away from the earth as the planet spins, and so there will be no measurable Doppler shift. Such a situation occurred in 1902. But about a quarter of an orbit later, Uranus' equatorial plane would almost pass through the earth, and then a clear Doppler effect should be observed. It was known that such a situation would occur in 1923.

Percival Lowell was anxious to measure the rotation period of Uranus as soon as possible, and was unwilling to wait until the optimum period of the early 1920s to do so. In 1909, therefore, he decided that the angle of Uranus' equatorial plane to the Uranus–Earth line may be small enough to yield a measurable Doppler shift. This turned out to be the case, and in 1912 he, assisted by Vesto Slipher, published their results, giving a rotation period of about 10 h 45 min, retrograde.

Five years later, E.C. Pickering announced that Leon Campbell, of the Harvard College Observatory, had found in 1916 that the light from Uranus varied in intensity by about 0.15 magnitudes with a period of about 10 h 49 min. In view of the similarity of this period to that deduced by Lowell and Slipher, Pickering concluded that Campbell had also measured the rotation period of the planet. Unfortunately, however, Campbell's subsequent measurements of 1917 and 1918 showed little or no evidence of intensity variations. Campbell suggested that this could be because the clouds responsible for the brightness variations that he had measured in 1916 may have dispersed in the meantime. Subsequently other visual observers, and those using photoelectric photometers, deduced a similar period to Campbell, although other observers could detect no brightness variations at all.

Over the period 1927–30, Joseph Moore and Donald Menzel repeated Lowell and Slipher's measurement of the Doppler shifts, and deduced a period of 10 h 49 min ±10 min, which was the same, within error, as that measured by Lowell and Slipher twenty years earlier and by other observers using intensity fluctuations. This agreement, between the periods calculated using two entirely different methods, seemed to be conclusive proof that this was the true rotation period of Uranus. As a result a period of about 10 h 50 min was accepted for many years. Then in the 1970s longer periods of 15 to 17 hours were deduced from intensity fluctuations and in 1984 R.G. French derived a period of about 15 h 35 min from Uranus' observed oblateness.

## Miranda

Gerard Kuiper photographed Uranus and its system of four satellites on 16th February 1948 to get more accurate data on the satellites' intensities. He instantly noticed when he developed the plate that there was an unexpected faint object close to the planet. At that stage, it was not clear if the object was a star or another satellite. But two plates exposed on 1st March clearly showed that the object had moved with Uranus, and so was the planet's fifth known satellite, now called Miranda.[58] Further plates, exposed on 24th and 25th March, showed that Miranda was orbiting Uranus in an approximately circular orbit with a period of about 33 h 56 min, in the same plane as the other four satellites.

## The rings

Gordon Taylor, of the Royal Greenwich Observatory, had predicted in 1973 that Uranus would occult the 8.8 magnitude star SAO 158687, as seen from the Indian Ocean, on 10th March 1977. Such an occultation would be important as it would provide valuable data on Uranus' atmosphere. In addition, if it was observed from more than one place on earth, the diameter and the oblateness of Uranus could be determined to an unprecedented accuracy. In fact, by the time of the occultation, observers from as far apart as Tokyo, Peking, Kavalur and Naini Tal in India, Cape Town and Perth had made arrangements to record it, along with astronomers from Cornell University flying on the Kuiper Airborne Observatory (KAO).

Unfortunately, the star SAO 158687 is about three magnitudes dimmer than Uranus, thus making it difficult to observe when it is very close to the planet. However, the intensity of Uranus can be very much reduced, relative to the star, if it is observed in the wavelengths of its methane absorption bands. As a result, these were chosen for the observations.

The Kuiper Airborne Observatory had only a limited fuel supply, and so it was essential that both the track of the eclipse, and its timing, should be known as accurately as possible beforehand. Fortunately, Taylor had calculated that the star and Uranus would pass within about 50″ of each other in late January 1977, before Uranus reversed its direction prior to the occultation. This would enable both objects to be photographed on the same plate, allowing their relative positions to be accurately checked.

In the event, the observations of Uranus and SAO 158687 in January 1977 were a big surprise. Not only was Uranus 0″.2 south of where it was expected to be, but the star was a very unexpected 1″.2 north of its catalogued position. This moved the northern limit of the occultation southwards on

---

[58] Kuiper chose the name Miranda from Shakespeare's *The Tempest*, in keeping with the names of the other four satellites which had been taken from either Shakespeare's or Pope's characters.

the surface of the earth, so that it would no longer be observed from Japan or India as originally expected.

As it turned out, the key occultation observations were made by James Elliot, Edward Dunham, and Douglas Mink, of Cornell University, on board the KAO high over the southern Indian Ocean. About forty minutes before the expected occultation of the star, they recorded a sudden dip in the star's intensity. A few seconds later, the star's intensity recovered to its previous value. Whilst they were still trying to understand what had happened, they observed another dip in intensity, followed by three more dips before the star's occultation by Uranus. They also detected a series of dips after the occultation.

Meanwhile at Perth Observatory, Robert Millis from the Lowell Observatory, with Peter Birch and Dan Trout from the Perth Observatory, had detected their first dip, which lasted for about eight seconds, about seventy seconds before the first intensity reduction observed by the KAO astronomers. Then in the next fourteen minutes they observed four more dips, each of which was shallower than the first, lasting about one second each. Unfortunately, Perth was too far north and so the observers there did not see the occultation by the planet itself.

After the occultation had finished, the KAO returned to Perth where it landed. There the KAO team met Millis, who was anxious to compare results. Although they briefly discussed the possibility that the intensity dips were caused by a series of rings around Uranus, they dismissed that idea as the rings would have to be very thin,[59] and there was no known mechanism to stop narrow planetary rings from expanding in space. Instead they attributed their observations to a swarm of small satellites around the planet. But when, on 14th March, Elliot and his wife carefully examined the 40 ft (12 m) long chart recording the star's intensity during the occultation, they noticed that the pattern of dips in intensity before the planetary occultation were matched, in reverse order, by the pattern after the occultation (see Figure 9.4). They immediately realised that these must, after all, be due to a system of very thin rings around Uranus.

Elliot and his colleagues designated their five newly discovered rings α, β, γ, δ and ε, in order of increasing distance from Uranus. The α, β, γ and δ rings were very thin and apparently circular in shape, whereas the ε ring was broader and eccentric. Then, after analysing their results, the Perth team announced that they had detected six rings, namely Elliot's ε, which they numbered 1, γ (2), β (3), 4, 5 and 6, with decreasing distance from the planet.

---

[59] At that time, the only planetary rings known were the broad rings of Saturn.

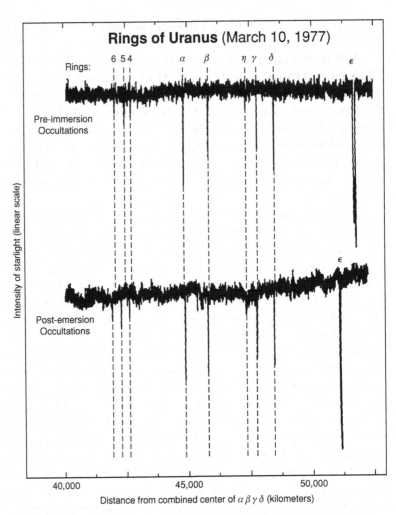

**Figure 9.4** Intensity traces recorded by James Elliot and colleagues using the Kuiper Airborne Observatory during the occultation of star SAO 158687 by Uranus. The top and bottom charts are before and after the occultation, respectively. The symmetry of the reductions in intensity due to the rings α, β, γ and δ is clearly seen, as well as the eccentricity of the ring ε. (With permission, from the *Annual Review of Astronomy and Astrophysics*, Volume 17 © 1979 by Annual Reviews www.AnnualReviews.org)

Later analysis showed evidence for the 4, 5 and 6 rings in the KAO data, together with an additional ring, between rings β and γ, which Elliot and colleagues designated the η ring.[60]

---

[60] For some reason, the sixth letter, ζ, in the Greek alphabet was omitted.

Initially, Bhattacharyya and Kuppuswamy, the observers at Kavalur, reported the stellar occultation by the ε ring as being caused by a new satellite. Two years later, however, after combining the results from Kavalur and Naini Tal, they and their colleagues identified the nine rings of Elliot, Millis, *et al.* In addition, they found evidence for another ring, which appeared to be about 3,000 km wide, nearer the planet than the others.

Numerous stellar occultations by the rings were observed over the next nine years, which separated the date of their discovery from that of the first spacecraft encounter with Uranus. These occultations showed that the nine 'classical' rings are very narrow, with all but the ε ring being 12 km, or less, in width. The η, γ, δ and ε rings are coplanar with Uranus' satellites, whereas the five inner rings have small but measurable inclinations. These inclinations, together with the marked eccentricity of the ε ring, causes these six rings to precess slowly around the planet.

## 9.12 Neptune

### Rotation period

In the 1880s Maxwell Hall had found that Neptune varied in brightness with a period of about 7.9 hours, but the variations died out after just over two weeks, so they could not be accepted as proof of the planet's rotation period. However, Hall was convinced that that was what he had measured; the variations having died out because the cloud structures on Neptune, that had presumably produced them, had changed. Then in March 1915 Hall undertook new photometric observations of the planet, finding a regular period of 7.835 hours over about a month, before the period became irregular, finally dying out completely in May.

Ernst Öpik and Liviänder, of the Tartu Observatory in Estonia, undertook photometric observations of Neptune in 1922–1923 and, in the following year, analysed their results alongside those of Müller (1884–1885), Baldwin (1908) and Maxwell Hall (1915), and concluded that they showed two superimposed periods of 7.707 and 7.836 hours. They suggested that this was because they were observing the drift of clouds at different latitudes, but with such fast rotation rates Neptune's polar flattening should have been easily observable. However Neptune, in the telescopes of the time, looked perfectly circular. It was then that Jackson suggested that we may be observing more than one spot at a particular latitude. In which case, the period may be a simple multiple of 7.8 hours. In fact, on a strictly theoretical basis,

Jackson concluded in 1926 that Neptune's period is of the order of 19 hours, assuming that its central condensation is similar to that of Jupiter, as their densities seemed to be about the same.

Jackson concluded in his 1926 paper that it would probably be impossible to measure such a long period of 19 hours using the differential Doppler shift of Neptune's surface. Joseph Moore and Donald Menzel of the Lick Observatory begged to differ, however, and in 1928 they were able to deduce a rotation period of $15.8 \pm 1$ hours from such a Doppler shift, or twice that observed from the light fluctuations, and the rotation was prograde. This 15.8 hour rotation period was generally accepted, until in 1977 Hayes and Belton deduced a period of 22 hours, based on Doppler shifts, whilst in 1981 Robert Hamilton Brown, Dale Cruikshank and Alan Tokunaga found a period of 17.95 hours photometrically.

Bradford Smith, Harold Reitsema and S.M. Larson were able to record clear cloud features in Neptune's atmosphere for the first time in 1979, using a CCD and a filter that transmitted light in the strong 890 nm methane absorption band. The images showed a broad, dark equatorial band, where methane deep in Neptune's atmosphere absorbs sunlight, and bright features in both the northern and southern hemispheres due to high altitude clouds.

The quality of this type of image improved dramatically over the next few years with better CCDs, and with the introduction of better computer-based processing. In particular, those images taken by Richard Terrile and Bradford Smith in May 1983, using the 2.5 m Irénée du Pont reflector at the Las Campanas Observatory in Chile, showed four clear atmospheric features, which allowed the planet's rotation period to be determined. The period of 17.83 hours, prograde, was virtually the same as that measured photometrically two years earlier by Hamilton Brown and colleagues.

Finally, Heidi Hemmel confirmed the changing nature of Neptune's atmosphere, which had been first reported a hundred years before by Maxwell Hall, when she found in 1986 and 1987 that the clouds, which had clearly been seen in the northern hemisphere just shortly before, had disappeared. In 1986 and 1987 she also noticed one very bright cloud at latitude $-38°$ that had a rotation period about the centre of Neptune of 17.83 hours. A year later, however, the only bright cloud was at $-30°$, with a rotation period of 17.67 hours.

## The satellites

In the nineteenth century, the only known satellite of Neptune was Triton, which orbits Neptune in a retrograde sense. This led astronomers to wonder if Neptune itself had a retrograde axial rotation, but this was disproved in 1928 when Moore and Menzel measured the differential Doppler shift of its surface. So Triton was seen to rotate in the opposite sense to the spin of its primary, Neptune, the first major satellite in the solar system to be observed to do so.

Dale Cruikshank and Peter Silvaggio detected the 2.3 µm absorption band of methane on Triton in 1978, indicating that Triton has a tenuous methane atmosphere. If that is the case, Triton would be only the third satellite in the solar system to be shown to possess an atmosphere. (The others are Titan and Io[61].) The 1.7 µm methane band was relatively weak, however, indicating that there is little or no methane frost or ice on the illuminated surface, which was assumed to be largely rocky.

Triton's orbit is inclined at 23° to Neptune's equator, and Neptune's equator is inclined at 29° to the plane of its orbit around the sun. Thus the sun is at the zenith on Triton at 52° latitude on midsummer's day. As Neptune's year is 165 years long, there is plenty of time for one hemisphere of Triton to heat up and the other to cool down, the poles each being without sunlight for 82 years. So Cruikshank and Silvaggio suggested that there may be methane ice or frost on Triton's unilluminated surface, even though they had not detected any elsewhere. Then in 1983 Apt, Carleton and Mackay analysed the visible spectrum of Triton, and concluded that there was clear evidence for methane frost or ice on its surface, although water ice appeared to be largely absent. In the same year Cruikshank, Clark and Hamilton Brown also detected nitrogen, possibly in liquid form, on the surface of Triton, implying that it would also be present in Triton's tenuous atmosphere, as nitrogen is highly volatile. So by the time that the first spacecraft arrived at Neptune in 1989, Triton was thought to possess an atmosphere of methane and nitrogen, with methane ice or frost and liquid nitrogen on its surface.

Gerard Kuiper discovered Nereid, Neptune's second satellite, in 1949, and found that it orbited Neptune in the opposite sense to Triton. Triton's orbit is almost circular, with a radius of 350,000 km, but Nereid's is highly elliptical, with an apogee of 9.7 and a perigee of 1.3 million km.

One would expect Triton, which is a large satellite, orbiting close to Neptune in an almost circular orbit, to orbit in a prograde sense, like the large satellites of Jupiter and Saturn. On the other hand, it would not be unusual for a small satellite like Nereid, which is orbiting at a large distance

---

[61] Io's very tenuous atmosphere had first been unambiguously detected by the Pioneer 10 spacecraft in 1973 (see Section 11.1).

from Neptune in a highly elliptical orbit, to orbit in a retrograde sense, like some of the small outer satellites of Jupiter and Saturn, which are probably captured asteroids. But it is Triton that has a retrograde orbit and Nereid that has a prograde one. So maybe Triton is the captured asteroid, and Nereid is a normal satellite. Maybe they have both been captured, as their orbits are inclined at more than 20° to Neptune's equator, and their orbits are also highly inclined to each other. It was hoped that the anticipated discovery of further satellites by the Voyager spacecraft, during its encounter with the Neptune system in 1989, would help to solve these questions.

### The rings

Saturn's ring(s) had been known for centuries. But rings had now been discovered around Uranus in 1977, and Jupiter in 1979 (see Section 10.1 later), so the obvious question was, does Neptune have rings?

Jupiter's rings had been discovered during a spacecraft intercept, and were not detectable from earth, whereas the rings of Uranus had been discovered from earth during a stellar occultation. Uranus' rings had been inclined at almost 70° to our line of sight when they were discovered, because of the 98° orientation of Uranus' spin axis, whereas the maximum inclination of Neptune's rings would only be about 30°, making them more difficult to detect. In addition, the maximum angular width of Neptune's rings would be about half that of Uranus' rings, as Neptune is almost twice as far away from earth. Neptune also moves more slowly against the stellar background. For all of these reasons, the chances of finding a suitable stellar occultation are clearly much worse for Neptune than for Uranus, although it was expected that they would occur about once per year, on average.

Orbital analysis showed that Neptune was due to almost occult a suitable star on 24th May 1981. This potential encounter was observed by Harold Reitsema, William Hubbard, Larry Lebofsky and David Tholen of the University of Arizona, who were hoping to find evidence of Neptune's rings. As Neptune approached the star, they found that the star was eclipsed for eight seconds, but there was no symmetrical eclipse on the other side of the planet. So the eclipse could not be due to a ring, as that should have produced a second eclipse as Neptune receded from the star (as happened with the occultation by Uranus in 1977). As a result, Reitsema and his colleagues concluded that they had probably discovered a new satellite of Neptune, designated 1981 N1, which they calculated was at least 180 km in diameter, and about 50,000 km from the centre of the planet.[62] They weren't completely sure, however, as the chance that a small satellite of Neptune had been just in the right place at the right time to occult the star was less than one in a thousand.

These results prompted astronomers at Pennsylvania's Villanova University to remember an occultation that they had observed in 1968. At that time Edward Guinan and J. Scott Shaw had been trying to measure the diameter of Neptune and observe its atmosphere from Mount John Observatory, New Zealand, when they had also observed a one-sided occultation of a seventh magnitude star. But that event had lasted for 2$\frac{1}{2}$ *minutes*. It could not have been due to a satellite, because the long period of the occultation implied that the satellite would have been large enough to have been easily visible. If it had been due to a ring or partial ring, however, it was not the same as the one seen in 1981, as the 1968 object was closer to the planet. So what had been detected in 1968 and 1981?

Another occultation opportunity occurred on 15th June 1983, but unfortunately no evidence of rings, or of a satellite, was found by the many teams of astronomers who observed the event. Then on 22nd July 1984 the first unambiguous discovery of a partial ring was found during the occultation of the star SAO 186001 by astronomers at both the European Southern Observatory (ESO) and 95 km to the south at the Cerro Tololo Inter-American Observatory (CTIO). The team at the ESO, who were observing the occultation at the request of André Brahic, of the University of Paris, found a 35% reduction in the intensity of the star that lasted for about one second. As before, the occultation was not observed on the other side of Neptune. It so happened, however, that William Hubbard and his team at the CTIO had actually recorded the same event, but they did not realise it because their quick-look data recording system did not have a fine-enough time resolution. Hubbard's team heard in October of the ESO discovery, and this caused them to re-examine all the data that they had recorded at the time which had a much finer time resolution. Their results turned out to be identical to those of the ESO.

These observations of an identical partial occultation at observatories spaced 95 km apart proved that the effect could not have been due to a satellite. It appeared to be due, instead, to a partial ring, or ring arc, about 20 km wide and 67,000 km $(2.71R_N)$ from the centre of Neptune. The ring arcs detected in 1968 and 1981 would be about 30,000 km and 70,000 km $(1.21$ and $2.83R_N)$, respectively, from the centre of Neptune, if ring arcs they

---

[62] The distance was later updated to 70,000 km.

were.[63] A further occultation in 1985 also showed that there was probably another ring arc 56,000 km from the centre of Neptune.

The question of how ring arcs could exist for, presumably, millions of years, without forming a continuous ring, or completely dissipating, was difficult to answer. Peter Goldreich, Scott Tremaine and Nicole Borderies suggested that each ring arc could be shepherded by a satellite, in a slightly inclined orbit. On the other hand, Jack Lissauer of the University of California suggested that each ring arc could be at a Lagrangian point 60° in front of or behind a satellite. The satellite would stop the arc spreading along its orbit, whilst a second satellite, in an orbit inside or outside the arc, would stop the arc spreading radially. More data would soon be available from spacecraft observations of Neptune's system to help solve this problem.

## 9.13 Asteroids

### Orbits

As the number of known asteroids continued to increase in the early twentieth century, it was natural to try to get an overall picture of these small members of the solar system, to see if the orbits are random or grouped in some way.

Daniel Kirkwood had identified gaps in the orbital periods of the asteroids (see Section 7.2), but it was not until 1918 that asteroid families were first identified by Kiyotsugu Hirayama, based on their orbital radius, eccentricity and inclination. Initially, Hirayama identified three families, namely Themis (22 members), Eos (21 members) and Koronis (13 members), followed a few years later by the families of Maria and Flora.[64] Hirayama believed that the existence of these families was no accident, but due, in each case, to the fracture of a larger asteroid, thus resurrecting, in modified form, the theories of Thomas Wright of Durham, Johann Lambert, and Wilhelm Olbers (see Section 6.4).

Over a thousand asteroids had had their orbits determined by the early 1920s, and by then asteroids were treated by most astronomers as a nuisance, appearing at random on images of other objects. Although finding new asteroids was an interesting project for amateur astronomers, it was rapidly loosing its appeal to the vast majority of professionals. So the number of

---

[63] Reitsema still thought that the 1981 occultation was due to a satellite, however.

[64] The families are named after the senior asteroid member of the family.

numbered asteroids discovered per year gradually reduced from a peak of ninety-nine in 1938, to a minimum of three in 1958.[65]

When the asteroid Eros had been discovered at the end of the nineteenth century, it was the only asteroid known to have an orbit well inside that of Mars; so far, in fact, that it came relatively close to the earth's orbit. Then in 1911 the asteroid Albert (719) was discovered by Johann Palisa that has a similar orbit, although it did not come quite so close to the earth as Eros.[66] Two other asteroids, Alinda (887) and Ganymed (1036), which also have orbits that come close to the earth's, were discovered in 1918 and 1924. Then on 12th March 1932, Eugene Delporte at Brussels discovered another asteroid that came even closer to the earth than Eros. Enough observations were obtained of this new asteroid,[67] now called Amor (1221), to enable its orbit to be calculated. As a result, Delporte was able to recover it in 1940 only a few degrees from its predicted position, after it had completed three orbits of the sun. With a diameter of only about one kilometre, Amor was, at the time, the faintest asteroid to be given a number. Its perihelion distance of 1.086 AU (see Figure 9.5), and orbital inclination of about 11°.9, brings Amor to within about 16 million km of the earth every 8 years. Asteroids in similar orbits, with a perihelion between 1.017 AU (the aphelion distance of the earth) and 1.3 AU are now called Amor asteroids. They have perihelia that almost reach to the earth's orbit, but do not cross it, at least, on average.[68] At the time of writing (early 2002) over 800 of these Amor asteroids have been discovered.

Amor held the record for the asteroid approaching closest to the earth's orbit for just six weeks, as on 24th April 1932 an asteroid was found that has an orbit that not only crosses that of the earth, but also that of Venus (see Figure 9.5). This new asteroid, with the provisional designation 1932 HA, was discovered by Karl Reinmuth of the Heidelberg Observatory and followed

---

[65] Only asteroids with clearly defined orbits are numbered. The number of asteroids discovered per year, quoted above, refers to the year in which the asteroids were discovered, not the year in which they were numbered, which could, in some cases, be many years later. The minimum numbers of 1 in 1944, 0 in 1945, and 1 in 1946 have been ignored, as they were clearly affected by war and early post-war conditions.

[66] The perihelion distances of Eros and Albert are 1.13 and 1.19 AU, and their orbital inclinations are 10°.83 and 10°.82.

[67] Including a pre-discovery observation from Japan near opposition, when the asteroid was moving at $1\frac{1}{2}°$ per day in declination.

[68] Because of gravitational perturbations, some Amor asteroids have a perihelion that oscillates across the 1.017 AU boundary, however.

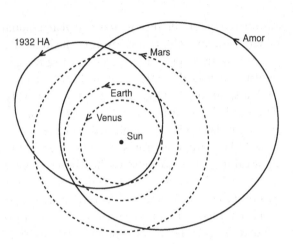

**Figure 9.5** This diagram shows the orbits of the asteroids Amor and 1932 HA, otherwise known as Apollo, relative to those of three of the inner planets. Amor's orbit is seen to stretch from just outside the Earth's orbit to well outside the orbit of Mars, whilst that of Apollo crosses the orbits of Venus, the Earth and Mars.

until 15th May when Van Biesbroeck last photographed it at the Yerkes Observatory, as it passed just 11 million km from earth. The preliminary orbit for 1932 HA indicated that it passed within just 0.65 AU of the sun on 7th July, but, because it is so small, it was thought that it would never be seen again. As a result, it was not given a name or asteroid number, although it was informally called Apollo.

In the early 1970s an attempt was made to recover Apollo. Unfortunately, the first searches, in 1971, by Charles Kowal, using the 48 inch (1.2 m) Palomar Schmidt telescope, and Paul Wild, using the 16 inch (40 cm) Bern University Schmidt, were unsuccessful. As a result it was decided to widen the search area at the next opportunity in 1973. Then, amazingly Richard McCrosky and Cheng-Yuan Shao were successful on their first night of 28th March, using the 61 inch (1.5 m) reflector at Harvard Observatory's Agassiz Station, at the edge of their search area. This asteroid is now called Apollo (1862).

Four years after the original discovery of Apollo, Eugene Delporte found another asteroid, designated 1936 CA, whose orbit crosses that of the earth. He first observed it on 12th February 1936, just a few days after it passed just 2 million km from the earth. This asteroid was intrinsically much fainter than Apollo, and so it could only be tracked for about two months even using the 100 inch (2.5 m) Mount Wilson reflector, which was then the largest telescope in the world. This asteroid, now generally called Adonis (2101), was, like Apollo, lost for about 40 years. Its perihelion turned out to be even closer to the sun at 0.44 AU.

It did not take long for the third member of this earth-crossing family, now called the Apollo asteroids, to be found. Like Apollo itself, this new asteroid, designated 1937 UB, was discovered by Reinmuth. Now called Hermes, it was found on 28th October 1937, two days before it was to pass within just 780,000 km of the earth, moving across the sky at an incredible 5° per hour. This was uncomfortably close to the earth as, although Hermes is probably only 1 km in diameter, the impact of such a body with the earth would cause a major catastrophe. It so happens that images of Hermes were accidentally recorded on plates taken at the Harvard Observatory, Sonneberg in Germany, and Johannesburg. But this fleeting visitor was seen for only five days in total, which was not enough to provide an accurate orbit. So it is unnumbered.

At the time of writing, over 800 of these Apollo asteroids have been found, which, by definition, have a perihelion within the 1.017 AU aphelion of the earth, and so cross the earth's orbit.

One of the most well-known Apollo asteroids is Icarus (1566), which was discovered by Walter Baade in 1949 as an object of the sixteenth magnitude. Herrick calculated its orbit, and found it to have a perihelion of about 0.19 AU, which is well within the orbit of Mercury, and an aphelion that lies outside of the orbit of Mars. Gilvarry pointed out a few years later that it has a relativistic advance of its perihelion of 11″/century (compared with 43″/century for Mercury), which gave further proof of Einstein's general theory of relativity.

Phaethon (3200), discovered in 1983 by J. Davis and S. Green, using the IRAS satellite, is about 5 km in diameter, and has a perihelion of 0.14 AU, which is even closer to the sun than that of Icarus. It was known in the nineteenth century that some meteor showers are associated with comets, but no comet had been discovered that has the same orbit as the Geminid meteors. It was, therefore, considered very significant when it was found that Phaethon's orbit is virtually the same as that of the Geminids. Whether Phaethon is the remains of an old comet is thought to be doubtful, however, but it does appear to be the source of the Geminid meteors.

Yrjo Vaisala had begun an asteroid survey programme at the University of Finland in 1935 which continued until 1957, whilst Frank Edmondson of Indiana University had started a similar survey programme in 1949. Gerard Kuiper had also undertaken such a programme in the early 1950s, but it was not until twenty years later that asteroid survey programmes were carried out specifically to find near earth asteroids. One of these was the Palomar Planet-Crossing Asteroid Survey carried out by Eugene Shoemaker and Eleanor Helin, starting in 1973, which used the 18 inch (46 cm) Palomar Schmidt.

One of Helin's first significant discoveries with the Palomar survey was asteroid 1976 AA, whose image appeared on a photograph taken on 7th January 1976. A further ten exposures were taken on the following night, which enabled James Williams of the Jet Propulsion Laboratory (JPL) to calculate a preliminary orbit. Additional observations were then made, not only at Palomar, but at other observatories in the USA and Japan. Then, using observations from 7th to 19th January, Brian Marsden, of the Harvard-Smithsonian Center for Astrophysics, calculated an accurate orbit that had a perihelion of 0.791 AU, an aphelion of 1.141 AU, and a period of just 346.8 days. Now called Aten (2062), this was the first near earth asteroid found that has a period of less than one year. At the time of writing a further 140 of these so-called Aten asteroids have been discovered.

Interestingly, Aten orbits the sun twenty times, whilst the earth orbits the sun nineteen times. However, this is by no means the only asteroid that has a resonance with planetary orbits. For example, the Apollo asteroid Toro (1685), that was discovered in 1948, orbits the sun five times, whilst the earth and Venus orbit the sun eight and thirteen times, respectively, and Icarus describes seventeen orbits for every nineteen of the earth. In addition, the Amor asteroid 1953 EA, now called Quetzalcoatl, was found in 1953 to have a four year orbit that puts it in resonance with Jupiter, making three orbits to Jupiter's one.

Finally, there is the question as to whether there are any asteroids with orbits well inside that of the earth, and, in particular, whether there are any with orbits inside that of Mercury. Intramercurial planets had been looked for in vain in the nineteenth century (see Section 7.1), but maybe there are many much smaller objects orbiting the sun inside the orbit of Mercury.

Attempts were made to photograph such intramercurial asteroids, or Vulcanoids as they are now known, during total solar eclipses in the first half of the twentieth century, but all failed. A few possible candidates were reported in the 1960s and 1970s, but subsequent investigations with the SOHO spacecraft have cast doubt on these observations. So at the time of writing there have been no unambiguous observations of such Vulcanoids.

## The Kuiper belt[69]

The Irish astronomer Kenneth Edgeworth pointed out in 1943 and again in 1949 that it seemed strange that the solar system seemed to abruptly stop at a

---

[69] This is sometimes called the Edgeworth–Kuiper belt.

large planet like Neptune at about 30 AU from the sun. There were the large planets of Jupiter, Saturn, Uranus and Neptune, and then nothing until the minuscule Pluto completed the solar system. Gerard Kuiper expressed similar views in 1951, both astronomers reasoning that there should be a large number of planetesimals in this apparent void, left over after the formation of the planets from the original solar nebula. The result should be a belt of planetesimals or asteroids, orbiting the sun from just outside the orbit of Neptune to well beyond the aphelion of Pluto (at 49 AU). Unfortunately, the technology of the time did not allow the detection of these distant objects.

Kuiper further proposed that Jupiter and Saturn's outer satellites had originated from near the inner edge of this belt, where they had had their orbits modified by the gravitational pull of the four giant planets. In addition, Kuiper and Fred Whipple suggested that this Kuiper belt, as it is now known, may also be the source of short-period comets.

## 9.14 **Comets**

### Chemical composition

In the 1860s, Huggins had shown that the spectra of gases in the heads of comets were the same as those produced by hydrocarbon compounds made luminous by electrical discharges and by meteorites heated in the laboratory. The exact hydrocarbon compounds involved were something of a mystery, however, until molecular spectra became better understood in the first half of the twentieth century. Molecular carbon, $C_2$, was identified in the head of a comet just after the turn of the century, followed by CN, or cyanogen, and CH, or methylidyne. In the case of Halley's comet, in particular, Nicholas Bobrovnikoff (1896–1988) identified bands of CN, $C_2$ and CH in the coma region during its 1909–1911 apparition. He also found bands at about 405 nm, but these were not correctly identified until 1951 when Alexander Douglas (1867–1962) showed that they were due to the $C_3$ molecule.

Henri Deslandres (1853–1948) and A. Bernard recorded the first clear spectrum of a comet's tail when they observed Daniel's comet (1907 IV) in August 1907, and found three weak bands between 400 and 458 nm. The same bands were observed in the tail region of Daniel's comet by John Evershed (1864–1956), from Kodaikanal Observatory in India, and similar bands were observed by Deslandres and Bernard in the tail of Morehouse's comet (1908 III), along with another band at 391 nm. These bands were identified two years later by Alfred Fowler (1868–1940). Those between 400 and

458 nm were due to singly ionised carbon monoxide, $CO^+$, and that at 391 nm due to the singly ionised nitrogen molecule, $N_2^+$. Bobrovnikoff also found the bands due to both of these molecules in the tail region of Halley's comet.

New information was forthcoming in 1941 when the Belgian astronomer Pol Swings (1906–1983) and his colleagues recorded the ultraviolet spectrum of the nucleus region of Cunningham's comet (1941 I), using a quartz optical system attached to the 82 inch (2.1 m) McDonald reflector. For the first time, ultraviolet bands were detected due to both the hydroxyl radical, OH, and NH. In addition, Swings and his team discovered the $NH_2$ band at 630 nm.

In the spring of 1948, Pol Swings and Thornton Page took high resolution spectrograms of Bester's comet (1948 I). They found ionised carbon dioxide, $CO_2^+$, in the tail region, together with a band at 405 nm in the head region, which they attributed to the $CH_2$ molecule. This latter deduction was based on Herzberg's similar attribution of a band at 405 nm in laboratory spectra. However, as mentioned above, Douglas found this in 1951 to be due to the $C_3$ molecule.

So, by the mid 1950s the molecules $C_2$, $C_3$, CH, CN, OH, NH and $NH_2$ had been found in the heads of comets, and the ionised molecules $CO^+$, $CO_2^+$ and $N_2^+$ had been found in their tails.

In the nineteenth century, sodium and iron spectral lines had been observed to replace the hydrocarbon lines when comets pass very close to the sun. The sungrazing comet Ikeya–Seki (1965 VIII) showed an even more marked effect in 1965 when it passed within two solar radii of the sun on 21st October of that year. Remarkably, the lines of neutral iron, nickel, potassium, chromium, calcium, cobalt, copper, vanadium and manganese were detected together with those of ionised calcium, $Ca^+$.


Density

Faint stars had been seen through the heads of comets in the nineteenth century, so their density was clearly very low, but how low? Karl Schwarzschild and E. Kron examined Halley's comet during its 1909–1911 apparition, and concluded that only 150 g/s of material was being emitted from the solid nucleus into the tail, giving a density in the tail of only one molecule per cubic centimetre. So when molecules are evaporated from the nucleus and stream away in the tail, they would generally not be subjected to collisions with other molecules.

This very low density of comets was demonstrated in 1957 when Pol Swings and Jesse Greenstein observed the forbidden lines of neutral oxygen in the spectrum of Mrkos' comet (1957 V). Under normal circumstances, the time required for oxygen molecules to spontaneously de-excite, and produce the so-called forbidden emission lines, is much longer than the time between molecular collisions, and so these spectral lines cannot be produced as the molecules lose their energy by collision first. Contrariwise, the presence of these lines clearly indicated that the densities of gases in comets must be very low, substantially reducing the probability of molecular collisions.

In 1957 William Liller observed that sunlight scattered by particles in Arend-Roland's (1957 III) and Mrkos' comets is redder than incident sunlight. This he attributed to the scattering of light by particles in the range 2.5 to 5 μm in diameter. This was the first time that a tolerably accurate estimate of the sizes of such particles had been made.

Tail structure

Bredikhin had suggested in 1877 that the different shapes of cometary tails could be explained by electrical repulsion between the sun and different chemical elements emitted by the cometary nucleus (see Section 7.3); the lighter the elements, the straighter the tail. This was indeed a plausible explanation, but at about the same time a new effect became available to explain cometary tails, namely radiation pressure.

In 1873 the brilliant physicist James Clerk Maxwell (1831–1879) published his theory of electricity and magnetism, which required light to exert what we now call radiation pressure. Unfortunately, ten years later, George Fitzgerald showed theoretically that such pressure acting on a hydrogen molecule would be only just enough to overcome the sun's gravitational attraction at the distance of the earth's orbit from the sun. In 1900, however, the Swedish physicist Svante Arrhenius (1859–1927) suggested that comets' tails could be caused by the effect of radiation pressure on cometary dust, rather than on cometary molecules. Shortly afterwards, Karl Schwarzschild (1873–1916) followed up this suggestion, and showed that the repulsive force on particles of the order of 0.1 μm could be as much as twenty times as strong as the sun's gravitational attraction. Unfortunately, it was known that in many comets the repulsive force is much greater than this.

It was well known that the sun has a vast corona which can be seen during a total solar eclipse. In 1893 John Schaeberle (1853–1924) suggested that it was particles emitted by this corona that were causing cometary tails. When these coronal particles impinged on the material coming from the cometary nucleus, they slowed down and became denser, whilst accelerating the cometary material. The coronal particles, together with the cometary material, formed the tail.

In spite of these promising ideas, there was no fully satisfactory theory in the first half of the twentieth century for the production of comets' tails. Most astronomers favoured the radiation pressure idea, even though Fitzgerald had shown it insufficient for molecules and Schwarzschild's analysis had shown it insufficient for dust, given the observed velocities of tail material. Then in 1951, Ludwig Biermann, of the Max Planck Institute in Germany confirmed that radiation pressure could not exert enough force to create the observed tail velocities. On the other hand, he showed that solar ions and electrons could do so, if their velocities were between 500 and 1,000 km/s, and their density was between 100 and $1,000/cm^3$ at the distance of the Earth.

At this stage, spacecraft were little more than a glint in the enthusiasts' eyes. But it was to require in-situ measurements in space before the mechanisms behind the production of comets' tails could be properly understood (see Section 11.7), and these various proposals put into perspective.

## Composition of the nucleus

In the mid 1930s, Karl Wurm had noted that many of the molecules observed in comets, such as $C_2$ and CH, were very active chemically, and so could not have been present in any particular comet for very long. As a result, he concluded that these active, so-called daughter molecules must have come from more stable parent molecules in the nucleus, such as $(CN)_2$, $H_2O$, and $CH_4$ (methane), which had been dissociated into the daughter molecules under the action of sunlight. Based on this idea, Pol Swings proposed in 1948 that the parent molecules were water, ammonia ($NH_3$), methane, molecular nitrogen ($N_2$), carbon monoxide and carbon dioxide, all of which were in the form of ice, before being heated by the sun.

The linkage between comets and meteor showers, that had been found in the nineteenth century (see Section 7.4), had led some astronomers to believe that the nucleus of a comet was nothing more than a swarm of small, solid particles, orbiting the sun in close order, in the same orbit. As the twentieth century progressed, however, this model fell out of favour, due mainly to the problem of explaining the large ratios of gas to meteoric material in large comets. Then in 1950 and 1951 Fred Whipple proposed his dirty snowball theory, in which the nucleus is a single body, composed of various ices, with meteoric material embedded within it. As the nucleus approaches the sun, the surface ices turn to vapour, releasing their entrapped meteoric material which, together with the vapour, streams away from the sun.

Whipple envisaged the meteoric material, embedded in the cometary ices, to consist of iron, calcium, magnesium, manganese, nickel, silicon, sodium and aluminium, in the form of free atoms or particles from a few microns up to at least a few centimetres in size. The free atoms had been responsible for the metallic emission lines detected in the nineteenth century when comets passed very close to the sun.

Whipple, like Swings, assumed that the nucleus consisted of ices, such as those of methane, which are necessary to explain the existence of some daughter molecules. Unfortunately, some of these parent molecules are also highly volatile, and it was difficult to understand how they could have survived in the comets for more than a few perihelion passes. Then in 1952 Armand Delsemme and Swings suggested that the highly volatile methane and other elements could be embedded within the crystalline structure of water ice as so-called clathrate hydrates.

## Size of nuclei

There had been considerable confusion, in the early twentieth century, about the diameter of a comet's nucleus. For example in 1926, Russell, Dugan and Stewart, in their book *Astronomy*,[70] referred to values of 1,500 km for Donati's comet of 1858, 2,900 km for of the Great Comet of 1882, and 800 km for Halley's comet in 1910. In 1951, Sir Harold Spencer Jones, in his *General Astronomy*,[71] was still talking about sizes of a few hundred kilometres, or several thousand kilometres in exceptional cases. But by 1963, Wurm in *The Solar System*,[72] mentioned that the nuclei of comets are probably only about a few kilometres in diameter.

The reason for the confusion was basically because a number of astronomers had thought that they had resolved the cometary nuclei with their telescopes, whereas they had not. In fact, it was impossible to resolve the nucleus of any comet in the ground-based telescopes of the period. It was only when estimates were made, starting in the late 1920s, that were based

---

[70] *Astronomy*, by Russell, Dugan and Stewart, Ginn, Vol. 1, p. 428.

[71] *General Astronomy*, by Sir Harold Spencer Jones, Arnold, third edn., 1951, p. 270.

[72] *The Solar System*, by Gerard Kuiper (gen. edn.), University of Chicago Press, Vol. 4, 1963, p. 574.

on the apparent intensity of the nucleus, that satisfactory dimensions, of the order of a few kilometres, could be established.

## Origin

The twentieth century opened with the question of the origin of non-periodic comets still unresolved. Carrington had provided evidence in the 1860s that they were part of the solar system (see Section 7.3), but other astronomers were not convinced. If comets had hyperbolic orbits before they entered the gravitational fields of the planets, they must have come from interstellar space, but were these original orbits hyperbolic?

There were two major problems that beset astronomers trying to answer this question. First, comets are only observed when they are relatively near to the sun, and so they are seen for a relatively small amount of one orbit. As a result, calculating an accurate orbit can prove to be very difficult, especially if it is highly eccentric and the comet has been seen only once. The method adopted was to try to fit the best parabolic orbit to the observations, and only declare that the orbit was elliptical or hyperbolic if the parabolic orbit did not match the observations, within error.

The second problem was trying to decide what the orbit of a comet was before it had entered the gravitational fields of the planets, from observations made after it had crossed most of the planetary orbits, and was close to the sun. These cometary orbits determined when the comets are near the sun are known as osculating orbits. That is, they are the orbits that the comet would take in leaving the vicinity of the sun if all the planets were to instantaneously disappear.

A survey of the osculating cometary orbits computed up to 1910 showed that about 100 were clearly elliptical, and about 20 appeared to be hyperbolic, but the majority, about 300, had orbits that did not significantly deviate from a parabola. As all these orbits had been calculated when the comet was near the sun, some of the hyperbolic orbits may originally have been elliptical or parabolic before the comet had been perturbed by the planets. On the other hand, some of the elliptical orbits may originally have been hyperbolic.

In 1914 Strömgren analysed those cometary orbits that appeared to be hyperbolic around perihelion, and concluded, in fact, that only eight were clearly not parabolic. Then, in collaboration with Fayet, he showed that, in each of the eight cases, the comet's velocity had been increased, as it approached the sun, by the gravitational attraction of Jupiter and Saturn. This had caused their original elliptical orbits to become hyperbolic. Then in 1927 G. Van Biesbroeck confirmed this for Delavan's comet, which followed a hyperbolic orbit near the sun, showing that it had originally moved in an elliptical orbit with a semimajor axis of 2.7 light years. So, at that time, it was not clear if any comet had originally come from interstellar space.

In 1932 Ernst Öpik (1893–1985) concluded, from an analysis of the effect of stellar perturbations on comets, that comets could remain bound to the sun at distances of up to about $10^6$ AU, or about 16 light years, which is about four times the distance to the nearest star. He found that stellar perturbations would tend to increase the perihelion distance of long period comets with time, causing them to eventually form a cloud or shell surrounding the sun at large distances. He also concluded that a minority of long period comets would have their perihelion distances reduced as a result of perturbations.

In 1948 the Dutch astronomer Adrianus Van Woerkom re-examined possible scenarios for the capture of comets by the solar system, and concluded that the velocity of the sun relative to that of interstellar comets, if the latter existed, would be far too large to allow capture. This was assuming that interstellar comets have similar velocities relative to the sun as do the stars in the solar neighbourhood. His conclusion indicated that long period comets could not originate in interstellar space.

Van Woerkom then showed that Hubert Newton's modelling of the effect of Jupiter on comets in near parabolic orbits was incorrect, as the principal effect of Jupiter was not direct capture of near parabolic comets, as Newton had shown (see Section 7.3), but small changes in the reciprocal semimajor axes, $a^{-1}$, of about 0.001 AU$^{-1}$ per encounter. These changes were cumulative for each comet after a number of Jupiter encounters, but they could go either way, increasing or decreasing $a^{-1}$ in steps of the order of 0.001 AU$^{-1}$. As a result, after about one million years, all long period comets would either have become short period comets in the inner solar system, or have been ejected into interstellar space.

In addition, Van Woerkom showed that there should be about an equal number of comets for an equal interval of $a^{-1}$. So if, for example, there were ten comets with $a^{-1}$ in the range 0.000–0.002 AU$^{-1}$ (having periods from $\infty$ to 11,200 years), there should be ten comets with $a^{-1}$ in the range 0.002–0.004 AU$^{-1}$ (11,200–4,000 years), and so on. Unfortunately, of the comets observed between 1850 and 1936, 177 were in the first interval (i.e. with $a^{-1}$ in the range 0.000–0.002 AU$^{-1}$), 10 in the second interval, down to just 1 in the seventh interval. This implied that there was a continuous source of new near-parabolic comets. Van Woerkom then went on to point out that a cloud of comets, moving with the sun through interstellar space, could be such a source. This would also overcome the problem of relative velocities mentioned above.

Two years later Jan Oort took Van Woerkom's analysis further. To start with, Oort only included the orbital statistics for nineteen comets with well-determined orbits. These orbits had all been calculated for the period before the comets were perturbed by the solar system's planets. In Oort's analysis he had ten comets in the $a^{-1}$ interval from 0 to $50 \times 10^{-6}$ AU$^{-1}$, four in the next interval, and so on. He further found that the average value of $a^{-1}$ for the ten comets in the first interval was $18 \times 10^{-6}$ AU$^{-1}$, equivalent to a semimajor axis of about 55,000 AU, or an aphelion distance of about 110,000 AU. Oort then took the definitive step of proposing that these, and all long period comets, had originated in a cloud of comets, now called the Oort cloud, about 50,000 to 150,000 AU from the sun.

Clearly the solar system planets could not have caused these comets in the Oort cloud to de-orbit and approach the sun in highly eccentric orbits. But Oort reasoned that, at the distances of 50,000 to 150,000 AU envisaged, the comets would be subject to perturbations by passing stars. As the nearest stars surround the sun in three dimensions, the comets perturbed by these stars would enter the solar system with a wide variety of orbital inclinations, as observed. Oort then calculated that about $2 \times 10^{11}$ comets would be required in the cloud to produce the observed number of comets with highly eccentric orbits. This number sounds too great but, in astronomical terms, comets weigh very little, and the whole cloud could well have a mass of less than that of the earth.[73]

Oort's theory satisfactorily explained the observed distribution of comets with periods down to about the 12 year period of Jupiter. But there were far too many short period comets to be explained by Oort's theory or any other theory of the time.

The question now arose as to where the Oort cloud came from in the first place. Oort suggested that it may have originated, together with the asteroids and meteorites, from an explosion[74] of a planet that was between the orbits of Mars and Jupiter. Those fragments that had had almost circular orbits, became members of the observable solar system, losing their gaseous constituents because of their continuous exposure to solar radiation, and

---

[73] The mass of the earth is about $6 \times 10^{27}$ g. A cometary nucleus of $3 \times 3 \times 1$ km, with a density of 1 g/cm$^3$, would have a mass of $9 \times 10^{15}$ g. So $2 \times 10^{11}$ of such comets would have a mass of $1.8 \times 10^{27}$ g, or less than that of the earth.

[74] This is a development of an idea originally proposed by Olbers in the early nineteenth century, to explain the origin of the asteroids (see Section 6.4).

becoming asteroids and meteorites. Those fragments with elliptical orbits, on the other hand, had their orbits perturbed by Jupiter and the other major planets. As a result a number of these fragments were given hyperbolic orbits, and were thus lost to the solar system, but a significant percentage were given orbits with aphelia of about 50,000 to 150,000 AU. Stellar perturbations then distorted these orbits near aphelia, making them more circular, thus producing the Oort cloud.

Kuiper proposed an alternative theory in 1951, in which comets result from the condensation of the original solar nebula outside the orbit of Neptune. These condensations have now lost their original, highly volatile material, but they are otherwise hardly changed since the origin of the solar system. Kuiper then suggested that the major planets had caused the orbits of these condensations to become highly elliptical, causing them to be injected into the Oort cloud, where their orbits had been made more circular by neighbouring stars.

# Chapter 10 | THE SPACE AGE – TERRESTRIAL PLANETS

## 10.1 **The earth**

### The ionosphere

It had been suggested in the 1930s that solar X-rays, particularly those emitted during solar flares, may be responsible for creating and maintaining the E and F layers of the ionosphere (see Section 9.6). At that time, however, the idea that the sun may be emitting X-rays was purely speculative, and most astronomers thought that the ionospheric levels were being created and maintained by ultraviolet solar radiation, possibly at the Lyman-α wavelength of 121.6 nm.

The first observational indication that the sun may be emitting X-rays was obtained by Robert Burnight, of the Naval Research Laboratory (NRL), on a V2 rocket flight of 5th August 1948. Unfortunately, a repeat experiment later that year failed to indicate any X-rays. At about the same time, Tousey, Watanabe and Purcell of NRL also found indications of solar X-ray emission with a V2 experiment, but Herbert Friedman (1916–2000), also of NRL, was the first to prove that X-rays detected in the earth's upper atmosphere came from the sun.

Friedman's first successful experiment, to prove the solar origin of X-rays, was flown on a V2 on 29th September 1949. The soft X-ray detectors started to respond at an altitude of about 85 km, and showed a gradual increase in X-ray intensity up to the peak altitude of about 150 km as the rocket traversed the E region of the earth's ionosphere. Moreover, the response of these detectors was higher when they pointed towards the sun, proving the solar origin of these X-rays. However, this initial, relatively crude experiment indicated that the intensity of the X-rays may not be high enough to sustain the E region. Eventually, after a number of rocket failures, Friedman was able to show that the X-ray intensity *was* sufficient to sustain the E region, using a Viking sounding rocket experiment on 15th December 1952.

Tousey, Watanabe and Purcell's V2 experiments, undertaken over the period 1948–1950, demonstrated that the Lyman-α emission line of hydrogen dominated the far ultraviolet solar spectrum. Herbert Friedman also made Lyman-α measurements on V2 and Viking flights in September 1949 and December 1952, respectively. In the event, both Tousey's and Friedman's experiments showed that the intensity of the Lyman-α solar radiation was enough to maintain the earth's ionospheric D region.

Over the next few years, Friedman's group continued to study solar X-rays, finding that the majority of the radiation in the range 0.8 to 2.0 nm was due to the solar corona. Then in 1956 they found that solar flares are the source of very high energy X-rays which disrupt the ionosphere and cause problems with radio communications.

## Van Allen radiation belts

At 22 h 48 min on 31st January 1958 Eastern Standard Time, the United States entered the space age with their first successful launch of an earth-orbiting spacecraft, the 14 kg Explorer 1. On board was a Geiger counter provided by James Van Allen of the University of Iowa.

When James Van Allen examined the data from his Geiger counter, he found that it showed an increase in radiation levels with increase in altitude, as expected, as the effect of the earth's atmosphere reduced. After saturating at an altitude of about 800 km, however, the intensity unexpectedly decreased to virtually zero at higher altitudes. It was difficult to analyse the data, because Explorer 1 did not have a tape recorder, and so the data could only be obtained when the spacecraft was within visibility of a ground station. Because of this, the altitude dependency could not be clearly separated from any latitude/longitude effect. The launch of Explorer 2 was a failure on 5th March 1958, but Van Allen was still trying to make sense of his Explorer 1 measurements when Explorer 3 was launched three weeks later.

The payload of Explorer 3 was almost identical to that of Explorer 1, except for the addition of a small tape recorder. This tape recorder was a real bonus, as it meant that measurements made when the spacecraft was out of visibility of the ground station could be stored and relayed back to earth when the spacecraft returned to within the ground station's visibility. Putting his Explorer 1 and 3 results together, Van Allen concluded that there

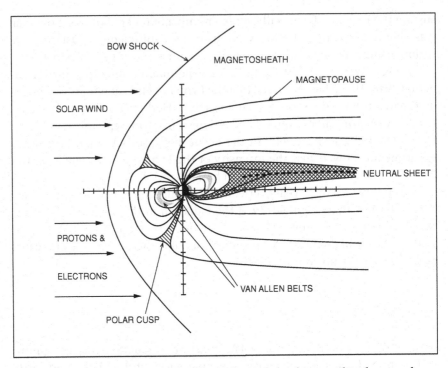

**Figure 10.1** The inner and outer Van Allen radiation belts shown within the general configuration of the earth's magnetosphere. The exact size and configuration of this system, which varies with solar activity, was gradually determined during the first ten years of the space age. The inner Van Allen belt is relatively stable and composed of high energy protons, whereas the outer belt is variable and composed of electrons and low energy protons. (The 'tick' marks on both axes are at intervals of two $R_E$.)

was a belt of elementary particles, probably in the form of low energy electrons, around the earth. This announcement, made on 1st May 1958, was a shock to everyone. Further results with Explorer 4, launched on 26th July, confirmed the existence of this so-called Van Allen belt (the inner belt in Figure 10.1).

The second successful Russian spacecraft, Sputnik 2, had carried a pair of Geiger–Muller tubes three months before Explorer 1 and they could have shown evidence of the inner Van Allen belt, had the Soviet scientists not had great difficulty in analysing their results. Similarly, Sputnik 3 would also have been able to map the inner Van Allen belt in May 1958 had its tape recorder not failed. It did, nevertheless, show that the inner belt was composed mostly of protons, rather than the low energy electrons suggested by Van Allen.

On 6th December 1958, the American Air Force launched Pioneer 3 in an attempt to reach the moon. This, their fourth lunar attempt, was no more successful than any of the previous three, reaching an altitude of only 107,000 km, or a little over a quarter of the way to the moon, before plummeting back to earth. Nevertheless, it enabled Van Allen to discover the second radiation belt around the earth (see Figure 10.1). In the following year, Pioneer 4, which missed the moon and went into orbit around the sun, provided additional information about the extent of the two Van Allen belts. In particular, it appeared as though the position of the outer edge of the outer belt varied over time.

### The magnetosphere

In 1941, the Swedish physicist Bengt Edlén (1906–1993) had found that the temperature of the solar corona was at least 2 million K. The corona could be seen during a total solar eclipse extending from the sun several million kilometres into space. However it was now recognised that, with such a high temperature, its influence must extend much further than that. In fact, in 1957 Sydney Chapman (1888–1970) calculated that such a corona, consisting mainly of electrons and protons, would extend beyond the earth's orbit, where its temperature would be about 200,000 K.[1]

A few years before Chapman published his theory, Ludwig Biermann (1907–1986) showed that ions and electrons emitted from the sun at velocities of the order of 500–1,000 km/s could produce comets' tails (see Section 9.14). Then in 1957 Eugene Parker of the Enrico Fermi Institute developed a new theory, based on those of Chapman and Biermann, of the so-called solar wind. In this, Parker envisaged charged particles emitted by the sun drawing the magnetic field lines in the corona out into the solar system, causing the lines to spiral out from the sun as the sun spins on its axis.[2]

---

[1] It may be wondered why, if the earth is surrounded by gas at a temperature of 200,000 K, the earth itself is not at such a high temperature. This is due to the very low density of the surrounding gas, which is no denser, in the region of the earth, than a good laboratory vacuum, so its energy density is rather low.

[2] The behaviour of the solar wind is fundamentally driven by the complex behaviour of the sun and, as such, is more appropriately covered in a book on the history of solar research rather than here. A good summary of the early work on it is given in *Exploring the Sun*, by Hufbauer, Johns Hopkins University Press, 1991.

The Russian Luna (or Lunik) 2 was, in September 1959, the first space-craft to measure the plasma, or gas of charged particles, away from the influ-ence of the earth's magnetic field, when it was *en route* to the moon. Unfor-tunately, although the ion trap on Luna 2 could measure the flux density of the charged particles, it could not measure their speed and direction. That was achieved by Explorer 10, Mariner 2 and IMP 1 (otherwise known as Explorer 18) in the early 1960s, which provided evolving evidence that Parker's theory of the solar wind was basically correct.

Explorer 10, which was launched on 25th March 1961, surveyed space on the night side of the earth, at an angle of about 130° to the earth–sun line. With an apogee at about $47R_E$, it was expected that this spacecraft would spend some time observing the solar wind well beyond the earth's magnetic influence. In the event, Explorer 10 appeared to cross the bound-ary between the relatively quiet region controlled by the earth's magnetic field, and the turbulent region outside of that, at least six times in its as-cent between $22R_E$ and $47R_E$. This appeared to be because the position of this boundary, now called the magnetopause (see Figure 10.1), varied with time.

As envisaged in 1961, therefore, there is a region around the earth, bor-dered by the magnetopause, that is dominated by the earth's magnetic field. Beyond that, there is a turbulent region containing plasma from the solar wind, which causes the earth's magnetic field there to vary also. One possi-bility was that this turbulent region could gradually become less turbulent, the further it is away from the earth, until the earth's influence is negligible and the undisturbed solar wind is observed. On the other hand, there may be a clear boundary, beyond the magnetopause, that separates the turbulent region from the undisturbed solar wind. This latter hypothetical boundary was called the bow shock.

The second hypothesis was proved to be correct when the bow shock was detected by magnetometers on board IMP 1 in 1963. This spacecraft was launched on 27th November into a four day orbit, with an apogee of $31R_E$. Initially, IMP 1 spent about 75% of each orbit, which initially made an angle of about 30° to the earth–sun line, beyond the bow shock at about $15R_E$. But as time passed, it spent less and less of each orbit beyond the bow shock, because of the changing orientation of the earth–sun line, as the earth orbited the sun, so that by February 1964, IMP 1 no longer crossed the bow shock. Inside the magnetopause IMP 1 found a well-ordered magnetic field of about 30 gauss intensity. Between the magnetopause and bow shock the field was found to be variable between about 5 and 20 gammas, and outside the bow shock it was generally steady at about 4 gamma.

So, in the period of a few short years, not only had the Van Allen belts been discovered, but the solar wind had been observed, and the basic configuration of the earth's magnetosphere determined.

## Plate tectonics

There was a resurgence of interest in the concept of continental drift in the 1950s, when new data on the earth's geomagnetic past was provided by the work of Stanley Runcorn (1922–1995), P.M.S. Blackett (1897–1974), and others. It had been known for many years that the direction of the magnetic field in magnetite and other ferromagnetic minerals is the same as that of the earth's magnetic field at the time that they crystallised. Such ferromagnetic minerals are found in igneous rocks, so measuring the direction of their magnetic field can provide the direction of the earth's magnetic field at the time of their formation. Sometimes, however, these igneous rocks have been broken down by weathering, and particles of magnetised materials have been released, later to form parts of sedimentary deposits. At that time the particles tend to realign themselves with the new direction of the earth's magnetic field. So a study of the so-called remanent magnetic field of ferromagnetic minerals, whether in igneous or sedimentary rocks, can enable the direction of the earth's magnetic field to be determined for previous eras.

In this way, it was found in the 1950s by Stanley Runcorn and colleagues that, over the last 600 hundred million years, the earth's north magnetic pole has moved, as seen from Europe, from near the present geographical coordinates of Hawaii to Japan, and then via Siberia, to its present position in Northern Canada. This could be because either the earth's magnetic pole has moved, or Europe has moved, or both. It was then found that the remanent magnetism of rocks in North America implied a route for the magnetic pole displaced mainly in longitude from the one based on rocks from Western Europe. The two routes could be made to coincide, however, if it was assumed that North America and Europe had been joined together until about 100 million years ago, when they started to drift apart, as proposed by Wegener. Similar paleomagnetic data from other continents indicated that they had also, as Wegener had proposed, once been joined together in one super continent. However, because of the difficulty of decoding the magnetic signatures from rocks, and a natural scepticism, most geologists of the time remained unconvinced as to the validity of these results.

Meanwhile evidence of a different kind for continental drift was being collected through a gradual understanding of the structure of the ocean floor.

The first deep-sea soundings had been made by Sir James Ross of the British Royal Navy in 1840, and by 1855 the presence of shoal water in part of the North Atlantic had suggested the presence of an ocean ridge. But deep-sea sounding was a laborious business in the nineteenth century, as it consisted of lowering a heavy lead weight on the end of a length of rope until it touched the bottom. As a result, only a few such soundings were made, and the ocean floor was thought to be generally flat and featureless, until echo sounding equipment was first used in 1920. Since then, a great central ridge has been found that runs the length of the Atlantic Ocean, and a mid-ocean ridge has also been found in the Indian Ocean. The Pacific Ocean was found to be by no means featureless, but the ocean ridge system there is more complex. The Pacific was also found to have a number of very deep trenches around its periphery, whilst more limited trenches were found in the Atlantic and Indian Oceans.

In the 1950s Marie Tharp, an American oceanographer, undertook a detailed examination of the mid-Atlantic ridge. As a result, she found evidence of a rift valley down its centre. In addition, she found that most of the earthquakes in the mid-ocean region had originated from just below the floor of this rift valley. Then, during the International Geophysical Year (IGY) in 1957–1958, work was undertaken on a world-wide scale, which uncovered a global system of oceanic ridges coinciding with major earthquake zones (see Figures 10.2 and 10.3).

There had been a number of problems with Wegener's theory of continental drift, probably the most serious of which was the lack of a driving force sufficiently powerful to move the continents. In 1929 the English geologist Arthur Holmes had suggested that convective currents in the mantle could rise up to the crust and carry the continents along with them. Then, as new evidence for continental drift was accumulated in the 1950s, the American geologist Harry Hess (1906–1969) used Holmes' concept as the basis for his own theory. This theory, which he wrote up in 1960, was not published until 1962, however, during which time additional supporting evidence of continental drift had been forthcoming.

Hess suggested, in his so-called seafloor spreading model, that the mid-ocean ridges are caused by mantle material breaking through the crust. This ex-mantle material then cools, subsides, and moves away from the ridges, finally being carried back to the mantle in the ocean trenches. This explained why earthquakes originate on the mid-ocean ridges, and why volcanoes also exist on those ridges, causing them to rise above the ocean surface in Iceland and the Azores, for example. It also explained why only relatively young rocks had been found on the ocean floor, compared with

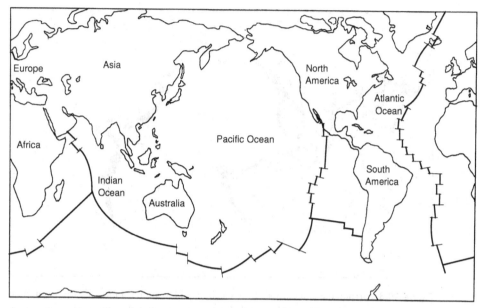

**Figure 10.2** Solid lines show the oceanic ridges in the Indian, Pacific and Atlantic Oceans. The main lines of the ridges are distorted by so-called transform faults, which create the broken structure shown.

on the continents. Hess' theory obviously predicted that the age of the ocean floor should increase with distance from the mid-ocean ridge, and this was confirmed almost immediately. But to understand the evidence, we need to go back to the early twentieth century, and the earth's magnetic field.

The French physicist Bernard Brunhes (1867–1910) had reported, as long ago as 1909, that he had found evidence in ancient lava flows that the earth's magnetic field had reversed direction in the past. Twenty years later, Motonori Matuyama (1884–1958) had made a similar discovery in Japan, but like that of Brunhes, it was ignored as it was not thought credible. After the second world war, however, a number of observers, such as Johnson, Graham, Kawai and so on, independently discovered similar behaviour in igneous and sedimentary rocks. But the scientific community were unsure as to whether this was due to a true reversal of the earth's field or to some other effect. For example, it was suggested that it could be due to a spontaneous tendency for the rock to become magnetised in a reverse direction to the earth's field. Then in 1963, Allan Cox, Richard Doell and Brent Dalrymple found clear, consistent evidence of at least three field reversals in the last four

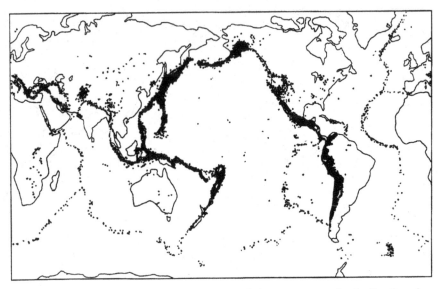

**Figure 10.3** Earthquake epicentres, shown as black dots, are not randomly distributed on the earth's surface, but are mainly found along clearly defined fault lines. Many are found along oceanic ridges (compare Figure 10.2) and subduction zones or oceanic trenches (see Figure 10.4 later).

million years, using data from rocks in North America, Europe, Africa and Hawaii, indicating that they are really due to reversals of the earth's magnetic field.

In 1961, the American geophysicists Arthur Raff and Ronald Mason measured a strange pattern of alternating stripes of material of high and low magnetic intensity in the eastern Pacific, off the coast of North America, running approximately north–south. A similar pattern was found the next year in the Indian Ocean. Then in 1966 two British geophysicists, Frederick Vine and Drummond Matthews, concluded, having analysed data from a number of oceans, that stripes of reversed magnetisation occur symmetrically on both sides of ocean ridges. This clearly showed that there was a lateral movement of the extruded mantle material from the ocean ridges in a magnetic field that reversed direction from time to time, giving powerful support to Hess' theory of seafloor spreading.

Further confirmation of Hess' theory was then obtained by the drilling ship *Glomar Challenger*, which took samples of the ocean floor in the Atlantic. The ages of these samples, determined from an analysis of their fossils, increased uniformly with distance from the mid-ocean ridge, and agreed with the ages derived from magnetic variations. The spreading rate

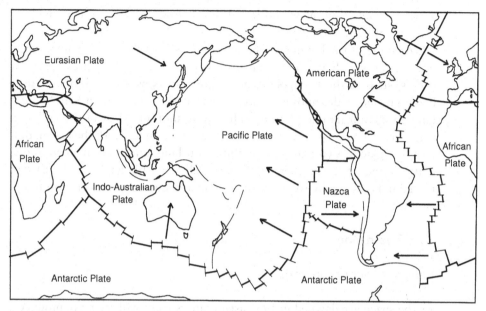

**Figure 10.4** The six major tectonic plates of Eurasia, Africa, Indo-Australia, Antarctica, Pacific, and America are shown, separated by ridges (solid lines) and subduction zones (light dashed lines). The largest of the smaller plates, called the Nazca plate, is also shown off South America. The Pacific plate is seen to be almost ringed by subduction zones, which are not present in the Atlantic. The movement of the plates is shown by arrows.

was found to be about 2 cm/year for the Atlantic. Later work in the other oceans produced rates of from 1 to 5 cm/year.

There was a serious problem with Hess' theory, however, as although there was an extensive network of trenches on either side of the Pacific, particularly on the western side, to allow subduction of the moving seafloor, there was no similar network on either side of the Atlantic. So where was the subduction region for the Atlantic? The only solution appeared to be that the adjacent continents of the Americas on one side, and Africa and Europe on the other, moved apart with the spreading seafloor.

The details of this theory of plate tectonics, as it is now called, were progressively worked out over the next few years by McKenzie and Parker in the UK, Morgan, Oliver, Isacks and Sykes in the USA, Le Pichon in France, and Wilson in Canada. In essence, it is believed that the earth's crust and upper mantle, down to about 60 km, are composed of relatively strong rock material in the so-called lithosphere. This layer is split into six major plates (see Figure 10.4), and a number of smaller ones, that float on the asthenosphere, which extends from about 60 to 250 km below the surface. It is made of material close to its melting point, which can flow under stress.

At some of the boundaries between plates, like the mid-Atlantic ridge, new material is being forced up from below, forcing the plates apart. At others, such as the Kermadec trench, north of New Zealand, one plate is being pushed under another plate, whilst at the well-known San Andreas fault in California one plate is sliding past another. Most of the collision zones between two plates occur at the edge of, or underneath the ocean. But, in a few places, two plates are colliding on land, and then, instead of one plate sliding under another, they push each other higher, as in the case of the Himalayas, where the Indo-Australian and Eurasia plates are colliding.

## 10.2 The moon

### Early spacecraft missions

As the space age dawned in 1957 astronomers were still trying to produce a satisfactory theory of the origin of the moon (see Sections 9.5 and 9.6). To do so, it was essential to obtain data on the age and type of lunar surface material. There was also the question about whether the moon still has a molten interior and, if so, what is its size and what is it made of? Are the lunar craters the result of meteorite impacts or volcanic activity? Another question that was of key interest to both astronomers and the engineers designing landing probes, was the nature of that surface. Are the lunar mare filled with volcanic ash or dust a few metres deep, as Thomas Gold had suggested, or is the dust only a few centimetres thick? If Gold was correct, it would be virtually impossible to land a spacecraft in the mare areas of the moon. Even though the moon was only 400,000 km away, we knew surprisingly little about it.

The Soviet Luna 1 spacecraft, which started man's exploration of the moon in January 1959, showed that it has, at most, a very weak overall magnetic field, as no field could be detected by the spacecraft's magnetometer. Some astronomers took this weak field as evidence that the moon has a solid core, as then there could be no dynamo effect producing a magnetic field. However, other astronomers suggested that the moon's core could still be molten if it rotated very slowly.

Later in 1959, Luna 3 took the first photographs of the far side of the moon. Although the quality of the images was very poor by modern standards, they appeared to show that the far side is broadly similar to the visible side, although it appeared to have fewer mare regions and more craters than the latter.[3]

The American Ranger 7, which was the first spacecraft to successfully impact the moon, did so on 31st July 1964, just 15 km from its target point.[4] Its last image, taken at an altitude of just 500 m, showed craters as small as one metre in diameter. Importantly, the images indicated that the surface dust was not very deep, and could probably support the manned Apollo lander, although it was going to be difficult to find a landing site free of craters.

Luna 9, which was the first spacecraft to soft-land a capsule on the moon, was launched on 31st January 1966. Its capsule, which came to rest on the inner slope of a 25 km diameter crater in the Ocean of Storms (Oceanus Procellarum), contained just two 'instruments', a panoramic TV camera and a radiation detector. Three panoramic photographs were transmitted, between each of which the lander settled slightly on the lunar soil, fortuitously allowing stereoscopic imaging. Much to the Russians' annoyance, however, the images had been received by Jodrell Bank in the UK and distributed to the world's press before the Russians were prepared to release them. Although Bernard Lovell of Jodrell Bank protested his innocence, he did not repeat this early release with later Russian probes.

As it turned out, the Jodrell Bank photographs exaggerated the vertical scale by a factor of 2.5, as the Russians had not provided information for correctly scaled images to be produced. As a result these British photographs were somewhat misleading, and it was not until the official Russian photographs were released a few days later that satisfactory interpretations could be undertaken. The surface appeared to have a rough texture like pumice or slag of volcanic origin, and to be covered with rocks of various sizes. There seemed to be no dust, thus confirming the conclusions from the earlier Ranger images.

In April 1966, the Russians were also the first to successfully put a spacecraft, called Luna 10, into lunar orbit. One of the instruments on board Luna 10, a magnetometer, detected a weak magnetic field of about 0.05–0.1% that of the earth. As this value did not change with increasing distance from the moon, however, it was thought that the Luna 10 magnetometer was not

---

[3] Later images from other spacecraft have shown that maria are virtually absent from the far side, covering just 2.5% of that surface, compared with about 16% for the moon as a whole.

[4] Although earlier spacecraft had impacted the moon, Ranger 7 was the first to do so with its instruments, in this case its cameras, working correctly.

measuring the true magnetic field of the moon, but rather that of the interplanetary magnetic field in the region of the moon.[5] Radio occultation measurements, made as the spacecraft just disappeared and just reappeared from behind the moon, also showed that the moon has no detectable atmosphere. In addition, Luna 10 showed that the moon is pear-shaped, with an elongation facing away from the earth. Shortly afterwards, the American Lunar Orbiter 1 showed that there are mass concentrations or 'mascons' under the moon's maria.

In the same year Surveyor 1, the first American spacecraft to soft-land on the moon, found that the moon's surface in the Ocean of Storms, where it landed, had a slight 'give' like uncompacted terrestrial soil. Surveyor 3,[6] which also landed in the Ocean of Storms, had a soil scoop which showed that the surface material was soft and clumpy. Both Surveyors 5 and 6 included an alpha-particle scattering experiment which found that the chemical composition of the lunar soil in their landing areas of the Sea of Tranquillity (Mare Tranquillitatis) and the Central Bay (of the Sinus Medii) was similar to that of terrestrial basalt, which is produced by volcanic activity.

Surveyor 7, the last Surveyor, was launched on 7th January 1968. It landed on the ejecta thrown over the lunar highlands when the meteorite impacted the moon to form the crater Tycho. This spacecraft had both a soil scoop and alpha scattering experiment. The scoop dug a number of trenches, and weighed a rock by measuring the current needed to operate its scoop arm. This enabled the density of the rock to be estimated at about 2.8 $g/cm^3$, compared with an average density of the moon of 3.3 $g/cm^3$. The alpha scattering experiment found that the iron content of the soil was lower than in the maria areas analysed by Surveyors 5 and 6.

So what had been achieved before the manned Apollo landings?

Prior to the launch of space probes it had been unclear as to whether the solid lunar surface was still relatively thin, with virtually all the lunar features (craters, mountains and maria) being caused by volcanic activity, or whether the moon was completely solid and had remained unchanged for most of its $4\frac{1}{2}$ billion year lifetime, except for the occasional impact of meteorites that had produced the maria and craters. In the latter cold moon theory (see Section 9.6), it was thought that the energy of the largest impacts would have been sufficient to melt the rocks locally and form large pools of molten rock which would then have cooled to form the flat maria.

---

[5] The value of the moon's general magnetic field has since been found to be about 0.01 % that of the earth.

[6] Surveyors 2 and 4 were failures.

Spacecraft had now shown that the moon has a very weak magnetic field, indicative of either a solid core or a liquid core with little internal motion. The moon's atmosphere seemed to be virtually non-existent, and there appeared to be relatively little dust on its surface. The maria, under which there are mass concentrations, appeared to have a basalt surface, whereas the highland material appeared to be different. Nevertheless it was still unclear, prior to Apollo, how many of the moon's surface features were due to volcanism and how many were due to meteorite impacts.

## Scientific results from Apollo

When Neil Armstrong and Buzz Aldrin landed on the moon on 20th July 1969 in the Sea of Tranquillity (Mare Tranquillitatis, see Figure 10.5), they collected 22 kg of moon rocks, which they brought back to earth. They also left a seismometer (see Figure 10.6) and laser reflectors behind on the moon; the seismometer to measure moonquakes and the reflectors to measure the earth–moon distance with high precision. The lunar material that they brought back was found to be generally basalt, confirming previous indications that the moon's maria are of volcanic origin.[7]

So the moon had once been geologically active, with a molten core. The Apollo 11 basalt samples also implied an age for the Tranquillity lavas of 3.65 billion years,[8] or about 1 billion years after the moon was thought to have been formed. Impact rocks called breccias were also found, many of which pre-dated the Tranquillity lavas. Interestingly, there was no trace of water in any of the rocks, even in the interstices, nor was there any organic material. So the moon appeared, from these limited samples at least,[9] to have had no opportunity to support even the most basic forms of life.

The basalt samples returned from Apollo 12 that landed in the Ocean of Storms (Oceanus Procellarum) were generally similar to those of Apollo 11,

---

[7] Although chemical analysis undertaken by the Surveyor probes could determine what chemicals were in the lunar soil, it was only the microscopic analysis of samples returned to earth by the Apollo and Russian Luna missions that could confirm the true nature of the rocks.

[8] The isotope potassium 40 decays to argon 40 with a half-life of 1.28 billion years. So the measurement of the proportions of potassium 40 to argon 40 in basalt samples enabled their ages to be determined. Abundance ratios of uranium 238 (half-life 4.5 billion years) to lead 206 were also used to estimate the ages of lunar rocks.

[9] No traces of water or organic material were found in any of the samples returned by any subsequent Apollo (or Luna) missions either.

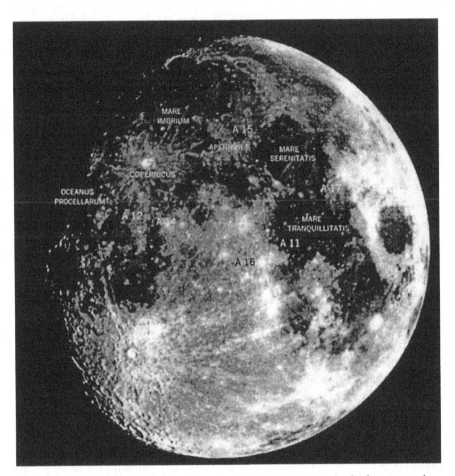

**Figure 10.5** Apollo landing sites on the moon (labelled A11 to A17). The first manned landing was that of Apollo 11 in the Mare Tranquillitatis.

but they were about 400 million years younger, showing that the maria did not all form at the same time.

The next successful mission, Apollo 14,[10] was the first manned mission to the lunar uplands. It landed at Fra Mauro, a region of hills and craters on the edge of the Mare Imbrium, with the astronauts being asked, on one of the two moon-walks, to climb to the rim of the 350 m diameter Cone crater. This was to find samples of rocks that had been thrown out of the crater when it

---

[10] Apollo 13 did not land, as an explosion on board early in its flight caused the mission to be aborted. Apollo 14 took over its mission.

**Figure 10.6** Buzz Aldrin is seen near to the seismometer that he and Neil Armstrong set up on the moon with Apollo 11. The surface in this mare area is seen to be remarkably flat. Their lunar module is seen in the background. (Courtesy NASA; AS11-40-5948.)

was formed. In fact, the rocks from Cone crater showed that it is only about 25 million years old. On the other hand, the breccias found by the astronauts in this general region indicated that the impact that had created the 1,140 km diameter Imbrium basin,[11] and the adjacent Fra Mauro hills, had occurred about 3.85 billion years ago. Rocks made of so-called KREEPy[12] lavas were also found of the type that had first been recognised in small quantities in the Apollo 12 samples. These lavas appeared to have been produced by partial melting of the primitive crust.

Six months after the launch of Apollo 14, Apollo 15 landed near the 1.5 km wide, 400 m deep Hadley Rille at the foot of the lunar Apennines (see Figure 10.7). Mission Control hoped that the astronauts would find samples here of the original lunar crust in the form of anorthosite, which would have risen to the surface of the moon whilst it was still molten. To assist the astronauts, a motorised vehicle called the Lunar Rover was provided for the first

---

[11] G.K. Gilbert had been the first person as long ago as 1893 to suggest that the Imbrium basin was an impact structure.

[12] So called as they are rich in potassium (K), rare earth elements (REE), and phosphorous (P).

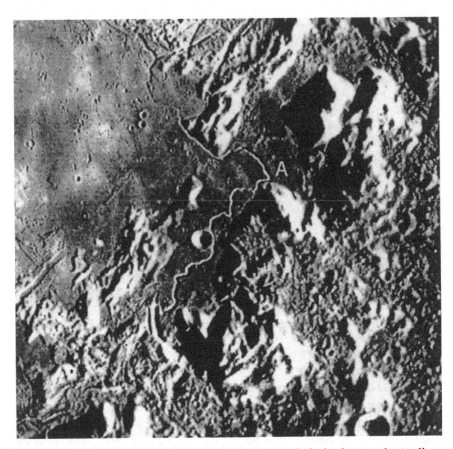

**Figure 10.7** Hadley Rille as seen by the Apollo orbiter, with the landing site for Apollo 15 indicated by the letter A. The mountains are part of the lunar Apennines (see Figure 10.5) which rise to a height of over 15,000 ft (4,500 m), and the flat area to the upper left is the edge of the Imbrium basin. (Courtesy NASA; AS 510.)

time. During their exploration the astronauts came across an unusual white rock which, once cleared of dust, was found to consist almost entirely of crystallised plagioclase, which is the main constituent of anorthosite. They had discovered what came to be known as the Genesis Rock, which was, at about 4.5 billion years old, an example of the original lunar crust, just 100 million years younger than the moon itself. Other rock samples confirmed the date of 3.85 billion years, deduced from Apollo 14 samples, for the Imbrium impact that had created the Apennines and the Fra Mauro hills. The astronauts also discovered a few examples of green-coloured rock, the colour of which was later found to be due to very small green-coloured glass spheres. These had apparently been brought to the lunar surface from hundreds of kilometres down by fire fountains. On the surface of the beads were volatile elements, like zinc, lead, sulphur and chlorine, which were the remains of the gases that had caused the eruptions. Surprisingly, these samples of green-coloured rock were only analysed after Apollo 17 had returned with samples of orange-coloured rock, which were also found to have been produced by fire fountains.

So the history of the moon was gradually being pieced together as the Apollo 16 mission was being planned. The moon had apparently formed about 4.6 billion years ago and, within about 100 million years, a crust had formed of light anorthosite rock as the magma ocean had cooled, with the heavier iron- and magnesium-rich materials sinking. The ALSEP packages,[13] left on the moon by the various Apollo missions, had shown that the loose lunar soil, or regolith, is about 2 to 8 m thick over the maria, and some 15 m or more over the uplands. The regolith covers the crust, which is now, on average, some 60 km thick, and below that is the mantle of iron- and magnesium-rich rocks which had been, about 4 billion years ago, the source of the maria (volcanic) basalts when the mantle was still molten. The numerous craters seem to have been formed by meteorite impacts, although the existence of some volcanic craters could not be ruled out. Some rock samples of about 3.8 billion years old were found to have a remanent magnetic field of over 100 times that of the moon today, indicating that the moon must have had, at that time, an appreciable magnetic field, possibly caused by dynamo action in a large liquid core. Temperature and heat conductivity measurements carried out by ALSEP packages showed that there was still an appreciable heat flow at the surface, which implied that the moon has, even today, a relatively large core, at a temperature of at least 1,000 K.

After Apollo 15, geologists wanted to find samples of volcanic rock from places other than the maria. They thought, for example, that the shape and relatively light colour of the Cayley plains, in the Descartes highlands, indicated that they were made of rhyolite, which is volcanic rock of higher silica content than basalt.[14] So the Descartes highlands was the target chosen for Apollo 16. In the event, most of the rocks that the Apollo 16 astronauts

---

[13] All the Apollo landing missions left behind experiments on the moon, varying from the two instruments left by Apollo 11 (see Figure 10.6) to much more sophisticated sets left by subsequent missions. These experiments, that had an autonomous power supply and antenna for communicating with the earth, continued to operate for some time after the astronauts had left the moon. They were known as the ALSEP packages.

[14] Rhyolite is also lighter coloured and more viscous than basalt.

brought back in 1972 were breccia or impact rocks, with a few anorthosite remnants of the ancient crust. There were no examples of the predicted rhyolite, so at least the area of the Descartes highlands where the astronauts had landed was not volcanic. Instead, it appeared as though the Cayley plains had been formed by molten ejecta from the Imbrium impact about 1,600 km away, whilst the older Descartes highlands had been produced by the impact that had created the Nectaris basin at a distance of about 600 km.

So far no geologists had been to the moon. NASA had relied, instead, on astronauts who had received special geological training for their missions. That was about to change with Apollo 17, however, which had the geologist Harrison Schmitt on board. The area selected for this last Apollo landing was the Taurus-Littrow valley on the edge of the Sea of Serenity (Mare Serenitatis), chosen because of its unusual very dark surface with what looked like very small volcanic craters nearby. In the event, the most interesting discovery was made on the edge of a 100 m diameter *impact* crater where Schmitt discovered a layer of orange soil under the surface dust. This spectacular find proved to be made of very small beads of glass, 3.7 billion years old, with a high titanium content (of at least 9%). The material had originated deep inside the moon, but it had been brought to the surface by a fire fountain, produced by the impact that had made the crater 19 million years ago. In addition, the black colour of the valley proved to have been due to volcanic glass generated in a similar way – the different colour being due to a different mineral content.

Whilst all this high profile lunar exploration was being undertaken by the Americans, the Russians were forced to attempt automatic sample return missions, because of serious problems with their new N-1 rocket. In all, only about 0.3 kg of surface material was returned, compared with 380 kg of selected material returned in toto by the Apollo astronauts.

The samples of surface material returned by Luna 16 from the Sea of Fertility (Mare Fecunditatis) were of basaltic composition, similar to the samples returned by Apollos 11 and 12, whereas those returned by Luna 20 from a highland region had a high concentration of anorthosite material like Apollos 16 and 17. Luna 21 found residual magnetism in lunar rocks and Luna 24, which landed near a 10 km diameter crater in the Sea of Crises (Mare Crisium), found evidence of an ejecta blanket caused by a meteorite impact about 3 billion years ago.

The ALSEP experiments had shown that the very tenuous lunar atmosphere consisted mostly of helium, which appeared to come from the solar wind, and argon, which appeared to come from the moon as its density correlated with seismic events. Then, in 1987, Drew Potter and Tom Morgan of

NASA found minute amounts of sodium and potassium in the lunar atmosphere. Four years later Mike Mendillo, Alan Stern, Jeff Baumgardner and Brian Flynn of Boston University found that the moon has a very tenuous sodium tail, similar to that of a comet, pointing away from the sun. Some process on the surface of the moon must be producing sodium to replenish the sodium in the 'tail', the most likely cause being the vaporisation of small amounts of the lunar surface by the continuous bombardment of micrometeorites.

## Clementine and Lunar Prospector

Although the Apollo and Luna missions provided substantial new evidence on the structure and composition of the moon, they only produced rock samples from a few small areas of the lunar surface. Equally, although the American Lunar Orbiters, and the astronauts in the Apollo command modules, produced unprecedentedly high resolution images of the lunar surface, they were unable to provide a detailed topographic or relief map of the moon. All this changed in 1994, however, when for $2^1/_2$ months the American Clementine spacecraft observed the moon in eleven different wavebands between 415 nm, at the blue end of the spectrum, and 2.75 μm in the infrared. In addition, it carried a laser altimeter that had a height resolution of about 40 m.

William Hartmann and Gerard Kuiper had suggested in 1962 that there is probably a large crater or basin on the far side of the moon that has produced the mountains seen on the moon's south-western limb. Evidence for something of the sort had been found in the 1970s, and by the Galileo spacecraft in 1990, but in 1994 Clementine showed the full extent of this so-called South Pole–Aitken basin for the first time. It turned out to have a diameter of about 2,500 km, stretching from the south pole to the Aitken basin at latitude 17° S, and to be about 12 km deep. Overall Clementine found that the range of the moon's topography (i.e. from the top of the highest mountain to the depths of the lowest crater) was a surprising 16 km on the far side, which is about the same as the range on the much larger earth, whereas on the near side it is only about 6 km.

Continuous measurement of changes in Clementine's orbital velocity, as the spacecraft orbited the moon, enabled a map to be drawn of gravitational anomalies, including a better delineation of the mascons (regions of high gravity) first identified by Lunar Orbiter 1. Clear evidence was found for mascons under the Imbrium, Serenitatis, and Crisium maria, for example, whilst the Mare Orientale, which is on the very edge of the moon as seen from earth, was found to have a central mascon surrounded by a ring of low gravity.

Putting the gravity anomaly data and topographic data together enabled a map to be produced giving an estimate of the thickness of the moon's crust all over its surface. This indicated that the crustal thickness ranged from about 120 km, for parts of the far side north of the South Pole–Aitken basin, to about 10 km in the maria on the near side. In fact, the centre of mass of the whole moon appears to be offset by about 2 km from the geometric centre, in the direction of the earth, probably because the relatively light crust is thicker on the far side than on the near side.

Analysis of the relative reflectance of the lunar surface in the various wavebands observed by Clementine confirmed that anorthosite is the dominant constituent of the lunar highlands, and that the near-side maria have been flooded with iron rich lavas. But the titanium content of the basalts was generally not as high as that of the Apollo samples.

It had been pointed out in 1961 by Kenneth Brown, Bruce Murray and Harrison Brown of Caltech that, as the sun never deviates by more than $1°.6$ from the moon's equatorial plane, the floor of some craters very close to the moon's poles could be in permanent darkness. As a result, they suggested that there may be water ice deposits in such regions, left there over time by cometary impacts. Clementine's highly inclined orbit was ideal to undertake a search for such deposits, and to do so it illuminated both lunar poles with radio energy which was picked up by earth-based radio dishes. The resulting back-scatter, which was higher than expected in the south polar region, in particular, was interpreted by some astronomers as indicating that there was probably ice at the south pole. Not all astronomers agreed, however.

In early 1998 the Lunar Prospector spacecraft was put into orbit around the moon by NASA. On board was a neutron spectrometer to detect neutrons of various speeds that are emitted from the moon when cosmic rays strike the lunar surface. This instrument showed an increase in the number of slower neutrons over both poles, together with a decrease in the number of medium speed neutrons (see Figure 10.8). This slowing of the neutrons was thought to indicate that the regolith over both poles contains a substantial amount of hydrogen in the form of water molecules. There was no change in the number of high speed neutrons over both poles, however, and so the water was thought to be in the form of ice crystals mixed in with the regolith, rather than in the form of ice sheets, which would have produced fewer fast neutrons. Unfortunately, there are other possible explanations for the

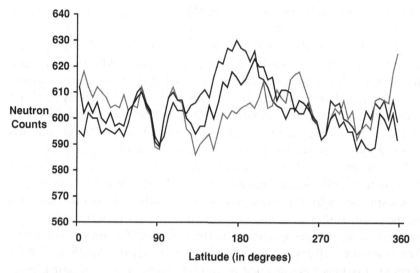

**Figure 10.8** The number of medium energy neutrons detected by the Lunar Prospector spacecraft as a function of latitude for a number of lunar orbits (one line per orbit). Consistent reductions at the north and south poles, at 90° and 270° latitude respectively, were attributed to the presence of water ice.

observed effect, and so the existence of water ice at the moon's poles is still somewhat speculative.

## Origin and subsequent history of the moon

In 1955, Gerstenkorn had suggested that the moon had been captured by the earth (see Section 9.5). Then in 1972/73 Alastair Cameron and Fremlin independently suggested that the moon had originally been orbiting the sun inside the orbit of Mercury. Fremlin proposed that a resonant interaction between Mercury and the moon had caused the orbit of Mercury to become smaller and more eccentric, whilst the moon had been ejected into an earth-crossing orbit where it was captured. Analysis of lunar rocks brought back by the Apollo astronauts showed, however, that the earth and the moon have similar relative amounts of the oxygen isotopes $^{16}O$, $^{17}O$ and $^{18}O$, indicating that they were both formed in the same part of the solar system. In addition, the relative amounts of these oxygen isotopes for the earth and moon were found to be different from those of Mars and the meteorites. This gave powerful evidence for the common origin of the earth and moon.

In 1975, William Hartmann and Donald Davis suggested that the moon had been formed as a result of the off-centre impact of a body the size of Mars with the earth about $4\frac{1}{2}$ billion years ago.[15] A similar theory had been proposed thirty years earlier by the geologist Reginald Daly of Harvard University, but his suggestion had been ignored at the time. Hartmann and Davis suggested that the material in the core of the impacting body had been incorporated into the earth's mantle, but that a cloud of mantle debris from both the impactor and the earth had been ejected by the collision, and this debris had subsequently aggregated to form the moon. This theory solved the angular momentum problem of spontaneous fracture (see Section 9.5), and was consistent with contemporary theories of the origin of the solar system. It also explained why the moon has proportionally less volatile elements than the earth, as they were lost by the mantle debris that formed the moon at the high temperatures generated by the collision. Theories of the origin of the moon are still speculative, however, although the Apollo and other spacecraft missions have enabled us to understand its subsequent history more clearly.

In 1988, the geochemists Richard Carlson and Gunter Lugmair dated one of the oldest samples of lunar crust brought back by the Apollo astronauts, showing that the crust solidified $4.44 \pm 0.02$ billion years ago, during a period of intense meteoritic bombardment. This bombardment appears to have gradually reduced in intensity about 3.8 billion years ago, when the last major lunar basin, Orientale, was formed. Over the period from 3.8 to 3.0 billion years ago, basaltic lavas from the molten region below the anorthositic crust poured onto the surface to fill the low-lying areas. Early on in this period the crust was not thick enough to support the extra weight and fractured, and these fracture features can still be seen today. Surface activity virtually ceased about 2.5 billion years ago, and the moon's surface has been modified since then only by the occasional meteoritic impacts.

## 10.3 Origin of the solar system

All early work on the origin of the solar system (see Sections 6.3 and 9.4) had been undertaken with little observational evidence from other stars to assist the astronomers and philosophers in their theories and speculations. But that changed in the latter part of the twentieth century when observations

---

[15] Alastair Cameron and William Ward independently came up with the same idea in the following year.

of other stars, particularly those known as T Tauri stars, began to help our understanding of how the solar system was probably formed.

T Tauri stars had been known for many years as variable stars associated with dark clouds. They have masses similar to that of the sun, and an excess of lithium, and this, together with their position on the H–R diagram (of temperature versus luminosity), indicated that they are stars in the early stages of development. They are apparently still contracting and have not yet reached the Main Sequence of the H–R diagram, where stars spend most of their observable lifetimes.

In the 1980s it was found that about 30% of T Tauri stars have an infrared excess; that is they emit more energy in the infrared than would be expected from their intensity in visible light. This infrared excess was attributed to energy being emitted by a dust cloud surrounding the central T Tauri star which is itself heating the cloud. However, the amount of infrared energy detected implied that the clouds are very dense. So dense, in fact, that we should not be able to see the central T Tauri stars in visible light, assuming, that is, that the clouds are uniformly distributed around the stars. So it was concluded that the dust clouds surrounding T Tauri stars are not uniform, but are in the form of a flattened disc, which, for many stars, is inclined to our line of sight, thus allowing us to see the central star.

T Tauri stars are not the only stars to have circumstellar discs or shells of dust. For example in 1983 the star Vega, a well-studied, young, stable star about 2.5 times the mass of the sun, was found by H.H. (George) Aumann and Fred Gillett, using the IRAS spacecraft, to be too bright in the infrared. This excess emission appeared to be coming from a thin shell of dust at a temperature of about 80 K. The surface temperature of Vega was known to be about 10,000 K, so the IRAS measurements implied that the dust was about 80 AU away from its surface, assuming that the dust was only being heated by the star. IRAS measurements also indicated that the dust grains were rather large, being at least 1 mm in diameter, which is at least 1,000 times larger than the size of interstellar dust grains. In fact, it was thought that we could be observing a cloud of dust which may be in the process of forming planets.

The discovery of Vega's dust cloud created a great stir in the astronomical community, and other similar stars were examined with IRAS to see if they too had an infrared excess. A number were found, of which the most interesting was β-Pictoris, which was found to have a shell about 800 AU in diameter that could be imaged by ground-based telescopes. Further ground-based observations showed that there appeared to be a region within about 30 AU of β-Pictoris which is dust free, possibly because the material that had been there had already formed into planets. Large amounts of gas were also found in the shell around β-Pictoris. In addition, the ultraviolet spectrum of the star, observed with the IUE spacecraft, showed that some of this gas is falling onto the star with velocities reaching 400 km/s, which is about the velocity of free fall.

Astronomers had known for some time that the Orion Nebula, which contains a number of very young stars, is probably still forming stars from the plentiful amounts of gas and dust that it contains. In 1967, for example, Eric Becklin and Gerry Neugebauer had discovered in the nebula an object, now called the BN object, that had a temperature of only 600 K. At first it was thought that it was a protostar, but radiation was soon found coming from hydrogen atoms at a temperature of 10,000 K, indicating that the BN object is a cloud of warm dust surrounding a hot, young star. Later George Herbig of the University of Hawaii showed that the density of stars in Orion's Trapezium Cluster at the centre of the nebula is about 10,000 times that in the vicinity of the sun. He also found that most of these stars were still contracting, not having yet reached the Main Sequence, so the cluster is a dense stellar nursery. Because of this the Orion Nebula was one of the first objects to be imaged by the Hubble Space Telescope (HST), following its launch in 1990. Unfortunately, these images were severely compromised by the HST's faulty mirror, and it was not until after the first servicing mission at the end of 1993 that satisfactory images could be made.

Within a few weeks of the first servicing mission, the HST returned stunning images of what Robert O'Dell of Rice University called *proto-planetary discs*, or proplyds, in the Orion Nebula. Some proplyds were expected in this region, but the sheer number was a big surprise, with over 30% of the stars imaged being surrounded by proplyds of one sort or another (see Figure 10.9).

The structure of the individual proplyds seen in the HST's high resolution images proved to be very interesting. They are basically small clouds of dust and gas whose outer parts, which face the star $\theta^1$C Orionis, are being ionised by its intense ultraviolet radiation, in the same way that the gas in the Orion Nebula is being ionised. Those proplyds nearest to $\theta^1$C Orionis have a bright cusp on the side facing the star, and have a comet-like tail, created by the strong stellar wind, trailing in the opposite direction.

So we now know that a substantial minority of very young stars, both T Tauri and others, are surrounded by large discs of dust. The question is, how do these discs develop and how do they condense to form planets?

The starting point in the development of the solar system is now thought to have been a large molecular cloud. It was at a temperature of the order of 10 to 50 K, and throughout it there were magnetic fields. As time progressed,

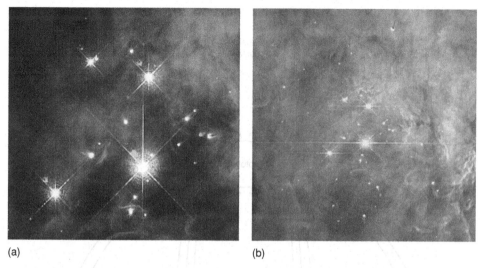

(a)                                                    (b)

**Figure 10.9** (a) The four young trapezium stars in the Orion Nebula, as seen by the Hubble Space Telescope, are strong emitters of ultraviolet radiation, but $\theta^1$C Orionis, the star to the lower right, outshines all the others. This has resulted in the proplyds, which are clouds of dust and gas surrounding very young stars, forming tails generally pointing away from $\theta^1$C Orionis. More of these proplyds are shown in (b), which gives a wider field of view. (Courtesy D. Johnstone (CITA), STScI and NASA.)

the initially small variations in gas density throughout the cloud resulted in gas concentrations that became favoured sites of gravitational collapse. In some cases of similar clouds, two or three major centres of collapse were created, which eventually resulted in a binary or multi-star system being created. But in the case of the solar system there was just one major centre of collapse which became the sun.

The gravitational attraction of this major centre of collapse, or cloud core, in the early solar system, eventually overwhelmed the resistance caused by the magnetic fields. The cloud material then cascaded onto the cloud core, turning it into a protostar. As a result, the protostar's pressure and temperature increased, eventually causing it to radiate a significant amount of energy. It is thought that typically about half of the energy of collapse would have been used to heat the stellar core, and about half would have been radiated away.

Initially the random motion of the gas molecules in the cloud would not have completely cancelled each other out, and there would have been a net angular momentum of the cloud. As more and more of the gas was drawn into the protostar, however, it and its surrounding gas cloud would

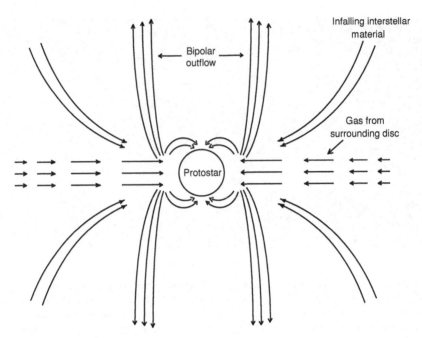

Infalling interstellar material

Bipolar outflow

Gas from surrounding disc

Protostar

**Figure 10.10** Gas motion around a protostar/protosun is shown in this schematic drawing, according to one model of the formation of the early solar system. Large amounts of material flow in towards the protostar, mainly from the surrounding disc, whereas other material, in the form of partially ionised gas, follows the field lines *away from* the protostar, creating a bipolar outflow. Observations of some protostars seem to indicate that the bipolar outflow may also come from regions much nearer to the poles.

have started to spin faster and faster, to conserve angular momentum. The natural consequence of this is that the cloud particles, as well as the local concentrations within the cloud, would eventually form a disc of material orbiting the central, major concentration or protostar.

Details of the above processes are not completely understood but, as mentioned above, very young stars are known to be surrounded by large dust clouds. In addition, radio telescopes have detected strong winds of material blowing away from such protostars in two opposite directions. These bipolar outflows, as they are called, eventually clear the cloud material from over both poles of the protostar (see Figure 10.10). So, when the star is about 1 million years old,[16] there is a disc of material flowing in towards the star

---

[16] This, and subsequent timescales quoted in this section, are for stars with approximately the same mass as the sun. Heavier stars develop more quickly, and lighter stars more slowly.

in the equatorial regions, together with gas and dust free regions over each pole, which allows us to see the star, if we are not in the disc plane. The star remains in this so-called T Tauri stage for about 10 million years, gradually accumulating material and becoming hotter, before its core temperature is high enough for hydrogen burning to commence.

As a consequence of the above, the solar nebula, just before the creation of the planets, was basically a disc of gas whose temperature, density and velocity were largest nearest the sun. In the inner region of the nebula, it was too warm for water to condense as ice, and so the planets forming there, the terrestrial planets, consist of silicate material and other elements like magnesium and iron, the so-called refractory elements, that condense into solids at high temperatures. In the middle region of the nebula, water ice was stable, and the large outer planets of Jupiter, Saturn, Uranus and Neptune were formed, possibly around a nucleus of water ice. Finally, the density of material in the outer part of the nebula was too small for sizeable planets to be produced. Instead, the planetesimals in that region generally remained small, and they are now what we know as the comets and asteroids of the Kuiper belt (see Section 9.13).

It is possible to see the results of some of the processes that were at work in building up the terrestrial planets by examining meteorites. In fact, as long ago as the nineteenth century, the British geologist Henry Sorby (1826–1908) undertook a detailed analysis of meteorites, concluding that they had been formed out of the same nebula that had produced the sun and planets. After the nebula had condensed into small particles he suggested, with a remarkable degree of foresight, that 'These... collected together into larger masses... and [were then] broken up by repeated mutual impact, and often again collected together and solidified.... The study of the microscopical structure of meteorites reveals to us the physical history of the solar system at the most remote period of which we have any evidence.'

Meteorites are now classified as irons, stony-irons and stones, with the latter category being divided into chondrites and achondrites. Chondrites have small spherical inclusions within them called chondrules, which may be of metal, silicate or sulphide materials. Chondrites have a very similar chemical composition to that of the sun, although they contain no free hydrogen and helium. The formation of the planets will now be examined, given what is now known about meteorites.

The spherical chondrules that are now in meteorites were clearly once molten, which required a temperature of at least 1,600 K, before they cooled, solidified and dispersed in the solar nebula. This high temperature limited their formation to the inner part of the solar nebula, however, where most of them remained, becoming coated with small dust particles.[17] How the chondrules then stuck together to form the larger, meteorite-sized bodies that we see today is unknown, but they clearly did so, as we now have the chondrites to prove it. Theoretical analysis shows that turbulence in the solar nebula may have played a significant rôle in this.

When the chondrites had grown to a few centimetres in size, friction with the residual gas in the solar nebula caused them to slow down in their solar orbits, and they spiralled inwards through the nebula, picking up more and more material on the way in the form of dust and other chondrites. But once they had reached about a kilometre or so in size, these bodies, now called planetesimals, would no longer be seriously affected by the depleted gas of the nebula, and their orbits would stabilise.

If a planetesimal now met a much smaller body in its orbit, it would collect it, because of the planetesimal's gravitational attraction. If, on the other hand, the planetesimal collided with another planetesimal, its fate would depend on their relative velocity. If both bodies were moving at about the same velocity in about the same orbit, their relative velocity would be low and the bodies would merge. But if the collision velocity was high, the planetesimals would shatter into many pieces, and growth would have to start all over again. Inevitably, however, those planetesimals that were able to grow by merging with similar sized bodies and the collection of much smaller ones, would grow rapidly at the expense of their other brethren, until all available material in their orbital region had been collected. They would then have become the planets that we see today, at least as far as their size is concerned.

Remarkably, theoretical analysis indicates that it would have taken only a few thousand years for kilometre size planetesimals to form. The terrestrial planets would have reached almost their current size after just ten million years, although it would probably have taken about ten times that time for them to have swept up all the planetesimals.

Jupiter and the other large planets probably grew in a similar way to the terrestrial planets by merging planetesimals, but in that part of the solar system the planetesimals would have been largely composed of water ice. Once they had reached a certain size, however, they would have started to accumulate hydrogen and helium, which was plentiful at that time, from the solar nebula. On the other hand, instability in this region of the nebula may have been such that Jupiter and the other large planets formed directly from

---

[17] Many chondrules in meteorites are, in fact, still coated with dust.

the nebula as large protoplanets. At present, it is not known which scenario is correct.

The above description is designed to give only a flavour of what we think we know about the formation of the solar system, and what we don't know. There is still a great deal of disagreement about the details, however, and it is not possible in a book of this nature to explore these here, and describe in detail how these ideas have been developed in recent years in this very complex subject area.[18]

## 10.4 Mercury

So far the Mariner 10 spacecraft that was launched in November 1973 is the only spacecraft to have visited Mercury. En route it flew by Venus, whose gravitational force redirected it to intercept Mercury in a so-called 'sling-shot' manoeuvre, thus enabling NASA to use a smaller launcher than would otherwise have been required. This arrangement also enabled the spacecraft to fly past Mercury on three consecutive orbits. Unfortunately, the orbital dynamics were such that these intercepts took place at intervals of exactly two Mercurian years. This meant that the spacecraft imaged the same illuminated half of the planet at each fly-by, because the planet rotated exactly three times between each intercept. As a result, there is about 50% of Mercury's surface that has still not been imaged.

Mariner 10 showed that Mercury has a surface that looks, at first glance, very much like the moon with a large number of impact craters, although extensive lunar-like seas are noticeably absent. Mercury's primary impact craters appear to be somewhat shallower than those on the moon, because of the planet's greater surface gravity. This greater gravitational force has also caused the distance of the ejecta blankets from the crater rims to be smaller than on the moon. There is no evidence of wind erosion of Mercury's craters, indicating that the planet has not had an appreciable atmosphere for most of its lifetime.

The most obvious large-scale feature on Mercury is the 1,300 km (800 mile) diameter Caloris basin, whose age, based on cratering density,

---

[18] For the reader wishing to go further, many more details are contained in *A History of Modern Planetary Physics*, in 3 vols., by Stephen Brush, Cambridge University Press, 1996. A simpler exposé of the current thinking is contained in John Wood's contribution to *The New Solar System*, 4th edn., by Beatty, Petersen & Chaikin (eds.), Sky Publishing and Cambridge University Press. A modified version of the latter contribution is given in *Sky & Telescope*, Jan. 1999, pp. 36–48, by John Wood.

has been estimated at about 3.8 billion years.[19] The basin is ringed by the Caloris mountains, which are about 2 or 3 km (7,000 to 10,000 ft) high and up to 100 km across. The floor of the basin seems to be composed of lava which has been modified to produce a pattern of both concentric and radial ridges and grooves. Robert Strom, Newell Trask and J.E. Guest suggested in 1975 that the ridges were produced by compression when the floor of the basin subsided. The grooves, which generally form polygonal patterns, are up to 10 km wide and 500 m (1,600 ft) deep, with generally steep sides and flat floors. They seem to be younger than the ridges as some grooves have modified the ridges, whilst the opposite is not observed.

An area called the 'weird terrain' was found on the other side of Mercury, consisting of hills and mountains up to 1,800 m high, arranged in a strange pattern (see Figure 10.11).[20] Interestingly, this weird terrain is at the antipodes of the Caloris basin, which led Schultz and Gault to suggest that the impact that had produced the Caloris basin had formed these strangely shaped hills and valleys. In 1975 they showed that this was quite possible, as the seismic energy from the Caloris event would have been focused by the large planetary core.

Scarps or faults up to 500 km long and 2 km high are found to be common on Mercury, providing evidence for the contraction of the whole planet by about 1 to 2 km radius. This could have been due either to cooling, or to the planet taking on a spherical shape as its rotation rate reduced, or both.

Mercury's atmosphere was found by Mariner 10 to consist mainly of atomic helium and hydrogen at a total pressure of only about $10^{-13}$ bar. So its atmosphere appeared to be nothing more than a local concentration of the solar wind caused by Mercury's gravity. This lack of a meaningful atmosphere, together with Mercury's slow axial rotation, means that the surface facing the sun is very hot, but that on the other side just before dawn is very cold. In fact the temperatures measured by Mariner of 430 °C (max.) and −173 °C (min.) show the largest temperature range of any planet in the solar system.

Astronomers received a big surprise in August 1991 when radar studies of Mercury by Duane Muhleman, Martin Slade and Brian Butler of Caltech indicated that there may be water ice at the planet's north pole. Later observa-

---

[19] The surface ages of solid bodies in the solar system can be estimated by observing the density of craters of various sizes and assuming a given impact rate.

[20] Similar terrain exists on the moon at the antipodes of the Mare Imbrium and Mare Orientale.

**Figure 10.11** A Mariner 10 image of Mercury showing the weird terrain which is on the opposite side of the planet to the very large Caloris basin. (Courtesy NSSDC, World Data Center-A for Rockets and Satellites, NASA.)

tions showed similar results for the south pole. This completely unexpected observation, of potential water ice on a planet so close to the sun, could be explained by the fact that the axis of Mercury is virtually perpendicular to the plane of its orbit around the sun. As a result, the sun, as seen from the poles, is permanently on the horizon, and so the floors of craters near the poles are in permanent shadow, and always very cold. David Page estimated in 1992 that the temperature could be as low as 60 K ($-213\,°$C), well below the temperature of 112 K ($-161\,°$C) required to retain water ice on Mercury for billions of years.

## The interior

Mercury is a much smaller planet than the earth and, because of this, it would have cooled much the faster after its initial formation. As a result, it was generally thought that Mercury would have an almost solid core. It was not surprising, therefore, when Mariner's images indicated that the surface is a single lithospheric plate which, although broken, shows no differential lateral movement.

The magnetic field on earth is generally thought to be due to a dynamo effect in its large liquid core. Mercury, on the other hand, is small, with a slow axial rotation and an apparently solid core, and so it was expected that there would be no magnetic field. It was something of a surprise, therefore, when Mariner 10 discovered that Mercury has a magnetic field which, although smaller than that of the earth, is still appreciable. Later analysis showed that Mercury has an exceptional amount of iron in its core, which explains the source of the field, although how the iron became magnetised is unknown.

## 10.5 Venus

### Early spacecraft results

The surface conditions on Venus were a complete mystery in 1960 before the first spacecraft visited the planet. In the early 1950s most astronomers had assumed that the surface temperature would be about 100 °C, but in 1956 the analysis of radio emissions indicated a surface temperature of about 300 °C (see Section 9.3). Likewise, surface atmospheric pressures of the order of a few bar had been assumed until Carl Sagan had estimated a figure of the order of 100 bar. So in one view, put forward by Menzel and Whipple, the surface is hot and humid, with oceans of water and clouds of water vapour. In an alternative view, Venus may be so hot that the surface is one great desert, with no water to be found anywhere. In 1960 either concept was considered possible.

The American Mariner 2 spacecraft, which was launched in 1962, was the first spacecraft to visit Venus. Flying past the planet at a distance of 35,000 km, it measured a surface temperature of 425 °C,[21] which is hot enough to melt lead, and a surface atmospheric pressure of about 20 bar. So the surface must be a desert, as liquid water could not possibly exist under these conditions. The spacecraft's infrared radiometer showed that the cloud tops were between 60 and 80 km above the surface,[22] with cloud-top

---

[21] This temperature was estimated assuming a surface emissivity of 90%, as deduced from earth-based radar measurements. Even if the emissivity was 100%, the surface temperature would still have been about 300 °C, however.

[22] These clouds are higher than the rain clouds on earth, which are generally below about 12 km in altitude.

temperatures of from $-30\,°C$ to $-50\,°C$. Furthermore, Mariner 2 found that Venus had no measurable magnetic field and no radiation belts.

Venera 4, the first successful Russian planetary spacecraft, consisted of a main spacecraft bus, with an ejectable probe that was designed to land on the surface of Venus. Launched in 1967, it found that the magnetic field of Venus was very weak, being not very much stronger than the interplanetary field near Venus. As a result the bow shock, where the solar wind is deflected by the Venusian magnetic field, was found to be only about 500 km above the planet's surface, compared with about 70,000 km for the earth. Data transmission from the landing probe stopped, at what the Russians believed to be the surface, when the probe measured an atmospheric pressure of 20 bar and a temperature of $270\,°C$. The atmosphere was found to be composed of about 96% carbon dioxide, with the remainder being nitrogen, water vapour[23] and oxygen. This was a surprise, as in 1960 Hyron Spinrad of JPL had reanalysed Adams and Dunham's spectrograms of 1932, in which carbon dioxide had first been detected, and had concluded that the atmosphere was composed of about 96% nitrogen and only 4% carbon dioxide. In fact, Joseph Chamberlain of the Kitt Peak National Observatory had even suggested that the carbon dioxide figure should be less than 4%.

The American Mariner 5, which flew past Venus just one day after Venera 4's capsule had landed, confirmed that the planet's magnetic field was very weak, and that the bow shock was only a few hundred kilometres above the surface. It also showed that the surface temperature was at least $430\,°C$, with a surface pressure in the range of 75 to 100 bar.

For some time there was concern about the apparent disagreement between the surface temperature and surface-level atmospheric pressure deduced from the Venera 4 landing probe ($270\,°C$ and 20 bar) and from Mariner 5 (at least $430\,°C$, and 75 to 100 bar). It was noted, however, that the two sets of measurements could be reconciled if the last data received from the Venera 4 probe was not from the surface, as had been assumed, but from an altitude of about 25 km. This was proved when earth-based radar measurements of the diameter of Venus were compared with the distance that the Venera 4 probe had been from the centre of the planet when it had stopped transmitting.

Although the results of Mariner 2, Venera 4 and Mariner 5 differed in detail, by 1968 a general idea of the Venusian environment had been deduced although, as yet, there had been no attempts to photograph the planetary surface as it was permanently covered in cloud. Venus was shown to have a surface temperature of at least $430\,°C$, with a surface atmospheric pressure

---

[23] The water vapour diagnosis was later found to be erroneous.

of about 75 to 100 bar. So Venus' surface environment is considerably more hostile than envisaged by pre-war astronomers. Instead, the surface temperature estimates made from radio measurements in 1956, and the atmospheric pressure estimates made by Carl Sagan, which had been disputed by many astronomers at the time, were largely correct. In addition, the Venusian atmosphere was found to consist mostly of carbon dioxide, and it was the greenhouse effect produced by this gas which was responsible for the very high surface temperatures. Venus was also found to have a very weak magnetic field with, consequently, no Van Allen radiation belts and a bow shock only about 500 km above the surface.

Veneras 7 and 8

The 1970s started with the first successful landing of a capsule on Venus, which enabled the surface conditions to be measured in situ. This capsule, which landed from the Russian Venera 7 spacecraft in December 1970 on the *night-time side* of the planet, confirmed earlier spacecraft findings that the atmosphere was about 97% carbon dioxide, with a ground-level temperature of $474 \pm 20\,°C$. Unfortunately, the lander's telemetry system malfunctioned so it did not transmit atmospheric pressure data from the surface. However, extrapolating from data transmitted earlier in the descent gave an estimated ground-level atmospheric pressure of $90 \pm 15$ bar.

About eighteen months later, a lander ejected from the Venera 8 spacecraft successfully soft-landed on the *daylight side* of Venus. It measured a surface temperature of $470 \pm 8\,°C$ and a surface pressure of $91 \pm 1.5$ bar. So, comparing these results with those from the Venera 7 lander, the surface conditions were seen to be the same, within error, on both the daylight and night-time sides of the planet. This is because of the very high density of the atmosphere, whose pressure at Venus' surface is about the same as that some 800 m below the surface of the earth's oceans. The atmosphere was found to be largely transparent below about 32 km altitude but, because of the overlaying clouds, only about 2 to 3% of the sun's light reached the surface.

The Venera 8 lander had included a number of new instruments compared with its predecessors, one of which was to measure concentrations in the atmosphere of ammonia, whose presence had been predicted theoretically, from evolutionary models of the atmosphere, and detected from earth. The concentrations measured of the order of 0.01 to 0.1% at an altitude of about 40 km were consistent with expectations. The lander also measured a wind velocity as it descended of from 100 m/s (225 mph) at 48 km altitude to only 1 m/s (2 mph) below 10 km. The probe landed relatively close to the terminator, and this very low surface-level wind velocity was consistent with surface temperatures that did not vary greatly between the day and night sides of the planet. A light meter showed that only 1.5% of the sun's illumination reached the ground where the probe landed. Measurement of γ-rays showed that the upper atmosphere shielded the planet from high energy cosmic rays, and γ-ray spectroscopy indicated that the surface at the landing site was probably composed of volcanic rock. This finding was important, as it suggested that Venus had once been hot enough for the heavier elements to fall towards the centre of the planet, thus leaving the lighter elements on top, and producing what is called a 'differentiated structure' like that of the earth and moon.

## Veneras 9 and 10

Venera 9 was launched to Venus by a Russian Proton vehicle on 8th June 1975, and six days later it was followed by its twin, Venera 10. The Venera 9 lander was released by the main spacecraft on 20th October as it approached Venus, and two days later the lander entered the Venusian atmosphere. At the same time, the orbiter undertook a braking manoeuvre to enter into Venus orbit, the first spacecraft to do so. The lander, meanwhile, had touched down about 2,500 m above the mean Venusian surface level on what is now known to be the eastern slope of the shield volcano Rhea Mons. Data was received on earth via the orbiter for 53 minutes until the orbiter disappeared below the lander's horizon. Three days later the Venera 10 lander also touched down about 2,200 km from Venera 9 at the foot of another shield volcano now called Theia Mons. Data was received on earth for 65 minutes until the Venera 10 orbiter also disappeared below the lander's horizon.

The atmospheric data received from the landers, as they descended through the atmosphere, showed no great surprises, and generally confirmed the data from Venera 8, although the new data was more detailed and accurate. In particular the vertical profile of the various cloud layers was more clearly defined, with three layers being detected centred on heights of 64, 55 and 51 km. This compares with the earth where most of the clouds (and most of the atmosphere) are below 12 km altitude. The clouds on Venus were found to be relatively transparent, and their opacity as seen from earth was found to be due to their great depth.

On the surface the γ-ray spectrometers measured the chemical composition of the rocks which appeared to be similar to slowly cooled basalt, confirming the volcanic origin deduced from the Venera 8 data. But the most excitement was generated by the receipt of a single monochromatic photograph of the surface from each of the landers.

It had generally been thought that the surface of Venus would probably look like a sandy desert produced by billions of years of wind[24] and heat erosion, but the photographs taken by the Venera 9 and 10 landers showed a very different scene consisting of numerous rocks with little or no sand. The rocks at the Venera 9 site had sharp edges, with no evident erosion, whereas those at the Venera 10 site appeared to be somewhat older as they showed clear evidence of erosion. Nevertheless, γ-ray analysis showed that the material at both sites is younger than at the Venera 8 site. Overall it appeared as though the surface of Venus is relatively young and still probably active.

In the meantime, the orbiters were continuing to transmit their own scientific data to earth. This showed that the cloud particles were definitely not water droplets, although they were liquid droplets of some sort and, somewhat bizarrely, the cloud-top temperatures on the night side of the planet were found to be somewhat higher than those on the day side. The Venera 10 radar altimeter showed that the surface of Venus, under its orbital path, was relatively flat, varying by only a few kilometres in elevation from a perfectly smooth surface.

Some years earlier the Venera 5 landing module had detected what was thought to be lightning on its descent through the Venusian atmosphere, but, as no such detection was observed with subsequent probes, the data was considered to be erroneous. However, both the Venera 9 and 10 landers detected radio signals during their descent which were tentatively attributed to lightning. In addition, a spectrometer on board the Venera 9 orbiter also detected what appeared to be very intense lightning. Whether these were true lightning observations or not is still not clear, however.

It had been known since the seventeenth century that, when the phase of Venus is new, the dark part of the disc sometimes appears to be slightly illuminated by what is called the 'ashen light' (see Section 4.3). It is thought that this ashen light, whose existence is still disputed, may be caused by some sort of auroral discharge, or by the scattering of sunlight in the Venusian upper atmosphere producing an extensive twilight. In 1967 an ultraviolet photometer on Mariner 5 had detected a faint ultraviolet glow on the night side of the planet which was thought by many to be the ashen light. Then the

---

[24] Although the wind velocity at the surface of Venus is very low, the density of the atmosphere is very high, so wind can still cause significant erosion.

Venera 9 and 10 orbiters showed that there was a night-side airglow caused by molecular oxygen in the upper reaches of the Venusian atmosphere. Whether this was the ashen light is still not clear.

## Pioneer-Venus

The design of the Pioneer-Venus 1 spacecraft, otherwise known as the Pioneer Venus Orbiter or PVO, showed a fundamental departure from that of previous spacecraft sent to Venus, as it included a radar altimeter that could 'see' through the clouds. It was designed to provide the first detailed map of the surface of Venus, with a maximum surface resolution of 20 km and an altitude resolution of 100 m.[25]

The PVO's sister spacecraft, variously called Pioneer-Venus 2 or the Pioneer Venus Multiprobe, consisted of a carrier spacecraft plus four atmospheric probes, none of which were designed to survive impact with the surface. The probes were targeted to enter the atmosphere at radically different locations, so that their results could be compared to give some idea of the atmospheric dynamics.

Both the Pioneer-Venus spacecraft were launched in 1978, with the PVO going into orbit around Venus a few days before the multiprobes began their descent through the atmosphere. In the event, the probes confirmed the theoretically based predictions made in 1973 by the Americans Godfrey Sill, Andrew Young and Louise Young that the Venusian clouds are composed mainly of corrosive sulphuric acid. So finally the true nature of the beautiful evening star as seen from earth had been revealed. Not only was its surface hot enough to melt lead, under an atmospheric pressure of ninety times that of earth, but its clouds were made of sulphuric acid! As was said about another dreadful environment 'Great God! this is an awful place'.[26]

Over the next two years the PVO radar mapped 93% of the surface. The result showed a relatively flat surface, with 65% being within ±500 m of the mean level, although there were exceptions like the 11,500 m (38,000 ft) high Maxwell Montes[27] and the 2,000 m (6,500 ft) deep Diana Chasma. Two

---

[25] Very crude maps had previously been made by earth-based radar (see Section 9.3) but they were very difficult to interpret as they only showed areas of different radar reflectivity, rather than height.

[26] Said by Captain Scott when he reached the South Pole in January 1912.

[27] The Maxwell formation had been detected by radar from earth (and called simply 'Maxwell') but its nature as a chain of high mountains was first resolved by the PVO.

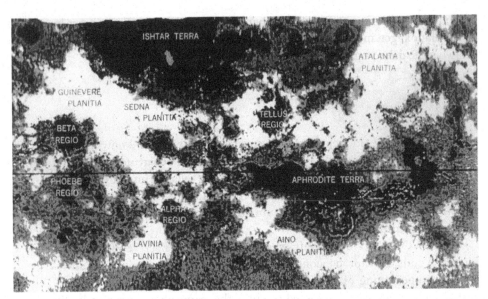

**Figure 10.12** Map showing the global surface features of Venus, as mapped by the PVO radar altimeter from about 75° N to about 65° S. Two large elevated plateaux are clearly seen, called Ishtar Terra in the north, and Aphrodite Terra centred just south of the equator (shown by the line). The lighter coloured areas, called planitia, are low-lying areas. (Courtesy MIT, US Geological Survey and NASA.)

large elevated plateaux were found, namely Ishtar Terra (named after the Babylonian goddess of love) in the far north, and Aphrodite Terra (named after the Greek fertility goddess) just south of the equator (see Figure 10.12). Ishtar Terra was found to have in its western half a 2,500 km plain, called Lakshmi Planum, that rises about 4,000 m (13,500 ft) above the rolling lowland plains. In many ways Lakshmi Planum, which was found to be bordered by towering mountain ranges including Maxwell Montes, resembles the Tibetan Plateau, which is bordered by the Himalayas including Mount Everest. Two shield volcanoes, called Rhea Mons and Theia Mons, which are much larger than any on earth, were also found in Beta Regio, isolated from the two elevated plateaux.

Infrared images from the PVO showed a circumpolar collar[28] of very cold air, about 2,500 km from the pole, 1,000 km across and 10 km deep. The collar, which is at an altitude of about 70 km, is just above the clouds and is about 30 °C colder than the atmosphere on either side. Surprisingly, at

---

[28] A similar dark collar had been detected previously in the ultraviolet by Mariner 10 and by some ground-based observers.

an altitude of 85 km the PVO found that the Venusian atmosphere is warmer at the poles than at the equator. The PVO also measured a slow reduction in sulphur dioxide in the atmosphere with time, which was attributed by some scientists to a major volcanic eruption, although there was no other evidence indicating that such an eruption had just taken place.

## Veneras 11 and 12

In 1978 the Russians also launched two spacecraft to Venus. Designated Veneras 11 and 12 they each included a lander with improved equipment to analyse the clouds. The landers also included a camera, but a common design fault meant that the camera covers could not be ejected on either lander, and so no images were obtained.

The landers found that the droplets in the upper-level clouds were composed mostly of chlorine, although those in the other clouds were confirmed as being largely sulphuric acid. Heat balance measurements indicated that the heat that Venus radiates into space comes mostly from the upper-level clouds. Sulphuric acid aerosols were found which form during daylight and disappear at night, and which inhibit heat flow into space during daylight hours from the lower levels of the atmosphere. These aerosols were the reason why the Veneras 9 and 10 had found that the cloud-top temperatures on the night side of Venus are higher than those on the day side

The relative proportions of argon 36 and argon 40 were recognised in the 1970s as key measurements in understanding the formation of planetary atmospheres. This is because argon 36 was thought to have come from the original solar nebula, from which the planets were formed, whereas on earth argon 40 is produced by the radioactive decay of potassium 40 deep inside the planet. Argon 40 is then brought to the surface by volcanoes or through fissures caused by tectonic motion. On earth there is much more argon 40 compared with argon 36 because of this volcanism and tectonic motion. It was, therefore, highly significant that the Venera 11 and 12 landers found that the ratio of argon 36 to argon 40 in their atmospheric samples was close to unity. This relatively high level of argon 36 on Venus was thought to indicate that its atmosphere was generally formed from the original solar nebula. However, the fact that Venus has a significant amount of argon 40 indicated that there must also have been some planetary outgassing produced by volcanism or tectonic motion.

In absolute terms there is more argon 36 on Venus than on the earth which, in turn, has more than on Mars. This is contrary to what had been expected before the launch of Veneras 11 and 12, as it was thought that the solar wind would have been strong enough to strip the inner planets of their original atmospheres, so the above Venus/Earth/Mars sequence should have been the reverse. Not for the first time had astronomers, trying to understand the formation of the solar system, had to tear up their theory and start again.

Veneras 13 and 14

Both Veneras 13 and 14, which were launched in 1981, consisted of a fly-by spacecraft and a landing module. The landers each contained a significant number of new instruments compared with the Venera 11/12 landers, including a drilling and analysis system to analyse surface material, and a new panoramic camera system with higher resolution, higher contrast, and a colour capability. Soil analysis of the drilled sample was undertaken by an X-ray fluorescence spectrometer, which was much more sensitive than the previously used $\gamma$-ray spectrometer.

En route to the surface, the gas chromatograph on board the landers detected both carbonyl sulphide (COS) and hydrogen sulphide ($H_2S$) in the Venusian atmosphere. The discovery of COS was particularly important as it provided a key to understanding the formation of the atmosphere. The three cloud layers previously observed were also detected down to an altitude of 49 km (Venera 13) and 47.5 km (Venera 14) where the atmosphere became relatively transparent.

The images transmitted of the Venera 13 landing site (see Figure 10.13), in the rolling foothills of Phoebe Regio, showed a sandy landscape littered with sharp rocks, with the dust that was blown onto the bottom of the lander moving between successive images. The site looked similar to those of Veneras 9 and 10, whereas the Venera 14 site, on the eastern flank of a 75 km diameter volcano in Phoebe Regio, looked different (see Figure 10.13 also), with much less fine grained material and with rocks of a more rounded shape, indicating that they were older. Both of the new sites showed layering of rocks but the evidence was clearer for the Venera 14 site.

Geologists concluded from an analysis of the surface material that the surface at both of the Venera landing sites is made of basalt. The material in the rolling plains region of Venera 13 appeared to be like leucitic basalt on earth, which is rich in potassium and is often found on the slopes of volcanoes, and that in the lowland region of Venera 14 appeared to be like terrestrial tholeiitic basalt, which is found on the earth's ocean floor.

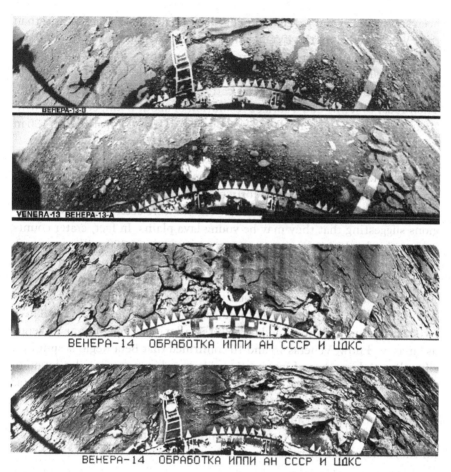

ВЕНЕРА-13-В

VENERA-13 ВЕНЕРА-13-А

ВЕНЕРА—14    ОБРАБОТКА ИППИ АН СССР И ЦДКС

ВЕНЕРА—14    ОБРАБОТКА ИППИ АН СССР И ЦДКС

**Figure 10.13** Images of the surface of Venus transmitted by the Venera 13 lander (top pair of images) and Venera 14 lander (bottom pair). Both landing sites look remarkably flat, with more dust and pebbles at the Venera 13 site, and large flat expanses of rock at the Venera 14 site. The images are distorted because the cameras (one on each side of the landers) panned in an arc-like motion. The crescent-shaped objects on the surface are camera covers. (Courtesy NASA and NSSDC; Venera 13 Lander, YG 06847, and Venera 14 Lander, YG 06848.)

## Veneras 15 and 16

The Americans had produced very useful data on the overall surface topography of Venus using a simple radar altimeter on the PVO in the late 1970s, but since then the performance of the earth-based Arecibo radio telescope had been substantially improved to produce higher resolution radar images than the PVO. In addition, the PVO had not imaged the polar regions, and these regions could not be satisfactorily resolved from earth, so there was a gap in our knowledge here. So in 1983 the Russians rectified this for the north polar region with the launch of two radar-carrying spacecraft, Veneras 15 and 16. Their radar systems had a surface resolution of about 2 km, compared with 20 km for the PVO, and an altitude resolution of 50 m, compared with 100 m for the PVO.

The images returned by Veneras 15 and 16 showed many ridges and narrow valleys which appear to have been formed by the horizontal motion of the crust. Several ten to forty metre diameter craters were found in the elevated Lakshmi Planum, but few craters were seen in the lowland regions suggesting that they may be young lava plains. In fact, crater counts in some lowland regions gave an age as young as about 300 million years. A 100 km diameter double ring impact crater, now called Cleopatra, was clearly seen on Maxwell Montes. In addition, two unusual circular features called Anahit Corona and Pomona Corona were observed, which appear to have been formed by a mixture of both volcanic and tectonic processes. These two large corona structures and over 300 smaller ones[29] are unique to Venus as far as we know. Veneras 15 and 16 confirmed that Beta Regio is split by a rift valley. A large area of intersecting ridges and grooves was also found to the east and north-west of Maxwell Montes, in the upland region of Ishtar Terra. At first this type of terrain, where the intersecting ridges and grooves produced both diagonal and random patterns, was called 'parquet',[30] but this term was thought not to be scientific enough and the term 'tessera' (Greek for tile) is now used instead.

## Vegas 1 and 2

In 1984 the Russians launched two spacecraft, Vegas 1 and 2, as part of an international fleet of spacecraft to fly past Halley's comet (see Section 11.7). However, the Vega spacecraft were unique in also being designed to intercept Venus on the way to the comet, with each spacecraft releasing a landing module as it flew past the planet. Interestingly, both landing modules also

---

[29] Most of these smaller coronae were subsequently found by the Magellan spacecraft.

[30] The first examples of this type of surface produced diagonal patterns reminiscent of parquet flooring. Later, the Veneras also found areas of more random patterns.

had an attached balloon system that was designed to drift in the Venusian atmosphere.

Balloon 1 was released from the Vega 1 lander at an altitude of 54 km and latitude of 7° N. During its 47 hours of operation it returned valuable data about the atmospheric circulation in the most active cloud layer of Venus. Its instruments included a nephelometer to measure cloud particles, two temperature sensors, an anemometer to measure vertical wind velocities, and a light detector to detect lightning.

The horizontal wind velocity turned out to be an average of 240 km/h (150 mph), with downdraft gusts of up to 12 km/h (8 mph), which were much stronger than expected. The light detector did not detect any unambiguous evidence of lightning, and the nephelometer found no clear areas in the clouds. The balloon drifted over one-quarter of the way around the planet in its two days of operation before the battery failed.

Two days later, Balloon 2 was released from the Vega 2 lander, as it descended through the atmosphere 7° south of the equator. It operated for exactly the same time as Balloon 1. The horizontal velocity was the same as for Balloon 1, but for the first 20 hours the downdrafts for Balloon 2 were very light. After 33 hours, however, Balloon 2 came into a very turbulent area after it crossed over a 5 km high mountain peak, and this turbulence continued for another 2,000 km. Meteorologists were very surprised that a 5 km peak could have such an effect on the atmosphere at an altitude ten times as high as the mountain and for such a great horizontal distance.

Neither of the Vega 1 or 2 landers contained cameras, because the timing of the missions dictated night-time landings, but both carried a drilling system and X-ray fluorescence spectrometer to analyse the surface material. During its descent through the atmosphere, Lander 1 found only two cloud layers, not the three previously observed, and found that the total cloud layer extended from about 60 to 35 km above the surface, rather than stopping at the altitude of about 48 km previously observed. Sulphur, chlorine and possibly phosphorus were found in the clouds, which contained an average of 1 milligram of sulphuric acid per cubic metre of atmosphere.

Unfortunately the drilling system failed on Lander 1. However, there were no such problems with Lander 2, which touched down successfully in the eastern part of Aphrodite Terra, between lowland plains to the north-west and mountains to the east. This was a new type of terrain as the previous landings had been in the lowland or rolling plain areas of Venus. The X-ray fluorescence spectrometer analysed the surface material excavated by the drilling mechanism, and found that it was similar in constituents to the material in the highland areas of the earth and moon.

## Magellan

On 4th May 1989 the American Magellan spacecraft was launched from earth towards Venus aboard the Space Shuttle Atlantis, the first planetary spacecraft to be launched by a shuttle, and the first American planetary probe to be launched for ten years. On board was an imaging radar that was designed to provide a surface resolution of about 120 m, and a radar altimeter with a height resolution of about 10 m. The images produced were a revelation.

The surface of Venus revealed by Magellan (see Figure 10.14) was found to be older than the average age of the earth's surface, as one would expect as there are no oceans, rivers or rain on Venus to modify its surface. But crater counts showed that it is appreciably younger than the surface of the Moon, Mars or Mercury. Most of the surface of Venus appears to be about 400 million years old, with no features older than about 900 million years. Although Magellan found thousands of volcanoes, none seemed to be active at the time of observation.

Most of the 900 impact craters imaged by Magellan were fresh-looking, suggesting that weathering[31] on Venus is a slow process. The largest crater imaged was a highly modified, double-ringed crater called Mead of about 275 km diameter. The lack of any larger craters, and the relative paucity of other large double-ringed craters, was due to the relative youthfulness of the planet's surface. Interestingly no craters were found of less than about 3 km in diameter because the small meteorites, that would normally have produced these small craters, would have been burnt up by the very thick atmosphere. Marks were seen on the surface, possibly caused by the hot gases hitting the surface from such burnt-up meteorites. Craters of between 3 and 20 km in diameter were found to have uneven floors with signs of rubble on them (see Figure 10.15), indicating that the meteorites that caused them were breaking up just before impact. Craters in the 20 to 70 km range were found to be generally pristine in appearance (see Figure 10.16), with a central peak and smooth floor of solidified lava. Some craters were found to have an

---

[31] The weathering on Venus is probably caused mostly by the wind, aided by the high atmospheric pressure and density and the high surface temperature.

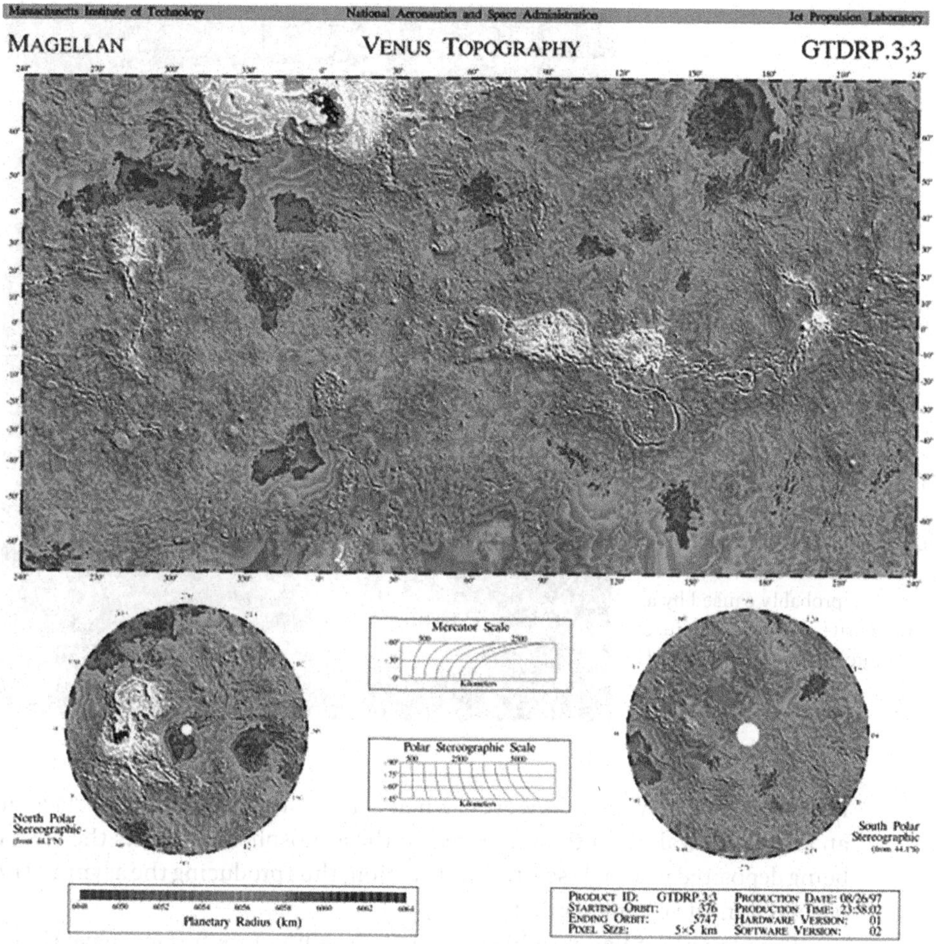

**Figure 10.14** The relative paucity of high and low areas on Venus is clearly seen in this map produced by Magellan's radar altimeter over two years of operation. The highland regions of Ishtar Terra (top left), Aphrodite Terra (right of centre) and Beta Regio (near the left edge) are seen as bright areas, and the low-level planitias are dark. The chasmas cutting through Beta Regio and south of Aphrodite Terra, for example, are clearly seen. **Notes** (a) The Maxwell mountain range in Ishtar Terra is also seen as dark in this map, as it is a black and white copy of a coloured original. (b) The PVO map of Figure 10.12 is a negative image compared with the above. (Courtesy NASA/MIT/JPL.)

asymmetry in their ejecta blankets (see Figure 10.17), unlike similar sized craters on the moon and Mercury. In the case of the moon and Mercury, when a meteorite hits the surface in all but the most glancing impacts, the ejecta is emitted equally in all directions, as there is no atmosphere to prevent

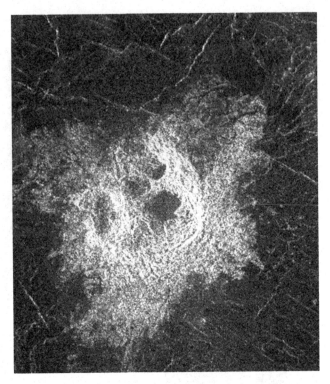

**Figure 10.15** This Magellan image shows a distorted 14 km diameter crater, now called Lillian, which is actually four craters in contact. It was probably caused by a meteorite that broke up just before impact. (Courtesy of NASA/JPL/Caltech; PIA 00476.)

it. In the case of Venus, however, when a meteorite strikes the surface at an angle, the wake[32] of the meteorite in the atmosphere prevents the ejecta being deposited in a back-scattering direction, thus producing the asymmetry observed in the ejecta blanket.

The plains[33] of Venus, which cover about 80% of its surface, have thousands of individual shield volcanoes scattered over them. Otherwise the surface of the plains is either smooth, reticulated, gridded or lobate in appearance. The reticulated plains have irregular grooves, the gridded plains have regular grooves, and the lobate plains are where various lava flows meet. All of the four types of plain were originally produced by lava flows, which in the case of the reticulated and gridded plains have since been modified. So the surface of Venus is relatively young and of volcanic origin. It was something of a surprise, therefore, when Magellan's gravitational measurements, made

---

[32] Because the atmosphere on Venus is so thick, it responds to impacts more like water would on the earth.

[33] The plains described in this section also include the lowland areas of Venus.

**Figure 10.16** Part of the Lavinia Planitia, showing three large impact craters with diameters ranging from 37 to 65 km. Each crater has an almost circular rim, a central peak and a lava-covered floor. (Courtesy NSSDC, World Data Center-A for Rockets and Satellites, NASA; Experiment Principal Investigator, Dr Gordon H. Pettengill, The Magellan Project; P-36711.)

later on in its mission, indicated that the rigid lithosphere of Venus appears to be at least 30 km thick.

The solidified lava flows imaged by Magellan are very interesting, often showing characteristics not seen on earth. In some places the lava had obviously been extremely fluid, running down slopes as shallow as 1°, whilst in other places it had been very viscous. In many ways the fluid lava behaved like water on earth, cutting channels in previous lava deposits to produce features that look more like rivers of water than lava flows (see Figure 10.18). The very viscous lava has produced a number of pancake-like volcanic domes about 20 to 60 km in diameter and about 100 to 750 m high

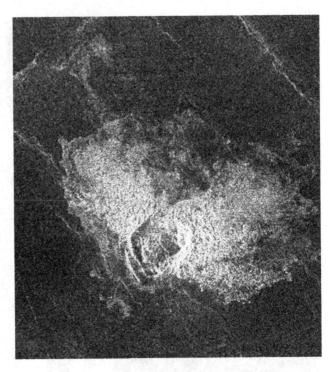

**Figure 10.17** This irregular 8 km crater has an asymmetric ejecta blanket, probably caused by the glancing impact of the meteorite that produced the crater. (Courtesy of NASA/JPL/Caltech; PIA 00474.)

(see Figure 10.19), some of which pre-date faults which have now split them, and some of which cover faults, obviously post-dating them. These pancake domes are not always connected with faults, however, often appearing on relatively smooth surfaces.

Venus shows clear signs of past tectonic activity in the highland regions. The deformational (tectonic) features showing the results of both compressional and extensional forces. Rifting of the crust has occurred to produce relatively shallow chasmas[34] and abundant faulting in the Aphrodite Terra and Beta Regio highlands (see Figure 10.20), whereas compressional forces have produced the mountain belts of Ishtar Terra. It is unclear if there is any limited resurfacing proceeding today due to local tectonic activity. There is no global network of faults like on earth, and no evidence of subduction, showing that there has been no obvious *plate* tectonic activity over the last few hundred million years.

---

[34] The deepest chasmas on Venus, which are similar in structure to the East African Rift Valley on earth, are only about 2 km deep. This is nothing like as deep as the ocean trenches on earth, which can reach a depth of about 8 km below the earth's mean datum level.

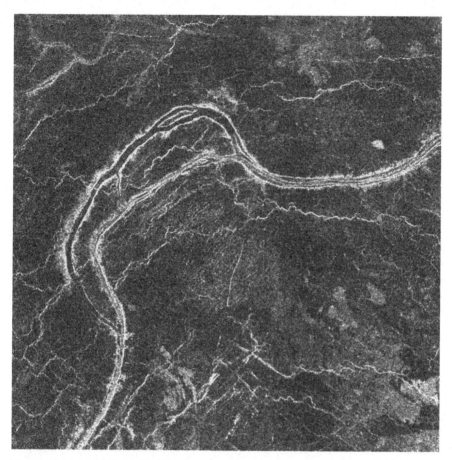

**Figure 10.18** This 2 km wide channel has been created by lava, not water. There are many such channels on Venus, which can be up to several thousand kilometres long. (Courtesy NASA/NSSDC/GSFC; Magellan F-MIDR 45N019;1.)

So ends this general overview of Venus as seen by Magellan, the last Venus probe of the twentieth century. At least 35 spacecraft had been launched towards Venus since 1961,[35] of which about 60% had been successful. This signified a major investment in planetary research but, at least in the early days, the incentive had been political rather than scientific. As high profile missions, they had been a convenient way during the cold war of demonstrating the relative technological capabilities of the two rival

---

[35] The exact number is unclear even today, as the Russians tried to cover up many of their early failures.

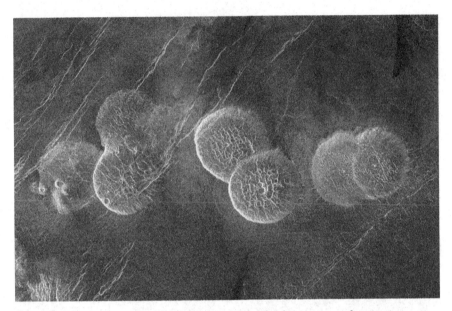

**Figure 10.19** These pancake-like domes are the result of very viscous lava oozing on to the surface. On average, they are about 750 m high and 25 km in diameter, with steep sides and flat tops. (Courtesy of NASA/JPL/Caltech; PIA 00215.)

superpowers.[36] It is no accident, therefore, that as the cold war began to thaw, the number of such missions began to reduce. So the last Venus-dedicated spacecraft of the twentieth century was launched in 1989, the year that the Berlin Wall came down.

## 10.6 Mars

### Pre-spacecraft concepts

For many years it had been known that Mars has dark areas on its surface which appear to change with the season. Before the first spacecraft reached Mars in the 1960s, it was thought that these dark areas were probably covered with lichens or moss, although this was by no means certain, with Dean McLaughlin suggesting that they were ash deposits from volcanoes (see Section 9.7). White polar caps, whose size also changed with the season, were

---

[36] The political background to these and other scientific satellite programmes is explained in *New Cosmic Horizons*, by David Leverington, Cambridge University Press, 2000.

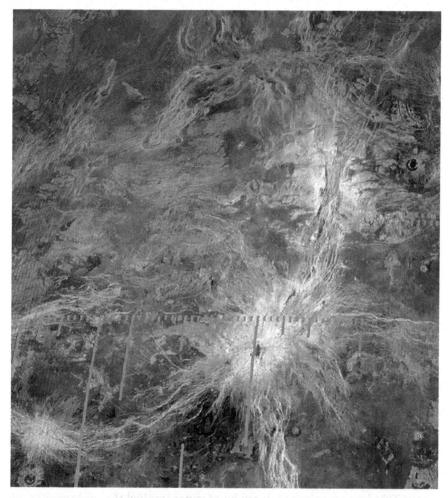

**Figure 10.20** This Magellan image of the Beta Regio region shows two radar bright peaks, Rhea Mons (top) and Theia Mons (bottom), with the Devana Chasma rift cutting through them both. The distance between the two peaks is about 1,000 km. (Courtesy NASA/NSSDC/GSFC; Magellan C3-MIDR 14N300;1.)

thought to consist of very thin water ice or a thick layer of hoar frost. The reddish coloured areas of Mars were variously thought to consist of limonite sand (a hydrated iron oxide) or igneous rock.

Mars' surface atmospheric pressure had been estimated by de Vaucouleurs in 1954 as being about 85 millibars, or a little less than 10% of that of the earth. Then in 1963 Kaplan, Münch and Spinrad deduced a figure of $25 \pm 15$ millibars from Mars' infrared spectrum. In the following year a

NASA sponsored Summer School, using all the most up to date evidence, settled on a range of 10 to 80 millibars as the likely surface atmospheric pressure.

As far as the constituents of the atmosphere were concerned, it was thought that it probably consists mainly of nitrogen or argon, with carbon dioxide as a minority constituent. Although the atmosphere is basically transparent, there appeared to be occasional clouds and dust storms.

Mariner 4

The American Mariner 4, which was the first successful spacecraft to fly-by Mars, provided a big surprise to planetary scientists when it discovered craters on Mars in July 1965. Instead of a planet with patches of vegetation, like moss or lichens, the spacecraft's images showed a cratered, desolate world resembling the moon, with the dark areas visible from earth being simply low albedo areas.

Whenever surprises occur, there is always a review to see if anyone had predicted what had been found. In this case, craters had apparently been seen on Mars by Edward Barnard in 1892 using the Lick 36 inch (91 cm) refractor, and by John Mellish in 1915 using the Yerkes 40 inch (102 cm) refractor, although not everyone believes that Barnard and Mellish saw craters.[37] However, their existence had been predicted by D.L. Cyr in 1944, Clyde Tombaugh, Ernst Öpik, Ralf Baldwin and, most recently, Fred Whipple of the Smithsonian Astrophysical Observatory. Nevertheless, the discovery of craters on Mars was a major surprise to the astronomical community as a whole. The atmospheric pressure measured by Mariner 4 was also a surprise as it turned out to be only about 4–7 millibars at ground level, rather than the 10–80 millibars anticipated, and the 1,000 millibars at the surface of the earth. This low pressure measured by Mariner implied that at least 50% of the atmosphere would be carbon dioxide, given Kaplan, Münch and Spinrad's model for the atmosphere of Mars.

The Mariner 4 images showed no evidence that water had ever existed on Mars, and water could not exist there now with such a low atmospheric pressure, and at the low surface temperatures anticipated from earth-based measurements.[38] The atmosphere was also so thin that it would not shield the surface from solar ultraviolet radiation, which would have killed off any

---

[37] See, for example, W. Sheehan, *The Immortal Fire Within; The Life and Work of Edward Emerson Barnard*, Cambridge University Press, 1995, pp. 258–259.

[38] Mariner 4 made no temperature measurements.

micro-organisms long ago, even if there had been water to sustain them. So conditions were highly unfavourable for life on Mars now, and it appeared probable that there had never been any form of life in the past. But without actually landing on Mars and examining and analysing the surface rocks, astronomers could not be sure.

Mariner 4 also found no measurable magnetic field on Mars, indicating that it has no molten magnetic core. The apparent lack of volcanoes also indicated that the layer of rock near the surface has not been molten for a long time. Finally Mariner found no Van Allen radiation belts, which is consistent with there being no measurable magnetic field, so charged particles emitted by the sun would impact the surface virtually unimpeded, again being a threat to life.

Although the Mariner 4 data indicated that Mars was unlikely to have supported life in the past, some astronomers pointed out that this spacecraft had imaged only about 1% of the surface at a resolution of only 3 km. The other 99% of the planet may look different, and evidence of the previous existence of water, for example, may be seen when the resolution was significantly improved. But the prevailing view following Mariner 4 was probably best summed up by the Mariner 4 imaging team, chaired by Robert Leighton of Caltech, who concluded, and I quote:[39]

- Reasoning by analogy with the moon, much of the heavily cratered surface of Mars must be very ancient – perhaps 2 to 5 billion years old.
- The remarkable state of preservation of such an ancient surface leads us to the inference that no atmosphere significantly denser than the present very thin one has characterised the planet since that surface was formed.
- Similarly, it is difficult to believe that free water in quantities sufficient to form streams or to fill oceans could have existed anywhere on Mars since that time. The presence of such amounts of water (and consequent atmosphere) would have caused severe erosion over the entire surface.

## Mariners 6 and 7

The next two spacecraft to successfully visit Mars were Mariners 6 and 7 which flew past the planet in 1969. Mariner 6, which passed over equatorial regions, measured surface temperatures ranging from 15 °C during the day to −73 °C at night, with a south polar cap temperature during the southern spring of about −125 °C. Similar temperatures had been measured using

---

[39] See NASA News Release 65–249 of 29th July 1965.

earth-bound radiometers, and the cratered terrain imaged by the television cameras was now no longer a surprise. The craters were much easier to see than in the Mariner 4 images, however, and it was clear that the larger craters were shallower than those on the moon with fewer central peaks, although the smaller craters had steeper sides and did resemble lunar craters. Mariner 6 also imaged a new type of terrain centred on about 40° W and 15° S which was almost craterless, with short jumbled ridges and depressions unlike anything seen on the moon. The surface atmospheric pressure was measured as about 6 millibars, about the same as measured by Mariner 4, and the atmosphere was found to be almost 100% carbon dioxide, which was a disappointment to those people hoping to find conditions suitable for life.

Mariner 7 flew over Mars one week after Mariner 6 in a trajectory that allowed better imaging of the south polar region. It again showed a cratered surface, except for the area called the Hellas plain which had no discernible craters. So this area was either very sandy, allowing the craters to be filled in shortly after they had been created, or was much younger than the other areas observed. The surface of the polar ice caps seemed to be covered with frozen carbon dioxide in the form of ice or snow.

Mariner 7 confirmed that the atmosphere was almost 100% carbon dioxide, with clouds of dust and carbon dioxide crystals. There were only minute amounts of water vapour, and the only ozone found was in the region of the south polar cap. The average ground-level atmospheric pressure was confirmed as being about 6 millibars. However, over Hellespontus, which had previously been thought to be a low lying area, the ground-level atmospheric pressure was found to be only 3.5 millibars, indicating that it was, in fact, an elevated area.

So the 1960s ended with the first substantial results from spacecraft sent to Mars. As a result most people had given up the idea of life on Mars, unless it was in a very simple form or not based on carbon and oxygen. The lack of a significant ozone layer added to the problems that living organisms would have to overcome, as the surface was bathed in almost unattenuated ultraviolet light.

The main difference between Mars as understood at the end of 1969, and Mars of before the space age, was that it was now seen to have a highly cratered and apparently dead surface, rather than one possibly covered by lichens or moss. The ground-level atmospheric pressure was only about 0.5% of that of earth, instead of about 1% to 8%, and the disputes on the atmospheric constituents had been largely settled by the discovery that it is almost 100% carbon dioxide, with clouds of dust and carbon dioxide crystals, and with only minute amounts of water vapour. There was also no planet-wide ozone layer, and the surface of the polar caps seemed to be composed of frozen carbon dioxide.

## Mariner 9, Mars 2 and 3, and the great Martian dust storm of 1971

Over the period from 1970 to 1974, when Veneras 7 and 8 and Mariner 10 were examining Venus, both the Americans and Russians were attempting to send spacecraft to Mars, five of which were launched in the 1971 launch window. The Russian spacecraft included landing modules, unlike the American spacecraft which were only designed to orbit Mars. Although two of the launches failed, three spacecraft successfully reached Mars in 1971, namely the American Mariner 9 and the Russian Mars 2 and 3.

Mariner 9 arrived at Mars on 14th November 1971 in the middle of a planet-wide dust storm. It was placed into an initial orbit that was trimmed two days later to be a $1,400 \times 17,100$ km orbit inclined at about $65°$. This allowed imaging of the same piece of ground every seventeen days, to enable study of transient features such as the seasonal colour changes seen from earth and the seasonal advance and retreat of the polar caps. Because of the dust storm, the initial surface images received from Mariner 9 showed limited surface detail but, whilst waiting for the storm to blow itself out, the spacecraft took the first detailed images of the two small Martian satellites, Phobos[40] (see Figure 10.21) and Deimos, showing them to be irregularly shaped and covered in craters. The largest crater on the $27 \times 21 \times 19$ km irregularly shaped Phobos was found to be a massive 9 km in diameter, although the craters on Deimos were significantly smaller. In fact the collision that produced the large crater on Phobos, called Stickney, must have come very close to shattering Phobos.

Although the Mars imaging mission had to be put largely on hold during the dust storm, other experiments produced useful results. In particular, the infrared spectrometer found water vapour over the Martian south polar cap, where it was summer. So maybe there was a substantial amount of water locked up in the south polar cap after all, which was partially released during summer. This discovery created great excitement, even though the previous Mariners had shown that the Martian surface was now apparently dry and

---

[40] Crude photographs had been taken of Phobos by Mariner 7 two years earlier, but no surface detail had been visible.

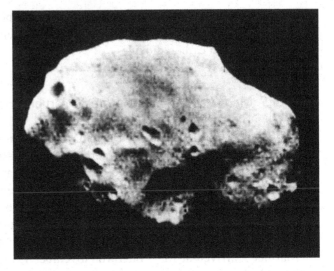

**Figure 10.21** Phobos, the larger of the two small satellites of Mars, imaged by Mariner 9. The notch in the upper right edge is the site of the 9 km diameter crater Stickney. (Courtesy NASA and NSSDC; Mariner 9, MTVS 4109-9.)

cratered. If there was some water vapour in the atmosphere now, maybe there had been water on the surface earlier in its history.

Later observations with Mariner 9 showed that the water vapour content of the atmosphere at mid-latitudes remained fairly constant at 10 to 20 precipitable microns.[41] However, the water vapour content over the south polar region varied from about 10 to 20 precipitable microns during the summer in the southern hemisphere, to zero in winter, whereas over the north polar region the content varied from about 20 to 30 precipitable microns in the northern hemisphere summer to zero in winter. So there appeared to be water ice at both poles that partially evaporated during the local summers. Alternatively, maybe the water vapour content of the atmosphere stayed approximately constant, and the concentrations simply moved from one hemisphere to the other with the seasons.

Glimpses of craters were seen as the dust storm gradually blew itself out towards the end of 1971 and the dust gradually settled. Routine imaging started on 2nd January 1972 and the excitement level reached fever pitch as a new Mars was slowly revealed. This new Mars was not the disappointing, crater-strewn planet revealed by previous Mariners, but an exciting planet with massive volcanoes and a giant canyon system. The largest volcano, Olympus Mons (see Figure 10.22) in the Tharsis region, was 25,000 m (80,000 ft) high and 500 km (300 miles) in diameter,

---

[41] This is the depth of water that would be on the surface if the whole of the water vapour in the atmosphere precipitated out.

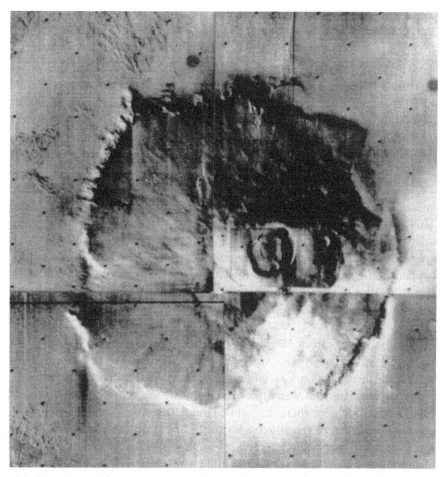

**Figure 10.22** The largest volcano on Mars, Olympus Mons, imaged by Mariner 9 in 1972, is 80,000 ft (25,000 m) high. Its 65 km diameter summit caldera is seen just right of centre. (Courtesy NSSDC, World Data Center-A for Rockets and Satellites, NASA; Principal Investigator, the late Dr Harold Masursky; P-12834.)

which is far larger than any volcano on earth.[42] What is even more remarkable is that this massive volcano is on a planet only half the diameter of the earth. Schiaparelli had, in fact, noticed a small whitish patch in this area in 1879, and had christened it Nix Olympica, the Snows of Olympia. Mariner 9 imaged three other large volcanoes in the Tharsis region, where

---

[42] Mauna Loa in Hawaii is the largest volcano on Earth, rising 9,000 m above the sea floor and being 120 km in diameter at its base.

a number of ground-based observers had also seen small whitish patches.[43] All these patches were now seen to be due to clouds near the tops of the volcanoes.

The large canyon system imaged by Mariner 9, now called the Valles Marineris,[44] was found to be vast (see Figure 10.25 later). It was seen to be 4,000 km (2,500 miles) long, up to 200 km (120 miles) wide, and up to 6,000 m (20,000 ft) deep.[45] The Hellas plain turned out to be a 2,300 km diameter impact crater or basin whose floor, at the lowest point, is some 5,000 m (16,000 ft) below the mean Martian surface level. The white deposits often seen in this Hellas basin were found to be carbon dioxide frost.

The discovery of huge volcanoes like Olympus Mons and the other three volcanoes in Tharsis indicated that there has been no plate tectonics on Mars. This was because, with no plate movement, hot spots in the underlying mantle had continued to eject lava through the same place in the crust year after year to build up these enormous volcanoes. On earth this does not happen as the crust is in motion relative to the mantle. It was also thought that the Valles Marineris could possibly be where two plates started to move apart, but were stopped as Mars cooled quickly, shortly after the crust was formed.[46]

As the images from Mariner 9 continued to be received on earth, astronomers started to wonder, since Mars looked much more interesting than before, and clearly has at least some water vapour in its atmosphere, whether there may be signs that water had existed on the surface of Mars in the past. Imagine their delight, therefore, when sinuous channels were discovered in an area called Chryse Planitia (see Figure 10.23). However, because of a combination of low atmospheric pressures and low surface temperatures, it was clear that water could not exist in liquid form on the surface of Mars today. But it could well have existed in the past when the atmosphere was probably denser.

As far as the possibility of life was concerned, Mariner confirmed that Mars does not have any ozone in its atmosphere between about 45° N and 45° S. However, Mariner did detect ozone over both polar regions down

---

[43] These volcanoes, now called Arsia Mons, Pavonis Mons and Ascraeus Mons, corresponded to the telescopic spots previously called Nodus Gordii, Pavonis Lacus and Ascraeus Lacus.

[44] After the Mariner spacecraft that discovered it.

[45] The Grand Canyon in the USA is only 150 km long, and at most 2,000 m deep.

[46] Mars would cool much more quickly than the Earth because it is only about half the Earth's diameter.

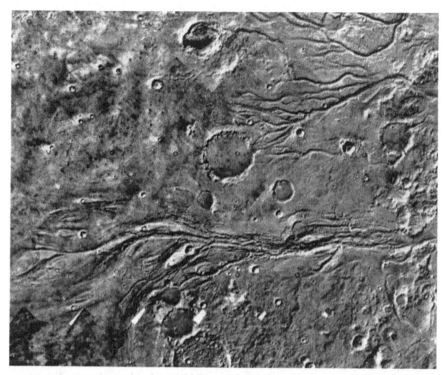

**Figure 10.23** A Mariner 9 image of channels in the Chryse region of Mars that appear to have been made some time ago by running water. To get an idea of scale, the large crater just above centre is about 75 km in diameter. (Courtesy NSSDC, World Data Center-A for Rockets and Satellites, NASA; Experiment Team Leader, Dr Michael H. Carr, US Geological Survey.)

to about 45° latitude during their respective winters. But in their respective springs, the ozone gradually started to disappear, so that by early summer it had completely disappeared from the appropriate hemisphere. It was thought that there may well be a direct linkage between the increase of ozone and the reduction of water vapour with season in high latitudes, as ozone was known to be destroyed by the products of water photolysis.

It had been known for many years that parts of Mars, as seen from earth, changed colour with season. The cause of this had long puzzled astronomers, some of whom had interpreted it as signs of plant life on the surface. Unfortunately, Mariner 9 was able to show that the colour change was due to nothing more than seasonal winds alternately covering and uncovering the darker substrate with lighter dust. In fact, wind velocities of up to 180 km/h

(110 mph) were measured by observing clouds in the lee of the Martian mountains.

Unfortunately, the landers from the Mars 2 and 3 spacecraft had landed during the Martian dust storm of late 1971. Probably damaged by dust, the Mars 2 lander never transmitted any data from the surface, and the Mars 3 lander only transmitted 20 seconds of test data before falling silent. The Russians had much more success with their Mars 2 and 3 orbiters, however, which measured a very weak Martian magnetic field of 30 gammas. This is not much above that of the interplanetary magnetic field, and is about three orders of magnitude less than that of the earth. The infrared detectors confirmed that there is a small amount of water vapour in the Martian atmosphere. The surface of Mars was found to cool quickly after sunset suggesting that it was covered with dust or sand, although the darker areas, which were warmer than the other areas during the day, generally cooled more slowly after dusk indicating that they were made of rock.

The Russian orbiters found that the dust particles in the dust storm were about 60% silicon, in agreement with Mariner 9 results, with sizes generally in the range from 2 to 15 μm. Although the largest particles precipitated out of the atmosphere quickly once the winds had died down, the smaller particles took months to reach the surface, as the dust had ascended, during the storm, to heights of more than 30 km (100,000 ft). During the dust storm the surface cooled by about 20 to 30 °C, as the dust obscured the sun, whilst the atmosphere warmed up, as the suspended dust absorbed solar energy.

In summary, Mariner 9 had shown us a much more interesting Martian surface than previously seen, because the previous three spacecraft had only imaged a total of 10% of the surface close-up which, as it happens, did not include the most interesting features. Mariner 9 had also found a small amount of water vapour in the atmosphere, which Mars 2 and 3 had confirmed. There is no liquid water on the surface today, however, but there were signs in some of the Mariner images that water may have flowed in significant amounts in the past.

If significant amounts of water had been present earlier in the history of Mars under a more substantial atmosphere, then maybe life could have been present. As the planet gradually lost its atmosphere, however, and the water evaporated, without a protective ozone layer, damaging solar radiation would probably have killed off all but the most resistant forms of life. But could such life still exist today? Maybe it could do so, shielded from the worst of the solar radiation under rocks or beneath the surface. After all, it is possible that there may be water still trapped beneath the surface. All of this may sound improbable, but our exploration of the universe has regularly shown us surprises, and in some cases things have been found which at the time had been thought impossible. Finding life on another planet would have such important psychological and religious consequences that we should leave no stone unturned (literally!) to find such evidence. This was the prevailing view in NASA in the early 1970s.

Clearly a lander would be required to analyse samples of the Martian surface to see if there was any evidence of biological activity past or present, and this is what was intended with the American Viking probes.

## Viking

The main purpose of the Viking lander missions, which were launched in 1975, was to search for life. So the landing sites chosen for the two spacecraft were those where there was a maximum possibility of finding water, which implied that they should be low lying. In fact, the landing site chosen for Viking 1 was in the Chryse region where sinuous channels had been discovered by Mariner 9 (see Figure 10.23) in an otherwise relatively flat region, and that for Viking 2 was in the Cydonia region near the southern fringe of the north polar hood, where it was hoped to find evidence of liquid water from the melting polar cap in the northern summer. Both were very low lying sites.

The Viking spacecraft each consisted of an orbiter and a landing module. The orbiter had a twin television camera system, a water detector, and an infrared instrument to map the surface temperature. Initially the television cameras and water detectors would try to locate a smooth, moist surface on which to land the landing module and, when that part of the mission had been successfully accomplished, the television cameras would produce high resolution images of the whole planet.

The most sophisticated instrument package on the lander was a miniature biological laboratory to test the soil, which was deposited in the analysers by a scoop, for signs of elementary forms of life. This laboratory package consisted of three different experiments, called the pyrolitic release, labelled release and gas-exchange release experiments.

In the pyrolitic release experiment, a soil sample was heated using a lamp that simulated Martian sunlight, minus the ultraviolet, in an atmosphere of carbon dioxide containing a small amount of radioactive carbon 14. If organisms ingested the carbon dioxide, they would also take in the carbon 14 tracer. After five days the chamber was flushed with an inert gas, and then the soil/organism mix was heated up to a temperature of about 625 °C to see if carbon 14 was released. If it was, this would be taken as an indication that there had been living organisms in the soil.

In the labelled release experiment, water and nutrients labelled with (i.e. containing) carbon 14 were fed to the soil sample. It was then expected that any organisms present would consume the nutrients and release gases labelled with carbon 14 which would be detectable.

Finally, in the gas-exchange release experiment unlabelled nutrients were fed to a soil sample in a humid carbon dioxide based atmosphere. Detectors then searched for hydrogen, nitrogen, oxygen and methane produced by the organisms.

Viking 1 arrived at Mars on 19th June 1976, and spent a month in orbit around the planet searching the chosen Chryse region for a suitable landing site. Eventually a suitable site was found at 22.5° N, 47.5° W, and on 20th July the Viking 1 lander separated from the orbiter and touched down close to its target. Touch down was on the western side of the Chryse region, 2.7 km below the datum representing Mars' average surface elevation.

Mariners 6 and 7 had shown that Mars' atmosphere was almost 100% carbon dioxide, but during Viking 1 lander's entry sequence its mass spectrometer, whilst confirming this result, also detected small amounts of nitrogen, argon, carbon monoxide, oxygen and nitric oxide in the upper atmosphere. The discovery of nitrogen was a great fillip to those exobiologists hoping to find evidence of life, as it was believed that nitrogen was an essential ingredient in any environment in which life had evolved.

Twenty-five seconds after landing, the first ever image from the surface of Mars began to be transmitted, and over the next few months numerous images of the red, rock-strewn, sandy surface and pink sky were received and analysed on earth (see Figure 10.24). It was early summer in the northern hemisphere when the Viking 1 lander touched down, and on the first day the atmospheric temperature varied from −86 °C at dawn to −33 °C in the early afternoon. The atmospheric pressure was 7.6 millibars, and the wind velocity gusted up to 52 km/h (32 mph). It was found that there was little change in the temperature cycles on a day to day basis, except during dust storms when the diurnal variation was much reduced.

Later in the northern summer a maximum atmospheric temperature of −14 °C was measured in the early afternoon at the Lander 1 site, with a dawn temperature of −77 °C, and an atmospheric pressure of 6.8 millibars. During the northern winter the atmospheric pressure increased to 9.0 millibars, as the southern polar cap released some of its carbon dioxide. Evidently the southern polar cap was releasing more carbon dioxide in the southern summer than the northern cap was collecting in the northern winter.

**Figure 10.24** Mars as seen by the Viking 1 lander, showing small wind-blown sand dunes and a plethora of rocks. The large boulder at left of centre is about 1 m high. The rocks show clear signs of weathering, most likely caused by wind-blown sand. (Courtesy NASA and NSSDC.)

The soil was analysed by an X-ray fluorescence spectrometer to be iron-rich clay with 21% silicon and 13% iron as its major constituents, and the atmosphere was found to consist of 95.3% carbon dioxide, 2.7% nitrogen, 1.6% argon and 0.13% oxygen. Interestingly, both the soil and the very fine dust were found to have an abundance of magnetic particles. On earth the most common magnetic particles are iron or magnetite, which is an iron oxide. The other common iron oxide on earth, haematite, which is red, is non-magnetic. So it was suggested that the red coloration and the magnetic properties of the particles on Mars could be explained if they were made of magnetite with a covering of haematite. A mass spectrometer found no evidence of organic matter in the surface sample tested, indicating that there was no life in it, but the biological experiments produced confusing results.

The first results from the biological experiments came from the gas-exchange experiment (GEX), where a large amount of oxygen was measured as soon as the surface sample was humidified. Then large amounts of radioactive carbon dioxide were produced in the labelled release experiment (LR) when the soil sample was moistened with nutrients. The results of the GEX and LR experiments were so clear and startling, particularly when compared with the lack of organics found by the mass spectrometer, that the experimenters began to doubt whether they were really measuring the effects of life, and various chemical reactions were suggested as a cause of the results. The problem was that, when the soil sample in the LR experiment was heated up to 170 °C to kill off any life and then re-tested, no radioactive gases were detected. So, either living organisms had been killed off, or the heat had broken up the chemical constituents so that they no longer reacted. A similar effect was also found with the pyrolitic release experiment (PR) which gave positive results with normal soil, but showed no activity after the soil had been heated. Later results of the PR experiment were not as clear-cut in showing this effect, however.

At face value the GEX, LR and PR experiments all showed that there was life on Mars yet that was inconsistent with the mass spectrometer results that showed that there was no organic material in the soil samples. It was possible that a small number of living organisms could have activated the biology experiments but not have been sufficiently numerous to be detected by the mass spectrometer. For every living organism on earth, however, there are thousands of dead ones, and there should have been enough dead organisms on Mars to have been detected by the mass spectrometer. Maybe there was life on Mars but the organisms were highly efficient cannibals, leaving few dead organisms, or maybe there was no life and the GEX, LR and PR results were all caused by chemical reactions. Either was possible.

Viking 2 arrived at Mars on 7th August, but imaging of its proposed landing site showed that it was too rough for a safe landing. After a number of orbit manoeuvres a new site was found in Utopia Planitia at 47.9° N, 225.8° W, some 3 km below the datum level. This was about 7,000 km from the Viking 1 lander and about 4° further north than originally planned. A more northerly position was chosen because it was hoped that the soil would have more moisture, being that much closer to the north polar cap. In the event the lander landed just 10 km (6 miles) from this chosen site.

The images sent back to earth by the Viking 2 lander showed a similar red, rock-strewn surface to that seen by Viking 1. This was something of a surprise as the scientists had anticipated seeing gently rolling sand dunes with more dust and less rocks than for the Viking 1 site. The eagerly awaited results of the biology experiments on Viking 2 were as inconclusive as those of Viking 1. Even today, in fact, the reasons for these strange results from the biology experiments are not clear, although most scientists think that they were produced by chemical reactions rather than living organisms.

The atmospheric temperature at the more northerly Viking 2 site was about 5 to 10 °C lower than those at the Viking 1 site. For example, the Viking 2 lander measured a minimum temperature of −87 °C during the northern summer, compared with −77 °C for Viking 1. During the winter the minimum temperature fell to −118 °C at the Viking 2 site. Frost was first observed on the surface near the Viking 2 lander in September 1977 when the temperature was −97 °C, after an overnight low of −113 °C.

Atmospheric pressure at the slightly lower Viking 2 site was somewhat higher than at the Viking 1 site, varying from 7.3 millibars in the summer to 10.8 millibars in the winter. Wind velocities were generally low at both sites, with typical figures of 7 km/h (4 mph) at night and up to 25 km/h (15 mph) during the day. Many astronomers were surprised to find that the maximum wind velocity measured at either site during dust storms was only 120 km/h (75 mph), compared with an expected value of two or three times higher.

Over their lifetimes, the Viking 1 and 2 orbiters imaged virtually 100% of the Martian surface at a resolution of at least 300 m (see Figure 10.25), with 2% being imaged at a resolution of about 25 m.[47] Later in the mission the periapsis of both the Viking orbiters was reduced to 300 km, allowing selected areas to be imaged at a resolution of just 8 m.

The Viking images showed clear signs of flash floods in the Chryse region (see Figure 10.26), and many landslides, some of which seem to have occurred in saturated soil. Dendritic or branching drainage features were also seen resembling terrestrial river systems. Temperature and water vapour sensors on the orbiters showed that the north polar cap (see Figure 10.27) is made of water ice in the northern summer at a temperature of about −65 °C. It is then covered by carbon dioxide ice in the northern winter. In the case of the southern cap, however, the carbon dioxide ice does not completely melt in the southern summer. Clearly some of the water ice in the northern polar cap evaporates in the northern summer, which explains why there is more water vapour in the northern hemisphere during its summer, than in the southern hemisphere during its summer. There was also evidence of extensive permafrost regions, with the permafrost around the polar caps possibly extending down to a depth of several kilometres in places.

Mariner 6 had shown that the south polar cap temperature was about −125 °C, which is the equilibrium temperature of carbon dioxide at the south polar atmospheric pressure of about 6 millibars. The infrared thermal mapper on the Viking orbiter indicated a south polar temperature as low as −139 °C, however, which was difficult to explain unless there was a higher

---

[47] A detailed comparison was made of the Mariner 9 images taken during the second half of 1972 with those from the Viking orbiter. They showed that the atmosphere had not fully settled after the large dust storm observed by Mariner 9 over six months earlier. In particular, the residual atmospheric dust had masked some surface features, particularly large lava flows, in the Mariner 9 images that were now clear in those from Viking.

**Figure 10.25** This global mosaic of Viking 1 Orbiter images shows the 4,000 km long Valles Marineris cutting across the centre, with the Tharsis volcanoes being the dark circular features on the left. (Courtesy NASA/NSSDC/GSFC; Viking 1 Orbiter, MG07S078-334SP.)

concentration of nitrogen and argon at the south pole than elsewhere on Mars. Nitrogen and argon would not condense at this temperature, and so they could easily provide the extra cooling, but it was not clear if there were higher concentrations of these gases than normal at the south polar region.

The Viking orbiters also measured much larger daily temperature ranges than previous spacecraft. In some places for example the temperature went from a daily maximum of 4 °C to a minimum of −133 °C, which was difficult to explain. The daily thermal profiles of other areas did not match

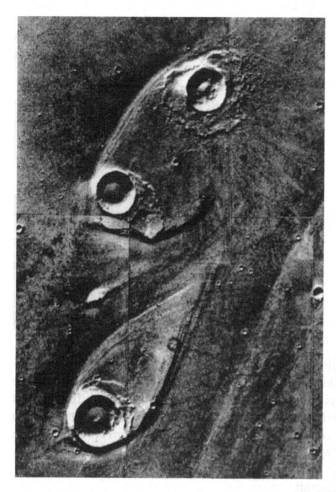

**Figure 10.26** Evidence of flash floods at the mouth of the Ares Vallis near the southern boundary of the Chryse Planitia. (Courtesy NSSDC, World Data Center-A for Rockets and Satellites, NASA; Experiment Team Leader, Dr Michael H. Carr; 76.H.480.)

predictions either, falling more rapidly than expected in the late afternoon, before slowing down to match the predicted temperatures during the night.

The water vapour in the Martian atmosphere was found to be highest in low lying regions and, as with Mariner 9, more water vapour was found during the summer than during the winter, when it was presumably frozen out of the atmosphere. In regions of rough terrain there were marked daily fluctuations in the amount of water vapour, possibly due to changing wind patterns. The atmosphere near the edge of the residual northern polar cap was found to be saturated with water vapour in the northern summer.

The images of both Phobos and Deimos from the Viking orbiters were much sharper than from Mariner 9 and, at one stage, Viking Orbiter 2 was

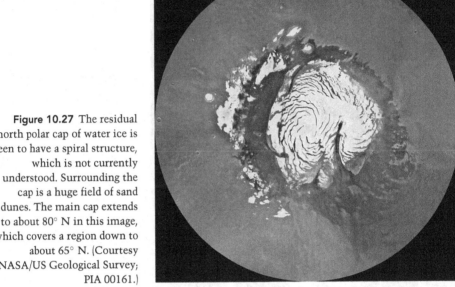

**Figure 10.27** The residual north polar cap of water ice is seen to have a spiral structure, which is not currently understood. Surrounding the cap is a huge field of sand dunes. The main cap extends to about 80° N in this image, which covers a region down to about 65° N. (Courtesy NASA/US Geological Survey; PIA 00161.)

manoeuvred to within only 30 km of the surface of Deimos, producing images with a resolution of 3 m. Phobos was found to have a network of linear grooves, which appear to be fractures caused by the collision that created the 9 km crater called Stickney. On the other hand, Deimos was seen to have a smoother surface than Phobos, and some of its craters appeared to be partially filled by material.

## The Phobos spacecraft

Mars' 27 × 21 × 19 km irregularly shaped satellite Phobos had been shown by earlier Mars spacecraft to be covered in craters (see Figure 10.21), and in some ways it looked a very uninteresting place to examine in any detail. But it was thought that Phobos and Deimos were probably captured asteroids and, as such, were probably bodies left over from the formation of the planets. This was why Phobos was of interest.

So in 1988 the Russians launched two spacecraft to fly-by Phobos and deploy landers. Unfortunately Phobos 1 failed, but Phobos 2 successfully reached the vicinity of Mars in January 1989 and fired its retrorocket to put it into orbit around the planet. Phobos 2 imaged Mars over the next few weeks, and on 18th February its orbit was circularised with imaging of

Phobos starting three days later from a distance of 860 km. The orbit of the spacecraft was then gradually modified ready for the intercept on 9th April but, unfortunately, near the end of March contact was permanently lost.

Before contact was lost, Phobos 2 took thirty-seven images of Phobos covering about 80% of the surface at resolutions of up to 40 m. This included some parts not previously imaged by the Viking orbiters. Before the Phobos 2 mission it was thought that Phobos was probably similar in composition to a carbonaceous chrondrite meteorite, but Phobos 2 showed that the surface colour was not as homogenous as previously thought, so reopening the question of its composition. The density of Phobos was measured as 1.95 g/cm$^3$, slightly less than anticipated and low enough to imply that it may be partially composed of ice. It was, therefore, a surprise to find that its surface temperature reached 27 °C in sunlight which, according to the spectrometer, produced a bone-dry surface. So Phobos 2 had produced some interesting data about Phobos but, because its mission was so cruelly cut short, it had raised more questions than it had answered.

## Meteorite ALH 84001

In 1985 a 1.9 kg meteorite known as ALH 84001, which had been picked up in the Allen Hills in Antarctica the previous year, was chemically analysed. As a result, it was put in the diogenite category of meteorites that are thought to come from the asteroid Vesta. There matters rested until 1993, when David Mittlefehldt of Lockheed found trace materials in ALH 84001 that do not exist in diogenites. Further analysis showed that ALH 84001 is one of a group of ten similar meteorites that are thought to come from Mars, having been thrown into space by a large impact millions of years ago. But this meteorite was different from the other Martian meteorites in one key respect; it was considerably older. In fact, it appears to have crystallised slowly from molten rock beneath the surface of Mars about 4.5 billion years ago.

Once in space, a meteorite is subject to cosmic ray bombardment, and this changes the isotopic balance of its material. In the case of ALH 84001, an analysis of such isotopic abundances showed that the meteorite had spent about 16 million years in space, before landing in Antarctica. There it had been covered in snow and ice for something like 13,000 years, until subsurface ice flows had brought it to the surface. As a result, we now have, at next to no cost, a 1.9 kg piece of the ancient crust of Mars. So what does it tell us?

A team of scientists lead by David McKay and Everett Gibson Jr., of NASA's Johnson Space Center, spent the next two years analysing ALH 84001. Then in August 1996 they made the startling declaration that they had found evidence in the meteorite for early life on Mars. The team acknowledged that each of their four pieces of evidence could be disputed, however, but claimed that, taken as a whole, it added up to strong circumstantial evidence for early life on Mars.

First, they found strange looking, tiny carbonate globules that were apparently formed about 3.6 billion years ago. These carbonate globules gave strong evidence for the presence of water at the time of their formation. Earlier in its life the rock had apparently been subjected, along with the rest of Mars, to a heavy bombardment. This had caused it to crack, which had allowed water to percolate through the rock and form the globules that we see today.

Second, Richard Zare and colleagues at Stanford University found organic molecules in ALH 84001, the first ever discovered in a Martian rock. Specifically, the molecules were polycyclic aromatic hydrocarbons, or PAHs as they are know. The evidence for previous water and now the evidence of carbon in this meteorite gave a powerful boost to the idea that life could have existed on Mars.

Third, the team found minute grains of magnetite, gregite and pyrrhotite inside and on the surface of the carbonate globules. On earth such small grains of these minerals are usually produced by bacteria and other microbes.

Then fourth and finally, images taken with a scanning electron microscope showed clusters of elongated, worm-like shapes (see Figure 10.28), no more than 0.1 µm long, in and near the carbonate globules. These elongated shapes appeared remarkably similar to those of so-called microfossils which have been found on earth and which had been formed about 3.5 billion years ago.

The announcement of these four strands of evidence for early life on Mars created a sensation at the NASA press conference at which it was released on 7th August 1996. As a result, the photographs of the hypothesised microfossils appeared in newspapers and technical magazines all over the world. Immediately, the sceptics weighed in, however.

There was disagreement about the temperature of the water from which the carbonate globules had been formed, with some scientists suggesting a figure well in excess of 100 °C, which would have made life virtually impossible. Then there was the question of contamination; maybe the PAHs were due to contaminants in Antarctica. But if the PAHs did come from space, they are known to be present in other meteorites and are not, on their own, taken as evidence of life. Unfortunately, the existence of magnetite and the other minerals does not prove the existence of microbes, either, as these minerals could have been formed in other ways. And then there was the

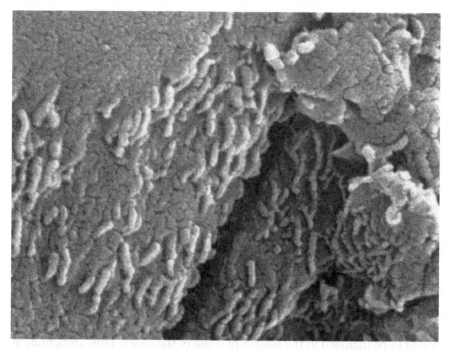

**Figure 10.28** A scanning electron microscope image showing clusters of elongated shapes associated with carbonate globules in the ALM 84001 meteorite, which is thought to have come from Mars. They have been interpreted as possible 3.6 billion year old microfossils from Mars. (Courtesy NASA.)

problem of the supposed Martian microfossils, as they are 100 times larger than microfossils on earth.

In the face of these criticisms, McKay and Gibson emphasised that the evidence has to be taken as a whole, as no individual piece is conclusive on its own. But their opponents argued that each of the individual pieces of evidence was flawed, and so the overall conclusion was flawed also. McKay and Gibson then went on to present more evidence in support of their argument, but today discussions are still ongoing as to whether ALH 84001 has provided evidence for early life on Mars or not.

## Pathfinder

NASA returned to the surface of Mars in 1997, after a wait of twenty years, with the Pathfinder spacecraft and its six-wheeled rover vehicle called Sojourner. This was the first vehicle to be placed on Mars, enabling ground-based observers to direct it and its alpha proton X-ray spectrometer to analyse the Martian soil and any interesting surface rocks in the vicinity of the lander. Its landing site had been chosen in the Ares Vallis, as this area seemed to be the site of an ancient flood plain (see Figure 10.26 earlier), which the Pathfinder was expected to confirm.

The first large rock to be observed and analysed by Sojourner was found to be covered in small depressions of the sort left by bubbles of gas whilst a rock is cooling. This suggested that the rock was volcanic but, much to everyone's surprise, the silica content was found to be unusually high, indicating that it was not a normal basalt. Instead it appeared to be from an underground magma ocean in which the heavier elements had time to separate out and fall towards the centre of the planet. Some other rocks were also found by Sojourner to have a high silica content, but others did not. Unfortunately, the rocks were covered with varying amounts of red Martian dust, which was lower in silica and higher in sulphur, which probably affected the analysis results.

Great excitement was caused amongst the ground-based scientists when Sojourner found a number of rounded pebbles, of the sort seen at the sea-shore on earth, together with a small sand dune. In addition, some pebbles were found to be included in the surface of some of the rocks. This configuration was reminiscent of that of so-called conglomerates on earth, which are the result of a long time exposure to running water. Furthermore, the large rocks in the vicinity were observed to be generally leaning in the same direction as each other, indicating that they had been subjected to a strong flow of water in the past. These discoveries taken as a whole provided powerful evidence for the previous existence of flowing water on Mars.

## Mars Global Surveyor

NASA had originally intended Pathfinder as an experimental 'proof of concept' spacecraft to which they later added scientific instruments, whereas the Mars Global Surveyor (MGS) was planned as a scientific spacecraft right from the start. Although launched a few days before Pathfinder, MGS took a more leisurely trajectory to Mars, arriving in September 1997. It then used aerobraking[48] in the Martian atmosphere to reach its final sun-synchronous orbit, in which it passed over the Martian equator at the same local time on each orbit.

---

[48] Aerobraking is the technique of using air resistance to slow a spacecraft down and so modify its orbit.

The first significant scientific results from the MGS were forthcoming during the aerobraking process, when it was found that Mars' magnetic field is not dipolar like the earth's. Instead there is a strong remanent magnetic field over only the older parts of the Martian surface. This was highly significant, implying as it did that Mars had had a molten core in the distant past, which had produced a magnetic field like the earth's, whereas today the core does not produce a magnetic field. So, either the core is made of very different materials to that of the earth, or its internal motion is very much slower, or possibly a mixture of both.

Another interesting finding made during the early aerobraking phase was the remarkable flatness of Mars' northern plains, which led some astronomers to believe that this area is the dried up bed of an ancient ocean. This concept was reinforced when it was found that much of the shelf defining the 'shoreline' is at the same altitude around this hypothesised former ocean.

Recently the picture of Mars has been changed significantly by an analysis of the MGS images. Instead of the concept of Mars as a planet that has been more or less dead geologically for well over the last billion years, it appears that significant changes have occurred over that timescale. The evidence is still not fully clear but, nevertheless, it is still fascinating.

For example, until ALH 84001 was discovered, the oldest known meteorite from Mars was about 1.3 billion years old. But recently it has become clear that these meteorites from Mars, excluding ALH 84001, and others discovered since, fit into two basic groups. One group consists of those that are made from lava that solidified about 1.3 billion years ago, and which were ejected from Mars about 11 million years ago. The second group consists of those that solidified on Mars only 100 to 700 million years ago. At one stage it had been thought that the relatively young age of the latter group was because these meteorites had been melted that long ago by the impact that threw them into space. But now it appears as though the 100 to 700 million years is the period in which they originally solidified on Mars, indicating that there had been geological activity on Mars until about 100 million years ago. Furthermore, like ALH 84001, carbonate deposits or clay materials have been found inside these meteorites, indicating that they had been subjected to a water environment before they were ejected from the planet. If liquid water or concentrated water vapour was present on Mars just a few hundred million years ago, it may well be present there today, possibly just under the Martian surface.

The young age of some of the lava flows on Mars has been confirmed by the MGS. Images of lava flows in the Elysium Planitia region, for example,

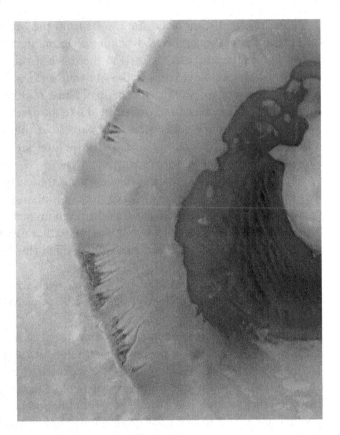

**Figure 10.29** This image of part of an impact crater in Noachis Terra shows possible water seepage from a sub-surface layer near the top of the crater walls. In the crater floor is a relatively smooth, dark region with 'islands' of brighter material. Whether the dark area is the remains of a lake is still uncertain. (Courtesy Malin Space Science Systems/NASA; MOC 7707.)

show that they have hardly been modified by impact craters, indicating that they are very young, probably only a few tens of millions of years old, and cratering counts of the youngest lava flows on Olympus Mons indicate that the age of these flows is less than 100 million years. Evidence is also being found that some of the channels and valleys on Mars were formed by water very recently. Interestingly, Michael Carr of the US Geological Survey has pointed out that the river channels often start very abruptly, with collapsed pits 'upstream' of the river head indicating that the channel has been fed by underground water.

Probably the best evidence yet produced by MGS of the seepage of underground water to the surface is that shown in an image of a 50 km diameter impact crater in Noachis Terra (see Figure 10.29). Not only is there evidence of seepage patterns at the top of the crater walls, but there is evidence of water, which may be frozen or covered with sand, on the crater floor.[49]

So ends this brief summary of our concepts of Mars as they have developed over the years. Clearly great strides have been taken since the start of the space age. But, although we know a great deal more than we did just forty years ago, there is still much that we do not know. In particular, has there ever been life on Mars, and are there really large amounts of water just under the surface? The second question could possibly be answered by a spacecraft in the near future, but to answer the first we will probably need to return rock samples to earth. However, whether automatic sample return missions currently under planning will be sufficient, or whether we will eventually have to send astronauts to Mars to answer this question, only time will tell.

---

[49] Evidence for substantial amounts of sub-surface water was announced by NASA whilst this book was in the process of publication. The newly arrived Mars Odyssey spacecraft found evidence for large quantities of sub-surface hydrogen, using its $\gamma$-ray and neutron spectrometers. This hydrogen was thought most likely to be in the form of water ice.

# Chapter 11 | THE SPACE AGE – THE OUTER PLANETS

## 11.1 Jupiter

Pioneer 10

Pioneer 10, which was the first spacecraft to visit the outer planets, was launched towards Jupiter by NASA on 3rd March 1972. It passed the orbit of Mars in May of that year and flew past Jupiter on 4th December 1973 after a journey of just 21 months. Meanwhile on 6th April 1973 Pioneer 11, its near twin, was also launched towards Jupiter, with the intention that it should then fly past Saturn if all went well.

Pioneer 10 first detected high energy electrons from Jupiter at about 300 Jupiter radii $(R_J)$ from the planet. It then passed through Jupiter's bow shock, where the solar wind interacts with Jupiter's magnetic field, whilst still about 7.7 million km, or about $108\,R_J$, from the planet. The discovery of high energy electrons from Jupiter in front of the bow shock was a surprise, as this meant that these electrons had escaped from the planet's magnetosphere and had somehow crossed the shock wave. The distance of the bow shock from Jupiter was also greater than expected, indicating that Jupiter's radiation belts may be more active than anticipated.

On crossing the bow shock, the magnetic field measured by Pioneer 10 increased from 0.5 to 1.5 gammas, the temperature of the solar wind increased dramatically and its velocity was reduced by a factor of two. On the following day the magnetic field suddenly increased to about 5 gammas when the spacecraft crossed Jupiter's magnetopause into its magnetosphere, a region protected from the solar wind and dominated by the planet's magnetic field. Then three days later, much to everyone's surprise, the magnetic field suddenly decreased back to its interplanetary level and the solar wind

Table 11.1 *Magnetic fields of the major outer planets compared with that of the earth*

|  | Earth | Jupiter | Saturn | Uranus | Neptune |
|---|---|---|---|---|---|
| Equatorial radius (km) | 6,380 | 71,400 | 60,200 | 25,500 | 24,760 |
| Inclination of equator to orbit | 23.4° | 3.1° | 26.7° | 97.9° | 28.8° |
| Angle between magnetic and spin axes | 11.5° | 9.6° | 0° | 58.6° | 47.0° |
| Dipole offset (km) | 460 | 7,000 | 2,400 | 7,700 | 14,000 |
| Dipole offset (radii) | 0.07 | 0.10 | 0.04 | 0.30 | 0.55 |
| Magnetic field at equator (gauss)[a] | 0.31 | 4.28 | 0.22 | 0.25 | 0.15 |
| Dipole moment (Earth = 1) | 1 | 19,000 | 540 | 50 | 28 |
| Av. sunward distance of magnetopause (planetary radii) | 10.4 | 70 | 22 | 20 | 27 |

[a] These magnetic field values are at the surface of the earth or at the cloud tops of the other planets.

was detected once more. Evidently an increased pressure of the solar wind had pushed the magnetopause closer to Jupiter and it and the associated bow shock had overtaken the spacecraft, but eventually the spacecraft crossed the bow shock and magnetopause once more. As Pioneer 10 flew past Jupiter it found that its radiation belts, where protons and high energy electrons are trapped in the planet's magnetic field, are about 10,000 times as intense as the earth's Van Allen radiation belts.

Later analysis of the data recorded during the spacecraft's fly-by showed that Jupiter's magnetic field seems to consist of two regions. The first being a dipole field with a strength of about 4 gauss ($4 \times 10^5$ gamma), at Jupiter's cloud tops,[1] that extended to about $20R_J$. Then there was a stronger field, outside the first, which was found to be more confined to Jupiter's equatorial plane because of the centrifugal forced caused by Jupiter's rapid rotation. The axis of the dipole field was found to be inclined at about 10° to Jupiter's spin axis, with the magnetic centre being displaced slightly from the geometrical centre of the planet (see Table 11.1). The radiation belts were found to be most intense inside the $20R_J$ boundary of the dipole field, and concentrated towards the planet's equatorial plane.

When the high energy electrons were found in front of Jupiter's bow shock, records of early earth-orbiting spacecraft were re-examined, and they showed an increase in the background level of cosmic ray electrons about

---

[1] This is about ten times the field strength at the surface of the earth.

every 13 months, which is the same as Jupiter's synodic period. Thus some of Jupiter's high energy electrons even reach the vicinity of the earth.

Prior to the Pioneer 10 encounter, there had been speculation about the effect of Jupiter's inner satellites on its radiation belts, as these inner satellites were thought, correctly as it turned out, to lie within the belts. The spacecraft found evidence of the sweeping effect of Io and Europa on the population of protons and electrons in the belts. Similar, but less pronounced effects were found for Amalthea, Ganymede and Callisto. However, not only do Amalthea and the four Galilean satellites trap particles from Jupiter's magnetosphere, but Io is unique in that it produces and accelerates them. Io was found to have an ionosphere extending up to about 100 km above its surface on the day side, which gradually decays during Io's 21 hour night. It also appeared to have a tenuous atmosphere, with a surface pressure of the order of $10^{-9}$ bar.

A torus of atomic hydrogen was found around Jupiter, at Io's distance from the planet, extending to $\pm 60°$ from Io. It was thought that it was produced by the sputtering of Io's surface by particles in Jupiter's magnetosphere. At about the same time, Robert Brown of Harvard University discovered, using a ground-based telescope, that Io was surrounded by a yellow sodium cloud, which was also thought to have been produced by surface sputtering.

Pioneer 10's ultraviolet spectrometer detected helium in Jupiter's atmosphere for the first time, and showed that Jupiter's atmosphere is about 99% hydrogen and helium. This, like the ratio of helium to hydrogen measured on Jupiter, is similar to that of the sun,[2] thus showing that Jupiter's atmosphere has apparently evolved relatively little since the planet was formed from the solar nebula. Jupiter's brightness temperature of 128 K measured by Pioneer's infrared radiometer was very similar to that measured from earth, confirming that Jupiter emits about twice as much energy as it receives from the sun. The day and night sides of Jupiter were found to be at the same temperature as each other because of the rapid rotation of the planet and its internal heat source. At a pressure level of 1 bar, Jupiter's atmosphere was found to have a temperature of 165 K. The temperature then went through a minimum of 108 K at 0.1 bar about 150 km higher in the atmosphere, before increasing to 150 K at 0.03 bar at cloud-top height. This increase was attributed to the absorption of sunlight by a thin haze of dust particles.

---

[2] Jupiter's helium to hydrogen mass ratio deduced from Pioneer 10 measurements was $0.28 \pm 0.16$, compared with the solar value which was thought at that time to be 0.22. This latter value was later revised to $0.28 \pm 0.01$.

The cloud tops of Jupiter's Great Red Spot (GRS) were found to be colder than those of the adjacent South Tropical Zone, so the GRS was thought to be an enormous high pressure hurricane rising about 8 km above its surrounding area. The scale of the GRS is quite unlike anything on earth, however, as it has a length three times the diameter of the earth, and it has existed for over a century at least (see Section 8.1). The temperature at the top of the bright zones in Jupiter's atmosphere was found to be about 9 °C lower than that of the adjacent dark belts, indicating that the bright zones are regions of rising gas which are about 19 km higher than the belts where the gas eventually descends.

On 3rd December 1973 Pioneer 10 passed within 1.4 million km of Callisto, and 90 minutes later it passed within 450,000 km of Jupiter's largest satellite, Ganymede. Their surface temperatures were found to be about 110 K and 148 K, respectively, but the resolution of the images was insufficient to show any clear surface features.

Pioneer 11

After evaluating the data from Pioneer 10, NASA announced in March 1974 that it would adjust Pioneer 11's intercept trajectory with Jupiter to allow it to fly past Saturn on 1st September 1979. The new trajectory would send Pioneer 11 just $0.6R_J$ above Jupiter's cloud tops, compared with $1.8R_J$ for Pioneer 10, as it traversed Jupiter flying from the south to north polar regions.

Pioneer 11 first crossed Jupiter's bow shock at about 6.9 million km from the planet, and over the next two days it crossed the bow shock and magnetopause three times as the magnetosphere was successively compressed and expanded in response to the varying pressure of the solar wind.[3] The spacecraft then took twenty-five images of Jupiter during its final encounter phase, together with one image of Io, Ganymede and Callisto, but, as with Pioneer 10, the images of the satellites had insufficient resolution to show any surface detail. As expected, Pioneer 11's magnetometers measured a higher maximum magnetic field of 1.2 gauss near closest approach than the 0.2 gauss of Pioneer 10, because Pioneer 11 had passed much closer to Jupiter than its predecessor. Similarly the peak counting rate of high energy protons was measured to be 40 times higher using Pioneer 11 than with Pioneer 10. Finally, Pioneer 11 found that Jupiter's magnetic field very close to the planet was not the dipolar field measured further out by the earlier Pioneer, but was

---

[3] The position of the magnetopause varied from about $100R_J$ to $50R_J$ from Jupiter during the Pioneer 10 and 11 fly-bys.

much more complex in structure, varying in intensity from about 3 to 14 gauss at the cloud tops.

In 1951 W.H. Ramsey had suggested that Jupiter consists mostly of hydrogen, the majority of which is so highly compressed that it behaves as a metal. This basic concept was confirmed by the magnetospheric, gravitational and other data returned by the two Pioneers, although the detailed structure was somewhat different to that proposed by Ramsey. It was now thought that Jupiter probably consists of a small rocky core, composed mainly of iron and silicates at a temperature of about 25,000 K, surrounded by a $0.50R_J$ thick layer of mostly liquid metallic hydrogen, followed by a $0.35R_J$ thick layer of liquid molecular hydrogen and a $0.01R_J$ or 1,000 km thick atmosphere. At the interface between the metallic and molecular hydrogen, the temperature was estimated to be about 11,000 K, at a pressure of about $3 \times 10^6$ bar. The liquid metallic hydrogen was thought to produce Jupiter's powerful magnetic field.

## Voyager 1

The much more sophisticated Voyager 1 and 2 spacecraft were the next two spacecraft to fly-by the outer planets. Weighing three times as much as the Pioneers, Voyager 1 was due to fly past Jupiter in 1979 and Saturn in 1980, but Voyager 2's itinerary included fly-bys of Jupiter in 1979, Saturn in 1981, Uranus in 1986 and Neptune in 1989.

On 20th August 1977 a Titan rocket launched Voyager 2 on its way to fly past Jupiter in July 1979. Then, two weeks later, Voyager 1 was also launched on a slightly different trajectory that would see it overtake Voyager 2 and fly past Jupiter in March 1979.

Before the Pioneer missions to Jupiter, astronomers had detected a pattern of alternating easterly and westerly winds in zones parallel to Jupiter's equator. The only exception to this so-called 'zonal flow' that had been observed was the anticyclonic flow around the Great Red Spot (GRS). Unfortunately the relatively poor resolution of the Pioneer images meant that they had provided little dynamic information on Jupiter's cloud system, but the Pioneers had shown a more complex pattern of clouds than seen from earth. The resolution of the Voyager images was far superior to those from the Pioneers, however, and even when millions of kilometres away the cloud motions became very clear. In place of the simple zonal motion, astronomers could see within the zonal system a planet-wide pattern of small-scale vortices (see Figure 11.1) that were forever changing on timescales of only

**Figure 11.1** This image of Jupiter, taken when Voyager 1 was still some 33 million km from the planet, shows detail never before seen from earth. The structure of the Great Red Spot, south of the equator, is clearly seen, together with the unexpectedly complex structure of Jupiter's zonal system. (Courtesy NSSDC, World Data Center-A for Rockets and Satellites, NASA; Team Leader, Dr Bradford A. Smith; P-20993.)

hours. The most spectacular pattern was that surrounding the GRS, where the clouds were ripped apart in regions of very high shear.

In January 1979 Voyager 1 detected low energy charged particles from Jupiter when it was still some $600\,R_J$ (about 40 million km) from the planet. Then in the following month the spacecraft crossed the bow shock at $86\,R_J$ and, as in the case of the Pioneers, crossed and recrossed that and the magnetopause several times over the next few days as the intensity of the solar wind varied.

One of the most dramatic discoveries of Voyager 1's fly past of Jupiter was made as the spacecraft flew through the planet's equatorial plane some 17 hours before closest approach. For centuries Saturn had been known to have rings and in 1977, six months before Voyager 1 took off for Jupiter, faint rings had also been discovered around Uranus (see Section 9.11). So did Jupiter have rings? Although the Pioneers had not seen a ring system around Jupiter, in December 1974 Pioneer 11 had detected a reduction of high energy particles near its closest approach. This led Mario Acuña and Norman Ness to suggest in 1976 that there was either an undiscovered ring or satellite there. Such an intriguing prospect could not possibly go unchecked, so Tobias Owen and Candy Hansen asked that an attempt be made using Voyager 1 to find the possible ring or satellite. As a result, 17 hours before closest approach, a single 11 minute exposure was made looking to the side of Jupiter, and much to most astronomers' surprise a faint, thin ring was seen.

Jupiter was known to possess thirteen satellites at the time of the Voyager 1 encounter, but because of the spacecraft's intercept trajectory only the four Galilean satellites and Amalthea would show significant discs in the spacecraft's images. Although these satellites were imaged from time to time as Voyager 1 approached Jupiter, the best images were generally received during the spacecraft's closest approach.

The first closest approach was to Amalthea, just 6 hours before Voyager 1's fly past of Jupiter. Amalthea at about $260 \times 145 \times 140$ km is about the size of a large asteroid, and is about ten times the size (in linear dimensions) of the two small satellites of Mars, but about ten times smaller than Jupiter's Galilean satellites. It is so small and the distance of closest approach was so great, however, that all that could be seen was a very blurred, irregularly shaped red image that showed evidence of cratering. One crater was seen to be enormous, however, compared with the size of Amalthea, being about 85 km in diameter. The red colour of the satellite's surface was attributed to sulphur particles from Io.

Prior to the Voyager 1 encounter it was thought that Io would look like a reddish version of our moon, but covered with sulphur-coated impact craters.[4] It was also thought that the other three Galilean satellites would have ice-covered surfaces with little surface relief and only a few impact craters, as the ice would have flowed and eliminated all but the most recent craters over time. Europa has the highest reflectivity and the highest density of these three icy satellites, so, although it was thought to have water ice on its surface,[5] it was thought that this is probably in a shell only about 100 km thick. Ganymede and Callisto, on the other hand, were thought to be richer in water ice than Europa because of their lower densities, with maybe 50% of their volume being composed of ice.

As mentioned above, it was generally thought, prior to Voyager 1, that Io's surface would be old and cratered like the moon. But in the 2nd March issue of *Science*, published just three days before Voyager 1's closest approach, Stanton Peale of the University of California and Patrick Cassen and Ray

---

[4] The presence of sulphur was deduced from Io's colour and from its ultraviolet reflectance spectrum.

[5] Water ice had been detected in Europa's near infrared spectrum.

**Figure 11.2** This Voyager 1 image of Io, taken from a distance of 860,000 km, shows its mottled, fresh-looking surface. The circular, doughnut-shaped structure at the centre is the volcano Prometheus, which was seen to be erupting in other images. (Courtesy NASA and NSSDC, Voyager 1 FDS 16368.36.)

Reynolds of NASA Ames came up with an alternative theory. They pointed out that since Io is subjected to resonant gravitational forces by the other Galilean satellites, its orbit would be eccentric, although over time it averages out as circular. As Io is so close to Jupiter, however, the planet's gravity would produce powerful tidal forces in Io's crust,[6] causing the satellite to heat up. As a result, these three astronomers suggested that there would be widespread volcanism on Io.

Initially as Voyager 1 closed in on Io a number of small dark dots were seen surrounded by faint rings. At first these were thought to be the generally expected impact craters, but later high resolution images showed no such craters. Even at closest approach, with a resolution of 0.6 km, no craters could be seen, implying a surface age of less than about 1 million years. What could be seen, however, was not the old, cratered surface generally expected, but a fresh-looking landscape (see Figure 11.2) showing numerous

---

[6] Io spins on its axis once as it orbits Jupiter once (that is its spin is synchronous with its orbital period). If Io's orbit is circular, its Jupiter-facing bulge caused by Jupiter's gravity would always be at the same place on Io's surface, so there would be no surface flexing. As the orbit is not always circular, however, this bulge would move on the surface of Io, causing it to flex and heat.

volcanic calderas and rivers of lava. Peale, Cassen and Reynolds had been absolutely correct in their predictions published just three days before. With such a young surface, and with so many volcanic caldera, it was possible that some volcanoes would still be still active, although the chances of seeing one erupting was thought to be very slim. But the astronomers were in for another shock.

On 8th March, just three days after closest approach, Linda Morabito of JPL was looking at images of Io, taken from a distance of 4.5 million km, that had been deliberately over-exposed to show stars as part of a navigational investigation. On one of these images she was astonished to see an umbrella-shaped plume on the limb of Io, reaching about 270 km above the surface. It was not an artefact, but was the plume of an erupting volcano! An immediate search of previous images was undertaken, and these showed even more plumes. Eventually it was found that there had been eight volcanoes active during Voyager 1's closest approach, producing plumes ranging from 70 to 300 km in height with vent velocities as high as 1.0 km/s (2,200 mph). Rather than being as dead as the moon, the surface of Io was found to be far more active than that of the earth.

In contrast to Io, Europa was found to be very bland, with surface markings of low contrast. Very little surface detail could be resolved, because the fly-by distance was so large, but numerous dark stripes were seen, tens of kilometres wide and up to thousands of kilometres long, criss-crossing the surface. It was thought that they may be faults or fractures caused by tectonic activity. These tantalising glimpses of Europa whetted the appetite of astronomers for the much closer fly-by of Voyager 2, just four months later.

Ganymede, the largest of Jupiter's satellites, was seen to have a complex intersecting pattern of parallel grooves and ridges, together with numerous rayed craters, on a basically two-toned surface. The largest area of dark terrain, called Galileo Regio, is an approximately circular feature of about 3,000 km in diameter on the anti-Jupiter facing hemisphere. Unlike on the earth's moon, however, the dark terrain on Ganymede is more heavily cratered than the lighter areas, and so it is clearly older. Many of the rayed craters were seen to be very light in colour, probably because the ice there is fresher and more powdery than on the surface as a whole. Interestingly, some of the grooves in the light coloured areas showed lateral offsets, indicating fault lines where the surface had moved laterally.

The large number of craters on Ganymede was a surprise, as it had been expected that the surface ice would have flowed, eliminating all but the most recent craters. So it was an even greater surprise when the relatively dark surface of Callisto was found to be almost saturated with what must be

**Figure 11.3** Evidence of turbulent motion is seen all over this Voyager 2 image of part of Jupiter. Structure is clearly seen in the Great Red Spot on the upper edge, and in a large white spot located just above the centre. (Courtesy of NASA/JPL/Caltech; PIA 00372.)

very old craters. These craters are quite shallow, however, with the limb of Callisto showing that there is virtually no surface relief. The largest feature was found to be the large bright impact basin, now called Valhalla, which is surrounded by a series of concentric rings or ridges spaced 20 to 100 km apart, extending to a radius of about 2,000 km. These were produced by shock waves caused by the impact that created the basin. Crater densities in the inner part of this ring system were found to be less than elsewhere, indicating that the large impact was not the earliest feature still visible on the surface. In common with the remainder of Callisto, Valhalla showed virtually no surface relief, presumably because water[7] and ice had flowed back after the impact to fill the depression.

## Voyager 2

Voyager 2 flew past Jupiter and its satellites four months after the Voyager 1 encounter. In general the fly-by distances were not as close as for Voyager 1, but this did not prevent it returning stunning images of Jupiter (see Figure 11.3). In fact, a comparison between the Voyager 1 and 2 images

---

[7] The heat caused by the tremendous collision would have melted some of Callisto's ice.

enabled the relative velocities of Jupiter's various belts and zones to be determined. The GRS, which appeared more uniform in colour than during the Voyager 1 intercept, was found to have drifted west by about 0.26° per day, relative to the rotation rate of Jupiter's core as deduced from variations in its radio emissions. The maximum wind velocity, relative to the core, was found to be about 150 m/s (340 mph) at low latitudes (see Figure 11.11 later).

Although Voyager 2 passed much closer to Europa than Voyager 1, the resolution was still not sufficient to show any structure in most of the numerous long dark stripes found all over the visible surface. However, some of the stripes, the so-called 'triple bands', were seen to have a light centre and dark edges. Bright mottled terrain was observed on Europa's leading hemisphere, and darker mottled terrain on the trailing hemisphere. Interestingly, Jupiter's magnetosphere rotates around the planet faster than do the Galilean satellites. So as a result, charged particles from Jupiter will impact the satellites' trailing hemispheres. Whether this was the cause of the darkening on Europa's trailing hemisphere was uncertain.

Voyager 2 imaged a few palimpsests[8] on Europa, with diameters of the order of 100 km, which appear to have been caused by large impacts. Near the terminator a few 20 m diameter craters were also observed. Analysing the size distribution of these various craters implied that, although the surface of Europa is younger than that of Ganymede or Callisto, it is still probably a few hundred million years old.

Voyager 2 also showed that the surface of Europa has virtually no surface relief above about 100 m high, which is consistent with the pre-Voyager concept of a 100 km thick ice layer covering the surface. Tidal heating and/or radioactive heating could be sufficient to melt the lower levels of this ice crust, however, and some astronomers wondered whether the surface of Europa may be composed of pack ice floating on water, with the observed linear patterns showing where the ice had fractured.

During the Voyager 2 fly-by the spacecraft found that Pele,[9] Io's most active volcano during the Voyager 1 intercept, had become dormant, although six of the seven other volcanoes[10] seen by Voyager 1 were still active, and two new volcanic vents were seen. Spacecraft images showed what appeared

---

[8] A palimpsest is a roughly circular spot on an icy surface that is the remains of a former crater.

[9] Pele was named after the Hawaiian volcano goddess.

[10] The seventh volcano was out of sight of Voyager 2.

to be clouds or white surface deposits along scarps or faults, particularly in Io's south polar region, and spectrophotometric analysis indicated that these white particles were composed of sulphur dioxide and sulphur. Sulphur dioxide gas was also discovered near the volcano Loki. So both sulphur dioxide and sulphur appear to be present on Io's surface.

Vent velocities of particles emitted by Io's volcanoes were found to be about a factor of ten higher than those on earth, indicating that Io's volcanoes are significantly different in nature. They are also more consistent in their output over time than those on earth, and this, together with their umbrella-shaped plumes, suggested to planetary geologists that they are more like geysers. The driving force for geysers is usually the transition from water to steam, but that on Io must be different because of the lack of water. The most likely material driving the plume eruptions of sulphur appeared to be sulphur dioxide or, in the case of very high plume velocities, possibly sulphur itself.

Voyager measured a maximum noontime temperature on Io's equator of about 120 K, whereas the temperature of a dark feature just south of the volcano Loki appeared to be about 300 K. This 200 km diameter feature was thought to be a lake of lava from the volcano. It was probably made of liquid sulphur and in the lake were lighter features that looked like icebergs probably of solid sulphur. Near Loki and the volcano Pele there were also very small hot spots with a temperature of about 500 K.

Voyager 2 imaged Jupiter's ring as the spacecraft approached the planet, and also repeated the observations as the spacecraft receded. Much to everyone's surprise the ring appeared more than twenty times brighter in forward-scattered light than in back-scattered light, implying that many of the ring particles are only about one to two microns in diameter. This was very interesting as such small particles can only exist in the ring for a short period of time, implying that there must be some resupply mechanism. It was thought that this could be volcanic dust from Io, or the result of interplanetary micrometeoroids colliding with larger ring particles. Four months after the Voyager 2 encounter Mark Showalter et al. found that they could distinguish three components in Jupiter's ring, namely a main ring, which was found to extend from 1.72 to 1.81 $R_J$ from Jupiter's centre, a broad faint 'halo' ring inside the main ring, and a much broader faint 'gossamer' ring outside it. Two previously undetected small satellites (now called Metis and Adrastea) were discovered near to the outer edge of the main ring, possibly constraining its outward expansion and/or helping to supply it with material.

The impact of Comet Shoemaker–Levy 9

Astronomers had been planning for many years to send a probe to impact Jupiter's atmosphere in an attempt to understand its very interesting dynamical and chemical structure. But before the spacecraft, named Galileo, could reach Jupiter, nature had intervened, in the guise of a comet that was itself heading straight for an impact with the planet. In fact there was not just one comet, but a number, as comet Shoemaker–Levy 9, the comet concerned, had broken into many pieces some months before the anticipated impact.

Comet Shoemaker–Levy 9 had been discovered in March 1993 when Carolyn Shoemaker noticed a strangely shaped comet on a photograph taken by Eugene Shoemaker and David Levy with the 0.4 m Palomar Schmidt telescope. Further work showed that the comet had been broken into a number of pieces by its July 1992 encounter with Jupiter, which had taken it to within just 20,000 km of the planet's cloud tops. The fragments, all of which were in virtually identical orbits around Jupiter, were then observed to break up even more in the months after discovery as they approached the planet once again. But this time they would not continue in their orbits, but would crash into Jupiter one by one over a few days in July 1994. The view in a telescope, as the fragments approached Jupiter in a linear array, was described as being like 'a string of pearls'.[11]

Unfortunately, the impacts of the fragments were calculated to all occur just beyond the limb of the planet, as seen from earth. However, the impact sites, all of which were estimated to be at about the same latitude and within about 5° longitude of each other, would gradually come into view on the limb a few minutes after the collisions. It would then take another 15 to 20 minutes for the limb to move into sunlight (as Jupiter was not completely illuminated as seen from earth at that time), although the impact sites should be visible in infrared during this unilluminated period. The Galileo spacecraft, which was still five months away from its own arrival at Jupiter, would have a better view of the collisions, however, as it was not in line with the earth and Jupiter.

Prior to the collisions, it was expected that the comet would disturb Jupiter's atmosphere in the immediate vicinity of the impact sites sufficiently for astronomers on earth to observe directly hydrogen sulphide ($H_2S$) and other gases from lower down in Jupiter's atmosphere, for the first time.

---

[11] A.A. Common had used a similar phrase over a hundred years earlier when describing the appearance of the fragments of the Great September Comet of 1882.

In the event, the collisions created much greater visible 'scars' in the planet's atmosphere than expected that lasted for well over a week.[12] They did enable hydrogen sulphide to be detected, but there was no evidence of water from lower down in Jupiter's atmosphere. Relatively small amounts of water were briefly detected at some impact sites, but it was thought to have come from the comet itself.

## Galileo

The first man-made object to enter Jupiter's atmosphere left earth on 18th October 1989 as part of the Galileo spacecraft. But, because of launcher constraints, multiple planetary fly-bys had to be used to generate enough velocity for the spacecraft and its attached entry probe to reach Jupiter. Bizarrely, therefore, Galileo was actually launched towards Venus in the inner solar system, even though it was eventually to reach Jupiter in the outer solar system! In fact, Galileo's planned itinerary was to take it via Venus in February 1990, followed by two encounters with the earth in late 1990 and 1992, before it arrived at Jupiter in December 1995. This extensive itinerary resulted in a trip time to Jupiter of 74 months, compared with just 21 months for Pioneer 10, and 18 months for Voyager 1. However, as some sort of compensation, Galileo would be directed to fly by the asteroids Gaspra in October 1991 and Ida in August 1993 (see Section 11.6) to give us our first close-up views of an asteroid.

The Galileo spacecraft consisted of the main spacecraft that would be put into orbit around Jupiter, and a smaller probe that would parachute into Jupiter's atmosphere. The main spacecraft would act as a data storage and relay facility during the probe's descent into Jupiter's atmosphere, but it would then undertake a mission of its own, surveying Jupiter and its Galilean satellites during numerous fly-bys. In the event, the entry probe was released from the main spacecraft on 13th July 1995, entering Jupiter's atmosphere some five months later on 7th December at an astonishing velocity of 170,000 km/h. Then an hour or so later the main spacecraft's retrorocket fired to place it into orbit around Jupiter.

Galileo's entry probe, which had started to measure Jupiter's particle environment about 3 hours before atmospheric entry, found two radiation

---

[12] The impact scars were even visible in quite modest amateur telescopes, which was quite something as the cometary fragments impacting Jupiter were each only a few kilometres in diameter.

belts, one inside and one outside of Jupiter's main ring. The outer belt was that previously discovered by the Pioneer spacecraft, but the inner one had not been seen before. Somewhat surprisingly, energetic helium ions were found in this inner belt.

It had been planned to start transmitting detailed atmospheric data from the entry probe at the 0.13 bar level in Jupiter's atmosphere, at an expected atmospheric temperature of 110 K, 42 km above the 1 bar reference level. This was after the radio black-out caused by the high entry speed of the probe, and just after deployment of the main parachute. Then, in the minimum expected mission, transmission would continue for another 36 minutes as the probe descended from this 0.13 bar level to the 10 bar level, at an expected atmospheric temperature of 335 K, 92 km below the 1 bar reference level. In the event, the probe continued transmitting for a further 22 minutes, by which time the atmospheric pressure was 23 bar at 425 K, 140 km below the 1 bar level.

The density of Jupiter's upper atmosphere, a few hundred kilometres above the 1 bar level, was found to be much greater than expected. In addition, the temperature was clearly above that solely produced by solar heating, strongly implying that Jupiter had an internal source of heat. Had Jupiter had no internal heat source, the wind velocity should then have decreased as the probe descended through the thicker atmosphere, as this would be less and less influenced by the sun. In fact the probe showed that the wind velocity increased (from 540 to 640 km/h) as the probe descended, confirming the existence of an internal source of heat.

One key part of the probe mission was to measure the helium to hydrogen mass ratio for Jupiter, to get a better understanding of the way that the planet had been formed. The Pioneer and Voyager spacecraft had only been able to measure this ratio remotely for the top of Jupiter's atmosphere, whereas the Galileo probe would be able to measure the ratio in situ for more than just the outside layer. Initially it was thought that the probe had measured a ratio of 0.14, which was far lower than expected, and this sent the theoreticians into something of a panic. But then it transpired that this value was erroneous, and the true value was a more reasonable 0.24 which was, as expected, almost the same as that of the sun. This figure seemed to confirm that Jupiter's atmosphere had evolved relatively little since the planet was formed from the solar nebula. However, the measured abundances of methane, ammonia and hydrogen sulphide clearly exceeded their solar values, showing that Jupiter had had significant enrichment due to the capture of planetesimals, asteroids and comets since its formation. This had been expected by a number of astronomers, but it was a big surprise when it was found that Jupiter was also enriched in the noble gases of argon, krypton and xenon, as that implied that either Jupiter or its captured objects had been formed at a maximum temperature of 30 K, significantly lower than previously expected. This has now caused planetary scientists to reconsider the detail of their theories of solar system formation, including the initial dynamics of the solar nebula in particular.

It is very dangerous to try to draw too many conclusions, about conditions on Jupiter as a whole, from measurements made by one probe that has sampled the atmosphere at only one site. This was particularly so in this case as earth-based telescopes had indicated that the probe had entered the atmosphere at the edge of an infrared hot spot. This local infrared excess suggested that there was significant heat escaping at that spot from the lower levels of Jupiter's atmosphere through, presumably, relatively large gaps in the clouds.

Before the Galileo mission it had been thought that there are three cloud layers on Jupiter, namely an outer layer of ammonia ice (which is seen from earth), followed by one of ammonium hydrosulphide crystals ($NH_4SH$), which is a compound of ammonia ($NH_3$) and hydrogen sulphide ($H_2S$), and finally one of water ice or possibly water droplets, at about the 8 bar level, 80 km below the 1 bar reference level. It was expected that the atmosphere below this last cloud layer would be relatively clear.

In the event, the probe's parachute deployment sequence started 53 seconds later than planned. So the probe was already 25 km further into the atmosphere than expected at the 0.35 bar level when it started making measurements. Fortunately this was still soon enough to detect the top layer of cloud at about 0.6 bar, at a temperature of 145 K, some 15 km above the 1 bar reference. The next level of cloud was detected at 1.6 bar and 200 K, 30 km below the reference level. Both of these cloud layers were thinner and more transparent than expected, with the lower cloud layer being no more opaque than a light fog or mist. Surprisingly, the expected third cloud layer of water ice or droplets did not appear to exist.

There then followed an extensive debate as to whether the probe site was exceptionally dry compared with the rest of Jupiter, or whether there was really relatively little water in Jupiter's atmosphere as a whole. Here the Galileo orbiter came to the rescue, however, showing, as it did, that the infrared hot spots are exceptionally dry parts of Jupiter, whilst there is ample water vapour elsewhere on the planet.

The two-tone Ganymede was the first of Jupiter's Galilean satellites to be observed close up by the Galileo orbiter, at a distance of just 850 km above

Figure 11.4 Bright impact craters are seen scattered over the two-toned surface of Ganymede in this Galileo image. Galileo Regio, which was first imaged by Voyager 1, is the large, dark, almost circular area at the top right. It is separated from the similarly toned Marius Regio on the left by the light, wide band of Uruk Sulcus. (Courtesy of NASA/JPL/Caltech; MRPS 76026.)

the satellite's surface.[13] This was some seventy times closer than during the Voyager fly-bys. Voyager had shown many impact craters and evidence of lateral surface movement, and it was now Galileo's job to produce high quality images to enable the surface processes to be better understood (including tectonics, cratering and cryovolcanism[14]), and to determine the nature of the surface materials.

Voyager had imaged the large, dark, almost circular Galileo Regio feature that is separated from the smaller, dark Marius Regio by the light, irregular young feature known as Uruk Sulcus (see Figure 11.4). Galileo now produced high resolution images of the surface of Uruk Sulcus, and showed that its ridges and troughs found by Voyager, which are up to some hundreds of kilometres long and tens of kilometres wide, actually consist of a much finer system of ridges, parallel to the main ridges. This appeared to be due to tectonic activity caused by the crustal water ice expanding. The variation in height between the ridges and troughs was found to be of the order of a few hundred metres, with the tops of the ridges appearing light and the troughs dark.

---

[13] Galileo had passed close to both Europa and Io during its initial approach to Jupiter, but no images had been taken because of a faulty tape recorder on board.

[14] Cryovolcanism is the resurfacing of an icy surface by ice–liquid mixtures.

The generally dark surface of Galileo Regio was found to be littered with craters of all sizes, with crater counts indicating a basic surface age of about 4.2 billion years. But its surface had also been clearly substantially modified since its formation to produce a hummocky terrain of bright hills and dark planes, caused by a mixture of tectonics and cryovolcanism. Sun-facing slopes were found to be generally dark, apparently because the sun had caused the ice to sublimate, leaving the darker substrate visible.

Spectroscopic analysis showed that the bright regions on Ganymede are rich in water ice with patches of carbon dioxide ice, whilst the dark surface material was found to contain abundant clays and tholins, which are rich in carbon, hydrogen, oxygen and nitrogen. Daytime temperatures were found to vary from 90 to 160 K depending on the time of day and latitude.

Surprisingly, Ganymede was found to posses a magnetic field, with a strength at the equator of about one percent that of the earth, implying that Ganymede has a molten or semi-molten core of iron and/or iron sulphide. This was surprising on two counts; first that the satellite was differentiated with a clearly defined core, and second that that core was still at least partly molten in such a relatively small body so far from the sun. A magnetosphere was also detected along with a thin ionosphere, indicating that Ganymede probably also has a tenuous atmosphere. This was apparently confirmed by earth-based infrared observations that detected molecular oxygen ($O_2$), and by Hubble Space Telescope observations in the ultraviolet that also detected signs of ozone ($O_3$), although it was not absolutely clear if this oxygen was on the surface or in the atmosphere. On the other hand, Galileo did find clear evidence of a tenuous atmosphere of atomic hydrogen,[15] with ionised hydrogen (that is protons) streaming away at high speed. Apparently, charged particles in Jupiter's magnetosphere are impacting the water ice on Ganymede's surface, driving off atoms of hydrogen and oxygen in a process called sputtering.

The second of Jupiter's satellites on the Galileo orbiter's itinerary was Callisto. Voyager had shown that this, the outermost of Jupiter's Galilean satellites, has a dark surface almost completely saturated with craters, the largest of which is the 4,000 km diameter Valhalla impact feature. It was generally expected that, with the higher resolution of Galileo, a large number of small craters would be seen, but astronomers were in for a shock. Much to everyone's surprise, Galileo showed that there was an almost complete lack

---

[15] Ganymede's surface temperature is far too high for hydrogen to be deposited on it in the form of a frost, and so it was clear that the hydrogen detected must be in the atmosphere. The same is not true of oxygen, however, as it has a far higher freezing point than hydrogen.

**Figure 11.5** The dark material that covers Callisto smoothes out small surface irregularities, as shown in this image of the Valhalla region. Side lighting shows up a fault scarp, running diagonally across the image, which was created by the Valhalla impact. (Courtesy of NASA/JPL/Caltech; PIA 00561.)

of craters of less than about one kilometre diameter. Callisto has virtually no atmosphere, so the surface must have been impacted by innumerable very small objects, so what has happened to their craters?

Callisto's second surprise was that its surface seemed to be covered with a dark material that created a very smooth surface (see Figure 11.5), quite unlike that seen anywhere else in the solar system. This dark material, which could be several metres thick, may well be covering the small craters, but where had this dark material came from and what is it made of? The central area of Valhalla was also covered in the dark material, so it could not have been ejecta from the Valhalla impact. Galileo's spectrometer showed that this 'soil' consists of clays and tholins, but so does the soil on Ganymede and it is completely different in appearance. Even now the reason for this difference in appearance is not clear.

Early, pre-encounter spectra showed that there is carbon dioxide frost on Callisto's surface but, strangely, the poles have less frost. Instead it tends to be concentrated on Callisto's trailing hemisphere, which is subjected to particle bombardment as Jupiter's magnetosphere overtakes Callisto in its orbit. Sulphur dioxide frost was also found on Callisto's surface, but its distribution was more patchy than that of the carbon dioxide. Also, unlike the latter, the sulphur dioxide frost showed no tendency to concentrate on Callisto's trailing hemisphere, suggesting that it does not originate from the sulphur ions in Jupiter's magnetosphere which come from Io's volcanoes. On the contrary, there was a tendency for the sulphur dioxide frost to be deposited on Callisto's leading hemisphere, which led to the suggestion that it came from the impact of numerous micrometeorites.

As in the case of Ganymede, Galileo detected hydrogen atoms escaping from Callisto's surface. This implies that there must also be oxygen somewhere on Callisto from the breakup of water molecules of its icy surface. As Callisto is further out from Jupiter than Ganymede, however, this breakup is probably caused by the bombardment of the surface with solar ultraviolet light, rather than by charged particles in Jupiter's magnetosphere, which was the case for Ganymede. In fact Galileo soon found evidence of the oxygen, probably both as a frost and as a gas, in Callisto's very thin atmosphere, which appears to consist of hydrogen, oxygen and carbon dioxide.

Dynamics data from the Galileo encounters indicated that, unlike Ganymede, Callisto does not have a central core. This was consistent with the fact that Galileo could find no evidence of an overall magnetic field on Callisto. Later analysis showed, however, that it has a variable magnetic field which is caused by Jupiter's magnetic field inducing subsurface electric currents on Callisto.[16] Calculations showed that these currents could be in a 10 to 20 km deep salty ocean, about 100 to 200 km beneath the surface. The heat to keep this ocean liquid, or at least slushy, must be coming from inside the satellite. Callisto has a cold, undifferentiated interior, however, and the gravitational stresses on its interior are relatively small at its distance from Jupiter, so it appears that the heat must be coming from radioactive decay.

Galileo's next target was Europa, which Voyager had shown to be a remarkably smooth, ice-covered satellite with numerous long dark stripes, some of them, the so-called triple bands, with lines down the middle. Most planetary scientists thought that these stripes must be the result of some sort of tectonic process, possibly like that in the mid-ocean ridges on earth. The exact processes were unclear, however, and so the high resolution images from Galileo were eagerly awaited to try to differentiate between the various possibilities. There was also the intriguing question to resolve, which had been left open since the Voyager mission, as to whether Europa has a sub-surface ocean.

One of the first things that was noticed on the Galileo images was that the edges of the stripes or bands on Europa were not clear-cut, as they had appeared to be in the Voyager images, but were rather indistinct. Close-up views under side lighting showed that some of the stripes consisted of

---

[16] This analysis was only undertaken after Europa had been found to have a similar magnetic field, see later in this section.

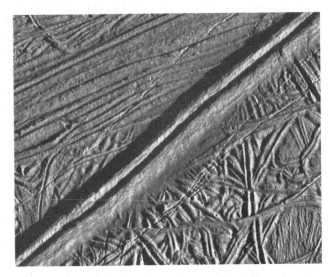

**Figure 11.6** This double ridge feature on Europa is about 2.6 km wide and just 300 m high. It is clearly relatively young as it cuts cleanly across all the other fractures. (Courtesy of NASA/JPL/ Caltech; PIA 00589.)

prominent ridges with a fracture running down the middle (see Figure 11.6). This appeared to be the result of repeated extrusions of ice or water along fractures in the surface ice sheet. The triple bands appeared to have a similar structure, although there were generally more dark and light bands in their cross sections. Some triple bands crossed other triple bands (see Figure 11.7), showing that they were clearly of different ages.

Galileo's second closest approach to Europa in February 1997 showed that the area of mottled terrain about 1,000 km due north of the 26 km diameter crater Pwyll was clearly chaotic in structure (see Figure 11.8). This area of chaotic terrain, now called Conamara, was found to consist of numerous polygonal rafts of ice each about 5 to 10 km across, whose top surfaces are about 20 to 200 m above the surrounding surface. Some of the rafts are tilted, as if they have been subjected to lateral compressional pressure at some stage. On the western fringe of this area, the rafts have been clearly powdered by ice particles ejected by the Pwyll impact. In addition, a number of small craters appear to have been produced by blocks of ice ejected by the same impact. It appears as though these rafts are nothing less than icebergs that at one stage were floating on ice or slush only about one kilometre or so below the surface. Some of the rafts are seen to carry segments of ridges, and reconstructions have shown that about half of the original surface is missing because, presumably, it melted during the event that caused this localised surface breakup.

Detailed observations of the relatively young crater Pwyll indicated that the collision that formed it was sufficient to puncture the icy crust of Europa,

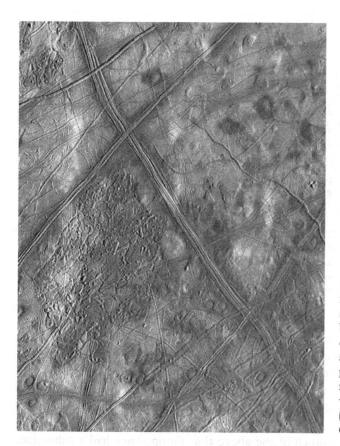

**Figure 11.7** This 200 km wide image of Europa's surface shows long, linear, triple bands, some of which intersect each other, indicating different ages. The light coloured patches are ejecta from the relatively young crater Pwyll, which is about 1000 km away. (Courtesy of NASA/JPL/Caltech; PIA 01296.)

releasing darker material from beneath the surface. The crater's floor lies at the same level as the surrounding terrain, suggesting that it filled immediately after formation with slushy material. The impact that created the 30 km crater Manann'an also appears to have punctured the surface, which was probably just a few kilometres thick at the time.

Crater counting statistics, which is an established technique for determining the age of a planetary or satellite surface, has proved problematic for Europa as it is not clear how its close proximity to Jupiter would have changed the density or velocity of impacting bodies over time. In addition, there was also confusion as to how many of the numerous small dark spots in the Galileo images are craters. The net result is a range of possible ages for the surface of Europa from about 10 million to 3 billion years!

Dynamics and Doppler measurements made particularly during Galileo's closest approaches indicated that Europa, like Ganymede, has a

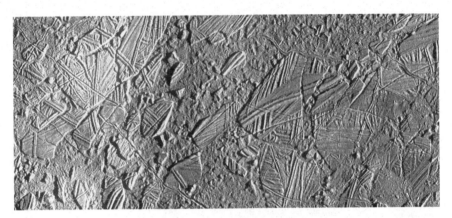

**Figure 11.8** The heavily fractured Conamara region of Europa, shown in this image, is believed to show icebergs that once floated on a subsurface ocean. The icebergs clearly moved laterally after each breakup, but they now appear to be frozen back into the surface ice. The area shown is a detail of that seen left of centre in Figure 11.7. (Courtesy of NASA/JPL/Caltech; PIA 01127.)

differentiated structure; the radius of the core in Europa's case being about half the total radius. Gravitational tidal action caused by Jupiter, and by Europa's orbital resonances with Io and Ganymede, will have not only kept Europa's interior warm, but will have caused flexing of its surface. This would have helped to create the fractures seen by Voyager and the Galileo orbiter.

It is clear from much of the above that Europa once had a subsurface ocean, but the key question is, does it still exist? Is tidal flexing, together with radioactive heating, sufficient to keep the subsurface ice liquid? A weak magnetic field had been detected early on in Galileo's mission, but scientists disagreed as to whether it was intrinsic to Europa, or whether it was induced by Jupiter's magnetic field. The question was finally resolved during Galileo's 343 km fly-by of 3rd January 2000.

As Jupiter's magnetic field is inclined to its spin axis, its direction at Europa changes sinusoidally with a period of 11 hours. As a result, if the measured magnetic field of Europa is induced, it should change with this frequency. In fact Galileo found that the field reversed direction every 5.5 hours. So it is clearly induced, and further analysis indicated that it was probably generated in a 100 km deep salty ocean only a few kilometres below Europa's surface.

Galileo had passed within about 900 km of Io on its initial approach to Jupiter, but the camera system was not operational at that time.

Nevertheless, the spacecraft's non-imaging systems were up and running to try to understand the complex electromagnetic relationship between Io, the innermost of the Galilean satellites, and Jupiter.

Voyager 1 had detected an aurora on Jupiter that extended in an arc around Jupiter's north pole. Surprisingly, this aurora did not appear to have been produced by particles in the solar wind, but by particles from Io. Further analysis indicated that Io appeared to be linked to Jupiter by a pair of flux tubes which run down Jupiter's magnetic field lines, carrying a current of the order of one million amperes.

As the Galileo spacecraft passed Io on its way to Jupiter, it made the first in situ measurements of electrons flowing up and down these Io/Jupiter flux tubes. The energy being dumped into each of Jupiter's polar regions, which was of the order of $10^{12}$ watts, was clearly enough to produce the observed aurora. In addition, as Galileo passed Io, the strength of Jupiter's magnetic field was observed to suddenly decrease by about 30%, and then increase again to about its original level. It appeared as though Io was in a bubble in Jupiter's magnetosphere where the planet's magnetic field was held at bay by Io's own magnetic field, which must be quite strong. Alternative explanations were put forward, however, that did not rely on Io having its own magnetic field.

Later analysis showed that the orbital behaviour of Galileo close to Io implied that Io has a clearly defined core, with a radius of about 50% of the satellite's radius, probably made of a mixture of iron and iron sulphide. The core seems to be surrounded by a mantle of partially molten rock, topped by a relatively thin crust. Unfortunately, such a structure can easily exist without creating an internal magnetic field, so this did not resolve the question as to whether Io has its own magnetic field.

Pioneer 10 had found that Io has an ionosphere up to a height of about 100 km on its day side, that gradually decays during Io's 21 hour night. Then Voyager found that volcanic plumes rise to a height of about 300 km above Io's surface. It was a surprise, therefore, when Galileo passed through Io's ionosphere at a height of 900 km, not the 100 km anticipated, detecting a very dense cloud of ionised oxygen, sulphur and sulphur dioxide, which clearly originated from Io's volcanoes. This discovery implied that the ionosphere varies not only on a daily basis, but over longer periods, presumably in response to changes in the level of volcanic activity.

It was difficult to determine the exact temperature of the surface around Io's volcanoes with the Voyager spacecraft, as the spatial resolution of the radiometers was relatively poor. Nevertheless, very small hot spots were found on or near the volcanoes Loki and Pele that had temperatures of about 500 K.

On earth sulphur is found in three forms, the $\alpha$ orthorhombic form, and the $\beta$ and $\gamma$ monoclinic forms. The melting point of the $\alpha$ form, which is by far the most common, is about 390 K. Sulphur exhibits clear changes in colour and viscosity with temperature. On melting it is yellow-orange, at 435 K it is pink, and at 465 K red. At 500 K sulphur becomes black and its viscosity increases, whilst at about 600 K its viscosity begins to reduce significantly. Pure sulphur's boiling point changes significantly with pressure, being about 720 K at 1 bar and 440 K at 0.001 bar, for example. It was natural to assume, therefore, that the yellow, red and black colours of the surface of Io are due to sulphur at various temperatures, although the sulphur probably includes significant impurities which affect its behaviour.

Voyager had found mountains up to 10 km high on Io, which implied that the upper crust of Io must be quite strong, and probably could not, therefore, be made of sulphur. In addition, the walls of some of the volcanic caldera appeared too steep to be made of pure sulphur, as this would have slumped after formation. So it was possible that the crust could be made of silicate rocks, and there may be some silicate/sulphur or silicate volcanism. The key to finding out was to measure the surface temperature across Io, as silicate rocks become molten at much higher temperatures than sulphur.

In 1986 a team of astronomers studying Io from the top of Mauna Kea in Hawaii observed a significant increase in its infrared brightness, indicating that there had been a brief eruption of lava with a temperature of at least 900 K. Since then other earth-based observations have detected similar events at even higher temperatures, indicating that at least some of Io's volcanic lava is silicate, as the temperature of 900 K is far too high for sulphur.

In 1979 Voyager had detected Pillan Patera as a simple caldera just over 10° away from Pele. Seventeen years later Galileo's initial observations showed that there had been no change in Pillan's appearance, but over the next few orbits Galileo detected that Pillan's albedo was variable, even though its temperature appeared to be approximately constant. Then on 5th July 1997 the Hubble Space Telescope discovered a 200 km high plume being emitted by Pillan Patera. Remarkably Galileo had imaged the plume a week earlier, during its ninth orbit of Jupiter, but the image had not been processed when the HST image was disseminated. In the Galileo image (see Figure 11.9), Pillan's plume was on the limb of Io, and there was another plume over the volcano Prometheus near the terminator. The magnitude of the change produced on Io's surface by the Pillan eruption became clear on

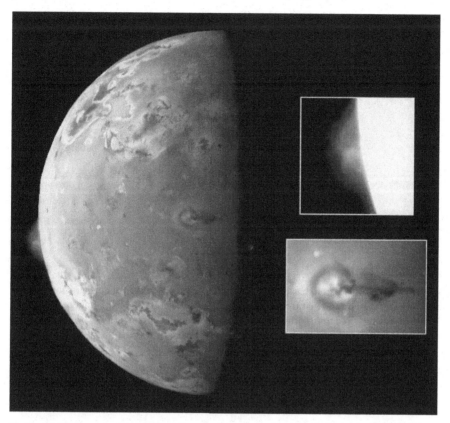

**Figure 11.9** The eruption of Pillan Patera is clearly seen on the left hand limb of Io in this Galileo image taken on its ninth orbit (see also insert upper right). Sidelighting of the plume from Prometheus near the terminator has cast a clear shadow to its right (see also insert lower right). Prometheus had been observed to erupt previously, but this is the first time that Pillan Patera had been caught in the act. (Courtesy of NASA/JPL/Caltech; PIA 00703.)

Galileo's next orbit, which showed that the volcano had produced a 400 km diameter, dark, pyroclastic blanket (see Figure 11.10) that covered part of the reddish halo around Pele. Unlike most other plume deposits on Io, however, which are white, yellow or red, Pillan's deposit was grey, resembling that of Babbar Patera nearby. At a temperature of at least 1,700 K it was clearly due to silicate volcanism.[17]

---

[17] The lava was so hot that its radiation saturated Galileo's detector. So it could be hotter than the 1,700 K maximum temperature that Galileo could measure. As a comparison, the temperature of the hottest terrestrial lava is only about 1,500 K.

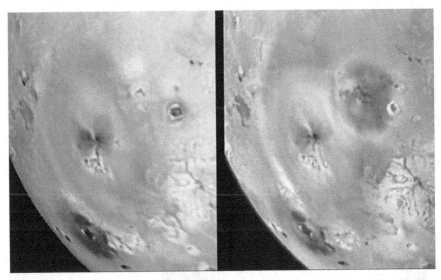

**Figure 11.10** Pillan Patera has dramatically changed the appearance of the region to the east of Pele, the volcano just to the left of centre, between these images taken on Galileo's seventh and tenth orbits (on 4th April and 19th September 1997). The volcano Babbar Patera, mentioned in the text, is the dark feature near the limb, to the southwest of Pele. (Courtesy of NASA/JPL/Caltech; PIA 00744.)

## 11.2 **Saturn**

Pioneer 11

In August 1979 Pioneer 11 closed rapidly on Saturn, following its successful encounter with Jupiter, crossing the orbit of Phoebe, Saturn's outermost satellite, at a distance of about 13 million km from the planet. Jupiter had been known to possess a very active magnetosphere and radiation belts before the Pioneer intercepts, because it had been found to emit radio waves. No radio emissions had been detected from Saturn, however, so it was not known if it possessed a magnetosphere and radiation belts before Pioneer 11 arrived. However, astronomers using intelligent guesswork, based on the Jupiter data, had estimated that Saturn would have a relatively strong field, but not as strong as Jupiter's, with a bow shock about 4 million km from the planet.[18]

---

[18] This value of 4 million km for Saturn, compared with 7 million km for Jupiter, may seem to indicate that Saturn's magnetic field was not thought to be very much less than that of Jupiter. But Saturn is almost twice as far from the sun as Jupiter, so the pressure of

In the event Pioneer 11 did not cross the bow shock until 31st August when the spacecraft was only 1.44 million km or $24 R_S$ from the centre of Saturn.

On the following day, Pioneer 11 crossed Saturn's ring plane about 30,000 km from the outer edge of the A ring. The spacecraft then passed under the rings, coming within 21,400 km of the planet's cloud tops, before recrossing the ring plane. On the next day Pioneer 11 flew past Titan, Saturn's largest satellite, at a distance of 354,000 km, but communications problems during this closest approach resulted in some of the Titan data being permanently lost.

The images of Saturn received during the Pioneer 11 encounter were disappointing, showing only the same subtle banded structure seen from earth. But the images of Titan were even more frustrating as they were completely featureless.[19] Saturn's atmosphere was found to be warmer than expected, and Iapetus, Rhea and Titan, its three largest satellites, were found to be of low density, indicating that they were mostly composed of ice. One new ring was found, the narrow F ring just 3,000 km outside the A ring, and a new small satellite, now called Epimetheus, was discovered orbiting about 14,000 km outside the A ring. In fact the spacecraft passed only about 2,500 km from this satellite, which was a close encounter in astronomical terms.

The strength of Saturn's magnetic field was calculated from Pioneer data to be about 0.22 gauss at the planet's equator, which is a similar value to that of the earth at its much smaller equator (see Table 11.1 earlier). But Saturn's magnetic moment is about 540 times larger than that of the earth because Saturn is so much larger. The magnetic axis of Saturn was found to be indistinguishable from its spin axis, and the magnetic field was dipolar. Although Titan was found to be generally within Saturn's magnetosphere, varying pressure from the solar wind was seen to drive the magnetopause back and forth across Titan, sometimes leaving Titan exposed to the solar wind.

Prior to the Pioneer encounter it had been anticipated that Saturn's ring system would prevent radiation belts from forming in the inner magnetosphere, and this is precisely what was found; the charge particle flux showing an abrupt cut-off as the spacecraft passed beneath the outer edge of the A ring. The newly discovered F ring and newly discovered satellite Epimetheus

---

the solar wind at Saturn compressing its magnetosphere is only about 30% of its value at Jupiter.

[19] This was not surprising to some astronomers, as Joseph Veverka of Cornell University had concluded in about 1972, from his polarisation measurements, that Titan is covered in clouds, although obviously the clouds may have had some structure.

were also found to have produced gaps in the electron population of Saturn's magnetosphere.[20] Pioneer 11 discovered similar features elsewhere in the magnetosphere, one of which was later found by Voyager 1 to be due to another ring (later called the G ring).

### Voyager 1

Combined data from both Voyagers 1 and 2 enabled the rotation rate of Saturn to be determined for the first time whilst Voyager 1 was still en route to Saturn. Previous estimates of Saturn's rotation period had been made by timing the movement of infrequent, temporary spots on the planet as seen from earth, but these produced results varying from 10 h 14 min at the equator to 10 h 40 min at mid- to high-latitudes (see Section 9.10). The new Voyager measurements were based on variations in Saturn's radio emissions, which were thought to be linked to the rate of rotation of Saturn's core. This produced a period of 10 h 39.9 min,[21] which implied that the 10 h 14 min spot previously seen had been moving around Saturn at the surprisingly high velocity of about 400 m/s (900 mph) relative to this rate of rotation.

Before the Voyager 1 fly-by of Saturn there was great a great deal of confusion as to how many satellites Saturn possesses. Phoebe, Saturn's ninth satellite had been discovered in 1899, and then in December 1966 Audouin Dollfus had photographed a tenth satellite, called Janus, from the Pic du Midi Observatory. Eleven years later Stephen Larson and John Fountain announced that they had found another satellite at about the same distance from Saturn as Janus, when re-examining plates taken in 1966. The exact orbits of both Janus and the eleventh satellite, now called Epimetheus, were uncertain because of limited data, but then in September 1979 Pioneer 11 accidentally flew within about 2,500 km of Epimetheus. It was not until March 1980, however, that it became evident that Janus and Epimetheus were in essentially the same 16.67 hour orbit.

That was not all, as in March 1980 Jean Lecacheaux and P. Lacques at the Pic du Midi Observatory found yet another satellite, which was recorded just hours later by astronomers at the Catalina Observatory in the United States. This satellite, initially designated 1980 S6, but now called Helene, was subsequently found to be near the 60° Lagrangian point ahead of Dione and co-orbiting with it. So as Voyager 1 closed in on Saturn for its November

---

[20] A similar effect had been observed by Pioneer 11 in 1974 at Jupiter, which Voyager 1 showed in March 1979 was caused by a ring around the planet (see Section 11.1).

[21] Later corrected to 10 h 39.4 min.

1980 fly past, it appeared as though Saturn had twelve satellites visible from earth with known orbits.[22] Of these it was planned to image eight of the nine largest during the Voyager 1 encounter and all nine during the Voyager 2 encounter. The first of these to be visited, Titan, was by far the most important because it has an atmosphere that some astronomers thought may resemble that of the early earth.

Systematic imaging of Saturn by Voyager 1 started on 25th August 1980, some 80 days before the close encounter and, much to the relief of the waiting astronomers, these early images started to reveal cloud features on the planet, after they had been computer-enhanced. So it looked as though the bland images returned by Pioneer 11 would give way to some that could be used to study Saturn's atmospheric dynamics. These and later images eventually showed that the wind velocity at cloud-top height reaches a maximum of almost 500 m/s (1,100 mph) practically on the equator (see Figure 11.11). This peak velocity is about three times that on Jupiter, and the equatorial jet stream on Saturn, of which this high speed movement is part, was found to cover the region from about 30° N to 40° S (where the velocity falls to zero), compared with Jupiter's jet that only covers the region from about 12° N to 16° S (see Figure 11.11). Voyager 1 showed that Saturn emits about 1.8 times as much energy as it receives from the sun, which is, within error, the same ratio as that deduced for Jupiter. So it was not clear why Saturn, which receives much less solar energy than Jupiter, should have much the more powerful equatorial jet. In fact the reason for this is still unclear.

The next surprise was the discovery of 'spokes' in Saturn's B ring (see Figure 11.12) by Richard Terrile whilst Voyager 1 was still over five weeks from closest approach. Radial shadings on the A ring, and occasionally on the B ring, had been observed from time to time in the last hundred years or so, but it had never been clear whether these shadings were real or not. About a week after the first images of the spokes had been received, Andy Collins and Richard Terrile found two new satellites, designated 1980 S26 and S27. Now called Prometheus and Pandora, one was just inside and one just outside the narrow F ring, apparently shepherding or stabilising it. In fact, shepherding satellites had first been proposed by Peter Goldreich of Caltech and Scott Tremaine of the University of Toronto in a 1979 *Nature* paper to explain the narrow rings of Uranus, but these two shepherding satellites of Saturn were

---

[22] Two other satellites, 1980 S13 and S25, now called Telesto and Calypso, had been imaged by Bradford Smith *et al.* in March and April 1980, but their orbits had not been well defined by the time of the Voyager 1 fly-by of Saturn. They were found to be at the Lagrangian points 60° in front of and 60° behind Tethys, and co-orbital with it.

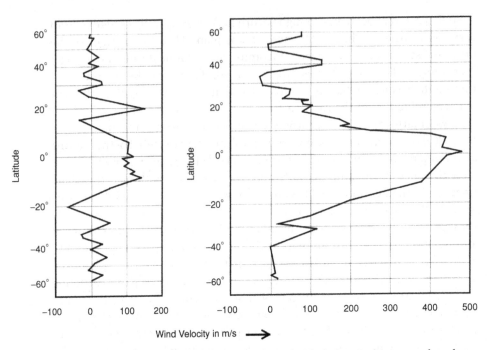

**Figure 11.11** Saturn's equatorial jet stream (right hand plot) is much stronger than that on Jupiter (left hand plot), reaching a maximum velocity of about 500 m/s (1,100 mph). The reason for this is unclear as Saturn receives appreciably less energy from the sun compared with Jupiter.

the first to be discovered for any planet. Then in early November Terrile found another small satellite, designated 1980 S28. Now called Atlas, this new satellite was orbiting Saturn just 800 km outside the outer edge of the A ring, and was apparently restricting the ring's outward expansion.

As Voyager 1 approached Saturn the smooth A, B and C rings seen from earth began to break up into more and more individual ring components. Then 18 minutes after its closest approach to Titan, the spacecraft passed through the ring plane, whilst still over one million kilometres outside the outer edge of the A ring. It then passed under the rings and recrossed their plane between the orbits of Dione and Rhea, about 340,000 km outside the outer edge of the A ring. This enabled the structure of the rings to be analysed in detail in both reflected and transmitted light.

The ring system of Saturn was found to consist of about 500 to 1,000 narrow rings, and even the Cassini division between the A and B rings, which is seen to be dark from earth and was assumed to be empty, contained over 100 narrow rings. The A ring was found to be bright in both reflected and

**Figure 11.12** The mysterious spokes of Saturn's rings are clearly seen in this Voyager image taken from a distance of 4 million km. Some of the spokes were observed to change their appearance over timescales as short as only 20 or 30 minutes. (Courtesy of NASA/JPL/Caltech; PIA 02275.)

transmitted light and to be relatively smooth and unstructured, whereas the B ring is bright in reflection but almost opaque to transmitted light. The B ring is also highly structured, consisting of hundreds of concentric rings with dusky radial, spoke-like markings. These spokes caused the theoreticians a real headache, as they appeared to contravene Kepler's second law, which requires that particles further away from Saturn orbit the planet at a slower angular rate than those nearer in. This would destroy the spokes very quickly. However it was soon realised that the spokes rotate around Saturn at the same angular rate as the planet rotates on its axis, and so they must be associated with its magnetic field.

The relatively dark C ring was brighter in transmission than reflection, and the radio occultation experiment showed that the particles in this ring were rather large, typically being about 1 m in diameter. There was a narrow gap between its outer edge and the inner edge of the B ring, which showed up bright in transmission. A very faint D ring was found inside the C ring, apparently confirming Guerin's observation of 1969, but it is unlikely that he saw the ring as it appears to be too dark to be seen from earth. The narrow F ring, discovered by Pioneer 11 just outside the A ring, was found to be three intertwined or braided rings, apparently contravening the laws of dynamics.

Like the spokes in the B ring, however, it was suggested that this braided ring could also be responding to Saturn's magnetic field. A very faint, relatively narrow G ring was found between the F ring and the orbit of Mimas, at the position predicted by the absence of charged particles as detected by Pioneer 11 (see above). Finally, the very large and faint E ring, first detected by Feibelman in 1966 (see Section 9.10), was recorded stretching 300,000 km from about the orbit of Mimas, past the orbits of Enceladus, Tethys and Dione to almost that of Rhea.[23]

Titan, by far the largest of Saturn's satellites, and the second largest satellite in the solar system (behind Ganymede), was the first on Voyager 1's trajectory during its fly-by of the Saturn system. In fact the spacecraft flew just 4,400 km above Titan's cloud tops some 18 hours before its closest approach to Saturn. Unfortunately, the images of Titan returned were featureless, with only the faintest shadings being visible after extensive computer processing, although images of the limb showed an extensive haze layer above the impenetrable clouds.

In 1944 Gerard Kuiper had detected methane lines in Titan's spectrum, but before Voyager 1's fly-by no other constituents of its atmosphere were known, although methane lines were known to suffer from pressure-broadening. Some astronomers thought that Titan's atmosphere probably consists of methane, with only traces of other gases, at a surface-level pressure of about 20 millibars. This would explain the observed pressure broadening, but Donald Hunten of the University of Arizona suggested that the same result could be obtained if the main atmospheric constituent was nitrogen, which was undetectable from earth, with methane being present at a much lower percentage. Hunten's model was somewhat flexible, however, depending on how deep the atmosphere was. Measurements by John Caldwell and colleagues in 1979, with the partially completed VLA radio telescope, implied that the surface temperature of Titan was about 87 K, and this led Hunten to conclude that the surface pressure would be about 2,000 millibars, or about twice that of the earth.

Voyager 1's radio occultation experiment showed that Titan's surface-level atmospheric pressure is actually about 1,500 millibars at a temperature of 94 K, thus vindicating Hunten's atmospheric model. The ultraviolet spectrometer also showed that Titan's atmosphere is approximately 90% nitrogen,[24] in line with Hunten's prediction. The minimum atmospheric

---

[23] So the order of the rings is (from Saturn outwards) D, C, B, A, F, G and E.

[24] It was originally thought that Voyager 1's measurements indicated a nitrogen concentration of 99%, but after Robert Samuelson had reanalysed the data he pointed out

temperature detected was about 68 K at a pressure of 380 millibars about 50 km above the surface. Interestingly, nitrogen could condense into droplets under those conditions and form clouds.

The orange colour of Titan seen from earth was found to be due to a layer of photochemical smog, which is produced when ultraviolet sunlight breaks up atmospheric methane and nitrogen molecules into their constituent atoms. These then recombine to form various hydrocarbon and nitrogen compounds, including ethane, ethylene, propane and hydrogen cyanide, which were found by Voyager 1 in small quantities in Titan's atmosphere. The surface temperature of $94 \pm 2$ K is tantalisingly close to methane's triple point of 90.7 K, at which methane can exist in solid, liquid and gaseous states. So it was surmised that the surface of Titan may have oceans of methane, with cliffs of methane in its polar region, with methane on Titan acting like water on earth in its various forms.

Titan, whose orbit generally lies just within Saturn's outer magnetosphere, was found to lose nitrogen gas from its atmosphere to Saturn's magnetosphere, where it is ionised and retained. Interestingly, as the plasma particles in Saturn's magnetosphere flowed past Titan their density increased, whilst their flow velocity slowed down. It was also found that Titan and the other large satellites inside the magnetosphere, namely Rhea, Dione, Tethys, Enceladus, and Mimas, absorb electrons and protons from the magnetosphere, in a similar way to the large inner satellites of Jupiter.

After the very close fly-by of Titan, Voyager 1 flew past most of Saturn's other large satellites at distances ranging from 73,000 km for Rhea, to 880,000 km for Hyperion and over 2 million km for Iapetus. The first to be intercepted was Tethys, but this was only seen at a distance of over 400,000 km. Little surface detail could be resolved, apart from its generally cratered appearance and a long north–south gorge about 4 km deep. Mimas, which was the next satellite on the itinerary, was found to have a 135 km diameter crater on its 390 km diameter surface. Not only is the crater very large compared with the size of the satellite, but it is an incredible 10 km (33,000 ft) deep. Enceladus was, like Tethys, poorly seen by Voyager 1, but even at a distance of 201,000 km it was expected that some craters would be seen. In the event, however, no surface topography could be observed on its very bright surface.

Dione was shown to have a dark and a light-coloured hemisphere, with broad white streaks criss-crossing the dark trailing side, that appeared to be relatively smooth (see Figure 11.13(a)). The light side, on the other hand, was

---

that Titan's atmosphere may contain up to about 12% argon, a gas that cannot be detected spectroscopically.

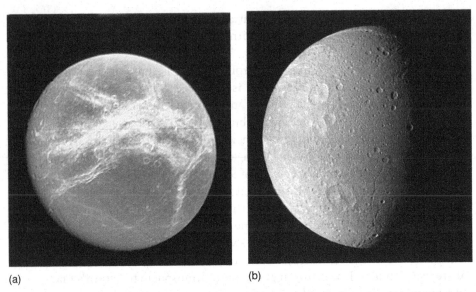

(a)                                                                (b)

**Figure 11.13** (a) High albedo streaks are seen crossing the dark, trailing side of Dione in this contrast-enhanced image. (Courtesy NASA and NSSDC; Voyager 1 FDS 34933.38.) (b) Dione's heavily cratered leading hemisphere, with an albedo of about 0.50, is about twice as bright as the dark hemisphere shown in the previous image. A long, sinuous valley system is seen near the day/night terminator. (Courtesy of NASA/JPL/Caltech; PIA 00028.)

seen to have a number of sinuous valleys on its heavily cratered surface (see Figure 11.13(b)). Rhea was found to be similar in appearance to its smaller neighbour Dione, having one heavily cratered hemisphere and a darker hemisphere that is less heavily cratered. Also like Dione it has broad white streaks on its surface, although the contrast difference between the streaks and the adjacent surface is not as high. Hyperion and Iapetus were not well imaged by Voyager 1, but the images were good enough to show that Hyperion is very irregular in shape. A satellite of its size (410 × 220 km) should be spherical, so it was thought to be part of a body that has been broken up by an impact. Iapetus had been known for some time to have one hemisphere with about ten times the reflectivity of the other, and Voyager 1 showed that the bright hemisphere is heavily cratered whilst the dark hemisphere appears to be featureless.

Voyager 1 now left the Saturnian system (see Figure 11.14) with no more planets on its itinerary, and on 19th December 1980 the two cameras were switched off. But some of the other instruments were left on to measure the interplanetary environment, as the spacecraft gradually traversed the outer realms of the solar system, to eventually end up in interstellar space.

**Figure 11.14** This image of Saturn and its rings was taken by Voyager 1, at a distance of 5.3 million km, four days after closest approach. Saturn's disc can be seen through parts of the rings, indicating their relative transparency, whilst their more opaque parts create a shadow on the planet. Such a view of the crescent Saturn can never be seen from earth, as from earth the planet is seen almost completely illuminated. (Courtesy of NASA/JPL/ Caltech; PIA 00335.)

## Voyager 2

In the meantime, whilst Voyager 1 had been undertaking its Saturn intercept mission, Voyager 2 had flown past Jupiter and was now on its way to fly past Saturn in August 1981. Before it could do so, however, it was necessary to decide on whether to continue Voyager 2 on its present course, which would allow it to visit Uranus and Neptune after its intercept with Saturn, or whether to modify its trajectory to fly closer to Titan and forego the Uranus and Neptune fly-bys. In order to make a decision, NASA analysed the performance of Voyager 1 at Titan and the health of Voyager 2, and in January 1981 announced that Voyager 2 would continue on its present course and visit Saturn, Uranus and Neptune.

Contrary to earlier proposals it was decided not to send Voyager 2 through the ring plane like Voyager 1, but to undertake the whole approach phase to

Saturn above the rings. The sun would illuminate the top of the rings at an angle of about 8° to the ring plane during the Voyager 2 encounter, compared with 4° for Voyager 1, and the Voyager 2 intercept trajectory would make an angle of about 15° to the ring plane, which was also higher than for its predecessor. This should produce better images of the rings. The Voyager 1 photopolarimeter had failed before the Saturn fly-by but that on Voyager 2 was still operational. So the fine structure of the rings should be determined to an even greater level of detail by analysing stellar occultations with the Voyager 2 photopolarimeter. Voyager 2 was also to pass closer to Iapetus, Hyperion, Enceladus, Tethys and Phoebe than Voyager 1, so producing better images, and it would image some of the smaller satellites that had been, at that stage, only recently discovered.

It is difficult to believe it now, but before Voyager 1 had arrived at Saturn it was thought that the ring system, which appeared to consist of three broad, featureless rings, was generally understood, with the gaps in the rings being caused by resonances with Saturn's satellites, Mimas in particular. For example, any particles in the Cassini division between the B and A rings would have a period of one-half that of Mimas (see Section 8.2), and so would suffer a pull from Mimas on alternate orbits. As a result these particles would gradually vacate that orbit to produce the empty Cassini division. There would also be similar effects for particles at the inner edge of the B ring, which has a period of one-third that of Mimas, or at the inner edge of the C ring, which has a period of one-quarter that of Mimas. However, the fine structure of the main A, B and C rings found by Voyager 1, the non-empty Cassini division, the spokes on the B ring, the braided F ring, and the existence of other new rings, demanded a fundamental rethink of the theory of Saturn's rings.

By the time that Voyager 2 arrived at Saturn, nine months after Voyager 1, there had been tentative explanations proposed for many of these new phenomena, with the new rôle of shepherding satellites receiving particular attention. Some astronomers thought that, even though the cameras on Voyager 2 had better detectors than those on Voyager 1, and the ring imaging conditions were better, the number of individual rings observed by Voyager 1 of 500 to 1,000 would not be exceeded. But this was not so. Voyager 2's images showed more fine structure in the main rings than its predecessor, and the Voyager 2 photopolarimeter showed even finer structure, with some rings in the main rings being as narrow as a few hundred metres. Such a fine structure implied that the rings must be very thin. The A ring, for example, which is about 15,000 km from its inner to outer edge, was estimated to have a thickness of only about 100 m at most.

The Voyager 1 images had shown the small satellites Prometheus and Pandora shepherding the F ring, and Atlas apparently preventing the A ring from expanding. So one of the key tasks to be carried out with Voyager 2 was to look for other shepherding satellites, to try to explain the fine structure of the rings. The fly-by period was going to be a hectic time, however, so the Voyager 2 investigators decided to concentrate, as far as the rings were concerned, on observing the fine detail of the Cassini division, where Voyager 1 had shown two clear gaps in the ring structure, and the highly structured B ring, including the spokes.

Surprisingly, Voyager 2 investigators found no shepherding satellites in the Cassini division. Had such satellites been found, of course, this would have given a great boost to the shepherding satellite theory of fine ring structure, and more shepherding satellites would have been looked for elsewhere in the system. A negative result did not disprove the theory, as the shepherding satellites may be too small to be detected by Voyager 2, but it did make astronomers look for other causes of the fine ring structure.

As mentioned above, the theory that the Cassini division is caused by a 2:1 resonance with Mimas had to be modified when Voyager 1 found that the Cassini division was not empty. But Carolyn Porco showed, using Voyager 2 data, that the outer edge of the B ring, which is the inner edge of the Cassini division, is not circular (see Figure 11.15) but slightly elliptical, with the major axis precessing once every 22.6 hours. This is the orbital period of Mimas, so the outer edge of the B ring is clearly controlled by this 2:1 Mimas resonance.

In the 1970s Peter Goldreich and Scott Tremaine had shown that spiral density waves, like those thought to occur in galaxies (producing their spiral arms), would also exist in circumplanetary discs of material. In the latter case the density waves would be produced by a satellite orbiting near the disc, producing a pattern of condensations and rarefactions in the disc. Voyager 2 found about fifty such spiral density wave trains in Saturn's A ring, partially explaining its complex structure.

In 1986 Mark Showalter and three colleagues published a paper in *Icarus* analysing the 325 km wide Encke gap in the A ring, and deduced the possible mass, size and orbit of the satellite responsible for producing the gap. This enabled a search to be made for the satellite using the Voyager images. In 1990 it was successful, with the discovery by Showalter, then of NASA-Ames, of a 20 km diameter satellite, now called Pan, orbiting within the Encke gap.

As far as the large satellites were concerned, the main aim of Voyager 2 was to obtain more data on Titan, particularly with the photopolarimeter,

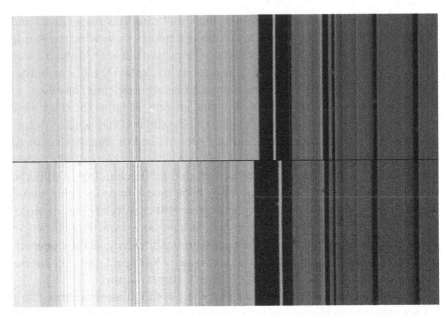

**Figure 11.15** These two images of the B ring (left 60% of image), from opposite sides of Saturn, are place side by side to show its lack of circularity. The images were both taken on the same day from a distance of about 600,000 km. Near the edge of the B ring, a separate thin ring in the Cassini division (right 40% of image) is also seen to be non-circular. The Cassini division is by no means empty, containing many, generally dark rings. (Courtesy of NASA/JPL/Caltech; PIA 01390.)

to provide higher-resolution images of Iapetus, Enceladus and Tethys than possible with Voyager 1, and to obtain the first good images of Atlas (the shepherd for the outer-edge of the A ring), the two shepherds (Prometheus and Pandora) for the F ring, the co-orbital satellites Janus and Epimetheus, the two Tethys trojans[25] (Telesto and Calypso), and the Dione trojan (Helene).

Voyager 2's polarisation measurements in the near ultraviolet and infrared bands indicated that Titan's atmosphere at an altitude of about 200 km contained droplets of maximum size 0.05 μm. However, some 30 km lower the droplets were larger, having a maximum size of 0.12 μm. It was hoped that a basic knowledge of these size distributions, linked to spectroscopic analysis of the atmosphere, would help scientists to disentangle the structure of Titan's clouds.

---

[25] The term 'trojan' is a general term referring to small bodies at the ±60° Lagrangian points co-orbiting with planets or satellites (see Section 7.2). The first trojans to be found were asteroids at the ±60° Lagrangian points in Jupiter's orbit around the sun.

**Figure 11.16** This Voyager 2 image of Enceladus shows that part of its surface is smooth, whilst part is cratered. Broad ridges and valleys are seen to cross both types of surface. (Courtesy of NASA/JPL/ Caltech; PIA 01367.)

The Voyager 2 images of Iapetus showed that there are numerous craters on its light-coloured hemisphere, with some of these craters near to the dark hemisphere having dark-coloured floors. Before Voyager 2 it was thought that, as the dark hemisphere is the leading side of Iapetus in its orbit around Saturn, the dark material could have been picked up as the satellite orbited the planet. The discovery of dark-floored craters in the light-covered hemisphere makes this unlikely, however. What is more probable is that the dark material has welled up from beneath the surface, but some astronomers feel uncomfortable with this explanation as Iapetus appears, from its density (of 1.1 g/cm$^3$), to be made of almost pure ice. Some of this could be in the form of methane ice, however, that could have come from the interior and turned black after its irradiation by ultraviolet sunlight. But that does not explain why the dark hemisphere happens to be the leading side in the satellite's orbit around Saturn, and why the reflectances of the two surfaces are so *radically* different. So a solution using both the welling up theory and the orbital intercept theory has been suggested as a possible answer.

Unfortunately the best Voyager 2 images of Enceladus were lost, owing to a spacecraft problem, but those that were returned showed remarkable detail on this the most reflective satellite in the solar system. Large areas were found to be craterless (see Figure 11.16), suggesting a surface age of less than 100 million years, but even the cratered areas showed relatively few large craters, suggesting that that surface is not as old as that of Saturn's other satellites. In addition there are numerous valleys and ridges up to 1 km high cutting across both the cratered and uncratered regions. All this indicates that Enceladus has been geologically active until the relatively recent past (in astronomical terms) and it may even be geologically active today. This is remarkable in a satellite of only 500 km in diameter, as its initial heat at formation should have been dissipated billions of years ago. Interestingly, neither Mimas nor Tethys, the satellites with orbits on either side of Enceladus, show anything like this level of activity. Enceladus is intermediate in size between these two satellites and all three have similar densities, so the cause does not appear to be due to flexure caused by its close proximity to Saturn. A more likely reason is thought to be the 2:1 resonance of its orbit to that of the larger satellite Dione, which will cause tidal flexing.

The Voyager 2 images of Tethys showed an enormous 400 km diameter crater called Odysseus which, relative to the size of the satellite, is even larger than the 135 diameter crater on Mimas. In addition Voyager 2 showed that the north–south gorge imaged by Voyager 1, that extends about two-thirds of the way around Tethys, has an average width of about 100 km and depth of about 4 km. It is unique in its size relative to that of Tethys for bodies in the solar system, as far as we know. It has been suggested that this gorge, now called Ithaca Chasma, could have been produced when the watery interior of Tethys froze, causing the satellite to expand (as water expands on freezing). But it is not clear why it appears to be the only large gorge on Tethys, and why similar gorges do not exist on the other icy satellites of Saturn.

Finally, one of Voyager's strangest discoveries involves the co-orbiting satellites Janus and Epimetheus. Their orbital periods were found to differ by only thirty seconds and, once every four years, they would approach each other so closely that they would swop orbits!

With two very successful planetary intercepts behind it, Voyager 2 was now on its way to Uranus and Neptune to give us, if all went well, our only close-up glimpses of these planets and their satellites of the twentieth century.

## 11.3 Uranus

Voyager 2 approached the Uranian system in a trajectory almost perpendicular to the orbital plane of its five satellites, as this is almost perpendicular to the ecliptic. This meant that the whole close encounter sequence in January

1986 would last for little more than 5 hours, compared with the 34 hours for the Jupiter system and an even longer time for Saturn. The trajectory of Voyager 2 would take it through the ring plane between the outermost ring and Miranda, the nearest known satellite to Uranus. It would then continue to pass behind both the ring system and the planet, as seen from earth, allowing more detailed measurements to be made of their constitution using radio occultation techniques.

No radio emissions had been detected from Uranus as Voyager approached the planet, suggesting that it did not have a magnetic field, although measurements of excess ultraviolet light by the earth-orbiting IUE spacecraft suggested the contrary.[26] In the event, it was not until five days before closest approach, much later than most astronomers had expected, that Voyager detected a strong burst of polarised radio signals, confirming the presence of a magnetic field. Analysis of timing variations in the radio emissions over the next few weeks showed that the interior of Uranus rotates with a period of 17 h 14 min. This was on the high side of previous estimates made from earth-based measurements of the planet's intensity fluctuations (see Section 9.11).

Ten and a half hours before closest approach the spacecraft crossed the bow shock some 600,000 km (or $23 R_U$) from Uranus. A day later it became evident that Uranus' magnetic axis was tilted at an amazing 60° to the planet's rotational axis, and that the magnetic centre was displaced from the geometric centre by about $0.3 R_U$ (see Figure 11.17). The magnetic field at cloud-top height at the planet's equator was calculated to be about 0.25 gauss, which is similar to that at the earth's equator (see Table 11.1, earlier), although because Uranus is larger, its magnetic moment is about 50 times that of the earth. Because of its intercept geometry, Voyager did not cross the bow shock as it left Uranus until 77 hours after closest approach, at a distance of 4.1 million km from the planet. Over the next day or so, the spacecraft crossed the bow shock three more times, as the shock contracted and expanded in response to the varying intensity of the solar wind.

All of Uranus' satellites were found to lie within the magnetosphere and, as at Jupiter and Saturn, the larger, inner satellites, namely Miranda, Ariel and Umbriel, and the particles in Uranus' rings, were found to absorb energetic particles from Uranus' radiation belts. The 98° orientation of the equator of Uranus to the sun–Uranus line, the 60° inclination of the magnetic axis to the spin axis, and the relatively fast axial rotation rate of

---

[26] This ultraviolet excess measured by IUE was attributed to auroral activity on Uranus, which implied that it has a magnetic field.

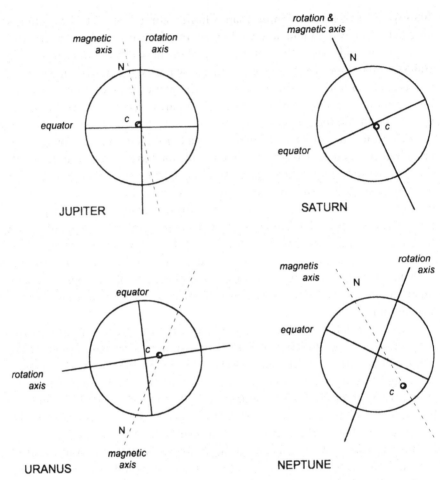

**Figure 11.17** The magnetic axis (dotted line) differs radically from the rotation axis for both Uranus and Neptune, and their magnetic centre, denoted by the letter 'c', is also far from their geometric centre. The situation at Jupiter and Saturn is more normal, like that of the Earth. (In this diagram the plane of the orbits of the planets around the sun is horizontal and perpendicular to the plane of the paper.)

just over 17 hours, were found to cause a complex diurnal variation of the belts.

The strange orientation and position of the magnetic axis of Uranus was immediately explained as being the result of a glancing impact early in Uranus' lifetime. This had not only caused the planet to tip on its side, as previously thought, but had also changed the position of its spin axis with respect to the structure of the planet. This theory could explain the observed axial orientations, at least in general terms. An alternative theory was also proposed in which Uranus was supposed to be in the process of a reversal in magnetic polarities, of the sort known to happen on earth about every 500,000 years. Unfortunately, both theories only lasted three years, until a similar magnetic/spin axis arrangement was discovered for Neptune. Clearly it was most unlikely that Voyager had visited two planets just when they were both in the process of reversing their magnetic polarities. Instead, these orientations are now thought to be produced, for both Uranus and Neptune, by convection in ionized water or ammonia in their mantles.

The cloud motion movies of Uranus produced in November 1985, as Voyager closed in on the planet, showed no clouds at all, even after computer-enhancement, and it was not until late December that computer-enhanced images began to show a banded pattern. But it was not until ten days before closest approach that individual clouds could be seen, enabling the wind velocity to be determined.

As the sun was virtually overhead at the south pole during the Voyager 2 fly-by, clouds could only be seen in the southern hemisphere, and even here they were few and far between. Nevertheless, some wind velocity data was produced between 25° and −71° latitude,[27] with an additional single data point at -6° produced by radio occultation measurements. The results showed a 100 m/s (220 mph) easterly[28] equatorial jet, with a strong westerly flow for most of the southern hemisphere. This is contrary to the case of both Jupiter and Saturn where the equatorial jets are both westerlies. In fact the strongest winds on Uranus were found at medium (southerly) latitudes, again unlike Jupiter and Saturn, where the strongest winds were found to be at or near the equator (see Figure 11.11, earlier).

Before the launch of Voyager 2 it was thought that the four large gaseous planets of Jupiter, Saturn, Uranus and Neptune would have a constitution basically unaltered since the time they were formed from the solar nebula. If that was the case, they should all have the same helium to hydrogen mass ratio, or helium abundance as it is called. It was known that the helium abundance of the sun was 0.28 ± 0.01, but nuclear reactions in the sun are continuously creating helium from hydrogen, so the helium abundances of the gaseous planets should be less than this.

Voyager measured the helium abundance of Jupiter's atmosphere to be 0.18 ± 0.04, which is less than that of the sun, but the value measured for

---

[27] The minus sign indicates a southern latitude.

[28] Westerly winds are defined as those that blow in the same direction as the planet rotates, as on the earth.

Saturn of $0.06 \pm 0.05$ is much less than the solar value. This value for Saturn caused theoreticians to reconsider their theories, and it was then realised that, at the very high pressure in the interiors of Jupiter and Saturn, liquid helium, which is heavier than liquid hydrogen, would gradually sink towards the interior. The effect would be greater for Saturn than Jupiter because of Saturn's lower temperature. This would cause a depletion of helium in the atmospheres of Jupiter and Saturn, with a larger depletion for Saturn, as observed. Uranus and Neptune are much smaller than Jupiter and Saturn, however, with consequently much lower internal pressures, so the above mechanism should not operate there, and their helium abundance should be that of the original solar nebula. The key question was, therefore, whether the helium abundances of Uranus and Neptune are similar to, but less than that of the sun, and higher than those of Jupiter and Saturn.

Just before the Voyager 2 encounter with Uranus, Glenn Orton of JPL deduced, using his analysis of its infrared spectrum observed from earth, that more than half of Uranus' atmosphere is helium. This caused great consternation and excitement as, if he was correct, there must be an unknown mechanism either increasing the amount of helium or reducing that of hydrogen in Uranus' atmosphere. In the event the theoreticians need not have worried, as the helium abundance for Uranus was measured by Voyager 2 as $0.26 \pm 0.05$, or about the same as that of the sun, within error.

The only gases detected by Voyager in the lower atmosphere of Uranus were methane, helium, and molecular and atomic hydrogen. The height of the methane cloud deck was determined by Voyager's radio occultation experiment as being at about the 1 bar level, and acetylene $(C_2H_2)$ was detected in the upper atmosphere. It is the methane which is responsible for Uranus' characteristic blue-green colour. In addition the acetylene was thought to be produced by the action of sunlight on methane, producing a photochemical smog, which raises the temperature of the upper atmosphere and reduces that of the lower one.

Voyager could not probe Uranus' deep atmosphere, but its temperature was calculated to reach 145 K (i.e. ammonia's melting point) at a pressure of about 8 bar, and 270 K at a pressure of about 120 bar. As a result, it was thought that liquid ammonia and liquid water, respectively, would exist below these levels in the planet's deep atmosphere.

As mentioned previously, Voyager 2 arrived at Uranus when the sun was almost directly overhead at the planet's south pole and, because of its long year of 84 earth years, it was estimated that the illuminated south pole would be about 5 to 10 K warmer than the dark north pole at cloud-top height. It was something of a surprise, therefore, when the temperatures over the two poles were found to be almost identical, indicating that there must be very good atmospheric mixing across the latitudes. Such mixing cannot be due to an internal heat source, as Voyager showed that, unlike Jupiter and Saturn, Uranus' internal heat source appeared to be very weak, if indeed it has one.

Before the Voyager intercept, due on 24th January 1986, it had been hoped that there would be pairs of shepherding satellites found for each of Uranus' nine known rings. On 30th December 1985 Voyager discovered the first new satellite of Uranus, designated 1985 U1 (now called Puck), and on 3rd January 1986 the next new satellites, 1986 U1 and 1986 U2 (Portia and Juliet), were discovered. Cressida, or 1986 U3, was found on 9th January, and 1986 U4, U5 and U6 (Rosalind, Belinda and Desdemona) were discovered four days later. None of these satellites were shepherds for the rings, as they all orbited Uranus outside its outermost ring, namely the ε ring. However, four days before Voyager's closest approach to Uranus, two small shepherding satellites 1986 U7 and U8 (Cordelia and Ophelia) were found on either side of the ε ring, and on the following day these two satellites were imaged together with the nine rings detected from earth (see Figure 11.18). On the same day a tenth new satellite, 1986 U9 (Bianca),[29] was added to the list of Voyager discoveries. A later review of earlier Voyager images showed that all of these satellites had been imaged previously, enabling their orbits to be calculated. Only Puck was large enough to have its diameter (of 155 km) measured directly, however, with all the other satellite diameters (ranging from 25 to 110 km) being estimated from their brightness, assuming that their reflectivity was the same as Puck (a dark 7%). No further shepherding satellites were found at that time, so only one of the nine previously known rings had been found to have a pair of shepherds.

The nine known rings of Uranus, which had first been imaged by Voyager 2 a few days before closest approach, looked narrow and dark, very much as expected. Then on the day before closest approach a very faint tenth ring, called the 1986 U1R or λ ring, was found between the ε and δ rings, and on the following day a faint ring (1986 U2R) about 2,500 km wide was found between the innermost ring (ring 6) and the planet. Over 200 images were taken of the rings after the spacecraft had passed through the ring plane, as

---

[29] It was suggested that it would be a fitting memorial to the seven astronauts who lost their lives in the Challenger Space Shuttle disaster, in same month as the Uranus encounter, to name seven of these newly discovered Uranian satellites after them. In the event, the International Astronomical Union (IAU) decided to follow the precedent established for previous space fatalities, and name some craters on the far side of the moon after them.

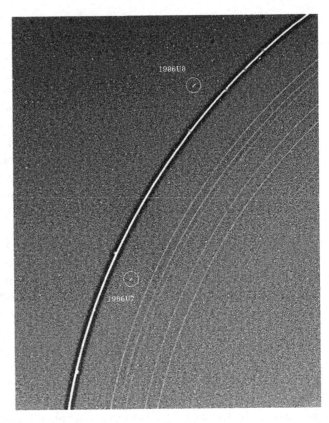

**Figure 11.18** Cordelia and Ophelia (1986 U7 and 8) are shown shepherding the bright ε ring of Uranus. All nine rings detected from Earth in 1977 are visible in the original image. (Courtesy of NASA/ JPL/Caltech; PIA 00031.)

it was expected that they would show up better in this forward-scattering arrangement, but all but one of these images were blank. Fortunately, the last exposure, with Voyager 2 just 8° from the sun line, showed a clear set of rings. Each of the ten known narrow rings were seen, with the newly discovered λ ring being the brightest, together with numerous new dust rings (see Figure 11.19).

Jupiter's rings, which had been found to be brightest in forward-scattered light, consist mostly of very fine dust of about 1 to 2 μm in diameter, whereas Saturn's rings had been found to consist of objects ranging from microns to tens of metres in diameter. The very low intensity of the Uranian rings in forward-scattering light indicated that they had relatively little dust in them compared with Jupiter's rings. Collisions of the larger particles in the Uranian rings and the impact of micrometeorites must produce dust. But Voyager found that there is probably enough atomic hydrogen at the altitude of the rings to limit the lifetime of these dust particles from less than a year

**Figure 11.19** Uranus' rings appear radically different in forward-lighting (top) and back-lighting (bottom), and the ε ring (at the left in both images) is clearly eccentric. Surprisingly, the λ ring (too faint to be seen in the top image) is the brightest ring in the back-lit configuration. (Courtesy NASA, composite of two images.)

for the inner rings to about 1,000 years for the outer rings, which is a very short time in astronomical terms.

The fine structure of the Uranian ring system was determined using the occultation of the star β Persei (Algol) as observed by Voyager's photopolarimeter. This provided a radial resolution of about 100 m for all of the rings. A similar occultation of the star σ Sagittarii (Nunki) provided a radial resolution of about 10 m for the ε, λ and δ rings. In addition, occultation of the radio signals from Voyager, when the spacecraft passed behind the rings as seen from earth, provided a radial resolution of about 50 m for all of the original nine rings. This data taken as a whole showed that the narrow rings have relatively well-defined edges, with the ε and γ rings, in particular, having sharp edges at both their inner and outer boundaries.

A density wave pattern, of the sort found in Saturn's A ring, was found in the highest resolution profiles of Uranus' δ ring. Unfortunately, the satellite causing this density wave pattern has not been found, presumably because it is too small to be imaged by Voyager. Depending on where it is, theory showed that the satellite could assist in constraining either the inner edge of the δ ring and the outer edge of the γ ring, or the outer edge of the η ring and the inner edge of the γ ring.

The ε ring was estimated to have a reflectivity of only 1.4%, compared with the reflectivity of Saturn's A and B rings of about 60%, and of the C ring and Cassini division of about 25%. The other Uranian rings were also very dark and, like the ε ring, colourless. No evidence of water was found in their spectra, and so it was concluded that they are probably composed of carbon, which may have been produced by the decomposition of methane ice by energetic protons.

Little was known about Uranus' five largest satellites that had been discovered using earth-based telescopes until the Voyager 2 intercepts. In fact, it was not until the early 1980s that their diameters had been measured with any certainty, as ranging from about 1,600 km for Oberon and Titania, to 1,300 km for Ariel, 1,100 km for Umbriel and 500 km for Miranda. This meant that they are smaller than Jupiter's Galilean satellites and about the same size as the large satellites of Saturn, excluding Titan. Because of their relatively small size and their great distance from the sun, it was anticipated that these five Uranian satellites would show little in the way of geological activity, being covered instead with craters dating back to their early formation.

The satellites of Saturn are generally less dense than those of Jupiter, because they contain more water ice, and it was generally anticipated that those of Uranus would contain even more water ice, and be even less dense, as they are further from the sun. They are darker than Saturn's satellites, however, and the ice spectral bands were not as clear. So it was assumed that the ice was more contaminated on the surface of the Uranian satellites. In the event Voyager 2 showed that the densities of the Uranian satellites were higher than those of the similar-sized satellites of Saturn, not lower, as they consisted of less ice, not more.

The outermost of the Uranian satellites, Oberon, was seen by Voyager 2 to have an ancient, heavily cratered surface, as anticipated, with some craters being more than 100 km in diameter. Some of the craters were seen to have smooth, dark floors, which is thought to be due to a mixture of melted ice and carbonaceous rock coming from beneath the surface. On the highest resolution image, an Everest-sized mountain, which is probably the central peak of an impact crater, was seen on the edge of the disc.

Titania, with a diameter of about 1,580 km, was found to be the largest of the Uranian satellites. It was seen to have fewer craters than Oberon, which was a surprise as the two satellites are of similar size and density, and are in adjacent orbits around Uranus. Two very large craters were imaged near Titania's terminator, however, one of which (Gertrude) has three concentric rings, and the other (Ursula) has two. A large rift valley system was found (see Figure 11.20(a)) that is at least 1,500 km long, with some of the valleys being up to 50 km wide and 5,000 m (17,000 ft) deep. It is thought that this rift valley system may have been produced when the ice in the interior froze, in the same way that the large rift valley was thought to have been produced on Tethys, one of Saturn's satellites. Clearly the ancient surface of Titania

(a)

(b)

**Figure 11.20** (a) Titania's large rift valley system is clearly seen near the day/night terminator in this Voyager 2 image. A large, double-walled crater (Ursula) is seen at the bottom of the image, and another, larger crater (Gertrude) is on the terminator, near the top. (Copyright Calvin J. Hamilton.) (b) Ariel's rift valley system, shown in this image, is more extensive that that of Titania, at least as far as the visible surface is concerned. Ariel's valleys are over 60,000 ft (20,000 m) deep, which is amazing for a satellite of less than 1,200 km diameter. (Copyright Calvin J. Hamilton.)

has been modified since formation by geological processes, although it is still relatively old.

Umbriel with an albedo of 0.20 is the darkest of the large Uranian satellites. It is somewhat surprising, therefore, that Ariel, whose size and density are very similar, and whose orbit is just inside that of Umbriel, has an albedo of 0.40 and is the brightest of the satellites. Umbriel and Ariel are of almost identical size and density, so why one is so much darker than the other is a mystery. It may be partially explained by the age of their surfaces, as Umbriel was seen to have an ancient, heavily cratered surface like Oberon, which is also quite dark. Umbriel was seen to have just one outstanding feature, a bright, doughnut-shaped ring called Wunda near the edge of the disc, although there also appear to be linear features and prominent cliffs on the visible surface.

Ariel was found to be similar to Titania in so far as both have large rift valleys, although those on Ariel cover more of the visible surface (see Figure 11.20(b)). Why Ariel and Titania should have extensive rift valleys, whilst Umbriel and Oberon do not, is not understood. Maybe there are rift valleys on the latter two satellites, but they are in the hemisphere not imaged by Voyager 2. We will have to wait a long time to find out, however, as there are no spacecraft currently approved to revisit Uranus.

Some of the valleys on Ariel, many of which are flat-bottomed, are over 20,000 m (60,000 ft) deep, which is astonishing for such a relatively small satellite. Parts of its surface have fewer craters than Titania, indicating that they have been resurfaced more recently by volcanic activity. In fact, there is a relative dearth of large craters on Ariel, and detailed crater counts seem to indicate that there have been several periods of partial resurfacing extending up to quite recent times. The reasons for this are unclear as, unlike the case of Io, there are no orbital resonances between Ariel and any of the other large satellites, although there may have been some resonances in the past.

By a stroke of good fortune the highest resolution images of all the Uranian satellites were those of Miranda, as Voyager passed far closer to Miranda than to any of the other satellites of Uranus. With a diameter of only 480 km, Miranda is the smallest and least dense of the five large satellites and the closest to Uranus. In the event, Miranda proved to be the star of the show, with its chevron-shaped feature, two large ovoids and enormous cliffs (see Figure 11.21). The light-coloured chevron shape, which was detected when Voyager was over 1 million km from Miranda, was seen to be surrounded by a large almost rectangular-shaped feature, now called the Inverness Corona, that has several parallel linear features that intersect at right angles. The two large ovoids, now called Elsinore Corona and Arden

**Figure 11.21** Miranda was the star of the show at Uranus. Its complex surface included Arden Corona, the large, dirt track like feature on the right hand side, and Elsinore Corona on the left. Inverness corona and its chevron-shaped albedo feature is at the upper left. The enormous cliffs of Verona Rupes, that range from 10 to 20 km high, are at the top left. (Courtesy of NASA/JPL/Caltech; PIA 01490.)

Corona, have concentric sets of ridges and grooves that make them look like large dirt racetracks. The Inverness and Arden coronae appear to be bordered by trenches often associated with parallel scarps. In the case of Inverness corona, these scarps continue to the terminator where the enormous cliff face called Verona Rupes is seen, ranging in height from 10 to 20 km (30,000 to 60,000 ft), which is incredible on such a small satellite. So Miranda was seen to be far removed from the simple cratered world that had been expected.

Voyager 2 was now, following this successful intercept with the Uranian system, on its way to fly past Neptune and its system of satellites, the last target of this highly successful spacecraft mission.

## 11.4 Neptune

Like the case of Uranus, our knowledge of Neptune and its satellites was very sketchy before the Voyager 2 encounter. Because of Neptune's and Uranus' similar size and mass and their adjacent orbits around the sun, they were often considered to be twins. But Neptune, like Jupiter and Saturn, but unlike Uranus, was known to have an internal heat source, discovered in the 1970s, and Neptune was also known to be significantly denser than Uranus. The atmospheres of both planets were known to contain both methane and molecular hydrogen (see Section 9.8), but as long ago as 1948 Bernard Lyot had detected faint markings on Neptune, whereas Uranus appeared completely featureless from earth. However, the most striking difference between the two planets is the angle between their spin axis and their orbit around the sun, which is an unusual 98° for Uranus but a relatively normal 29° for Neptune. So, although Uranus and Neptune were seen to be similar in some ways, they were clearly different in others.

Before the Voyager intercept, Neptune was known to have two satellites, Triton and Nereid (see Section 9.12). Triton, which is the closer to Neptune, orbits the planet in a retrograde sense, suggesting that it may have been captured by Neptune. On the other hand its orbit is almost circular, whereas that of Nereid, which is prograde, is much further away from Neptune[30] and is highly elliptical, with an eccentricity of about 0.75. The satellites' orbits are inclined at 23° and 29°, respectively, to Neptune's equatorial plane, which are unusually large inclinations, so maybe both satellites have been captured, Triton because of its retrograde orbit and Nereid because of its large orbital eccentricity. It was hoped that the discovery of new satellites by Voyager 2 would help to clarify this.

By the time of the Voyager intercept in 1989, there was clear evidence of methane and nitrogen in Triton's atmosphere and on its surface. The exact amounts and condition of each constituent would depend critically on Triton's surface temperature, however, which was thought to be in the range 50 to 65 K, depending on latitude and on the seasons. As Neptune's equator is inclined at 29° to its orbit around the sun, and Triton's orbit is inclined at 23° to Neptune's equatorial plane, the sun can reach 52° altitude at Triton's poles at mid-summer in the extreme case, during a complex cycle of seasons that covers a period of about 680 years. In fact the Voyager fly-by would take place in early summer in the southern hemisphere.

In 1986 Cruikshank and colleagues found, using earth-based instruments, that the methane feature in Triton's spectrum was weakening compared with earlier years. Similarly whilst observations in 1977 and 1981 had shown that the intensity of Triton varied by 6% as it rotated, those in 1987 showed that the intensity variation was less than 2%. So it was thought that the methane atmosphere was becoming more hazy as the sun was gradually

---

[30] Nereid's mean distance from Neptune is about 5.5 million km, compared with about 350,000 km for Triton.

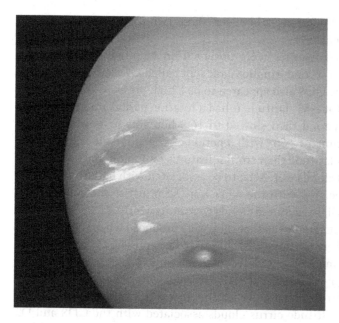

**Figure 11.22** This image of Neptune shows the Great Dark Spot left of centre, the triangular Scooter, and the small dark D2 spot near the bottom. All these features moved around Neptune at significantly different velocities, and so they could only occasionally be seen relatively close together, as here. Prominent white methane clouds are seen just to the south of the GDS and in the centre of D2. (Courtesy of NASA/JPL/Caltech; PIA 01142.)

heating up the southern hemisphere, and there were fears that the atmosphere may be too hazy during the Voyager fly-by for the spacecraft to image the surface.

Although the encounter with Neptune was not due to start until June 1989, images returned as early as January 1989 gave astronomers a taste of what to expect. These images, taken when Voyager 2 was still over 300 million km from Neptune, showed clear evidence of cloud structure with both a light and a dark spot visible. By April the light spot had faded but the dark spot was still evident, and as the spacecraft closed in on the planet this and other features became clear.

Voyager observations during the encounter phase showed that the dark spot, now called the Great Dark Spot or GDS (see Figure 11.22), was centred at about $-22°$ latitude, rotating around the planet once every 18.3 hours. It also oscillated in shape with a period of 8 days. Although at 12,000 × 8,000 km the GDS was physically smaller than Jupiter's Great Red Spot (GRS), it was about the same size as the GRS relative to the size of their respective planets. Also like the GRS, the GDS rotated counter-clockwise about its centre south of the equator, making it a high pressure feature. On the other hand, whilst the GDS shared in the general planetary circulation at its latitude, the GRS is in a shear region on Jupiter, where winds blow in one direction to its north and the other direction to its south. So there are similarities and differences between these two prominent features on Neptune and Jupiter.

Numerous other features were observed on Neptune during the encounter, including a second, smaller dark spot called D2 at about −53° latitude, which like the GDS also appeared to be a high pressure system. When D2 was discovered at −55° latitude it had a rotation period around Neptune of 16.0 hours. Its period then slowed to 16.3 hours as it moved north to −51°, before returning to −55° with a period of 15.8 hours. It seemed to be less constrained in latitude than similar features on Jupiter and Saturn.

Light-coloured clouds were seen to be associated with both the GDS and D2, but those associated with the GDS were seen just to the south of the spot, whereas D2's light-coloured clouds were at its centre. Detailed Voyager imaging of these clouds showed that they were produced when methane was forced into the higher atmosphere by these two high-pressure systems, where it condensed to form cirrus clouds. A small, fast-moving, light-coloured feature called the Scooter was seen at −41° latitude, which changed shape with time from a circular to a triangular feature. It was not as high in the atmosphere as the high-altitude cirrus clouds associated with the GDS and D2 spots, however. More high-altitude cirrus clouds were seen near the terminator of the planet, casting shadows on the blue atmosphere about 60 km below, and many other small features were seen which appeared and disappeared in one planetary revolution in Neptune's dynamic atmosphere.[31] Voyager's radio occultation experiment showed that Neptune's methane cloud layer was at an atmospheric pressure level of about 1.2 bar, as expected.

The GRS rotates around Jupiter at about the same rate as the planet's interior, as deduced from the variation in its radio signals. As a result it was assumed that Neptune's interior would be found to rotate every 18.3 hours, which was the rotation period of the GDS. It was something of a surprise, therefore, when eight days before closest approach Voyager 2 detected radio signals from Neptune[32] that were varying with a period of 16.11 hours,[33] which was assumed to be the rotation period of Neptune's interior. Relative to this, Neptune's clouds showed that there is an easterly equatorial jet, like

---

[31] The changing nature of Neptune's atmosphere has since been demonstrated by the Hubble Space Telescope which showed that both the GDS and D2 had disappeared by 1995, hence the use of the past tense in describing these features above.

[32] Later analysis showed that Neptune's varying radio signals had been detected by Voyager 2 some 30 days before its closest approach.

[33] It may or may not be a coincidence that this was about the period of the D2 spot, although, as explained above, the latter was not constant.

that on Uranus, only that on Neptune is five times as fast at an incredible 500 m/s (1,100 mph). These Neptune winds are, along with Saturn's equatorial jet, the fastest atmospheric winds in the solar system, which is surprising considering that Neptune receives such a small amount of radiation from the sun. So the winds are thought to be driven by Neptune's significant internal heat source. The maximum westerly winds were found to be at about −70° to −80° latitude where they reach about 200 m/s (440 mph).

Voyager found that Neptune's magnetic axis was even more bizarre than that of Uranus as, not only was Neptune's magnetic axis inclined at 47° to the spin axis, its magnetic centre was over half-way from the centre of the planet (see Figure 11.17 and Table 11.1 earlier). As a result, two of the theories explaining the strange orientation of Uranus's magnetic axis had to be abandoned (see Section 11.3). Not only was Neptune not spinning on its side like Uranus, so the impact idea had to be dismissed, but it was also not credible to suggest that Uranus and Neptune had been visited by Voyager just as both were suffering a reversal of their magnetic fields. Instead astronomers concluded that, unlike the cases of Jupiter and Saturn which have normal magnetic fields caused by a central dynamo, there is no such dynamo in the cores of Uranus and Neptune. Rather, Uranus and Neptune appear to have magnetic fields produced by convection in a highly compressed ionised water or ammonia layer, which is thought to exist a few thousand kilometres below their cloud tops. However, unlike Uranus, Neptune was known to have a significant heat source, so this would be expected to produce more convection, which in turn would produce a larger magnetic field than on Uranus. It was surprising, therefore, that Neptune's magnetic field is less intense than that of Uranus (see Table 11.1 earlier).

In mid June, about two months before closest approach, Neptune's first new satellite, initially designated 1989 N1 and now called Proteus, was discovered by Voyager 2. It was about 117,600 km from the planet's centre, in a circular, prograde orbit in Neptune's equatorial plane. This compares with the case of Triton which is about 355,000 km from Neptune in a circular, retrograde orbit, inclined at 23° to the planet's equatorial plane. Although Proteus, with a diameter of about 420 km, is much smaller than Triton, it is larger than Nereid (at 340 km). It had not been seen from earth, however, as its albedo is appreciably lower than that of Nereid, and observations are hampered by its relative closeness to Neptune's glare.

Over the next two months five other new satellites were discovered,[34] all of which are in circular, prograde orbits between 48,000 km (for 1989

---

[34] Called Larissa, Galatea, Despina, Thalassa and Naiad.

N6 or Naiad) and 73,600 km (for 1989 N2 or Larissa) from the centre of Neptune, with diameters ranging from about 50 km (for Naiad) to 200 km (for Larissa). Proteus and Larissa were the only new satellites to be imaged at high resolution, showing highly cratered surfaces with reflectivities of only 6%. Proteus, like Mimas, was found to have a very large crater on its relatively small surface. In Proteus' case the crater is about 160 km diameter on a satellite of 420 km diameter, which compares with Mimas' 135 km and 390 km, respectively.

Contrary to expectations, the presence of these six small satellites in circular, prograde orbits did not help to resolve the question as to whether Triton is an original or captured satellite of Neptune, as it was, at first, difficult to see how Triton could have been captured without disturbing the orbits of these much smaller satellites. In addition, these smaller satellites must have formed further out from Neptune than they are at present, as tidal disruption would have prevented them forming in their current orbits so close to the planet. This would have made it more likely that their orbits would have been disturbed by Triton. So, if Triton was a captured satellite, it must have been captured after the orbits of the smaller satellites had migrated towards Neptune.

Peter Goldreich and colleagues then developed a scenario in which Triton was captured by Neptune (that already had a satellite system) following Triton's collision with one of Neptune's larger satellites. In their theory, the collision slowed down Triton enough for it to be captured into a highly elliptical, retrograde orbit that, over time, gradually became circular, with a reduced inclination. The initial impact would have had a devastating effect on the orbits of all but the nearest and smallest satellites to Neptune. Triton would have probably captured most of the other satellites over time, but some of them may have been ejected from their orbits around Neptune, or put into a highly eccentric orbit of the sort currently occupied by Nereid.[35]

In early August, about three weeks before closest approach, Voyager confirmed that Neptune apparently had ring arcs as deduced from previous earth-based observations of stellar occultations (see Section 9.12). But, as the spacecraft drew closer to Neptune, it became clear that the ring arcs were just the brightest segments of a complete ring, designated 1989 N1R and later called Adams,[36] about 62,900 km from the centre of Neptune

---

[35] It is not yet clear if Nereid's orbit was changed in this way, or whether it is a satellite that was captured by Neptune after Triton.

[36] Named after John Couch Adams.

(a)

(b)

**Figure 11.23** (a) Neptune's Adams and Le Verrier rings were the only rings seen in front-lighting as Voyager approached the planet, with the clumpy nature of the former clearly evident. (Courtesy of NASA/JPL/Caltech; PIA 01493.) (b) In back-lighting the fainter, more diffuse Galle ring was discovered between the Le Verrier ring and the planet, and the broad Plateau ring was found extending half-way from the Le Verrier to the Adams ring. This image is a composite of two images taken at different times. The clumps of the Adams ring cannot be seen as they were on the opposite side of Neptune during each exposure. (Courtesy of NASA/JPL/Caltech; PIA 01997.)

(see Figure 11.23 (a)). The Adams ring was remarkably clumpy, with density variations of about a factor of ten along its length. A second complete ring, 1989 N2R, called Le Verrier,[37] was also seen inside the first and about 53,200 km from the centre of Neptune. After its closest approach to Neptune the spacecraft imaged the rings in back-lighting (see Figure 11.23 (b)), showing a third, diffuse ring, 1989 N3R, called Galle,[38] about 42,000 km from the

---

[37] After Urbain Le Verrier.
[38] After Johann Galle.

centre of Neptune, and a broad band of relatively large particles and dust, 1989 N4R, called the Plateau ring, between 1989 N1R and N2R. In addition, as Voyager passed through the ring plane, just outside the outer ring, a large number of dust impacts were recorded, starting about one hour before the crossing. The impact rate reached 300 impacts per second at its peak, compared with 400 per second when Voyager crossed Saturn's ring plane, and only 50 per second when it crossed that at Uranus. Voyager passed much closer to the outermost ring of Neptune than to that of Uranus, however, which could possibly explain the higher figure at Neptune. Evidently, in addition to the visible rings, there was also a thick cloud of dust orbiting Neptune.

Further analysis showed that the ring arcs discovered in 1984 and 1985, at 67,000 and 56,000 km from Neptune, were part of the Adams ring at 62,900 km. The reason for the discrepancy in distance was because of a slight error in the earlier estimation of the angle of Neptune's spin axis to its orbital plane. Interestingly, however, the stellar occultation observed by Harold Reitsema, William Hubbard, *et al.* in 1981, at about 70,000 km from Neptune was not due to a ring arc, but to the 200 km diameter satellite designated 1989 N2. Reitsema had been right all along when he maintained that the occultation was due to a satellite, not a ring arc, and in recognition of his team's prior discovery of this satellite, the IAU invited them to name the satellite from names connected with Neptune mythology. They chose Larissa, as both Reitsema and Hubbard had daughters named Laurie, a near match.

The satellite Galatea (or 1989 N4) orbits Neptune about 900 km inside the clumpy Adams ring, and the satellite Despina (1989 N3) orbits Neptune about the same distance inside the Le Verrier ring, helping to shepherd the rings. Galatea could not in itself cause the clumping of its associated ring, and neither Galatea nor Despina could stop either ring from expanding outwards. To do this a further satellite was required just outside each ring, but no such satellites were found.

Voyager 2 flew only about 37,000 km above Triton, some five hours after the spacecraft's closest approach to Neptune. The satellite was found to have a highly reflective surface with a diameter of about 2,700 km, near the low end of the expected range.[39] Its density was estimated as 2.05 $g/cm^3$, greater than that of any of the satellites of Saturn or Uranus, and somewhat higher than generally expected. This indicated that Triton has less ice and more rock in its interior than anticipated. Interestingly, Triton's density and size

---

[39] Prior to Voyager, diameter estimates for Triton ranged from about 2,200 to 5,000 km.

are almost identical to those of Pluto, adding to the idea that Triton is a captured asteroid.

Because of its highly reflective surface, with an average reflectivity of about 85%, Triton absorbs little solar energy, so its surface temperature was found to be lower than anticipated at 38 K. This was by far the lowest surface temperature yet measured in the solar system, and is far too cold for nitrogen to exist in liquid form on the surface, except in possible 'hot' spots. Because of the lower than expected temperature, the surface-level atmospheric pressure of 15 microbar was considerably less than expected, as most of the atmosphere was 'frozen out' on the surface. Fortunately, although there was a haze layer about ten kilometres above the surface, the surface could be clearly seen. The very thin atmosphere was found to consist almost completely of nitrogen, with only trace amounts of methane, carbon monoxide, carbon dioxide and water vapour.[40]

Voyager arrived at Triton during early summer in its southern hemisphere and revealed a very bright, pinkish-white, southern polar cap of nitrogen ice 'contaminated' with trace amounts of methane, carbon monoxide, carbon dioxide and water ice extending three-quarters of the way from the pole to the equator. The ice cap appeared to be slowly melting at its edges to reveal a darker, redder surface underneath. The edge of the ice cap was ragged, and at one place there was a small group of circular dark features with bright skirts.

The retreating and advancing polar ice caps have ensured that the surface of Triton is still subject to change, even though it is the coldest planet or satellite known in the solar system. Dark streaks up to 150 km long were seen all over the south polar cap and, as the ice retreats every summer, those streaks near the edge of the cap must have been produced recently. Their appearance, and the fact that all of these dark streaks point in approximately the same direction, suggest that nitrogen or methane had been ejected at one end of the streak in some sort of explosion, and carried by the wind in Triton's tenuous atmosphere.

Robert Hamilton Brown proposed a mechanism for these eruptions, likening them more to geysers than volcanoes. In particular, he suggested that sunlight penetrates the almost-transparent nitrogen ice of the polar ice cap, where about two metres down it is absorbed by frozen methane, which has been darkened by exposure to ultraviolet light. The heat is trapped by the nitrogen ice, as it is a poor conductor, and as the heat builds up the subsurface nitrogen ice is turned into gas. Eventually the gas pressure is too much

---

[40] The maximum abundance of any of these minor constituents was only about 0.1%.

**Figure 11.24** The paucity of impact craters in this image of Triton attests to the relative youth of its surface. This image, which is about 450 km across, shows two depressions, possibly old impact features, that have been subject to flooding by cryovolcanic fluids and subsequent freezing. There is also evidence of remelting, faulting and collapse. (Courtesy of NASA/JPL/Caltech; PIA 01538.)

to be resisted by the surface ice, and the nitrogen gas explodes through the surface, carrying with it the sun-darkened methane, to produce a geyser-like eruption.

A month after the encounter, Lawrence Soderblom and Tammy Becker at the US Geological Survey were stereoscopically examining Voyager images of the dark streaks on the south polar cap, when they were amazed to discover an image of an eruption in progress. It produced a plume 8 km high, which extended horizontally for about 150 km. They then found an image of a second eruption in progress. These two eruptions, and three other suspected eruptions, were all near to the sub-solar point, indicating that they are caused by solar heating, giving some support to Robert Hamilton Brown's theory of geyser-like eruptions.

There were only a relatively small number of craters seen on Triton (see Figure 11.24, for example), and the largest crater was found to be only 27 km in diameter, indicating that none of Triton's original surface has survived. So there must have been considerable geological activity in its early lifetime, possibly caused by gravitational stresses as a result of its hypothesised capture by Neptune. In addition, resurfacing must have continued almost up to the present in geological terms, and some resurfacing may still be occurring today.

In the equatorial regions there are large areas of relatively young, dimpled terrain, quite unlike anything seen elsewhere in the solar system. This so-called cantaloupe terrain (dimpled like orange peel) was criss-crossed in places by shallow linear ridges up to 30 km wide and 1,000 km long, probably caused when water ice in Triton's mantle froze and expanded. Within this cantaloupe terrain and to its north there are areas of frozen lakes and calderas, showing evidence of liquid flows, possibly as a result of volcanic activity. Some of these frozen liquid flows are relatively old with many craters, but some look fresh and are almost crater-free. These lakes and calderas cannot be made of nitrogen and/or methane, as the surface temperature is too near to their melting points for these ices to be able to support the crater walls, which, in some places, are over 1,000 m (3,000 ft) high. Instead it appears as if the subsurface here is made of a mixture of water ice, nitrogen and/or methane, with the water ice providing the strength, and the nitrogen and/or methane reducing the freezing point of water to enable it to flow at low temperatures.

So ended the remarkable mission of Voyager 2, without doubt the most successful spacecraft to investigate the solar system thus far.

## 11.5 Pluto and Charon

Walker and Hardie had detected regular variations in Pluto's brightness in 1955 (see Section 9.1) with a period of about 6.387 days, which was clearly due to Pluto spinning on its axis. Over subsequent years, however, Pluto's maximum intensity during its spin period decreased, even though it was approaching perihelion and so getting more light from the sun. In addition, the variation in the light curve over each spin cycle increased, so the dark part became even darker. Hardie attributed these effects to the sublimation of surface frost as Pluto's surface got warmer, so uncovering a darker surface. On the other hand, Leif Andersson and John Fix suggested that they could be due to a large tilt in Pluto's axis, thus allowing us to see different aspects of the surface as Pluto orbited the sun. Assuming Pluto's polar regions are brighter than its equator, maybe we have seen less and less of the pole pointing in our general direction and more and more of the equator.

In 1976 Dale Cruikshank, Carl Pilcher, and David Morrison detected, using the 4 m Mayall telescope on Kitt Peak, that Pluto has methane ice on its surface. This meant that Pluto has a much higher albedo than previously thought, which, in turn, meant that a smaller surface area was required to produce the observed intensity. As a result, Cruikshank, Pilcher and Morrison estimated that Pluto's diameter was probably between about 5,300 and 1,600 km, or from a little larger than Mercury to less than that of the moon,[41] depending on how much the reflectivity of the methane had been affected by impurities or by solar ultraviolet light.

A little later on 22nd June 1978 James Christy, of the US Naval Observatory, was examining plates of Pluto when he noticed that its images were not circular, whilst the stellar images were circular. So the cause of Pluto's non-circular images could not be a telescope tracking problem. In fact, Pluto appeared to have a large lopsided bulge, which Christy found to be in the same southerly position on plates taken on 13th and 20th April. But the bulge was in a northerly position on plates taken on 12th May. Christy wondered if he had found a satellite of Pluto, as the bulge was too large to be a mountain. If it was a satellite, however, its orbit was probably highly inclined. On the following day, he examined a series of five plates taken during one week in June 1970, and these showed the bulge moving clockwise around Pluto with a period of about six days. This appeared to prove that Pluto has a satellite, with an orbital period similar to Pluto's axial rotation period of 6.387 days.

Christy's colleague, Robert Harrington, computed the orbit of the new satellite, and found that all the observations were consistent with an orbital period exactly the same as Pluto's axial rotation period. As a result, the Pluto system is unique in the solar system as not only does the satellite presumably have one face permanently facing its planet, like many other planetary satellites, but in this case the planet has one face permanently facing its satellite. The centre-to-centre distance of Pluto to its satellite, now called Charon,[42] turned out to be only 20,000 km (or 1/20th of the distance of the earth to the moon), and the diameters of Pluto and Charon were estimated at about 3,000 and 1,000 km, respectively. Amazingly the total mass of the Pluto–Charon system is less than that of the earth's moon alone. Furthermore, the orbit of Charon showed that the angle of Pluto's equator to the plane of Pluto's orbit around the sun is similar to that of Uranus, with both planets almost spinning on their side.

---

[41] Mercury's and the moon's diameters are 4,880 and 3,480 km, respectively.

[42] Christy wanted to call the satellite Charon after his wife Charlene, but that was not allowed according to IAU regulations. Fortunately, however, he found that Charon was the name of the ferryman who took the souls of the dead across the River Styx to Pluto's underworld in Greek mythology. Because of that the name was officially recognised by the IAU in 1985.

The discovery of Charon made it less likely that Pluto was once a satellite of Neptune, as some astronomers had believed, as any disturbance powerful enough to eject Pluto from Neptune would almost certainly have separated Pluto and Charon. However, Pluto may have been in orbit around Neptune without Charon. A planetesimal could then have collided with Pluto, ejecting Pluto and the resultant debris from Neptune's orbit, with Charon forming from the debris.

However, there had been another blow to the idea that Pluto had once been in orbit around Neptune, when in 1965 Hubbard and Cohen had discovered that Pluto's orbit is in a 2:3 resonance with Neptune's; that is when Pluto orbits the sun twice, on average, Neptune orbits it exactly three times. This resonance, and their radically different orbital inclinations, means that, if the resonance had always been present, Pluto could never have come closer than 2.2 billion km to Neptune, and so could never have been part of the Neptune system. At face value this appears to have written off the idea that Pluto had once been in orbit around Neptune, although Pluto's orbit may have been disturbed since Pluto left Neptune's orbit.

The general view today is that Pluto, Charon and Triton are planetesimals left over from the formation of the solar system, and that Triton was captured by Neptune. There are some, however, who favour a modification of this idea, as proposed by William McKinnon and Steve Mueller to explain the large size of Charon relative to Pluto, and the angular momentum of the Pluto-Charon system. In their theory, which was based on Hartmann and Davis' theory of the origin of the moon (see Section 10.2), Charon was formed from the debris after a large planetesimal had collided with Pluto, which was in orbit around the sun.

In April 1980 Alistair Walker, a South African astronomer, was hoping to observe an occultation of a star by Pluto, and so determine Pluto's size, and hopefully detect any atmosphere. Unfortunately, Pluto did not quite occult the star, but Charon occulted it instead. This enabled him to estimate Charon's diameter which turned out to be about 1,200 km.

It so happens that Charon had been discovered just before it and Pluto were due to undertake a series of mutual eclipses, as they orbited the centre of mass of their system. This was fortunate, as these mutual eclipses require an orbital alignment that occurs only every 124 years (half Pluto's year). Unfortunately, the exact year for the start of this mutual eclipse period was not known, because the geometrical arrangement of the planet and its satellite were not known with sufficient accuracy. As a result the eclipses were looked for from 1979, but it was not until 1985 that the first eclipse was detected. In the event they lasted from 1985 to 1991, allowing Pluto's diameter to be accurately determined, for the first time, as 2,300 km, so Charon is the largest satellite, relative to its primary, of any satellite in the solar system. Charon was found to be appreciably darker than Pluto, with an albedo of about 0.35 in the visible waveband, compared with about 0.55 for Pluto. The angle between Pluto's spin axis and its orbit around the sun was found to be about 123°, so its spin was retrograde like that of Uranus. In addition Charon's orbit was found to be almost exactly circular.

Spectroscopic systems of the time were not good enough to separate the spectrum of Pluto from that of Charon, as their images were too close together. But the total eclipses of Charon by Pluto in 1987 allowed the spectrum of Pluto and Pluto plus Charon to be determined and, by subtraction, that of Charon alone. This showed that the previously observed methane feature of their joint spectrum is due to Pluto, and that Charon has water ice on its surface.

Pluto's high reflectivity of about 0.55 presented something of a problem, as the methane ice on its surface should have been darkened by the solar ultraviolet. If, however, the methane sublimated (to gas) around perihelion, it could condense as pure white methane frost every time Pluto approached aphelion. In that case, Pluto should have an atmosphere for a few decades at least around perihelion, which was due in 1989. A number of attempts had already been made, like Walker's in 1980, to detect Pluto's atmosphere by observing the planet occulting a star. But all had failed, until on June 9th 1988 a number of teams of astronomers were successful during a carefully planned observing programme. The star concerned was observed to fade gradually in intensity, and increase gradually again, indicating the presence of a tenuous atmosphere. Analysis of the light curve showed that, although the atmosphere has only a density of the order of 10 microbar, it was surprisingly large, being over 100 km deep.[43] It is thought that the atmosphere, which is probably produced by the surface ice sublimating, may exist for only a few decades around perihelion.

In 1993, Tobias Owen, Leslie Young and Dale Cruikshank made the surprising discovery that the ice on Pluto's surface, as on Triton, is mainly nitrogen, with small amounts of carbon monoxide and methane, rather than mainly methane as previously thought. In fact they concluded that the atmosphere must be about 98% nitrogen, with about 1.5% methane and 0.5% carbon monoxide.

---

[43] 99% of Pluto's atmosphere is below about 300 km altitude. This compares with the earth where 99% of its atmosphere is below about 40 km.

Table 11.2 *Triton, Pluto and Charon compared*

|  | Triton | Pluto | Charon |
|---|---|---|---|
| Diameter (km) | 2,700 | 2,300 | 1,200 |
| Density (g/cm$^3$) | 2.05 | 2.0 | 1.7 |
| Atmospheric pressure (microbar) | 15 | ~10 | ?[a] |
| Atmospheric constituents | Mainly nitrogen | Mainly nitrogen | ? |
| Surface ices | Mainly nitrogen | Mainly nitrogen | Water |

[a] No atmosphere has been detected on Charon.

All of this work indicated that Pluto is very similar to Triton (see Table 11.2). Their diameters and densities are similar, with similar atmospheres and similar ices on their surface. Charon is, however, very different, as not only is it significantly smaller, its surface ices are completely different. In fact Charon appears to much more like the larger satellites of Uranus than Pluto.

## 11.6 Asteroids

### Spacecraft intercepts

Radar had first been used in the early 1960s to try to understand the surface conditions of Venus and Mercury (see Sections 9.2 and 9.3), but it was not until 1968 that the first radar contact was made with an asteroid, in this case Icarus. Icarus is so small, however, that it was not possible to resolve any surface detail. All that could be determined from an analysis of the returned signals was that the asteroid was about 1 km in diameter with a rough surface. Because of technology limitations, subsequent radar work on various asteroids was of limited use until in 1983 Stephen Ostro found the first indications that an asteroid, Oljato (2201), probably consisted of two objects joined together. Six years later Ostro found more convincing evidence of such a structure for Castalia (4769). Slightly later, Scott Hudson reprocessed Ostro's data on Castalia and showed that it consisted of two lobes, each about 1 km in diameter, joined by a 100 m thick 'waist'. It appeared as though the asteroid had originally been two asteroids that had become joined together in a low-speed collision.

In October 1991 the Galileo spacecraft flew past Gaspra (951) on its way to Jupiter, producing the first close-up images of an asteroid. Prior to the

Galileo intercept, a young amateur astronomer, Claudine Madras, had found Gaspra to be an egg-shaped asteroid with a 7 hour rotation period. Then infrared observations made in 1990 by Jeffrey Goldader of the University of Hawaii implied that metallic iron exists on the asteroid's surface. This seemed to indicate that Gaspra had once been part of the lower mantle of a differentiated larger body that had been shattered by a collision. In fact, Noriyuki Namiki and Richard Binzel of MIT hypothesised that Gaspra would be found to be irregular in shape, consistent with this idea, and less than one billion years old.

In the event, the Galileo spacecraft, which flew to within about 1,600 km of Gaspra, confirmed that it is highly irregular in shape, with dozens of small craters on its surface. Crater counts confirmed an age of the order of a few hundred million years, consistent with Namiki and Binzel's prediction. At $18 \times 11 \times 9$ km in size, it is slightly larger than expected and similar in size to Mars' satellite Deimos. A number of grooves were seen, about 300 m across and 10 to 20 km deep, which cross the surface for up to a few kilometres, resembling the fractures seen on Mars' larger satellite, Phobos.

Two years later, Galileo imaged a second, larger asteroid, the 56 km long Ida (243). Although the intercept took place in August 1993, spacecraft problems meant that it was many months before all the data had been relayed satisfactorily to earth. By February 1994 only part of the data had been received, but with this Ann Harch noticed that there was an unexpected object close to Ida. The possibility that it was a star or another asteroid accidentally imaged close to Ida was quickly dismissed, as further images showed it to be a small spherical companion to Ida (see Figure 11.25). This one kilometre diameter satellite, which was given the initial designation 1993 (243) 1, was eventually named Dactyl.

The discovery of Dactyl ended a long argument about whether asteroids were large enough to retain satellites. In 1802 William Herschel had suggested that asteroids were not large enough to do so, and calculations in the twentieth century seemed to confirm this. But in the 1970s and 1980s a number of secondary intensity dips were observed when some asteroids occulted stars, and these dips were attributed by some astronomers to unseen asteroid companions. We still do not know whether this was so or not for the asteroids concerned, but Dactyl showed that even a relatively modest sized asteroid could retain a companion.

The fortuitous discovery of Ida's satellite had the great advantage of allowing Ida's mass and density to be determined. Although Ida's highly elongated shape was clearly seen, being 15 km across and 56 km long, Dactyl's

**Figure 11.25** Ida and its tiny satellite Dactyl are seen in this image taken by the Galileo spacecraft about 14 minutes before closest approach. The maximum separation between the asteroid and its satellite was larger than apparent here. In fact Dactyl's exact orbital parameters are unclear because its orbit was seen almost edge on. (Courtesy of NASA/JPL/Caltech; PIA 00136.)

orbit could only be determined within broad limits, as the orbit was seen nearly edge-on. As a result, Ida's density could only be determined to within the range of 2.0 to 3.1 g/cm$^3$. Nevertheless, these low densities indicated that Ida, which is thought to be a stony asteroid, probably has many voids inside it.

Richard Binzel had analysed the Koronis family of asteroids, of which Ida is one, before the Galileo intercept, and had concluded that these asteroids were the result of the breakup of a 200 km diameter body less than one billion years ago. Initially it was thought that such a fracture would eject the resulting asteroids with such a velocity, and in such different directions, that no two significant pieces would stay together in the same orbit. But the discovery of Dactyl brought this logic into question. Evidently the velocity of Dactyl relative to Ida in their orbits of the sun must have been very slow to allow the satellite's capture, so both objects must have been in virtually the same orbit.

Two other facts needed explaining about Ida and Dactyl. Firstly, from a crater count both objects appear to be much older than the one billion years maximum previously suggested by Binzel. Secondly, Dactyl should have been knocked out of its orbit around Ida long ago. In addition, Dactyl is spherical, which is almost impossible if it was produced following the violent disintegration of a far larger body.

The first of these problems is easy to solve, as if Ida and Dactyl are fragments of a larger body, the debris from the breakup would most likely have been impacting Ida and Dactyl for millions of years afterwards. This would have produced more craters than normal for objects of their age, giving the impression that they are older than they really are. The continued existence of Dactyl as a satellite of Ida could be explained if Dactyl is a fragment of the original explosion that has only recently been captured by Ida. Dactyl's spherical shape is more difficult to explain, however. It could be that Dactyl is only a loose collection of rocks that formed relatively recently, in which case it could be spherical. Maybe the original Dactyl was destroyed by an impact, and those pieces still in orbit around Ida recently came together to form the satellite that we now see.

Gaspra and Ida are so-called silicaceous, stony or S-type asteroids, which are reddish in colour and common in the inner regions of the asteroid belt. Their albedos typically range from 0.15 to 0.25. But the next asteroid, Mathilde (253), to be visited by a spacecraft was a very dark carbonaceous or C-type, which is more common in the outer regions of the asteroid belt.

In June 1997 the Near Earth Asteroid Rendezvous (NEAR) spacecraft flew within about 1,200 kilometres of Mathilde, whilst NEAR was en route to intercept the asteroid Eros (433) in 1999. Mathilde is a medium sized asteroid, being about 60 km in diameter according to ground-based observations. Its rotation period of 17.4 days makes it one of the most slowly spinning asteroids known, and with an albedo of just 0.03 it is very dark – much darker than charcoal.

The images returned to earth by NEAR showed that Mathilde has a remarkably battered surface, with at least four 20 to 30 km diameter craters on the 65% of the surface that was visible during the encounter. It is amazing that an object that was found to be just $59 \times 47$ km in size could have survived such a battering, but survive it it did. Interestingly, the albedo and colour of Mathilde's surface was the same at the bottom of the large craters as elsewhere on the surface, indicating that it is a uniformly dark, undifferentiated asteroid, that crater counts indicated is at least 2 billion years old.

Mathilde's gravitational pull on the NEAR spacecraft, and the observed size of the asteroid, produced a density estimate of about 1.3 g/cm$^3$, which is less than half the density of carbonaceous chondrite meteorites. This suggests that Mathilde is extremely porous. At first sight this seems to be highly unlikely, bearing in mind that Mathilde has clearly survived several massive impacts. But it transpires that porous bodies are more likely to survive an impact than more rigid ones, as the porosity allows more 'give' than a rigid surface, soaking up more energy. So smaller objects than normal could have made the large craters.

After the fly-by of Mathilde, it had been originally planned to put NEAR into orbit around the S-type asteroid Eros in January 1999, but spacecraft problems caused NEAR to fly past the asteroid. However, quick action on the ground by spacecraft controllers enabled NASA to have one more attempt to orbit Eros just over a year later. In spite of this problem, images were taken

**Figure 11.26** This is the last image to be transmitted by the NEAR spacecraft as it descended to the surface of Eros. It shows objects as small as one centimetre across, but few craters are seen. An irregular channel, just below left centre, is thought to indicate a subsurface crack. (The stripes in the bottom third of the image were caused by a breakage in transmission when NEAR touched down.) (Courtesy of NASA/JPL/ Caltech; PIA 03148.)

of Eros during the unplanned fly-by that showed that it is a 33 × 13 × 13 km object with one large impact crater.[44] Interestingly, there were fewer smaller craters compared with Ida, indicating that Eros has a younger surface, and spacecraft trajectory changes showed that the asteroid's density was about 2.7 g/cm$^3$.

The second attempt to put NEAR into orbit was successful in February 2000. Over the next year the spacecraft transmitted thousands of images, before NASA decided to end its life with a previously unplanned landing on Eros. The spacecraft had not been designed to do this, and it had no landing legs or air bags to cushion its fall. Nevertheless this amazing manoeuvre was a complete success with the spacecraft continuing to transmit from the asteroid's surface, whilst resting on two of its solar wings.

Images from orbit showed a surprising variety in geological features, ranging from odd-looking, squarish craters, to flat 'ponds' of material in low-lying areas, and boulders of various sizes, some of which had small craters on their surface. For some reason there was a relative scarcity of very small craters, possibly because they had been filled in by loose surface material every time EROS suffered a large impact that caused it to vibrate.

By good fortune NEAR landed on one of the flat 'sand ponds' and, in its last image (see Figure 11.26), it confirmed that the lack of small craters

---

[44] The actual size is not all that different from the 22 × 6 km deduced as long ago as 1931 by van den Bos and Finsen (see Section 7.2). They had also correctly estimated its rotation rate as about 5.3 hours.

extended down to those of centimetre size. It also showed parts of two small channels, which indicated some sort of slumping of surface material, possibly due to underground cracks.

## Distant asteroids and the Kuiper belt

In October 1920 Walter Baade discovered an object, later called Hidalgo (944), in an orbit with a perihelion of 2.0 AU, aphelion of 9.6 AU, and an inclination of about 42°. Such a highly eccentric, highly inclined orbit is typical of comets, but the object has never shown any evidence of cometary activity. So it was not clear if it was an asteroid or a comet that is no longer active.

For over fifty years no other asteroid was known with such a distant aphelion, until in October 1977 Charles Kowal, of the Hale Observatories, discovered such an object using the 48 inch (1.2 m) Palomar Schmidt. The object, initially called 1977 UB, but now called Chiron (2060),[45] was at that time about 17.8 AU from the sun. From its intensity it appeared to have a diameter in the range 100 to 400 km. Further photographs in October and November allowed a very preliminary orbit to be determined, which allowed a search through earlier photographs looking for pre-discovery images. The first to come to light was, in fact, on a plate taken by Kowal himself in September 1969.

Eventually pre-discovery images were found going back all the way to 1895, allowing a much better orbit to be defined, with a perihelion of 8.5 AU, aphelion 18.9 AU, inclination of 6°.9, and period of 51 years. In fact Chiron had been virtually at perihelion when it had first been imaged in 1895. So this, the first so-called Centaur asteroid to be discovered, had an orbit that crossed Saturn's orbit, and almost reached that of Uranus.

There has been a considerable discussion over the years as to whether Chiron is an asteroid or a comet, as its orbit is highly eccentric, like that of a comet. It may be thought that Chiron could not be a comet, as it did not have a tail, but comets do not have tails so far from the sun. What makes it less likely that it is a comet, however, is the fact that its diameter, which was thought to be in the region of 200 km, was ten times larger than that of any known cometary nucleus.

---

[45] Chiron was one of the wild, lawless, half horse/half man Centaurs of Greek mythology. He was the son of the god Kronos, or Cronus, who the Romans knew as Saturn.

The question of Chiron's identity became even more interesting in February 1988 when David Tholen of the University of Hawaii and colleagues found that Chiron had brightened unexpectedly. This still did not make it a comet, as no coma could be seen, and the spectrum contained no emission features. But on 9th April 1989 a coma of ice and dust was detected by Karen Meech of the University of Hawaii and Michael Belton of the American National Optical Astronomy Observatories. Then in 1990 Bobby Bus and Ted Bowell of the Lowell Observatory and Mike A'Hearn of the University of Maryland detected cyanogen (CN) gas surrounding Chiron to a distance of 50,000 km. Cyanogen was known to be a constituent of the ionised gas tails of comets, but this was the first time that it had been detected at such a large distance of 13 AU from the sun.

It seemed natural for Chiron to start emitting gas and dust as it approached perihelion, due in 1996, if it was a comet. But Bus and colleagues, on reexamining some pre-discovery images, found that Chiron had been brighter around *aphelion* in 1970 than it had been since, completely unlike the behaviour of a comet. Then in 1993, Meech, Belton, and Marc Buie found evidence, using the Hubble Space Telescope, of a dust atmosphere in the inner 1,200 km of its coma. At this stage it seemed that Chiron was neither an asteroid nor a comet.

Chiron's diameter was still uncertain in the early 1990s, with estimates ranging from 160 to 370 km. However, in 1994 there was a possibility of obtaining a much better estimate, as it was thought that Chiron may occult a 12th magnitude star on 9th March of that year. Observing the occultation should also provide more information about Chiron's coma. Unfortunately, Chiron just missed the star, but observations on board the Kuiper Airborne Observatory showed a number of dips in the star's intensity, the deepest and sharpest of which occurred just before closest approach.

Because Chiron missed the star, its diameter was still not known accurately. But a new estimate was made shortly afterwards by Wilhelm Altenhoff and Peter Stumpff, of the Max Planck Institute, based on Chiron's brightness at a wavelength of 1.2 mm, as at this wavelength asteroids are almost perfect thermal emitters. In the event, the calculated diameter of $170 \pm 20$ km was near the bottom end of previous estimates.

So, with all this information, is it possible to determine whether Chiron is an asteroid or a very large comet? Oddly, Chiron's activity decreased markedly as it approached perihelion in early 1996, so it is not a comet as we normally understand them. On the other hand, with its large coma, it is not a normal asteroid either. Whatever it is called, its solid structure is about 170 km in diameter and nearly spherical in shape. It is also thought to have a dark, dusty surface crust of silica or organic compounds. Underneath this crust are thought to be low temperature ices which, from time to time, erupt through the surface, projecting ice, dust and gas into space through geyser activity. In many ways it is more like a smaller version of Triton than anything else, although the dark surface of Chiron was thought to be due to dust, rather than darkened methane as on Triton.

Chiron's physical characteristics are not its only interesting attributes, however, as its orbit goes so close to that of Saturn and Uranus that it is unstable. This is clear, not only from an orbital analysis, but also from the fact that Chiron's subsurface ices would have evaporated long ago, if it had spent very long in its present orbit.

Although it had taken only just over a year after the discovery of Ceres for the second asteroid to be discovered, it took fifteen years after the discovery of Chiron before the second Centaur was found. This was the asteroid 1992 AD or Pholus (5145) that was discovered in January 1992 by David Rabinowitz, who found it with the Spacewatch telescope on Kitt Peak, Arizona. Within a few weeks Pholus, which was estimated to have a diameter of about 160 km, was found to be in an even more distant orbit than Chiron, with a perihelion of 8.7 AU, aphelion of 32 AU, inclination of 25°, and period of 93 years. Although it was discovered only a few months after perihelion, Pholus showed no evidence of a coma. It was found to be very red for an asteroid (or comet), being over three times as bright in the near infrared waveband as in visible light. Pholus' orbit was, like Chiron, also chaotic, crossing, as it does, the orbits of Saturn, Uranus, and Neptune.

Kenneth Edgeworth and Gerard Kuiper had suggested some time ago that there should be many planetesimals left over from the formation of the planets, orbiting the sun outside the orbit of Neptune (see Section 9.13). Kuiper and Whipple had also proposed that this Kuiper belt, as it is now called,[46] could be the source of short period comets. Then in 1980 Julio Fernandez, of the University of Uruguay, produced supporting evidence when he showed that the Kuiper belt would be far more efficient in producing short period comets than the much more distant Oort cloud (see Section 9.14). A few years later Martin Duncan, Thomas Quinn and Scott Tremaine, of the Canadian Institute for Theoretical Astrophysics, showed that the orbital inclinations of any comet coming from the Oort cloud should be random, whilst those from the flattish Kuiper belt would remain low. This is consistent with the observed orbital inclinations of short period comets, which are

---

[46] It is also sometimes called the Edgeworth–Kuiper belt.

generally less than about 35°. Clearly Chiron and Pholus are not planetesimals in the Kuiper belt now, but they may well have come from there in the past. What was really required now to confirm the existence of such a belt beyond Neptune was the direct observation of planetesimals still in it.

So it was that David Jewitt, of the University of Hawaii, and Jane Luu, of the University of California, started in 1987 to search for Kuiper belt objects using the University of Hawaii's 2.2 m telescope on Mauna Kea. For five years they found nothing, then in August 1992 they found what they had been looking for. Initially it was impossible to be sure that the orbit of 1992 QB$_1$ or Smiley, as the new object was informally called,[47] was wholly outside the orbit of Neptune. At that stage the reddish object appeared to be about 43 AU from the sun, but its orbital eccentricity could not be determined until it had been followed over more of its orbit. Eventually, however, it was confirmed that 1992 QB$_1$ was the first trans-Neptunian or Kuiper belt object to be found. It was always well outside Neptune's orbit, as 1992 QB$_1$ had a perihelion of 40.9 AU, aphelion 47.3 AU, orbital inclination 2°.2, and period 290 years. Assuming a typical albedo of 0.04, its diameter would be about 240 km, or very similar to that of Chiron and Pholus, although, unlike these, 1992 QB$_1$'s orbit is stable.

Whilst Jewitt and Luu were making further observations of 1992 QB$_1$, in order to determine its accurate orbit, they also continued their search for more Kuiper belt objects. In this they were quickly successful, for in March 1993 they found the second such object, called 1993 FW or Karla,[48] in a similar orbit to Smiley. Also like Smiley, 1993 FW was reddish in colour.

Whilst the first two Kuiper belt objects had orbits that came nowhere near the orbit of Neptune (see Table 11.3), the next four to be discovered (1993 RO, RP, SB and SC) had orbits that came much closer. More interesting, however, was the fact, first pointed out by Brian Marsden, that their orbital periods, like that of Pluto, were in a 2:3 resonance with that of Neptune. It is this resonance that ensures that their orbits are stable, ensuring that the objects themselves never come near to Neptune and are therefore not disturbed by it.

---

[47] Jewitt and Luu had called the object Smiley, after the name of the British spymaster in the Le Carré novels. Although they did not submit the name to the IAU for ratification, it would not have been ratified as there was already an asteroid called Smiley, after Charles Smiley, an expert in celestial mechanics.

[48] Like the name Smiley, Karla was also an informal name, again taken from the Le Carré novels. Karla was the Soviet spymaster, and Smiley's opponent.

Table 11.3 *The first Kuiper belt objects in order of discovery*

| Name | Discovered | Diameter (approx.) (km) | Perihelion (AU) | Semi-major axis (AU) | Aphelion (AU) | Orbital inclination (degrees) | Orbital eccentricity |
|---|---|---|---|---|---|---|---|
| 1992 QB$_1$ | Aug. 1992 | 240 | 40.9 | 44.1 | 47.3 | 2.2 | 0.08 |
| 1993 FW | Mar. 1993 | 260 | 41.7 | 43.6 | 45.6 | 7.7 | 0.05 |
| 1993 RO | Sept. 1993 | 160 | 31.5 | 39.4 | 47.2 | 3.7 | 0.20 |
| 1993 RP | Sept. 1993 | 100 | 34.9 | 39.3 | 43.8 | 2.6 | 0.1 |
| 1993 SB | Sept. 1993 | 160 | 26.7 | 39.5 | 52.4 | 1.9 | 0.32 |
| 1993 SC | Sept. 1993 | 240 | 32.3 | 39.7 | 47.0 | 5.1 | 0.19 |
| Pluto | | 2,300 | 29.7 | 39.5 | 49.3 | 17.1 | 0.25 |
| Neptune | | | 29.8 | 30.1 | 30.3 | 1.8 | 0.01 |

In addition to the discovery of the third to sixth Kuiper belt objects, the year 1993 also saw the discovery of the third Centaur. Called 1993 HA$_2$ Nessus (7066), this Centaur appeared to be somewhat smaller than the previous two, assuming that they all had the same albedo. It was even redder than Pholus and orbited somewhat further out than either Chiron or Pholus, with an orbit reaching well beyond Neptune's.

So at the end of 1993 three Centaurs were known, all with perihelia well within Neptune's orbit, together with four Kuiper belt objects, with mean orbits in a 2:3 resonance with Neptune, and two Kuiper belt objects in approximately circular obits at least 10 AU beyond Neptune. It was beginning to look as though there were just three categories of these distinct objects to consider, and over the next two or three years that was the case. But Nature is always ready to destroy our neat categorisations, whether they be of rocks, stars or asteroids, and in 1996 an object was found that did not fit into any of these three categories of distant asteroids.

Called 1996 TL$_{66}$, the new object was discovered in October 1996 by Luu, Jewitt and colleagues using a new, larger CCD array attached to the 2.2 m Mauna Kea telescope. Follow up observations by Carl Hergenrother, of the University of Arizona, indicated that its orbit was peculiar, however. As a result further observations were undertaken at Brian Marsden's request[49] by Warren Offutt of the Cloudcroft Observatory, to enable an unambiguous orbit determination to be made. In the event, 1996 TL$_{66}$'s perihelion distance was found to be an unremarkable 35 AU from the sun, but its aphelion was at the astonishing distance of 135 AU, giving it an orbital period of about 790 years. Since then a number of similar objects, now called scattered disc objects, have been found with aphelia well beyond the outer edge of the Kuiper belt.

Today (early 2002), therefore, we know of about 30 Centaurs, 80 scattered disc objects, and over 500 Kuiper belt objects, a significant percentage of the latter being in a 2:3 orbital resonance with Neptune. These are now often called Plutinos, following a suggestion by David Jewitt, as the first object to be found in such a resonance orbit was Pluto.

This discovery that Pluto is not alone has recently led to a somewhat sterile debate as to whether Pluto should now be struck off the list of planets. In the event, Pluto has retained its planetary status, at least for the present, as it has always been known as a planet, and it is far larger than any of the

---

[49] Brian Marsden was the director of the Minor Planet Center, and as such was responsible for issuing notices of asteroid discoveries to the astronomical community.

currently known Centaurs, scattered disc or Kuiper belt objects. What will happen, however, if and when a similar sized or larger body than Pluto is discovered much further out orbiting the sun, is anyone's guess.[50]

## 11.7 Comets[51]

Our knowledge of comets has been greatly enhanced by the international fleet of six spacecraft that examined Halley's comet after it had swung by the sun in 1986, 76 years after its previous visit. The fleet consisted of two Russian spacecraft Vega 1 and 2, two Japanese spacecraft Suisei and Sakigake, a European spacecraft Giotto[52] and the American International Cometary Explorer or ICE spacecraft. All these spacecraft were due to fly by the comet on the sunward side, at closest approach distances ranging from 28 million km for ICE to just 600 km for Giotto.

Some time previously NASA had decided to use the ISEE-3 spacecraft, which was already in space, to fly by the comets Giacobini–Zinner and Halley. As a result, on 11th September 1985 ISEE-3, which had been re-named ICE, flew through the tail of Giacobini–Zinner about 7,800 km from the nucleus. Although the spacecraft did not have a camera or spectrometers to analyse the cometary dust, it was able to measure the electromagnetic environment of Giacobini–Zinner and to observe the interaction between it and the solar wind. Somewhat surprisingly, no clear bow shock[53] could be detected by ICE, but the turbulence produced by the comet in the solar wind was found to extend further from the nucleus than expected.

The most accepted theory of comets at the time had been proposed by Fred Whipple of the Harvard College Observatory in 1950 (see Section 9.14). In this so-called 'dirty snowball' theory, Whipple proposed that a cometary nucleus is a relatively loose mixture of various ices and dust. As the nucleus begins to approach the sun it starts to heat up, and when its temperature

---

[50] This discussion of the rapidly expanding field of distant asteroids will be terminated here so as to retain the book's balance. For the reader wishing to go further, however, many more details are given in *Beyond Pluto: Exploring the Outer Limits of the Solar System*, by John Davies, Cambridge University Press, 2001. Also, up to date lists of all asteroids are currently to be found on http://cfa-www.harvard.edu/iau/lists.

[51] The impact of the comet Shoemaker–Levy 9 with Jupiter is covered in section 11.1.

[52] Named after the Florentine painter Giotto di Bondone who in 1303 painted a comet, thought to be Halley's comet, on a fresco in the chapel of the Scrovegni in Padua, Italy.

[53] It was anticipated that comets would behave in a similar way to planets in creating a bow shock in the solar wind.

exceeds the sublimation temperature of the ices, they start to evaporate and form the coma or head of the comet. These molecules are then ionised to form an ion tail. The evaporation of the ices releases the entrapped dust that also enters the head of the comet and forms a separate dust tail. Delsemme and Swings had suggested that highly volatile methane and other elements could be embedded within the crystalline structure of water ice. So it was of some considerable interest when the ICE spacecraft detected many water ions in the tail of the comet Giacobini–Zinner.

Dale Cruikshank suggested, using ground-based infrared measurements in 1985, that the nucleus of Halley's comet had an albedo of 0.04, whilst Neil Divine proposed an albedo of 0.06. Whipple's dirty snowball would have an albedo higher than either of these estimates, so was it correct? Prior to the Halley spacecraft encounters, no-one had imaged the nucleus of a comet nor knew its size for certain. David Jewitt and Edward Danielson estimated a diameter of 8 km for the nucleus from Halley's intensity at recovery in October 1982, Dale Cruikshank suggested 20 km, and Neil Divine 6 km.

In the event, the Giotto and Vega spacecraft found that the dirty snowball theory was broadly correct for Halley's comet. The inner coma had a very high water content, and the nucleus had a density of about 0.2 $g/cm^3$, very much the same as a snowball. The inactive part of the nucleus had an albedo of only 0.03, however, so it was a very dirty porous snowball, being blacker than coal. This dark or black surface was a good absorber of solar radiation, producing a temperature of 330 K, which is about 100 K greater than the sublimation temperature of water ice in space. So the water ice in the nucleus must sublimate below its dark insulating surface.

Halley's nucleus was found to be an irregular potato-shaped mass (see Figure 11.27), $16 \times 8 \times 8$ km in size, covered with bumps and hollows, some of which looked like craters. Bright jets could be clearly seen streaming towards the sun. At the time of the spacecraft encounters the water vapour production rates were measured to be about 40 tons/s (by both Vegas 1 and 2) and 15 tons/s (by Giotto) and the dust production rates to be about 10 tons/s (Vega 1), 5 tons/s (Vega 2) and 3 tons/s (Giotto).[54] Although these figures are many of orders of magnitude higher than Schwarzschild and Kron's estimate for the 1910 apparition (see Section 9.14), they are still low enough to allow about 1,000 more orbits of the sun before Halley's comet runs out of material.

Observations made by the American spacecraft OAO-2 in 1969, at the Lyman-$\alpha$ wavelength of 121.6 nm, had shown that comet Tago–Sato–Kosaka

---

[54] As a comparison, at one stage in 1998 comet Hale–Bopp was found to be losing about 1,000 tons of dust and 130 tons of water per second

**Figure 11.27** Ejecta are seen to be coming from the sun-facing surface of Halley's comet nucleus, as imaged by the Giotto spacecraft. On the original image a number of craters can be seen on the left hand side of the nucleus. (Courtesy MPAE, Dr H.U. Keller and ESA; 88.03.015-004.)

was surrounded by a spherical cloud of neutral hydrogen over a million kilometres in diameter. At a distance of 1 AU from the sun, this comet was producing hydrogen atoms at the rate of $10^{29}$ per second. Similar clouds were also found around Bennett's comet in 1970, and Kohoutek's comet in 1973, which, in the latter case, was found to be over 10 million km in diameter.

Halley's comet was also found by the Suisei spacecraft to be surrounded by a spherical corona of neutral hydrogen atoms, extending some 10 million km from the nucleus at the time of the spacecraft intercepts. These neutral hydrogen atoms become ionised when they have travelled about 10 million km, so at that distance they are controlled by the electric and magnetic fields associated with the solar wind.

The fleet of spacecraft found that Halley's bow shock was 400,000 km from the nucleus at its closest point, almost exactly as predicted. The solar wind was found to decrease in velocity from 400 km/s in interplanetary space, to 60 km/s about 150,000 km from the nucleus. Giotto was the only spacecraft to penetrate the ionopause, which was found to be some 4,700 km from the nucleus. At the ionopause the interplanetary magnetic field dropped abruptly to zero, the solar wind disappeared, and a stream of neutral molecules and cold ions were found flowing away from the nucleus.

It is these neutral molecules which drag dust particles away with them to form the coma or head of the comet.

A few months prior to these spacecraft encounters, Zdenek Sekanina of JPL and Stephen Larson of the University of Arizona had deduced a rotation period for Halley's nucleus of 2.2 days, by carefully analysing photographs taken during the 1910 appearance of the comet. This period was confirmed by Kaneda, using data from the Suisei spacecraft by measuring the brightness variation of the hydrogen corona. Various other observers confirmed this period using data from the other spacecraft to intercept the comet. Then Robert Millis of the Lowell Observatory and David Shleicher estimated a period of 7.4 days, using a different technique for analysing the variations in the corona's brightness. This period was also confirmed by other astronomers using ground-based observations and information provided by the IUE spacecraft. Unfortunately, the data is not clear-cut, and it is possible that both the 2.2 and 7.4 day rates are correct, caused by different rotation rates about two different axes.

Halley's inner coma was found by the Vegas and Giotto spacecraft to consist mainly of water ($H_2O$), with lesser amounts of carbon monoxide ($CO$), carbon dioxide ($CO_2$), methane ($CH_4$), ammonia ($NH_3$), and polymerised formaldehyde ($H_2CO)_n$. It was thought that the polymerised formaldehyde may be the reason why the nucleus is so dark. These spacecraft found that the dust particles consisted of carbon, hydrogen, oxygen and nitrogen, the so-called CHON elements, and simple compounds of these elements, or of mineral-forming elements such as silicon, calcium, iron and sodium. The relative abundances of key elements in the material emitted by Halley's comet are close to those in the sun, rather than in the earth or in meteorites, indicating that comets consist of very primitive material, which is depleted only in the volatile elements of hydrogen and nitrogen.

On 12th February 1991, Halley's comet suddenly increased in brightness by a factor of 300. It was too far away from the sun for this to be due to a spontaneous explosion caused by solar heating, so the sudden increase was generally attributed to the impact of a large meteorite or small asteroid, although such a collision would be very rare, considering how small Halley's nucleus is. Calculations showed that the impacting body need only be about 30 m in diameter.

It had been anticipated that the Giotto spacecraft may not survive its very close fly-by of Halley's comet, because of the battering it was expected to take from cometary dust so close to the nucleus. But now that Giotto had done so, the question arose of what to do with it. Orbital specialists had already calculated what other comets Giotto could visit, and had suggested the comet Grigg–Skjellerup which it could fly-by on 10th July 1992, via an earth fly-by in July 1990. However, only 60% of the experiments were still operational, with the camera being one of those that was inoperative. Nevertheless, after six orbits of the sun Giotto was retargeted to Grigg–Skjellerup for a fly-by in 1992.

Halley's comet visits the sun every 76 years and it has an impressive tail during its period in the inner solar system. Grigg–Skjellerup, on the other hand, orbits the sun every 5 years and never gets further from the sun than the orbit of Jupiter. Because of this its nucleus has suffered much more erosion than the relatively pristine Halley nucleus, and it is now much less active. So it was anticipated that the nucleus of Grigg–Skjellerup would be much smaller than the $16 \times 8 \times 8$ km nucleus of Halley, and that it would emit much less gas and dust and have less of an effect on the solar wind in its locality than does the larger comet.

The first direct evidence of Grigg–Skjellerup was found by Giotto when it was about 600,000 km from the nucleus. There pick-up ions were detected that were caused by contamination of the solar wind by cometary gas. Then about 25,000 km from the nucleus, spacecraft instruments detected a remarkable series of waves in the magnetism of the solar wind with a wavelength of about 1,000 km. These waves were much stronger and more clearly defined than those for Halley. Just before the bow shock, water, carbon monoxide and other ions were detected, and then about 17,000 km from the nucleus Giotto passed through the bow shock itself. This was much closer than the distance of Halley's bow shock from its nucleus, indicating that Grigg–Skjellerup was emitting only about 1% of the gas that Halley did during the Giotto fly-by. Although the camera was no longer operational, and so the nucleus of Grigg–Skjellerup could not be imaged, analysis of the results from those experiments that were still working indicated that Giotto had passed within about 200 km of the nucleus, even closer than to Halley.

As Grigg–Skjellerup was emitting less dust than Halley, a smaller number of impacts were expected than the 12,000 recorded by Giotto during the Halley fly-by. It was, nevertheless, something of a surprise when only three dust impacts were recorded during the whole encounter phase with Grigg–Skjellerup. The three particles were relatively large, however, and confirmed the result from the Halley intercept that the ratio of large particles to smaller ones is higher than originally expected, indicating that comets have a higher dust to gas ratio than originally thought. So rather than cometary nuclei being 'dirty snowballs' they appeared to be more like 'icy mudballs'.

The only other comet that has been visited by spacecraft at the time of writing was comet Borrelly, that was observed from a distance of just

2,000 km by NASA's Deep Space 1 (DS 1) in September 2001. Discovered in 1904, comet Borrelly, with a period of 6.9 years, is one of the most active comets that regularly visit the inner solar system. Its period had been increased to 7.0 years following its moderately close approach to Jupiter in 1936, and then reduced to 6.8 years following a similar close approach in 1972.

As DS 1 approached Borrelly, it confirmed that its nucleus was about 8 km long by 4 km wide,[55] with a shaped waist very similar to that of a bowling pin. The spacecraft also detected three jets of vaporised ice and dust extending towards the sun, together with smaller, non-sun-pointing jets from other parts of its surface. The main jets, which were up to 60 km long, appeared to come from relatively bright, smooth patches on the comet's surface. Elsewhere the surface, that has an average albedo of 0.04, was seen to be mottled with many small, very dark patches. These were thought to be made of an organic rich residue left behind after some of the comet's ices had sublimated into space.

Interestingly, the DS 1 intercept produced one completely unexpected result as, although the solar wind was found to flow symmetrically around the comet's coma, as expected, the nucleus appeared to be off-centre, with the outward streaming cloud of ionised gases offset from the nucleus by about 2,000 km. The cause of this asymmetry has still to be explained.

## 11.8 Other solar systems

Not all stars are, like the sun, single stellar systems. For example, in 1782 the young English astronomer John Goodricke (1764–1786) was studying the star Algol, that had been known to vary in brightness since ancient times, when he found that the intensity fluctuations had a regular period of 2 days 21 hours. As a result he concluded, correctly as it turned out, that the visible star was being eclipsed by a darker stellar companion, in what is now called an eclipsing binary pair.

Sixty years later, Friedrich Bessel found that the stars Sirius and Procyon each had a secondary oscillatory motion superimposed on their linear proper motion. In this case there appeared to be no intensity variations, but the oscillatory motion indicated that both stars had an unseen stellar companion; the visible star and its companion orbiting around their common

---

[55] Five years earlier such a size had been deduced by Philippe Lamy and colleagues using images from the Hubble Space Telescope.

centre of mass. Binary stars discovered in this way are now called astrometric binaries.[56]

Finally in 1889 Edward Pickering of the Harvard College Observatory found that some of the spectral lines of the star Mizar A were not only split in two, but the separation between the two components varied cyclically. He attributed the two sets of lines to two different stars, and the variation in the line separations to changes in their Doppler shifts as the two stars orbited around their common centre of mass. So Mizar A is what is now called a spectroscopic binary. In the same year Hermann Vogel at Potsdam found that the spectral lines of Algol oscillated slightly in frequency, with the same period as the variation in its light curve. So Algol was not only an eclipsing binary, but also a spectroscopic binary, although in this case only the spectrum of one of the stars was bright enough to be detected.

The discovery of these eclipsing, astrometric and spectroscopic binaries shows how the presence of a companion star can be deduced without actually observing or imaging it. Clearly each of these techniques could also be used, in principle, to detect a planet orbiting a star, but in the case of a planet the effects are naturally very much smaller, and are consequently much more difficult to detect. For example, if the sun was viewed from a position 10 light years away, which is very close in stellar terms, Jupiter would cause it to oscillate in position by just $1.6 \times 10^{-3}{''}$, or 1.6 milliarcseconds.

These three different techniques have different limitations. In particular astrometric techniques are clearly limited to examining nearby stars,[57] as they, in general, would exhibit the largest angular oscillations due to a companion planet. So it was that in 1938 Peter van de Kamp started to observe Barnard's star, the star with the largest proper motion and, at 6.0 light years, one of the closest stars to the sun. By the 1960s van de Kamp had accumulated over 2,000 photographic plates taken with the Sproul Observatory's 24 inch (61 cm) diameter telescope and, much to his delight, these plates showed that the star oscillated in position by 24.5 milliarcseconds relative to its linear proper motion. At first he attributed this to a planet 1.6 times the mass of Jupiter ($M_J$) in a highly eccentric orbit, but by 1969 he had concluded that there were two planets, with masses similar to those of Jupiter and Saturn, causing the oscillation.

---

[56] Sirius remained an astrometric binary for just 18 years as its dark companion, now called Sirius B, was observed directly in 1862 by Alvan and Graham Clark. Procyon's companion was first imaged in 1896.

[57] This is not so for spectroscopic binaries, provided the spectrum is clear.

The first evidence that there was something wrong with van de Kamp's results came from Nicholas Wagman, who found no such oscillations in the position of Barnard's star when imaged with the Allegheny Observatory's 30 inch diameter refractor. Then in 1971 George Gatewood began to photograph Barnard's star as his PhD project, using telescopes at the Allegheny Observatory in Pittsburgh and the Van Vleck Observatory in Connecticut. Two years later it became clear to Gatewood and his supervisor Heinrich Eichhorn that the trajectory of Barnard's star was, as Wagman had found it, linear with no oscillations. Others astronomers, starting with Heintz in 1976, came to the same negative conclusion from other observations, but van de Kamp refused to believe them. Similarly, initially positive results from observations of other nearby stars like 61 Cygni were also rejected after further investigations. So by 1990 no clear examples of planets around other stars had been found.

Although no clear evidence had been provided by 1990 for the existence of even one planet around a star other than the sun, evidence was beginning to be found of dust rings around stars. These rings, which were thought to be the remains of the nebula from which the star had been formed, could, like the solar nebula, eventually condense to form planets.

The IRAS spacecraft provided the first clear evidence for a dust ring or disc around a star when in 1983 it detected an excess of infrared energy coming from Vega. More careful measurements showed that the disc extended up to about 80 AU from the star. Further work with IRAS showed evidence of dust discs around other stars, including β Pictoris, Fomalhaut and ε Eridani.

Bradford Smith of the University of Arizona was intrigued by the IRAS results and wondered if he could image the dust disc around each of these stars using a coronagraph. A coronagraph is an instrument that was originally designed to block out the light of the sun and enable astronomers to see the sun's corona without having to wait for a total solar eclipse. Clearly a similar instrument could be used to block out the light of a star and enable its surrounding dust disc to be seen, if it is bright enough. Smith, working with Richard Terrile of JPL, was unsuccessful with Vega, Fomalhaut and ε Eridani but, much to his surprise, he was able to detect a dust disc around β Pictoris, which had been thought to be the most unlikely candidate of the four. Equally surprising they had detected the dust in visible light, rather than in the infrared where they had expected the disc to be prominent, because of its low temperature. The disc was much larger than that thought to be around Vega, extending some 400 AU from β Pictoris. Buoyed by their success Smith and Terrile then tried to detect dust discs around over a hundred other nearby stars, but without success.

As often happens in astronomy, evidence of the first planets outside the solar system came from a most unexpected quarter, from someone who was not looking for them. In 1990 Alex Wolszczan was looking for new examples of that rare breed of very fast pulsars, called millisecond pulsars, with the Arecibo radio telescope, when he found one, PSR B1257 + 12, with a period of 6.2 milliseconds. Analysis of its pulsations indicated that something was not quite right, however, and so Wolszczan asked Dale Frail, of the National Radio Astronomy Observatory, to try to observe it with the VLA radio telescope in New Mexico. Frail was unsuccessful at first, but in August 1991 he was able to detect and locate the pulsar. Analysing Frail's data, Wolszczan came to the surprising conclusion that the pulsar's signals were being distorted by two planets a few times more massive than the earth, with orbital periods of 67 and 98 days.[58] Their existence was confirmed two years later when Wolszczan detected the gravitational interaction between the two planets. At the same time he also found that there was a third, much less massive planet, inside the orbits of the other two, with a period of 25 days.

It seemed odd that the first planets to be found outside the solar system should be found around such an unusual object as a millisecond pulsar. But that happened because it was possible to deduce their presence from a detailed analysis of the pulsars pulses. To date only one more planet has been found around a millisecond pulsar. At face value this sounds suspicious until it is realised that millisecond pulsars are rare, with only a few dozen examples known currently.

The discovery of the first planet around a 'normal' star was made in 1995 by Michel Mayor and Didier Queloz of Geneva Observatory in Switzerland. In April 1994 Mayor and Queloz had started a programme to detect and measure the variation in Doppler shifts of a number of yellow and orange stars like the sun over time, looking for evidence of planets. At first they found nothing, but in September 1994 they measured the spectrum of the yellow, G-type star, 51 Pegasi, for the first time. Early on they saw that the Doppler shift was varying, but it was not until January 1995 that they realised that it was varying regularly with the ridiculously short period of just 4 days. The notion that a planet was orbiting 51 Pegasi with such a short period was difficult to accept, and it made them suspicious. What

---

[58] Bailes, Lyne and Shemar of Jodrell Bank were the first to announce the discovery of a pulsar planet, just before Wolszczan's discovery. But six months later they had to retract their announcement, as the effect that they thought they had detected was found to be spurious.

was doubly perplexing was that the planet appeared to have a mass of at least $1/2 M_J$.

A basic limitation of the Doppler technique is that it only allows a minimum mass of the disturbing planet to be determined. That mission is for the case where the planet's orbit is edge-on to us. If the orbit is inclined, then the mass of the disturbing planet is higher for a given Doppler shift. Unless the planet can be seen, or unless the planet eclipses the star, its orbital inclination cannot be determined, and so only its minimum mass can be determined. Statistically the most likely mass, which is often quoted, is 1.27 times this minimum mass.

The idea of a planet with a mass of at least half the mass of Jupiter orbiting a sun-like star in 4 days, at a distance of about one-eighth the distance of Mercury from the sun, caused many astronomers to doubt its existence. After all, how could a gas giant retain its integrity when subjected to a temperature of about 1,300 K? Other possible explanations of the observed variable Doppler shift were examined and rejected one by one,[59] until the idea of a planet seemed to be the only plausible answer. So on 6th October 1995 Mayor and Queloz announced their discovery to a sceptical audience. But just eleven days later the Americans Geoffrey Marcy and Paul Butler announced corroborative measurements from the Lick Observatory.

Marcy and Butler had been looking for extrasolar planets for eight years without success, whilst Mayor and Queloz had made their announcement after just one year's work. Naturally Marcy and Butler were somewhat 'miffed' to have been scooped by these newcomers. Instead of giving up, however, this galvanised them to increase their computational resources and work flat out to find planets around other stars. They were quickly successful, as in December 1995 they found that another solar type star, 47 Ursae Majoris, was oscillating in position. This appeared to be caused by a planet of minimum mass $2.4 M_J$ at a distance of 2.1 AU, with a period of 3 years; a much more reasonable set of parameters than those of the planet around 51 Pegasi.

Since these early discoveries there has been an escalation in the discovery rate of extrasolar planets, so that at the time of writing (early 2002) the number of confirmed planets exceeds 70. Inevitably the vast majority of these planets have masses in excess of that of Jupiter, as the heavier planets are the easiest to detect, but as the detection techniques become more sensitive a few planets with a mass as low as that of Saturn ($\sim 0.3 M_J$) have been found. Hopefully it will not be too long before much smaller planets

---

[59] These included starspots, stellar pulsations or the possibility that the stellar companion was a failed star called a brown dwarf, rather than a planet.

are discovered, enabling astronomers to observe individual stellar systems each containing many planets.

With the exception of the small planets Mercury and Pluto, the orbits of the planets of our solar system have eccentricities less than 0.1. Surprisingly, however, the orbital eccentricities of the extrasolar planets found so far are, in general, much larger than these, with about 60% having eccentricities greater than 0.1, and about 20% having eccentricities greater than 0.5. So theories of the origin of solar systems in general are having to be modified to take these statistics into account. Maybe our system is unusual in having most of its planets in almost circular orbits, or maybe not. Only time and the observation of many more planetary systems will tell.

An interesting new dimension was recently given to the characteristics of extrasolar planets when the atmosphere of such a planet was detected for the first time. A planetary transit had first been detected in front of the star HD 209458 by David Charbonneau, of the Harvard Smithsonian Center for Astrophysics, and colleagues in 1999. This enabled the radius of the planet to be accurately determined as $1.27 \pm 0.02\,R_J$, as well as its mass which turned out to be $0.69\,M_J$. But two years later Charbonneau and his colleagues observed the spectrum of the star using the Hubble Space Telescope during four transits of the planet. In particular they studied the spectral absorption lines of sodium, and found that these absorption lines were slightly enhanced during the transits. This indicated that the planet has sodium in its atmosphere which is absorbing part of the star's light but, more importantly, it showed that the planet has an atmosphere. Sodium happened to be chosen for Charbonneau's research as its lines are very distinctive. Now the search is on for other elements in planetary atmospheres.

## 11.9 Concluding remarks

'The progress of astronomy during the last hundred years has been rapid and extraordinary.' So said Agnes Clerke in 1885 in the preface to her excellent history of astronomy of the nineteenth century.[60] We could say the same today about planetary astronomy in the twentieth century and, no doubt, the same will probably be said a hundred years from now about planetary astronomy in the twenty-first century. One thing is certain, however, and that is that when discoveries of the twenty-first century are added to the story, our solar system will not be the only solar system under detailed review.

---

[60] Clerke, A.M., *A Popular History of Astronomy During the Nineteenth Century.*

Earth-sized planets will have been detected around many stars, and large extrasolar planets will have been imaged and their atmospheres analysed. In fact it is quite possible that our descendents in the year 2100 will know some of these extrasolar planets better than we knew Jupiter and Saturn just fifty or a hundred years ago.

As far as our own solar system is concerned, objects will be imaged over the next hundred years at much greater distances than is presently possible, so giving a new perspective on our system's size and configuration. Man will land on Mars, but what he will find there is anyone's guess. Whatever it is, it will no doubt be surprising.

In fact surprise has been one of the key features of astronomy over the centuries, as this book has hopefully shown. All the time, astronomical observers and theoreticians have been operating at the limits of what has been technically possible and, because of this, many false trails have inevitably been followed. In fact, most analysis has been based on what were thought to be the most likely scenarios, using the most eloquent models. But nature is full of the extraordinary and the unreasonable, and so we are certain to have more surprises in the future. That is what makes astronomy so interesting. Will we have discovered life in the solar system and/or elsewhere in the universe in the next hundred years? Only time, and man's enthusiasm, determination and commitment will tell.

# Glossary

*aberration of light*  The apparent displacement of the position of a star caused by the relative velocity of the earth in its orbit around the sun to the finite velocity of light. This results in a so-called annual aberration. There is also a much smaller diurnal aberration due to the spin of the earth.

*aerobraking*  Using the resistance caused by the atmosphere of a planet to slow down a spacecraft.

*albedo*  Reflectivity of an astronomical object. The units used are either on the scale 0 to 1 or 0% to 100%.

*annual equation*  The effect caused by the varying gravitational attraction of the sun on the moon as the earth's distance from the sun changes over the course of a year. As a result, the moon is slightly behind its expected position in the spring and slightly ahead in the autumn.

*annual parallax*  The effect of the orbital motion of the earth around the sun on the apparent positions of objects in the sky.

*anomalistic year*  The time between successive passages of the earth through perihelion.

*anorthosite*  An igneous rock consisting mainly of plagioclase feldspar, with much smaller amounts of pyroxene and olivine.

*ansae*  Those parts of Saturn's rings on either side of the planet as viewed from earth. ('Ansae' is the Latin word for 'handles', as that is how the rings appeared in early telescopes.)

*aphelion*  The point in the orbit of a planet, comet or asteroid where it is furthest away from the sun.

*apogee*  The point in the orbit of the moon or earth-orbiting spacecraft where it is furthest away from the earth.

*apsides*  See *line of apsides.*

*ascending node*  The point where an orbiting body crosses its orbital reference plane, travelling from south to north. The descending node is where the orbiting body crosses its orbital reference plane going from north to south. For an object orbiting the sun the orbital reference plane is the ecliptic.

*asteroid types*  Asteroids are classified according to their spectra, with the majority falling into one of the following three groups:

Most common are the very dark, carbonaceous, or C-type asteroids, that are very common in the outer regions of the asteroid belt. They appear to resemble the carbon-rich chondrite meteorites.

Next most common are the reddish or greyish, silicaceous, or S-types, with albedos typically in the range of 0.15 to 0.25. They are most common in the inner regions of the asteroid belt, and appear to resemble stony or stony-iron meteorites.

Least common of the three groups are the metallic or M-types, which appear to resemble nickel–iron meteorites, and which, like the S-types, are most common in the inner parts of the asteroid belt. They are strong radar reflectors.

*astronomical unit*   The mean distance of the earth from the sun.

*basalt*   A volcanic igneous rock consisting mainly of pyroxene and plagioclase, with smaller quantities of ilmenite.

*bow shock*   The boundary around a planet's magnetosphere where the solar wind is deflected.

*breccia*   A rock made of broken fragments held in a fine-grained matrix. They are commonly the result of impacts of, for example, meteorites on the moon.

*brown dwarf*   A star whose mass is too low to cause hydrogen fusion in its core.

*caldera*   A large volcanic crater created by the collapse of the surface and/or by an explosive eruption.

*celestial equator*   The projection of the earth's equatorial plane into space.

*celestial poles*   The two points on the celestial sphere about which the stars appear to rotate daily. They are 90° away from the celestial equator.

*Cepheid variable*   A type of variable star. Cepheid variables are important because their period of variability correlates with their absolute intensity, so allowing the latter to be determined from earth. Knowing their absolute intensity, their distance from earth can be estimated.

*chasma*   A steep-sided canyon.

*chondrites*   Stony meteorites that have small spherical inclusions in them called chondrules. Chondrites have a very similar chemical composition to that of the sun, although they contain no free hydrogen and helium.

*chondrules*   Spherical inclusions in chondrites. Chondrules were clearly once molten and can be of metal, silicate or sulphide materials.

*chromatic aberration*   The aberration produced by a lens that has different foci for light of different colours.

*coma of a comet*   Otherwise called the head. It is the diffuse luminous sphere that surrounds the comet's nucleus.

*concave lens*   A lens that is thinner at the centre than the edge, causing a parallel beam of light to diverge.

*conic section*   Various geometric figures are produced if a cone is cut by a plane at various angles to the cone's axis. These figures are called conic sections. Depending on the angle of cut, they are a circle (if the plane is perpendicular to the axis), an ellipse, a parabola and a hyperbola. A circle's eccentricity is 0, an ellipse's is greater than 0 but less than 1, a parabola's is exactly 1, and a hyperbola's is greater than 1.

*conjunction*   An alignment of three bodies. So, for example, Mercury can be in line with the earth and sun, and be either between the earth and sun, in an arrangement called *inferior conjunction,* or on the other side of the sun to the earth, when it is at *superior conjunction.*

*convex lens*   A lens that is thicker at the centre than the edge. It causes a parallel beam of light to converge to a focus.

*corona*   An oval-shaped feature on a planetary surface. See also *solar corona.*

*coronagraph*   An instrument containing an occulting disc that creates an artificial eclipse of a bright source like the sun, and allows observers to detect faint objects close to the bright source.

*crust*   The outermost solid layer of a planet or moon, usually consisting of rock and/or ice. It is the outermost part of the lithosphere.

*cryovolcanism*   The resurfacing of an icy surface by ice–liquid mixtures.

*declination*   (A) The angular distance of a celestial object on the celestial sphere, measured in degrees, north or south of the celestial equator.

(B) The angle between the direction of a magnetic field in a horizontal plane and true north for a planet or moon.

*deferent*   The circle on which the centre of an epicycle moves.

*degenerate matter*   Matter at very high pressures where the normal atomic structure is destroyed. For example, atoms are stripped of their electrons at the very high pressures within white dwarf stars. As a result, these stars consist almost entirely of degenerate matter in the form of a highly compressed mixture of atomic nuclei and electrons. In neutron stars, where the internal pressures are even higher, the degenerate matter consists almost entirely of neutrons.

*descending node*   See *ascending node.*

*diatomic molecule*   A molecule consisting of two atoms bound together, e.g. $H_2$ or CO.

*dipole*   A system of two equal and opposite charges situated a short distance apart. The dipole moment is the value of one of the charges times the distance between them. A small bar magnet is a magnetic dipole that produces a dipolar magnetic field.

*direct or prograde motion*   Orbital or axial motion of a body in the solar system that is counterclockwise as seen from north of the ecliptic. The term is also used to describe the motion of an object on the celestial sphere in a west–east direction.

*Doppler shift*   The shift in frequency of spectral lines due to the motion of a light source towards or away from an observer. A similar effect occurs to the frequency of sound waves when their source is moving towards or away from the listener.

*ecliptic*   The mean plane of the earth's orbit around the sun.

*elongation*   The angular distance between the sun and a planet or the moon as viewed from earth.

*epicycle*   A circle whose centre moves around a larger circle called the deferent.

*evection*   A periodic perturbation of the moon, most evident at quadrature, caused by the variation in the gravitational attraction of the sun as the moon orbits the earth.

*first point of Aries*   Where the earth's equatorial plane cuts the ecliptic at the spring equinox. Although this was originally in the constellation of Aries several thousand years ago, it is now in Pisces because of the precession of the equinoxes.

*geocentric*   Earth-centred.

*heliacal rising*   The first visibility of a star or planet in the morning sky just before dawn.

*heliacal setting*   The last visibility of a star or planet in the evening sky just after dusk.

*heliocentric*   Sun-centred.

*hippopede*   The figure-of-eight planetary track described by Eudoxus. It is the product of the motion of a planet fixed on the equator of one sphere, which has its poles fixed on the surface of a second, concentric sphere, with the earth at their common centre. Both spheres rotate uniformly with equal and opposite angular velocities about different poles.

*homonuclear molecule*   A molecule that contains atoms of only one element, e.g. $O_3$.

*H–R diagram*   See *main sequence stars*.

*Hubble constant*   The rate of expansion of the universe, currently thought to be about 70 km/s per megaparsec.

*ilmenite*   An iron titanium oxide, $FeTiO_3$.

*inclination*   (A) The angle between the orbital plane and a reference plane for an orbiting body. For an object orbiting the sun the reference plane is the ecliptic.

(B) The angle between a body's spin axis and a reference plane, which is usually the body's orbital plane.

(C) The angle between the direction of a magnetic field and the local horizontal for a planet or moon.

*infrared*   Electromagnetic radiation with a wavelength between about 0.8 and 300 μm.

*intensity of stars*   See *stellar magnitude*.

*intercalary month*   An extra month added from time to time into a calendar, to correct for the fact that there are not an integral number of synodic months in a tropical year.

*ionopause*   The region close to the nucleus of a comet where the interplanetary magnetic field drops abruptly to zero and the solar wind disappears. It is dominated by neutral molecules and cold ions flowing away from the nucleus.

*ionosphere*   A region of a planetary atmosphere where the atoms and molecules are ionised.

*Lagrangian points*   Points in the orbital plane of two objects that orbit around their common centre of mass, where a particle of negligible mass can remain in equilibrium. There are five such points for two bodies in circular orbits, three of which are unstable to small perturbations. The other two, which are 60° in front of and behind the less massive body, and in the same orbit, are stable.

*libration*   The effect of being able to see more than 50% of the moon's surface from earth, even though the moon's spin rate is the same as its orbital period. Libration is mainly due to the fact that the moon's orbit is not circular (producing a libration in longitude) and its spin axis is not perpendicular to its orbital plane (producing a libration in latitude). There are also smaller librations due to the fact that the moon's spin axis is not perpendicular to the ecliptic, and its spin is not absolutely regular about a fixed axis. There is also a diurnal libration caused by observing the moon from either end of the earth's diameter at 12 hourly intervals.

*line of apsides*   The major axis of an elliptical orbit joining the furthest and nearest orbital positions from the centre of mass. In the case of a planet's orbit around the sun, the line of apsides joins the aphelion and perihelion, the aphelion and perihelion being the apsides.

*line of nodes*   Where the plane of an inclined orbit intercepts a reference plane. For example, the line of nodes of the moon's orbit around the earth is where the plane of its orbit intercepts the orbit of the earth around the sun, the latter being the reference plane.

*lithosphere*   The rigid outer layer of a planet or moon, including the crust and part of the upper mantle.

*Lyman-α*   A line at 121.6 nm which is produced when an electron in the first excited state of a hydrogen atom decays to its ground state.

*magnetopause*   The boundary of the magnetosphere.

*magnetosphere*   The region around a planet in which its magnetic field is constrained by the solar wind.

*magnitude*   See *stellar magnitude*.

*main sequence stars*   Stars on a band of the H–R (Hertzsprung–Russell) diagram running from the very luminous, hot, white O-type stars through B, A, F, G and K stars of progressively lower luminosities and temperatures to low intensity, cool, red, M-type stars.

*mantle*   The layer of a planet or moon between the crust and the core.

*mare (pl. maria)*   The extensive dark areas on the moon that were originally thought to be seas of water. They are now known to be solidified lava.

*mascons*   Areas of 'mass concentration' or high density on the moon.

*Metonic cycle*   A period of 19 tropical years, which is almost exactly the same as 235 synodic months. So after 19 years the moon's phases recur on exactly the same days of the year (within 0.09 days).

*molecular energy levels*   Molecular spectral lines can be the result of electronic, vibrational or rotational energy level transitions. Typically the gaps between the energy levels are largest for electronic transitions, second largest for vibrational transitions, and smallest for rotational transitions.

*oblateness*   A measure of the amount by which the shape of a planet or other body differs from a perfect sphere. This non-sphericity is normally caused by axial rotation.

*obliquity of the ecliptic*   The angle between the earth's equatorial plane and the ecliptic.

*occultation*   The passage of one astronomical object in front of another so that an observer can no longer see the more distant object.

*octant*   The position of a moon or planet when half-way between conjunction or opposition and quadrature. In the case of the earth's moon, for example, the octants are at 45° before and after new and full moons.

*olivine*   A magnesium or iron silicate, $(Mg,Fe)_2SiO_4$.

*opposition*   The position of a superior planet when it is opposite to the sun in the sky, so crossing the meridian at local midnight. At that time the planet's elongation is 180°, and it is at its highest point in the sky.

*palimpsest*   A roughly circular spot on an icy surface that is the remains of a crater.

*perigee*   The point in the orbit of the moon or earth-orbiting spacecraft where it is closest to the earth.

*perihelion*   The point in the orbit of a planet, comet or asteroid where it is closest to the sun.

*plagioclase*   A feldspar of sodium or calcium aluminium silicate.

*planetesimal*   A body of rock and/or ice, typically 100 m to 10 km or so in diameter, that was formed in the early solar nebula.

*planum*   A plateau or high plain on a planetary surface.

*plate tectonics*   The mechanism that causes the structural plates of a planet's surface to move together or apart. On the earth this has produced continental drift.

*precession of the equinoxes*   The movement of the equinoxes with respect to the stars caused mainly by the gravitational effects of the moon and sun on the non-spherical earth. These gravitational effects cause the earth's axis to precess (trace out a conical motion) in space with a period of about 25,800 years. This results in the precession of the equinoxes with the same period.

*prograde motion*   See *direct motion*.

*proper motion*   The motion of a star across the celestial sphere, as seen from earth, that is a combination of its true motion and that of the solar system through space. It excludes the effect of the annual motion of the earth around the sun.

*pulsar*   A rotating neutron star which emits a beam or beams of radio and/or other waves. These are detected from earth as pulses at the rotational frequency of the star, usually of the order of seconds, but sometimes of milliseconds. Neutron stars are ultra-dense stars with a typical diameter of only about 10 km.

*pyroclastic blanket*   The blanket created around a volcano that is the result of the explosive eruption of fragmentary volcanic rocks and dust, rather than molten lava.

*pyroxene*   A calcium or sodium and magnesium, iron or aluminium silicate, i.e. $ABSi_2O_6$, where A is Ca or Na, and B is Mg, Fe or Al.

*quadrant*   An instrument for measuring the altitudes or angular separation of astronomical objects.

*quadrature*   The position of the moon or a planet when its angular distance from the sun, as viewed from the earth, is 90°.

*quadrupole*   A set of four electric charges.

*radiant*   The point on the celestial sphere from which meteors appear to radiate, giving the names to the various meteor showers, e.g. the Leonids that have a radiant in the constellation of Leo, etc.

*radiation pressure*   The pressure exerted by a stream of photons.

*reciprocal mass of a planet*   The mass of the sun divided by the mass of the planet.

*reflectivity*   See *albedo*.

*regio*   Means 'region' in the names of some planetary surface features.

*regolith*   The top layer of a solid planetary or similar body (e.g. the moon) covering the crust and composed of dust and loose fragments of rock.

*relativistic electrons*   Electrons moving at velocities close to the speed of light.

*resonance*   Vibrational resonance occurs when the frequency of application of periodic forces acting on a body matches the natural oscillation frequency of that body, causing a large increase in vibration amplitude. Orbital resonances occur when the orbital period of one body is a simple multiple of that of another body which is orbiting the same primary.

*retrograde motion*   Orbital or axial motion of a body in the solar system that is clockwise as seen from north of the ecliptic. The term is also used to describe the motion of an object on the celestial sphere in an east–west direction.

*rhyolite*   A volcanic igneous rock of relatively high silica content. Lighter in colour and more viscous than basalt.

*right ascension*   The coordinate used to define the position of a celestial object on the celestial sphere, that is the equivalent of longitude on the earth. It is measured along the celestial equator from the first point of Aries. The other coordinate, which is the equivalent of latitude on the earth, is called the declination.

*saros*   The period of 223 synodic months over which the pattern of solar and lunar eclipses repeats itself.

*second anomaly*   Another term for *evection* (see above).

*secular acceleration*   The acceleration of the moon in its orbit around the earth.

*seeing*   The effect of atmospheric turbulence on the quality of an astronomical image. When turbulence is low, and the seeing is good, the images appear sharp. But when turbulence is high, and the seeing is poor, the images are blurred and appear to be in constant motion.

*shield volcanoes*   Gently sloping volcanoes built up from successive lava flows from a single vent.

*sidereal year*   The orbital period of the earth around the sun relative to the stars.

*solar corona*   The bright outermost part of the sun that becomes visible during a total solar eclipse. It extends to many solar diameters.

*solar parallax*   The difference in the apparent position of the sun when observed by an observer on a line linking the centre of the earth to the sun, and when observed by an observer one earth radius to one side of that line.

*solar wind*   A stream of particles, primarily protons and electrons, flowing from the sun.

*spherical aberration*   The effect caused when a lens or mirror produces a separate focus for on-axis and off-axis light.

*star types*   See *main sequence stars.*

*stellar magnitude*   The magnitude or intensity of stars, as seen from earth, is given on a logarithmic scale, such that adjacent magnitudes have an intensity difference of about 2.5. So a magnitude 1 star, for example, is about 2.5 times as bright as a magnitude 2, and so on. The dimmest stars visible to the naked eye, under good observing conditions, are of about magnitude 6.

*sublimation*   The act of a material changing from the solid to the vapour phase without going through an intermediate liquid phase. So, for example, at certain temperatures and pressures water ice can evaporate without passing through the water phase.

*sun-synchronous orbit*   An orbit in which a satellite crosses the equator of a planet at the same local time of day on each orbit.

*synchronised rotation*   The exact coincidence between the spin rate of an orbiting body and its orbital period, causing the same face of the body to face its parent planet.

*synodic arc*   The average movement in ecliptic longitude of a superior planet over one synodic period.

*synodic month*   The mean interval between two successive new moons.

*synodic period*   The mean interval for a planet between successive inferior or superior conjunctions with the earth and sun, or for a planetary satellite between successive conjunctions of the satellite with the sun as observed from its parent planet.

*telluric spectral lines*   Spectral lines caused by molecules in the earth's atmosphere.

*terminator*   The boundary between the illuminated and non-illuminated parts of a planet or moon.

*terra*   An extensive land mass on a planetary surface.

*thermocouple*   A temperature measuring device in which a voltage is produced by the junction of two dissimilar metals, and where the magnitude of the voltage is dependent on the temperature of the junction.

*Trojan asteroids or satellites*   Small bodies at the $\pm 60°$ Lagrangian points and co-orbital with a planet or one of its satellites.

*tropical year*   The year measured from equinox to equinox.

*ultraviolet*   Electromagnetic radiation with a wavelength between about 9 and 380 nm.

*Van Allen radiation belts*   Two ring-shaped belts around the earth of electrically charged particles trapped by the earth's magnetic field.

*variation*  A periodic perturbation of the moon, most evident at the octants, caused by the variation in the gravitational attraction of the sun as the moon orbits the earth.

*visible light*  Electromagnetic radiation with a wavelength between about 380 nm (blue) to about 780 nm (red).

*zodiacal light*  A faint cone of light extending partly along the ecliptic on clear moonless nights in the west just after sunset, and in the east just before sunrise.

# Bibliography

This bibliography includes the main, general sources used in the preparation of this book, which can be consulted if the reader requires more detail than I have been able to give. The book publishers mentioned below are generally those for copies available in the UK. In other countries the publishers may differ. Earlier and/or later editions are sometimes available.

## General

Ashbrook, J., *The Astronomical Scrapbook*, Cambridge University Press, 1984, ISBN 0-521-30045-2.

Audouze, J., and Israël, G., (eds.), *The Cambridge Atlas of Astronomy*, Cambridge University Press, Third edition, 1994, ISBN 0-521-43438-6.

Baum, R., *The Planets: Some Myths and Realities*, David & Charles, 1973, ISBN 0-7153-6055-8.

Beatty, J.K., Petersen, C.C., and Chaikin, A., (eds.), *The New Solar System*, Cambridge University Press, Fourth edition, 1999, ISBN 0-521-64587-5.

Bell, L., *The Telescope*, McGraw-Hill, 1922 (Dover reprint, 1981, ISBN 0-486-24151-3).

Briggs, G.A., and Taylor, F.W., *The Cambridge Photographic Atlas of the Planets*, Cambridge University Press, 1982, ISBN 0-521-23976-1.

Brown, L.M., Pais, A., and Pippard, B., (eds.), *Twentieth Century Physics, Vol. III*, Institute of Physics and American Institute of Physics, 1995, ISBN 0-7503-0355-7.

Burrows, W.E., *This New Ocean: The Story of the First Space Age*, Random House, 1998, ISBN 0-679-44521-8.

Cornell, J., and Gorenstein, P., (eds.), *Astronomy from Space: Sputnik to Space Telescope*, MIT Press, 1985, ISBN 0-262-53061-9.

Doel, R.E., *Solar System Astronomy in America: Communities, Patronage, and Interdisciplinary Research 1920–1960*, Cambridge University Press, 1996, ISBN 0-521-41573-X.

Gingerich, O., *The Great Copernicus Chase*, Cambridge University Press, 1992, ISBN 0-521-32688-5.

Greeley, R., *Planetary Landscapes*, Chapman and Hall, Revised edition, 1987, ISBN 0-04-551081-4.

Henbest, N., *The Planets: Portraits of New Worlds*, Penguin, 1994, ISBN 0-14-01.3414-X.

Heppenheimer, T.A., *Countdown: A History of Space Flight*, Wiley, 1997, ISBN 0-471-14439-8.

Herrmann, D.B., (Krisciunas trans.), *A History of Astronomy from Herschel to Hertzsprung*, Cambridge University Press, 1984, ISBN 0-521-25733-6.

Hoskin, M., (ed.), *Cambridge Illustrated History of Astronomy*, Cambridge University Press, 1997, ISBN 0-521-41158-0.

Hufbauer, K., *Exploring the Sun: Solar Science since Galileo*, Johns Hopkins University Press, 1993, ISBN 0-8018-4599-8.

Jaki, S.L., *Planets and Planetarians: A History of Theories of the Origin of Planetary Systems*, Scottish Academic, 1978.

Kaufmann, W.J., *Planets and Moons*, W.H. Freeman, 1979, ISBN 0-7167-1040-4.

King, H.C., *The History of the Telescope*, Griffin, 1955 (Dover reprint, 1979, ISBN 0-486-23893-8).

Kluger, J., *Journey Beyond Selene: Remarkable Expeditions to the Edge of the Solar System and its 63 Moons*, Little Brown, 1999, ISBN 0-316-64842-6.

Kolb, R., *Blind Watchers of the Sky: The People and Ideas that Shaped our View of the Universe*, Oxford University Press, 1999, ISBN 0-19-286203-0.

Kraemer, R.S., *Beyond the Moon: A Golden Age of Planetary Exploration, 1971–1978*, Smithsonian Institution Press, 2000, ISBN 1-56098-954-9.

Kuiper, G.P., and Middlehurst, B.M., (eds.), *The Solar System Vol. III: Planets and Satellites*, University of Chicago Press, 1961.

Lang, K.R., and Whitney, C.A., *Wanderers in Space: Exploration and Discovery in the Solar System*, Cambridge University Press, 1991, ISBN 0-521-42252-3.

McDougall, W.A., *The Heavens and the Earth: A Political History of the Space Age*, Johns Hopkins University Press, 1997, ISBN 0-8018-5748-1.

Malphrus, B.K., *The History of Radio Astronomy and the National Radio Astronomy Observatory*, Krieger, 1996, ISBN 0-89464-841-1.

Middlehurst, B.M., and Kuiper, G.P., *The Solar System Vol. IV: The Moon, Meteorites and Comets*, University of Chicago Press, 1963.

Moore, P., and Hunt, G., *The Atlas of the Solar System*, Mitchell Beazley, 1983, ISBN 0-85533-468-1.

Newell, H.E., *Beyond the Atmosphere: Early Years of Space Science*, NASA SP-4211, 1980.

North, J., *The Fontana History of Astronomy and Cosmology*, Fontana, 1994, ISBN 000686177-6.

Pannekoek, A., *A History of Astronomy*, Allen & Unwin, 1961 (Dover reprint 1989, ISBN 0-486-65994-1).

Payne-Gaposchkin, C., *Introduction to Astronomy*, Eyre & Spottiswoode, 1956.

Porter, R., (consultant editor), *The Hutchinson Dictionary of Scientific Biography*, Helicon, Second edition, 1994, ISBN 0-09-177179-X.

Ronan, C.A., *Their Majesties' Astronomers*, Bodley Head, 1967.

Rousseau, P., (Bullock, M., trans.), *Man's Conquest of the Stars*, Jarrolds, 1959.

Russell, H.N., Dugan, R.S., & Stewart, J.Q., *Astronomy I: The Solar System*, Ginn, 1926.

Schorn, R.A., *Planetary Astronomy: From Ancient Times to the Third*

*Millennium*, Texas A & M University Press, 1998, ISBN 0-89096-807-1.

Sheehan, W., *The Immortal Fire Within: The Life and Work of Edward Emerson Barnard*, Cambridge University Press, 1995, ISBN 0-521-44489-6.

Sheehan, W., *Worlds in the Sky: Planetary Astronomy from Earliest Times through Voyager and Magellan*, University of Arizona Press, 1992, ISBN 0-8165-1308-2.

Spencer Jones, H., *General Astronomy*, Arnold, Third edition, 1951.

Stern, S.A., *Our Worlds: The Magnetism and Thrill of Planetary Exploration*, Cambridge University Press, 1999, ISBN 0-521-64440-2.

Struve, O., and Zebergs, V., *Astronomy of the 20th Century*, Macmillan, 1962.

Tatarewicz, J.N., *Space Technology and Planetary Astronomy*, Indiana University Press, 1990, ISBN 0-253-35655-5.

Taton, R., and Wilson, C., (eds.), *The General History of Astronomy, Vol. 2, Planetary Astronomy from the Renaissance to the Rise of Astrophysics: Part A. Tycho Brahe to Newton*, Cambridge University Press, 1989, ISBN 0-521-24254-1.

Taton, R., and Wilson, C., (eds.), *The General History of Astronomy, Vol. 2, Planetary Astronomy from the Renaissance to the Rise of Astrophysics: Part B. The Eighteenth and Nineteenth Centuries*, Cambridge University Press, 1995, ISBN 0-521-35168-5.

Urey, H.C., *The Planets: Their Origin and Development*, Yale University Press, 1952.

Weissman, P.R., McFadden, L-A, & Johnson, T.V., (eds.), *Encyclopedia of the Solar System*, Academic, 1999, ISBN 0-12-226805-9.

## Nineteenth century sources

Ball, R.S., *The Story of the Heavens*, Cassel, Revised edition, 1897.

Berry, A., *A Short History of Astronomy*, Murray, 1898 (Dover reprint, 1961, ISBN 486-20210-0).

Clerke, A.M., *A Popular History of Astronomy During the Nineteenth Century*, Black, Fourth edition, 1908.

Flammarion, C., (Gore, J.E. trans.), *Popular Astronomy*, Chatto & Windus, 1907.

Guillemin, A., ed. by Lockyer, J.N., rev. by Proctor, R.A., *The Heavens*, Bentley, Fourth edition, 1871.

Herschel, J.F.W., *Treatise on Astronomy*, Longman etc., 1839.

Humboldt, Alexander von, (Otté, E.C., trans.), *Cosmos: A Sketch of the Physical Description of the Universe, Vol. 1*, Bohn, 1849; reprinted by Johns Hopkins University Press, with an introduction by Rupke, N.A., 1997, ISBN 0-8018-5502-0.

Humboldt, Alexander von, (Otté, E.C., & Paul, B.H., trans.), *Cosmos: A Sketch of the Physical Description of the Universe, Vol. IV*, Bohn, 1852.

Lebon, E., *Histoire Abrégée de l'Astronomie*, Gauthier-Villars, 1899.

Ledger, E., *The Sun: Its Planets and Their Satellites*, Stanford, 1882.

Newcomb, S., *Popular Astronomy*, Macmillan, Second edition, 1898.

## Books concentrating on classical and early planetary astronomy

Armitage, A., *British Men of Science: William Herschel*, Nelson, 1962.

Buttmann, G., (Pagel, B., trans.), *The Shadow of the Telescope: A Biography of John Herschel*, Lutterworth Press, 1974, ISBN 0-7188-2087-8.

Caspar, M., *Kepler*, Abelard-Schuman, 1959, (Dover edn., with additions by Gingerich, O., 1993, ISBN 0-486-67605-6).

Christianson, J.R., *On Tycho's Island: Tycho Brahe and His Assistants 1570–1601*, Cambridge University Press, 2000, ISBN 0-521-65081-X.

Drake, S., *Essays on Galileo and the History and Philosophy of Science, Vol. 1*, Selected and introduced by Swerdlow, N.M., and Levere, T.H., University of Toronto Press, 1999, ISBN 0-8020-7585-1.

Drake, S., *Essays on Galileo and the History and Philosophy of Science, Vol. 2*, Selected and introduced by Swerdlow, N.M., and Levere, T.H., University of Toronto Press, 1999, ISBN 0-8020-8164-9.

Drake, S., *Galileo at Work: His Scientific Biography*, University of Chicago Press, 1978 (Dover reprint, 1995, ISBN 0-486-28631-2).

Dreyer, J.L.E., *History of the Planetary Systems from Thales to Kepler*, 1906 (Dover reprinted as *A History of Astronomy from Thales to Kepler*, 1953, ISBN 486-60079-3).

Fantoli, A., *Galileo for Copernicanism and for the Church*, Second edition, Vatican Observatory Publications, ISBN 0-268-01032-3.

Gingerich, O., *The Eye of Heaven: Ptolemy, Copernicus, Kepler*, American Institute of Physics, 1993, ISBN 0-88318-863-5.

Grant, E., *Planets, Stars & Orbs: The Medieval Cosmos 1200-1687*, Cambridge University Press, 1996, ISBN 0-521-56509-X).

Hall, A.R., *From Galileo to Newton*, Harper & Row, 1963 (Dover reprint, 1981, ISBN 0-486-24227-7).

Hall, A.R., *Isaac Newton: Adventurer in Thought*, Cambridge University Press, 1996, ISBN 0-521-56669-X.

Heath, T., *Aristarchus of Samos*, Clarendon, 1913, (Dover reprint, 1981, ISBN 0-486-24188-2).

Koestler, A., *The Sleepwalkers: A History of Man's Changing Vision of the Universe*, Penguin, 1986, ISBN 014-01.9246-8.

Koyré, A., (Maddison, R.E.W., trans.), *The Astronomical Revolution: Copernicus–Kepler–Borelli*, Cornell University Press, 1973 (Dover reprint, 1992, ISBN 0-486-27095-5).

Machamer, P., (ed.), *The Cambridge Companion to Galileo*, Cambridge University Press, 1998, ISBN 0-521-58841-3.

Ronan, C.A., *Edmund Halley: Genius in Eclipse*, Macdonald, 1970, SBN 356 02942 5.

Sidgwick, J.B., *William Herschel: Explorer of the Heavens*, Faber & Faber, 1953.

Small, R., *An Account of the Astronomical Discoveries of Kepler*, Gillet, 1804 (Wisconsin reprint, 1963).

Walker, C., *Astronomy Before the Telescope*, British Museum Press, 1996, ISBN 0-7141-2733-7.

## English translations of key historical books

### Ancient Greek texts

Heath, T.L., (trans. & notes), *Greek Astronomy*, Dent, 1932 (Dover reprint, 1991, ISBN 0-486-26620-6).

### Ptolemy

Toomer, G.J., (trans. & notes), *Ptolemy's Almagest*, Princeton University Press, 1998, ISBN 0-691-00260-6.

### Copernicus

Rosen, E., (trans. & notes), *Nicholas Copernicus: On the Revolutions*, Johns Hopkins University Press, 1992, ISBN 0-8018-4515-7.

Wallis, C.G., (trans.), *On the Revolutions of the Heavenly Spheres: Nicholas Copernicus*, Prometheus Books, ISBN 1-57392-035-5.

### Kepler

Wallis, C.G., (trans.), *Epitome of Copernican Astronomy and Harmonies of the World: Johannes Kepler*, Prometheus Books, ISBN 1-57392-036-3.

### Galileo

Crew., H., and Salvio, A. de, (trans.), *Dialogues Concerning Two New Sciences: Galileo Galilei*, Macmillan, 1914 (Dover reprint, 1954, ISBN 486-60099-8).

Drake, S., (trans. & notes), *Discoveries and Opinions of Galileo*, Anchor, 1957, ISBN 0-385-09239-3.

Drake, S., (trans. & notes), *Galileo: Dialogue Concerning the Two Chief World Systems*, University of California Press, 1962.

Van Helden, A., (trans. & notes), *Sidereus Nuncius or The Sidereal Messenger: Galileo Galilei*, University of Chicago Press, 1989, ISBN 0-226-27903-0.

### Newton

Cohen, I.B., and Whitman, A., (trans. & notes), *Isaac Newton: The Principia: Mathematical Principles of Natural Philosophy*, University of Chicago Press, 1999, ISBN 0-520-08817-4.

## Terrestrial planets

Butrica, A.J., *To See the Unseen: A History of Planetary Radar Astronomy*, NASA SP-4218, 1996.

Frankel, C., *Volcanoes of the Solar System*, Cambridge University Press, 1996, ISBN 0-521-47770-0.

Reeves, R., *The Superpower Space Race: An Explosive Rivalry through the Solar System*, Plenum, 1994, ISBN 0-306-44768-1.

Surkov, Y., *Exploration of the Terrestrial Planets from Spacecraft: Instrumentation, Investigation, Interpretation*, Wiley, Second edition, 1997, ISBN 0-471-96429-8.

## Outer solar system

Davies, J., *Beyond Pluto: Exploring the Outer Limits of the Solar System*, Cambridge University Press, 2001, ISBN 0-521-80019-6.

Fimmel, R.O., *et al.*, *Pioneer: First to Jupiter, Saturn and Beyond*, NASA SP-446, 1980.

Littmann, M., *Planets Beyond: Discovering the Outer Solar System*, Wiley, 1990, ISBN 0-471-51053-X.

## Vulcan

Baum, R., and Sheehan, W., *In Search of Planet Vulcan*, Plenum, 1997, ISBN 0-306-45567-6.

## Mercury and Venus

Antoniadi, E.M., (Moore, P., trans.), *The Planet Mercury*, Gauthier-Villars 1934, Reid 1974, ISBN 0-904094-02-2.

Dunne, J.A., and Burgess, E., *The Voyage of Mariner 10*, NASA SP-424, 1978.

Fimmel, R.O., *et al.*, *Pioneer Venus*, NASA SP-461, 1983.

Grinspoon, D.H., *Venus Revealed*, Perseus, 1998.

Maor, E., *Venus in Transit*, Princeton University Press, 2000, ISBN 0-691-04874-6.

Marov, M.Y., and Grinspoon, D.H., *The Planet Venus*, Yale University Press, 1998, ISBN 0-300-04975-7.

Maunder, M, and Moore, P., *Transit: When Planets Cross the Sun*, Springer-Verlag, 2000, ISBN 1-85233-621-8.

Moore, P., *The Planet Venus*, Faber & Faber, Second edition, 1959.

Sandner, W., (Helm, A., trans.), *The Planet Mercury*, Faber & Faber, 1963.

Roth, L.E., and Wall, S.D., (eds.), *The Face of Venus: The Magellan Radar-Mapping Mission*, NASA SP-520, 1995.

NASA SP-59, 1965, Jet Propulsion Laboratory, *Mariner-Venus 1962, Final Project Report.*

NASA SP-190, 1967, *Mariner Venus 1967.*

## Earth

Burgess, E., and Torr, D., *Into the Thermosphere: The Atmosphere Explorers*, NASA SP-490, 1987.

Jeffreys, H., *The Earth: Its Origin, History and Physical Constitution*, Cambridge University Press, Fourth edition, reprinted with additions, 1962.

Kuiper, G.P., (ed.), *The Solar System, Vol. II: The Earth as a Planet*, University of Chicago Press, 1954.

## Moon

Allday, J., *Apollo in Perspective: Spaceflight Then and Now*, Institute of Physics, 2000, ISBN 0-7503-0645-9.

Brooks, C.G., Grimwood, J.M., and Swenson, L.S., *Chariots for Apollo: A History of Manned Lunar Spacecraft*, NASA SP-4205, 1979.

Chaikin, A., *A Man on the Moon: The Voyages of the Apollo Astronauts*, Penguin, 1995, ISBN 0-14-024146-9.

Collins, M.J., *Space Race: The US–USSR Competition to Reach the Moon*, Pomegranate, 1999.

Compton, W.D., *Where No Man Has Gone Before: A History of Apollo Lunar Exploration Missions*, NASA SP-4214, 1989.

Cortright, E.M., (ed.), *Apollo Expeditions to the Moon*, NASA SP-350, 1975.

Hall, R.C., *Lunar Impact: A History of Project Ranger*, NASA SP-4210, 1977.

Mellberg, W.F., *Moon Missions: Mankind's First Voyages to Another World*, Plymouth Press, 1997, ISBN 1-882663-12-8.

Schefter, J., *The Race: The Definitive Story of America's Battle to Beat Russia to the Moon*, Random House, 1999, ISBN 0-7126-8471-9.

NASA SP-184, 1969, Surveyor Program Office, *Surveyor Program Results.*

NASA-214, 1969, NASA Manned Spacecraft Center, *Apollo 11 Preliminary Science Report.*

NASA-235, 1970, NASA Manned Spacecraft Center, *Apollo 12 Preliminary Science Report.*

NASA-272, 1971, NASA Manned Spacecraft Center, *Apollo 14 Preliminary Science Report.*

NASA-289, 1972, NASA Manned Spacecraft Center, *Apollo 15 Preliminary Science Report.*

NASA-315, 1972, NASA Manned Spacecraft Center, *Apollo 16 Preliminary Science Report.*

NASA-330, 1973, NASA Lyndon B. Johnson Space Center, *Apollo 17 Preliminary Science Report.*

## Mars

Carr, M.H., *et al.*, *Viking Orbiter Views of Mars*, NASA SP-441, 1980.

Collins, S.A., *The Mariner 6 and 7 Pictures of Mars*, NASA SP-263, 1971.

Corliss, W.R., *The Viking Mission to Mars*, Revised edition, NASA SP-334, 1975.

De Vaucouleurs, G., (Moore, P., trans.), *The Planet Mars*, Faber & Faber, Second edition, 1951.

Ezell, E.C., and Ezell, L.N., *On Mars: Exploration of the Red Planet, 1958–1978*, NASA SP-4212, 1984.

Hartmann, W.K., and Raper, O., *The New Mars: Discoveries of Mariner 9*, NASA SP-337, 1974.

Perminov, V.G., *The Difficult Road to Mars: A Brief History of Mars Exploration in the Soviet Union*, NASA NP-1999-06-251-HQ, 1999, ISBN 0-16-058859-6.

Raeburn, P., and Golombek, M., *Mars: Uncovering the Secrets of the Red Planet*, National Geographic, 1998, ISBN 0-7922-7373-7.

Sheehan, W., *The Planet Mars: A History of Observation and Discovery*, University of Arizona Press, 1996.

NASA SP-329, 1974, *Mars as Viewed by Mariner 9.*

NASA SP-425, 1978, *The Martian Landscape.*

*Mars: The NASA Mission Reports*, Apogee Books, 2000, ISBN 1-896522-62-9.

## Jupiter

Harland, D.M., *Jupiter Odyssey: The Story of NASA's Galileo Mission*, Springer-Praxis, 2000, ISBN 1-85233-301-4.

Hockey, T., *Galileo's Planet: Observing Jupiter before Photography*, Institute of Physics, 1999, ISBN 0-7503-0448-0.

Morrison, D., and Samz, J., *Voyage to Jupiter*, NASA SP-439, 1980.

Peek, B.M., *The Planet Jupiter*, Faber & Faber, 1958.

## Saturn

Alexander, A.F.O'D., *The Planet Saturn: A History of Observation, Theory and Discovery*, Faber & Faber, 1962.

Morrison, D., *Voyages to Saturn*, NASA SP-451, 1982.

Spilker, L.J., (ed.), *Passage to a Ringed World: The Cassini–Huygens Mission to Saturn and Titan*, NASA SP-533, 1997.

## Uranus

Alexander, A.F.O'D., *The Planet Uranus: A History of Observation, Theory and Discovery*, Faber & Faber, 1965.

Bergstralh, J.T., Miner, E.D., and Matthews, M.S., *Uranus*, University of Arizona Press, 1991.

Miner, E.D., *Uranus: The Planet, Rings and Satellites*, Wiley, Second edition, 1998, ISBN 0-471-97398-X.

## Neptune

Miner, E.D., and Wessen, R.R., *Neptune: The Planet, Rings and Satellites*, Springer-Praxis, 2002, ISBN 1-85233-216-6.

Moore, P., *The Planet Neptune: A Historical Survey Before Voyager*, Second edition, Wiley-Praxis, 1996, ISBN 0-471-96015-2.

## Pluto

Levy, D.H., *Clyde Tombaugh: Discoverer of Planet Pluto*, University of Arizona Press, 1991.

Stern, A., and Mitton, J., *Pluto and Charon: Ice Worlds on the Ragged Edge of the Solar System*, Wiley, 1999, ISBN 0-471-35384-1.

## Asteroids, comets, meteors and meteorites

Burnham, R., *Great Comets*, Cambridge University Press, 2000, ISBN 0-521-64600-6.

Calder, N., *Giotto to the Comets*, Presswork, London, 1992, ISBN 0-9520115-0-6.

Peebles, C., *Asteroids: A History*, Smithsonian Institution Press, 2000, ISBN 1-56098-389-2.

Watson, F.G., *Between the Planets*, Churchill, 1947.

Yeomans, D.K., *Comets: A Chronological History of Observation, Science, Myth and Folklore*, Wiley, 1991, ISBN 0-471-61011-9.

## Other solar systems

Croswell, K., *Planet Quest: The Epic Discovery of Alien Solar Systems*, Oxford University Press, 1997, ISBN 0-19-850198-6.

## Magazines and journals – further reading

In addition to books there are several magazines and professional journals that carry various articles and papers for further reading. For example, for amateur astronomers the American magazines *Sky and Telescope* and *Astronomy*, and the British magazine *Astronomy Now*, are of interest. The latter two magazines are relatively new (*Astronomy* started in the 1970s and *Astronomy Now* in the following decade). But *Sky and Telescope* started in the 1940s and in the early post-war decades, in particular, it contained many unique insights into astronomical research written by experts in the field. Secondhand copies of these magazines can often be purchased via advertisements in the magazines themselves, and occasionally via secondhand booksellers on the Internet.

In fact the advent of the Internet has been a boon to students of the history of astronomy who do not have ready access to astronomical libraries. For example it is now possible to download copies of the *Philosophical Transactions of the Royal Society* from 1757 to 1777 (via http://www.bodley.ox.ac.uk), and the *Monthly Notices of the Royal Astronomical Society* from 1827, the *Astronomical Journal* from 1849, *Publications of the Astronomical Society of the Pacific* from 1889, and the *Astrophysical Journal* from 1895 (all via NASA's Astrophysical Data System at http://adsabs.harvard.edu), all free of charge. In addition, via the latter website, it is possible to access the *Annual Reviews of Astronomy and Astrophysics* from 1963 and the *Annual Reviews of Earth and Planetary Sciences* from 1973.

# Units

I have tried to use a set of units which would be acceptable to both amateur and professional astronomers. Some of these people normally use SI units, but some feel more familiar with cgs or imperial units, so my selection has been a compromise. The units used are as follows:

| Type of unit | Unit | Abbreviation | Equivalent |
|---|---|---|---|
| Distance | megaparsec | Mpc | $10^6$ pc |
| | parsec | pc | 3.26 light years or $3.1 \times 10^{16}$ m |
| | light year | | $6.3 \times 10^4$ AU or $9.5 \times 10^{15}$ m |
| | astronomical unit | AU | $150 \times 10^6$ km |
| | kilometre | km | 0.62 miles (1 mile $\equiv$ 1.6 km) |
| | metre | m | 39.4 inches |
| | centimetre | cm | $10^{-2}$ m (1 inch $\equiv$ 2.54 cm) |
| | millimetre | mm | $10^{-3}$ m |
| | micron | $\mu$m | $10^{-6}$ m |
| | nanometre | nm | $10^{-9}$ m or 10 Ångströms |
| Angle | arc minute | $'$ | 1/60 degree |
| | arc second | $''$ | 1/3600 degree |
| Mass | kilogram | kg | 2.2 lb (1,000 kg $\equiv$ 1 metric ton) |
| | gram | g | |
| Density | grams per cubic centimetre | g/cm$^3$ | $10^3$ kg/m$^3$ |
| Pressure | bar | | 1.01 atmosphere or $1.0 \times 10^5$ Pa |
| Temperature | kelvin | K | $^\circ$C $+ 273$ |
| Energy | electron volt | eV | $1.6 \times 10^{-19}$ J |
| | erg | | $10^{-7}$ J |
| Frequency | gigahertz | GHz | $10^9$ Hz |
| | megahertz | MHz | $10^6$ Hz |
| Magnetic field | gauss | | $10^{-4}$ tesla or $10^5$ gamma |

# Name index

# Subject index

*Main references are in bold type*